AF362052

Innovations and Perspectives of Industrial and Bioenergy Crops for Bioeconomy Development

Innovations and Perspectives of Industrial and Bioenergy Crops for Bioeconomy Development

Editor

Mariusz J. Stolarski

MDPI • Basel • Beijing • Wuhan • Barcelona • Belgrade • Manchester • Tokyo • Cluj • Tianjin

Editor
Mariusz J. Stolarski
University of Warmia and Mazury in
Olsztyn
Poland

Editorial Office
MDPI
St. Alban-Anlage 66
4052 Basel, Switzerland

This is a reprint of articles from the Special Issue published online in the open access journal *Agriculture* (ISSN 2077-0472) (available at: https://www.mdpi.com/journal/agriculture/special_issues/Industrial_Bioenergy_Crops).

For citation purposes, cite each article independently as indicated on the article page online and as indicated below:

LastName, A.A.; LastName, B.B.; LastName, C.C. Article Title. *Journal Name* **Year**, *Volume Number*, Page Range.

ISBN 978-3-0365-3519-7 (Hbk)
ISBN 978-3-0365-3520-3 (PDF)

Contents

About the Editor

Mariusz Jerzy Stolarski is a professor of agricultural sciences, working as a professor at the University of Warmia and Mazury in Olsztyn (UWM), conducting research into the following: bioresources, bioeconomy, renewable energy sources (RES), biomass growing, productivity, quality, the use of perennial and annual alternative crops for energy and industrial purposes, production technology and logistics of biofeedstock acquisition and processing to obtain higher added value bioproducts and conversion to solid, liquid, and gaseous biofuels, the cascaded use of various agricultural- and forest-origin biofeedstock types, the bioconversion of various types of biomass residues by insects, the assessment of the economic, energetic, and environmental effectiveness of cultivation, production, logistics, and processing of biomaterials, And the characterisation and assessment of the usability of biomass and other RES on individual, local, regional, national, and international levels.

He is the co-author of over 80 original publications with the IF, and the co-author of 11 *Salix* spp. varieties and 1 patent. He has been a member of the Polish Agronomic Association since 2001 and a member of the Association for the Advancement of Industrial Crops since 2014. https://orcid.org/0000-0003-0712-9678; https://www.researchgate.net/profile/Mariusz-Stolarski.

agriculture

MDPI

Editorial

Industrial and Bioenergy Crops for Bioeconomy Development

Mariusz Jerzy Stolarski

Department of Genetics, Plant Breeding and Bioresource Engineering, Faculty of Agriculture and Forestry, Centre for Bioeconomy and Renewable Energies, University of Warmia and Mazury in Olsztyn, Plac Łódzki 3, 10-719 Olsztyn, Poland; mariusz.stolarski@uwm.edu.pl

Citation: Stolarski, M.J. Industrial and Bioenergy Crops for Bioeconomy Development. *Agriculture* **2021**, *11*, 852. https://doi.org/10.3390/agriculture11090852

Received: 29 August 2021
Accepted: 31 August 2021
Published: 7 September 2021

Publisher's Note: MDPI stays neutral with regard to jurisdictional claims in published maps and institutional affiliations.

The production of industrial and bioenergy crops has been the subject of scientific research for many years; however, the implementation of previously proposed solutions for commercial production is still at an early stage. It should be emphasized that when developing the production of industrial and bioenergy crops on agricultural lands, it is important to avoid land-use competition with the production of food and feed. It is well justified, for initiating the sustainable production of industrial and bioenergy crops, to promote efficient species for growing on marginal lands, which are unsuitable or less suitable for food or feed production. Another possibility is restoring fallowed land for agricultural production, including the production of biomass for non-agricultural purposes. Agricultural land abandonment is a process observed in most European countries. In Central and Eastern Europe, it was initiated with the political transformation of the 1990s. For example, in Poland, it concerns over 2 million ha of arable land. Such a large acreage constitutes a resource of land that can be directly restored to agricultural production or perform environmental functions. Therefore, a new concept for management of these lands is to produce biomass for bioeconomy purposes [1].

Industrial and bioenergy crops should include nonfood and nonfeed crops and generate agricultural products categorized as commodities and/or raw materials for industrial goods and bioenergy. Therefore, the research is mainly focused on the following groups of crops: SRC—short rotation coppice (willow, poplar, black locust, eucalyptus, etc.), grasses (miscanthus, giant reed, switchgrass, reed canary grass, etc.), herbaceous crops (Virginia mallow, Jerusalem artichoke, etc.), fiber crops (hemp, etc.), oil crops (camelina, crambe, castor, cardoon, etc.), and other alternative crops and their residues that are suitable for the industry or energy sectors. Research with some of the most common willow varieties in Europe has shown that it is possible to effectively use marginal land for the cultivation of willow intended for industrial purposes. However, it must be underlined that the key element that determines the production effects is the appropriate selection of varieties. Varieties with high production potential developed fewer shoots, but were taller and larger in diameter than other varieties [2]. Additionally, in another study, where 15 genotypes were grown at two different sites and harvested in two consecutive three-year harvest rotations, the very high impact of the genotype (81%) on the yield of willow was demonstrated. Therefore, the choice of a willow genotype is of key importance to success in willow production, since single-genotype monoculture on a commercial plantation may be a significant source of future problems with disease development or pest infestations [3].

It must be underlined that yields of *Miscanthus* × *giganteus* clones were also comparable, if not slightly better than other lignocellulose energy crops (poplar, willow, or Schavnat) in Czech conditions. *Miscanthus* × *giganteus* clones have good potential for commercial production of energy biomass, especially in warmer regions of Central and Eastern Europe with an annual sum of precipitation above 500–550 mm. The results show that the current economic conditions favor annual crops over *Miscanthus* (for energy biomass) and that this new crop shows very good adaptation to the effects of climate change. Selected clones of *Miscanthus* × *giganteus* reached high biomass yields despite very dry and warm periods and low-input agrotechnology, and they have good potential to become important biomass crops for future bioenergy and the bioeconomy [4]. Additionally, the cup plant (*Silphium perfoliatum*) is a new and promising bioenergy crop in Central Europe that can achieve high yields, especially on moist soils. However, spontaneous spread of this crop has already been documented, and especially valuable moist ecosystems could be at risk for becoming invaded by cup plant. Hence, fields for cultivating cup plant should be carefully chosen, and distances to such ecosystems should be held. If precautionary measures are observed, cup plant can take a place in the Central European agricultural landscape and make a valuable contribution to the conservation of biodiversity [5]. It was also shown that other perennial industrial crops did not cause a decline in wild biodiversity in comparison with unmanaged marginal land. Nevertheless, the cultivation of some crop species can cause a decrease in diversity of flora and fauna in the long term. For example, miscanthus and black locust cultivation were linked with a decrease in the number of plant species. On the other hand, the greatest biodiversity of plants and animals among crops was linked with the cultivation of willow; however, other crops also provided a good habitat for arthropods. No significant decrease in abundance of pollinators or natural enemies of pests were found in any perennial industrial crop [6].

These industrial and bioenergy crops can become an important source of biomass. Of course, the concept of their cultivation for nonfood (and/or nonfeed) uses is not new but, despite considerable investment in research and development, little progress has been made with regard to the introduction of such crops and their products into the market. For example, it is known that vegetable oil can surrogate petroleum products in many cases, as in cosmetics, biopolymers, or lubricants production. However, the cultivation of oil crops for the mere production of industrial oil would arouse concerns regarding competition for land use between food and non-food crops. Additionally, the economic sustainability is not always guaranteed, since the mechanical harvesting, in some cases, is still far from acceptable. The research underlines that the mechanical harvesting of sunflower, canola, and cardoon seeds is performed relying on specific devices that perform effectively with a minimum seed loss. Crambe and safflower seeds can be harvested through a combine harvester equipped with a header for cereals. On the other hand, camelina and castor crops still lack the reliable implementation of combine harvesters [7].

It must be underlined that the bio-based economy needs a sustainable supply of biomass for bioproduct generation and multiple uses. In this concept, no biomass should be used for energy generation unless the other options have been considered of using it to produce higher-value-added products. Noteworthy is that valuable substances in biomass are also used in the pharmaceutical, cosmetics, chemical, food, and feed industries. Hence, these substances should be extracted first, and after that production residues (post-production biomass) from the above industries should be used for production of biofertilizers, substrates, or bioenergy. Therefore, it is very important to analyze the possibility of obtaining bioactive compounds from lignocellulosic biomass before its transformation into biofuel. The research indicates varying contents of polyphenolic compounds in the biomass extracts of perennial herbaceous crops (*Helianthus salicifolius*, *Silphium perfoliatum*, *Helianthus tuberosus*, *Miscanthus* × *giganteus*, *Miscanthus sacchariflorus*, *Miscanthus sinensis*, and *Spartina pectinata*) depending on the harvest term in the growing period [8]. This information can enable the utilization of the studied biomass for not only the production of bioenergy but also to obtain valuable components of foodstuffs, medicines, and cosmetics. Additionally, another study suggested that supercritical extracts obtained from the aerial parts of *H. salicifolius* and *H. tuberosus* appeared to be a promising source of natural compounds with biocidal effect. They possessed antibacterial activity against Gram-positive and Gram-negative species, as well as antifungal activity against yeasts from the *Candida* genus. It is worth noting their antistaphylococcal activity. These extracts may be also regarded as natural potential antioxidants. The obtained data suggest that these extracts and their isolated bioactive compounds may be used as conservants in cosmetics and/or natural preservatives in food [9].

It is important to add that the research presented in this Special Issue "Innovations and Perspectives of Industrial and Bioenergy Crops for Bioeconomy Development" is the first small step towards identifying industrial and bioenergy crops that will be important for bioeconomy development. In total, 22 papers are published in this Special Issue, including three papers related to land availability, activities towards the development of the bioeconomy, and renaturation of a former quarry area [1,10,11]. Among perennial industrial crops, three papers about Miscanthus [4,12,13], five papers about SRC (willow, eucalyptus, poplar, Siberian elm, black alder, white birch, boxelder maple, silver maple) [2,3,14–16], and two papers about cup plant [5,17] are reported. Studies on tobacco, oil crops, and agricultural residual biomass are also presented [7,18,19]. The next two papers describe the biodiversity of weeds and arthropods in different perennial industrial crops [6,20]. Finally, three papers on the topic of functional compounds and the content of other substances in perennial industrial crops biomass [8,9,21] and one paper about the growth potential of yellow mealworm reared on industrial residues [22] are reported. Of course, only a small excerpt is presented here from the point of view of: (i) the number and diversity of plant species; (ii) their productivity and biomass properties; (iii) their impact on broadly understood biodiversity and the environment; and (iv) the possibility of multi-directional and cascade use of their biomass. These issues will be expanded upon by further interdisciplinary research.

Conflicts of Interest: The authors declare no conflict of interest.

References

1. Kozak, M.; Pudełko, R. Impact Assessment of the Long-Term Fallowed Land on Agricultural Soils and the Possibility of Their Return to Agriculture. *Agriculture* **2021**, *11*, 148. [CrossRef]
2. Matyka, M.; Radzikowski, P. Productivity and Biometric Characteristics of 11 Varieties of Willow Cultivated on Marginal Soil. *Agriculture* **2020**, *10*, 616. [CrossRef]
3. Stolarski, J.M.; Krzyżaniak, M.; Załuski, D.; Tworkowski, J.; Szczukowski, S. Effects of site, genotype and subsequent harvest rotation of willow productivity. *Agriculture* **2020**, *10*, 0412. [CrossRef]
4. Weger, J.; Knápek, J.; Bubeník, J.; Vávrová, K.; Strašil, Z. Can Miscanthus Fulfill Its Expectations as an Energy Biomass Source in the Current Conditions of the Czech Republic?—Potentials and Barriers. *Agriculture* **2021**, *11*, 40. [CrossRef]
5. Ende, L.M.; Knöllinger, K.; Keil, M.; Fiedler, A.J.; Lauerer, M. Possibly Invasive New Bioenergy Crop Silphium perfoliatum: Growth and Reproduction Are Promoted in Moist Soil. *Agriculture* **2021**, *11*, 24. [CrossRef]
6. Radzikowski, P.; Matyka, M.; Adam Kleofas Berbeć. Biodiversity of Weeds and Arthropods in Five Different Perennial Industrial Crops in Eastern Poland. *Agriculture* **2020**, *10*, 636. [CrossRef]
7. Pari, L.; Latterini, F.; Stefanoni, W. Herbaceous Oil Crops, a Review on Mechanical Harvesting State of the Art. *Agriculture* **2020**, *10*, 309. [CrossRef]
8. Ostolski, M.; Adamczak, M.; Brzozowski, B.; Stolarski, M.J. Screening of Functional Compounds in Supercritical Carbon Dioxide Extracts from Perennial Herbaceous Crops. *Agriculture* **2021**, *11*, 488. [CrossRef]
9. Malm, A.; Grzegorczyk, A.; Biernasiuk, A.; Baj, T.; Rój, E.; Tyśkiewicz, K.; Dębczak, A.; Stolarski, M.J.; Krzyzaniak, M.; Olba-Zięty, E. Could Supercritical Extracts from the Aerial Parts of Helianthus salicifolius A. Dietr. and Helianthus tuberosus L. Be Regarded as Potential Raw Materials for Biocidal Purposes? *Agriculture* **2021**, *11*, 10. [CrossRef]
10. Jezierska-Thöle, A.; Rudnicki, R.; Wiśniewski, Ł.; Gwiaździńska-Goraj, M.; Biczkowski, M. The Agri-Environment-Climate Measure as an Element of the Bioeconomy in Poland—A Spatial Study. *Agriculture* **2021**, *11*, 110. [CrossRef]
11. Cossel, M.v.; Ludwig, H.; Cichocki, J.; Fesani, S.; Guenther, R.; Thormaehlen, M.; Angenendt, J.; Braunstein, I.; Buck, M.-L.; Kunle, M.; et al. Adapting Syntropic Permaculture for Renaturation of a Former Quarry Area in the Temperate Zone. *Agriculture* **2020**, *10*, 603. [CrossRef]
12. Gołąb-Bogacz, I.; Helios, W.; Kotecki, A.; Kozak, M.; Jama-Rodzeńska, A. Content and Uptake of Ash and Selected Nutrients (K, Ca, S) with Biomass of Miscanthus giganteus Depending on Nitrogen Fertilization. *Agriculture* **2021**, *11*, 76. [CrossRef]
13. Gołąb-Bogacz, I.; Helios, W.; Kotecki, A.; Kozak, M.; Jama-Rodzeńska, A. The Influence of Three Years of Supplemental Nitrogen on Above- and Belowground Biomass Partitioning in a Decade-Old *Miscanthus × giganteus* in the Lower Silesian Voivodeship (Poland). *Agriculture* **2020**, *10*, 473. [CrossRef]
14. Šiaudinis, G.; Karčauskienė, D.; Aleinikovienė, J.; Repšienė, R.; Skuodienė, R. The Effect of Mineral and Organic Fertilization on Common Osier (*Salix viminalis* L.) Productivity and Qualitative Parameters of Naturally Acidic *Retisol*. *Agriculture* **2021**, *11*, 42. [CrossRef]
15. Berbeć, A.K.; Matyka, M. Planting Density Effects on Grow Rate, Biometric Parameters, and Biomass Calorific Value of Selected Trees Cultivated as SRC. *Agriculture* **2020**, *10*, 583. [CrossRef]
16. Palmieri, N.; Suardi, A.; Latterini, F.; Pari, L. The Eucalyptus Firewood: Understanding Consumers' Behaviour and Motivations. *Agriculture* **2020**, *10*, 512. [CrossRef]

17. Peni, D.; Stolarski, M.J.; Bordiean, A.; Krzyżaniak, M.; Dębowski, M. *Silphium perfoliatum*—A Herbaceous Crop with Increased Interest in Recent Years for Multi-Purpose Use. *Agriculture* **2020**, *10*, 640. [CrossRef]
18. Berbeć, A.K.; Matyka, M. Biomass Characteristics and Energy Yields of Tobacco (*Nicotiana tabacum* L.) Cultivated in Eastern Poland. *Agriculture* **2020**, *10*, 551. [CrossRef]
19. Bernas, J.; Konvalina, P.; Burghila, D.V.; Teodorescu, R.I.; Bucur, D. The Energy and Environmental Potential of Waste from the Processing of Hulled Wheat Species. *Agriculture* **2020**, *10*, 592. [CrossRef]
20. Piotrowska, N.S.; Czachorowski, S.Z.; Stolarski, M.J. Ground Beetles (*Carabidae*) in the Short-Rotation Coppice Willow and Poplar Plants—Synergistic Benefits System. *Agriculture* **2020**, *10*, 648. [CrossRef]
21. Budny, M.; Zalewski, K.; Lahuta, L.B.; Stolarski, M.J.; Stryiński, R.; Okorski, A. Could the Content of Soluble Carbohydrates in the Young Shoots of SelectedWillow Cultivars Be a Determinant of the Plants' Attractiveness to Cervids (Cervidae, Mammalia)? *Agriculture* **2021**, *11*, 67. [CrossRef]
22. Bordiean, A.; Krzyżaniak, M.; Stolarski, M.J.; Peni, D. Growth Potential of Yellow Mealworm Reared on Industrial Residues. *Agriculture* **2020**, *10*, 599. [CrossRef]

Review

Herbaceous Oil Crops, a Review on Mechanical Harvesting State of the Art

Luigi Pari *, Francesco Latterini and Walter Stefanoni

Consiglio per la Ricerca in Agricoltura e l'analisi dell'Economia Agraria (CREA)—Centro di Ricerca Ingegneria e Trasformazioni Agroalimentari, Via della Pascolare, 16, 00015 Monterotondo (RM), Italy; francesco.latterini@crea.gov.it (F.L.); walter.stefanoni@crea.gov.it (W.S.)

* Correspondence: luigi.pari@crea.gov.it; Tel.: +39-0690675250

Received: 12 June 2020; Accepted: 22 July 2020; Published: 23 July 2020

Abstract: The sustainable production of renewable energy is a key topic on the European community's agenda in the next decades. The use of residuals from agriculture could not be enough to meet the growing demand for energy, and the contribution of vegetable oil to biodiesel production may be important. Moreover, vegetable oil can surrogate petroleum products in many cases, as in cosmetics, biopolymers, or lubricants production. However, the cultivation of oil crops for the mere production of industrial oil would arise concerns on competition for land use between food and non-food crops. Additionally, the economic sustainability is not always guaranteed, since the mechanical harvesting, in some cases, is still far from acceptable. Therefore, it is difficult to plan the future strategy on bioproducts production from oil crops if the actual feasibility to harvest the seeds is still almost unknown. With the present review, the authors aim to provide a comprehensive overview on the state of the art of mechanical harvesting in seven herbaceous oil crops, namely: sunflower (*Heliantus annuus* L.), canola (*Brassica napus* L.), cardoon (*Cynara cardunculus* L.), camelina (*Camelina sativa* (L.) Crantz), safflower (*Carthamus tinctorius* L.), crambe (*Crambe abyssinica* R. E. Fr.), and castor bean (*Ricinus communis* L.). The review underlines that the mechanical harvesting of sunflower, canola and cardoon seeds is performed relying on specific devices that perform effectively with a minimum seed loss. Crambe and safflower seeds can be harvested through a combine harvester equipped with a header for cereals. On the other hand, camelina and castor crops still lack the reliable implementation on combine harvesters. Some attempts have been performed to harvest camelina and castor while using a cereal header and a maize header, respectively, but the actual effectiveness of both strategies is still unknown.

Keywords: harvesting; work productivity; supply chain; harvesting efficiency

---

1. Introduction

The European Directives on Renewable Energy (RED I and RED II) [1] aim to increase the share of renewable energy up to 32% of overall domestic energy production by 2030. Agriculture plays a key role in such context [2]; indeed, this sector can contribute to bioenergy production both with the exploitation of agricultural residues [3–6] and with dedicated energy crops [7–10]. Concerning the latter strategy, biodiesel production from oil crops is crucial [11]. Oil crops are able to synthesize highly complex molecular structures that can be used to displace significant amount of petroleum oil derived compounds [12] and they can significantly contribute to reach the above mentioned European Directives about renewable energy [13]. Moreover, oil crops contribution to bioeconomy concept is not only limited to bioenergy production, while taking into account the suitability of vegetable oil to be used as feedstock for the production of several bioproducts, such as surfactants, plasticizers, emulsifiers, detergents, lubricants, adhesives, and cosmetics [12,14]. However, two main issues are related to sustainable cultivation of oil crops. The first one is linked to avoid the competition with food crops considering the global increasing demand for food [15]. Scientific community has been working hardly to address this issue for years, mainly evaluating the possibilities of cultivating such crops in polluted soils [16,17] or in marginal lands [18–20]. The second one regards the costs of production costs of biodiesel, which is 30% more costly than petroleum-based diesel [21]. In particular, 60–80% of the biodiesel production cost is linked to the raw materials [11]; hence, there is a strong need to reduce the supply chain costs.

A possible strategy to reach this goal is to improve the harvesting operations. Harvesting is the key stage of the supply chain that strongly affects both costs and biomass quality and it plays a key role in the three pillars of the sustainability [22–31]. In a developed country, it is practically impossible to set up a sustainable supply chain for a given crop without effective mechanical harvesting [32]. The labor costs would be too high to bear. The mechanical harvesting of a given crop can be performed either using a dedicated harvester or borrowing it from other crops and apply specific modifications, at different levels, in order to limit the loss and damage to kernels [32]. The availability of dedicated machines is not always guaranteed, since different challenges arise according to the phenology of the plant and the seeds. Oil crops suffer the lack of availability on the market of dedicated machines for the harvesting stage, and the present review aims to provide the reader with the state of the art of mechanical harvesting, focusing on large scale oilseed production of some herbaceous oil crops, relying on the references that are produced in the last two decades (reference period 2000–2020).

Seven herbaceous oil crops have been taken into account, namely: sunflower (*Heliantus annuus* L.), canola (*Brassica napus* L.), cardoon (*Cynara cardunculus* L.), camelina (*Camelina sativa* (L.) Crantz), safflower (*Carthamus tinctorius* L.), crambe (*Crambe abyssinica* R. E. Fr.), and castor bean (*Ricinus communis* L.). The common feature to all these species is that all of them are herbaceous oil crops whose oil is the main product.

After the present introduction section, materials and methods for the review are reported. Subsequently, each crop is treated in a dedicated paragraph, in which, as a first step, a general view about the species is given (i.e., main agricultural features, actual and possible uses of oil, possibility of exploitation of by-products), then the mechanical harvesting topic is reported in detail. Eventually, the discussions and conclusion section are reported.

2. Materials and Methods

The bibliographical search was performed through Boolean operators. The use of Boolean searching to carry out a systematic review allows to analyze all studies in a given research topic through the use of specific database [33], in particular Scopus repository, was used for the present work. A papers search within Scopus database was done looking for both the common name and binomial one for each crop along with the word "harvest *". Table 1 provides details of the research keys and of relative findings.

Table 1. Research keys and findings on Scopus database.

Research Key	Overall Findings	2000–2020 and English Language Findings	2000–2020 Papers Actually Dealing with Mechanical Harvesting	Reference n.
TITLE-ABS-KEY ("sunflower" AND "harvest *")	1020	773		
TITLE-ABS-KEY ("Heliantus annuus" AND "harvest *")	427	323	7	[34–40]
TITLE-ABS-KEY ("canola" AND "harvest *")	633	380		
TITLE-ABS-KEY ("Brassica napus" AND "harvest *")	1244	1016		
TITLE-ABS-KEY ("oilseed rape" AND "harvest *")	553	269	16	[41–56]
TITLE-ABS-KEY ("rapeseed" AND "harvest *")	742	577		

Table 1. *Cont.*

Research Key	Overall Findings	2000–2020 and English Language Findings	2000–2020 Papers Actually Dealing with Mechanical Harvesting	Reference n.
TITLE-ABS-KEY ("cardoon" AND "harvest *")	59	58		
TITLE-ABS-KEY ("Cynara cardunculus" AND "harvest *")	143	142	7	[32,57–62]
TITLE-ABS-KEY ("camelina" AND "harvest *")	67	66	4	[63–66]
TITLE-ABS-KEY ("castor" AND "harvest *")	245	178		
TITLE-ABS-KEY ("Ricinus communis" AND "harvest *")	156	112	2	[67,68]
TITLE-ABS-KEY ("safflower" AND "harvest *")	165	83		
TITLE-ABS-KEY ("Carthamus tinctorius" AND "harvest *")	114	97	1	[69]
TITLE-ABS-KEY ("crambe" AND "harvest *")	42	25	3	[66,70,71]

TITLE-ABS-KEY: title, abstract, keywords. Thus the research was performed looking for the written above words within title, abstract and keywords of the papers in Scopus database. "harvest *": every word which includes "harvest", thus: pre-harvest, post-harvest, harvesting, etc

Findings reported in Table 1 were analyzed to further refine the search and approximately 200 papers were used to build up the present review.

3. Sunflower

3.1. Sunflower Main Features

Sunflower is an annual plant belonging to *Asteraceae* family and it is one of the most important oil crop worldwide [72]. Main producing countries are Ukraine and Russian Federation, which approximately account for 50% of global sunflower production [73]. Contrary to the majority of the other oil crops, sunflower oil is mainly used for human consumption, but it can also be used for biodiesel production [74]. When considering the great importance at global scale of this crop, many different varieties have been selected in order to achieve the possibility of cultivation in different climatic conditions and with different crop management (rainfed, minimum tillage, no tillage, various fertilization rates) [75–77]. Oil content in the seeds ranges between 38% and 53% by weight [75,78–81]. Sunflower is commonly considered an "environmental friendly" crop [82]; indeed, it generally requires low level of fertilization and the row spacing of about 50 cm allows for mechanical control of weeds [82]. On the other hand, ploughing resulted in being helpful to ensure the correct development of the tap root [83]. In the last years, there has been growing interest for sunflower cultivation in no tillage or reduced tillage systems [75]. Interesting results under this point of view were showed by Sessiz et al. (2008), who found that seed yield and oil content did not vary between no tillage and conventional tillage management [84].

Sunflower by-products show interesting features for energy production [85–89].

In particular, sunflower cake could be used as feedstock for anaerobic digestion, when considering its C/N ratio over 10 [88], for biofuel production in co-pyrolysis [90] or for the production of biomaterials through extrusion process [91]. Sunflower stalks can be used for bio-ethanol and biogas production [92,93] and seed hulls are suitable for energy production in co-firing [94,95]. On the contrary, the sunflower cake did not exhibit interesting characteristics for animal feeding [96,97].

3.2. Sunflower Mechanical Harvesting

Sunflower harvesting is performed with specific headers (Figure 1) that are designed for collecting sunflower seeds and transporting them to the inclined chamber of the combine harvester. Sunflower headers generally consist of the frame, auger, seed conveyors, choppers stems, dividers, and cutting units [34]; suitable combine settings for sunflower are given in Table 2.

Figure 1. Sunflower header working on cardoon (photo by Authors).

Table 2. Suitable combine harvester setting for sunflower [98].

Parameter	Values
Threshing speed (rpm)	300–400
Concave clearance (mm)	30–40
Fan speed (rpm)	650–850
Upper sieve clearance (mm)	13–15
Lower sieve clearance (mm)	7–9

The use of sunflower headers is generally associated with values of seed loss of 2% [34], if considering the average global yield for sunflower in 2018 of 1.97 Mg ha^{-1} [73], which corresponds, approximately, to 39 kg ha^{-1} of lost seed. The working speed of the combine harvester generally ranges from a minimum of 3.80 km h^{-1} to a maximum value of 9.60 km h^{-1} [34]. The field capacity ranges from 1.50 ha h^{-1} to 3.70 ha h^{-1} in the case of headers with eight rows; and, from 4.30 ha h^{-1} to 5.40 ha h^{-1} for headers having 12 rows. Fuel consumption varies from 6.71 L ha^{-1} up to 11.50 L ha^{-1} [34].

Three main aspects of the sunflower harvesting stage should be further investigated: the possibility to reduce the loss of seeds, to reduce the impurity of the seeds collected and to manage properly the crop residues. They are mainly made of talks that are usually shredded and buried into the soil in a second passage operation which has economic and environmental impacts (for example causing higher soil compaction) [35,36]. Nalobina et al. 2019 developed a sunflower header to further reduce seed losses and it is able to cut the stalk at a lower height, avoiding clogging of the header through the appliance of a conical rotor for the collection and the cut of stems. This modified header allowed for a reduction of seed loss, even if the working speed was consistently lower (1.00–2.50 km h^{-1}) [35].

Startsev et al. (2020) also tried to reduce seeds loss in sunflower mechanical harvesting. Authors developed an auger-reel applied to the header of the combine harvester that allowed for the reduction of seeds loss from the typical value of 2% up to 0.63% [37]. In case of dwarf sunflower plants, Shaforostov et al. (2019) designed and presented another prototype whose innovative features permitted the regulation of the inclination (in relation to the soil plate) of the screws along the dividers. In such way, it was possible to adjust the inclination according to the characteristics of the crop (dwarf or tall sunflower), thus allowing to reduce the length of the stem entering the threshing apparatus, which is the main responsible of clogging [38]. The stem of dwarf and tall sunflower measured 15 cm and 25 cm in length, respectively, whilst seeds loss of 0.25% in dwarf plants and 0.98% in tall plants were reported. The working speed of the combine harvester was set at 5 km h^{-1} for both sunflower varieties [38].

The third issue in sunflower mechanical harvesting, i.e., avoiding/limiting extraneous matter entering in the header, was addressed by Startsev et al. (2020). In particular, the authors developed the design of a sieve with additional holes that can be installed as additional cleaning stage in the cleaning shoe of the combine harvester. Those sieve holes were longitudinally shaped, just like sunflower seeds, and they also permitted the adjustment the area according the characteristics of the seeds of cultivar harvested. Such improvements permitted the significant reduction of extraneous material in the hopper heap by 38–42% by weight, in comparison with a combine harvester not equipped with such particular sieve [39].

However, in the same reference period, the impact of mechanization on agriculture was also investigated. Particularly, some studies focused on soil compactness due to machinery traffic, which is a major concern in both agriculture and forestry mechanization activities [99–104]. Among them, Dalmis et al. (2013) designed a chopper unit that can be installed under the header of a combine harvester and driven by via transmission chain connected to the header. The chopper unit consisted of a main body, a main shaft, three bevel gears mechanisms, four bearings, and three blade modules. The chopping occurred simultaneously to harvesting operation; therefore, avoiding the need for a second operation also saving time, labor, and energy. The height of the stalks on field measured 15 cm in height on average, which was similar to the average size of the chopped material in typical second passage chopping operation. The additional fuel consumption of the combine harvester due to the chopping unit was reported in $1.00\,l\,h^{-1}$ [40].

4. Canola

4.1. Canola Main Features

Canola (or rapeseed or oilseed rape) belongs to *Brassicaceae* family. It is the most cultivated oil crop worldwide and there has been growing interest on it during the last two decades [105]. The main producing countries are Canada and China, which together account for the 38% of global production [73]. However, canola is also gaining interest in Mediterranean zone [106]. Like sunflower, canola oil is suitable for human consumption [107,108], although energy production [109–111] is the most relevant application [106,112]. The main features that drive the growing importance of canola is the high oil content (about 45–50%) [113] and suitability for phytoremediation in soils polluted with diesel and heavy metals [114,115]. Moreover, rapeseed by-products can be also exploited. The cake is suitable for fermentation [116], while shoots and leaves are edible for humans, moreover nectar production is suitable from flowers, flavonoids and amino-acids extraction from pollen and fodder can be obtained from straw and seed meal [117]. Scientific research is focusing on the development of varieties and cultivation techniques to improve the possibility of mechanical collection. Rapeseed phenotypes that are suitable for mechanization should exhibit specific traits as: tightness, lodging resistance, collective flowering, and maturation, and produce more pods on the main stem [118,119]. Scientists are tackling the problem by following two strategies: testing higher plantation densities, which have positive effects on those desired traits in mechanical harvesting [120,121] and via developing dwarf cultivars (plant height lower than 160 cm) [122,123].

4.2. Canola Mechanical Harvesting

Two different options are available for canola harvesting: "direct cut", performed with a combine harvester in a single operation when seed ripeness is reached, or cutting and swathing of the crop at an early stage of ripening in order to limit seed shattering [41,42], and then trashing is performed. Seed loss is generally higher in direct cut method [49], although working performance is better [50] and the soil disturbance is lower. However, differences in seed loss among the two harvesting methods are not always found and a mean value 10% of seed lost is to be accepted [51–53]. As a general conclusion, it can be stated that combine harvesting is taking over and, currently, it is most studied method reported in the literature. In fact, only Irvine et al. (2010) studied the swathing method applied to canola harvesting [50]. The authors developed an alternative system to swath the crop in order to avoid windrowing operation and limiting the seed loss, which can occur under the strong wind condition.

The device is made of curved front surface that mechanically bends the plants forward causing the lodging at a height of 10–20 cm above the soil surface. After maturity, the lodged crop is harvested according to the direct-cut methods by operating against the lodging direction. Advantages from such strategy are found in the high working speed of the device and its low purchase cost, particularly if compared with those of a windrower [50]. On the other hand, the possible drawbacks of applying such system could be the delay in the harvesting period as well as the reduction of the speed of the combine harvester due to unavoidable need to cut the plants closer to the soil [50]. However, it is important to underline that, to the best of our knowledge, such a method is not actually applied to substitute the typical swathing one.

Several studies have been focusing on the effects of combine harvesting on seed loss in canola (suitable combine settings are reported in Table 3), and possible solutions to limit such problem are provided accordingly.

Table 3. Suitable combine harvester setting for canola [54,124].

Parameter	Values
Threshing speed (rpm)	550–580
Concave clearance (mm)	35–40
Fan speed (rpm)	400–600
Upper sieve clearance (mm)	4–16
Lower sieve clearance (mm)	3–4

It is important to highlight that, when it comes to mechanically harvesting the seeds of a given crop, some compromises must be accepted. Firstly, timing is a really important factor that could trigger high seed loss during the harvesting, due to both the natural dehiscence and presence of immature seeds [55]. Moreover, the latter ones are more susceptible to damages [56] and less suitable for high quality biofuel production [43]. Secondly, canola seeds size has negative impact on seeds loss, since smaller seeds can either be lost during impact of the header with tips of the plants, or during the trashing due to airborne effect of the fan [44]. Lastly, because plants are prone to entangle each other, pulling and tearing movements are generated during the harvesting and this can reduce the working efficiency of the machine [45].

So far, three main options for canola combine harvesting are available: using a conventional wheat header [46] without modifications; using wheat header specifically modified [42,47] or using a dedicated header that is characterized by the presence of vertical separators on both sides and by the continuous regulation of the blade-conveyor screw distance [48] (Figure 2). The last two options are meant to reduce pulling and tearing among plants and help to reduce the loss of seeds. When considering the differences in seed morphology between wheat and canola, it is fundamental to adjust the combine harvester set up finely, with particular attention to the fan speed, rotor speed, concave clearance, and sieves choice [46].

Figure 2. Modified wheat header for canola harvesting (photo by Authors).

Chronologically, the first study dealing with canola loss in combine harvesting reported in the reference period (2000–2020) is Hobson et al. (2002) [42]. Authors modified a wheat header adding a conveyor between the cutterbar and the auger, which gently conveyed the cut plants to the feeder conveyer, thus reducing the harsh movements partially responsible of seed loss. This modified header showed seed loss of 4.0% w/w, while the standard header tested on the same field showed a value of 6–8% w/w. According their preliminary economic evaluation, the estimated costs for modifying the header can be leveled off if more than 171 ha year^{-1} over five years period can be harvested, in comparison with using a standard header of the same size [42].

Tests on specific rapeseed header working performance were conducted by Pari et al. (2010) in 2007 and 2009 [48]. In 2007, a comparison between conventional wheat header and specific rapeseed header was performed, whilst in 2009 a wheat header was compared to a modified header bearing a vertical blade on its right side. The lowest value of losses was reached while using a specific rapeseed head in 2007 (2.76%), while wheat head in the same year showed 3.84% w/w of seeds loss. In 2009, the wheat head showed 9.33% w/w seeds loss while the modified wheat head with vertical blade showed only 6.84% w/w of seeds loss. Referring to the whole width of the combine harvester pass, the quantity of seeds loss found in the middle part (just below the swath) were similar in both treatments. However, the vertical blade applied on the modified header helped to decrease the loss of seeds in the lateral part of pass up to more than 35%. With respect to the combine harvester settings for wheat harvesting, the following tuning has to be applied: 20–30% lower rotor speed, 15–20% lower fan speed, 100–150% higher concave clearance, top riddle screen clearance of 8–10/7–9 mm, and lower riddle screen one of 3–3.5/3.5–4 mm [48]. A similar experiment was conducted by Asoodar et al. (2012), which tested a modified wheat header that was equipped with a vertical cutting bar for seeds loss. The Authors confirmed the previous finding, also reporting a reduction in seeds loss reduction from 14% w/w to 2% w/w [47].

Other tests were further conducted on specific canola header in 2012 by Pari et al. (2012) [54] setting the combine parameters, as explained hereafter: rotor speed 440 rpm, fan speed 430 rpm, concave clearance 37 mm, upper sieve clearance 4 mm, and lower sieve clearance 3 mm. Such a header showed lower losses in comparison to wheat header (0.97% and 1.63%, respectively) [54]. This study also highlighted the importance of the combine harvester speed: when the speed is decreased, the combine harvester is fed more smoothly and the impact of the header on the crop weakens [54]. On the contrary, reducing the speed implies higher costs, since both field capacity and material capacity are lower.

5. Cardoon

5.1. Cardoon Main Features

Cardoon belongs to the *Compositae* family, which bears flowers in inflorescences, called capitula or heads. The name of the fruit is cypsela (an achene from an inferior ovary) crowned by plumose filaments, called pappi [125]. Cardoon is cultivated as a multiannual crop and it can reach 2 m in height. The ecological requirements of *C. cardunculus* suit the environmental conditions of the Mediterranean zone, it requires 450–1000 mm year^{-1} of annual rainfall [125], and the cultivation is possible in poor soils [18,126–128]. Cardoon also performs well as bioremediation crop for polluted soils [17,129]. In the Mediterranean area, the establishment of cardoon crops is performed either during autumn or spring [130]. Ploughing and harrowing are recommended in order to allow for the proper development of the root system [131], even if Fernando et al. (2018) highlighted the possibility of cultivating cardoon in no tillage system also demonstrating a reduction in the environmental impact [132]. Cardoon can be either sowed or transplanted, but the first strategy is the most common [133–135]. Plant density is generally 1–2 m^{-2} [136,137]. The harvest is performed in summer from July to September [134,138,139], the yield of whole biomass on dry basis ranges between 10–20 Mg ha^{-1} year^{-1} with about 500 mm annual rainfall [125,135]; seeds oil content ranges from 185 g kg$_{dm}$$^{-1}$ to 253 g kg$_{dm}$$^{-1}$ [140].

The cultivation costs of *Cynara cardunculus* are reported between 329.19 € ha^{-1} and 477.37 € ha^{-1}; a consistent share of such costs is represented by biomass collection stage, whilst the harvesting operation accounts for 35–45% of the total costs [141].

The most interesting aspect of cardoon is the possibility to exploit its biomass in several ways. In the energy sector, the oil seed is suitable for the production of biodiesel [142–144]. The lignocellulosic biomass of the stems can be used as solid biofuel [145] or as substrate for gasification [146], pyrolysis [147], bioethanol production [148], and biomethane [149]. The pappi are particularly suitable for the production of paper pulp [138,150,151]. Finally, cardoon exhibits interesting features as a medicinal plant [152,153] or source of food for both humans and animals [154].

5.2. Cardoon Mechanical Harvesting

The mechanical harvesting of cardoon can be performed in two different ways. The first one involves the collection of all the aboveground biomass produced and afterward, the separation in different fractions. The second strategy, which is currently the most adopted, involves the use of a combine harvester for separating the seeds from the residual biomass, which is collected successively via baler.

First attempts to apply the first method, e.g., (collection of the whole aboveground biomass), are dated back to the end of 1990's, in Spain. A self-propelled mower-baler and choppers, previously applied to *Miscanthus* spp harvesting, were tested on cardoon. Unfortunately, the low bulk density of the material (mostly due to the chopper) and the high quantity of pappi scattered around represented a serious problem for the safety and the proper functioning of the machines. Filters and radiators of the machines clogged with a high risk of spontaneous combustion of the biomass [32].

Biomass baling is a more feasible alternative to undertake for a whole biomass harvesting. Here, two passes are needed: the first operation is mowing the plants by using a drum mower, while the second is the baling of the biomass with a round baler. However, during the baling, excessive quantity of soil was picked and included into the bales, thus causing slagging in phase of combustion. A mower-baler towed by a tractor was tested in 2006 in Spain in order to avoid this problem. However, the machine did not perform successfully, mostly due to the very low bulk density of the obtained round bales [32]. The development of high performing machines is a step-by-step process that, relying on specific trials, aims to address all the aspects of the harvesting. First trials were performed in 1990's with using a combine harvester equipped with conventional wheat header, which reported very high loss of seeds, mostly due the different architecture of the cardoon inflorescence as compared to cereal's spikes. Further trials were conducted in 2004 trying to: adjust the height of the header according to the height of the plants, add screens on the header, and lower the speed of the moving parts of the combine harvester. All of the results improved, but the loss of seeds that remained high (35% approximately) with high percentage of broken achenes (16%) [31]. Two years later, in Portugal, maize header and a wheat header were tested. Wheat header resulted in a high loss of achenes, but the high cutting height permitted obtaining a satisfactory size of residual biomass for the collection. On the other hand, the maize header showed better performance, but the cutting height of the stalk resulted in being too high and a second pass of lignocellulosic biomass mowing was necessary to collect it, obviously with negative influence on costs [32].

As a consequence of the previous tests that were performed using both maize header and wheat header, in 2007 the first attempt to develop a dedicated header for cardoon mechanical harvesting was made. The prototype derived from the merge of the two: the maize header was brought higher detached the capitula, whilst the wheat header below cut the stalks at the ground level. The residues of the threshed capitula were discharged on the swath, in order to collect them together with the lignocellulosic biomass. The combine harvester that were equipped with this header showed good working productivity with 3.24 km h^{-1} working speed and 1.57 ha h^{-1} field capacity, but no data concerning seed loss were reported by the authors [57]. The main problem with such header regarded the lower part of the machine, which was developed for a row cultivation, like maize and wheat.

Cardoon instead is a long-standing crop that is initially set up as row cultivation, but, after the first cut, it loses the row setting thanks to offshooting. Further improvements of the header (Figure 3) aimed to guarantee the cut of the stalk throughout the length of the cutting bar [58]. Working speed of the combine harvester increased to 4.41 km h^{-1} with a field capacity of 2.10 ha h^{-1} and a material capacity of 1.50 Mg h^{-1} of harvested seeds and of 13.05 Mg h^{-1} of lignocellulosic biomass. Biomass loss (seeds included) was 10.70% and damaged achenes were 3.25% [155].

Figure 3. Dedicated header for cardoon harvesting (photo by Authors).

The performance of the prototype was also tested against the performance of two wheat headers: one of the two was modified by adding triangular plates under the cutting bar in order to reduce seeds loss. Therefore, three headers were tested. The field capacity resulted lower in cardoon header (1.27 ha h^{-1}) than in modified wheat header (1.74 ha h^{-1}) and in classical wheat header (1.67 ha h^{-1}). The loss of biomass was lower in the case of cardoon header in comparison with the others. The loss of seed triggered by the cardoon header was 86% lower in comparison with the conventional wheat header and 65% lower than the modified header. Moreover, lignocellulosic biomass losses were significantly lower with the cardoon header prototype (about 8%) than in the other two headers (more than 50%) [59].

When it comes to cultivate cardoon on marginal Mediterranean soils, some problems may arise in the case of presence of stones. Stony soils represent a real challenge for mechanical harvesting, since the rocks can clog the combine harvester and cause serious damages to the mechanisms. Additionally, such soils undergo neither tillage nor levelling. Therefore, the driver has to keep the cutting bar distant from the soil in order to avoid unwanted external material within the header. This cautionary maneuver triggers the increase residual biomass loss. The development of a new flexible bar, driven by a system for sensing and signaling the presence of obstacles during the forward of motion of a combine harvester, was realized and tested. This flexible bar performed well, but it is important to highlight that the combine harvester speed was rather low, i.e., 0.70 km h^{-1} [60,61], so further tests and adjustments are encouraged in order to reach a working speed as much as possible close to the working speed of other dedicated headers.

The last mechanical improvement found in literature concerning cardoon crops is the development of a specific system that is able to collect the pappi from the cardoon's flower. During threshing, pappi are discharged along with the residual biomass in swaths for being included in the bales during the baling (within 3–4 h after harvesting). Because of the light weight, they can be quickly blown away from the straw. Thus, in order to limit such phenomenon, the combine harvester was equipped with a wetting system made of a water tank, an electric pump, a flow regulator, and three pairs of nozzles that sprayed the threshed material with a wetting solution. Four adjuvants in two different concentrations were used to extend the retention time. The amount of pappi remaining on the windrow proved the efficacy of one of the adjuvants (alkyl polysaccharide at 0.20% concentration) in limiting dispersal (52.8 kg ha^{-1}), bearing in mind that, without wetting (untreated), the amount remaining was seven times less (7.51 kg ha^{-1}). Additionally, preliminary economic analysis was conducted, thus proving the economic feasibility of such system [62].

However, it is important to highlight that, notwithstanding the growing interest for cardoon cultivation, no data regarding the setting of the combine harvester are available in literature.

6. Camelina

6.1. Camelina Main Features

Camelina belongs to *Brassicaceae* family [156]. This species originated from South-East Europe and South-West Asia [157]. Plant height generally ranges from 65 to 110 cm [158]. Seeds are contained within silique, commonly called seed capsules or pods, which measure from 5 to 14 mm in length, pear-shaped, slightly flattened, and contain eight to 15 seeds [159]. They are very small (0.7 mm × 1.5 mm), with a 1000-seed weight ranging between 0.8 and 1.8 g, depending on cultivar and growing conditions [160,161]. Seed oil content has been reported to range from 30% to 49% [160–163].

Thanks to the availability of both winter and spring varieties and the relatively short life cycle, camelina is particularly suitable for double cropping with small grain cereals, soybean, and sunflower [63,164–169]. Concerning the crop establishment, sowing occurs at a depth of 6–13 mm [159] applying 4 and 6 kg ha^{-1} of seeds [170].

The current growing interest in camelina cropping is linked to several factors. Apart from the double cropping, camelina shows high seeds oil content (30–49%) [171] and multiple uses of it, i.e., biodiesel [172,173] and jet fuel production [174,175], animal feeding [176–178], aquaculture [179,180], raw material for agrochemical products [181], and medical and veterinary applications [182,183]. Moreover, camelina is a low input crop in comparison with most of the commodity crops cultivated for biofuel production [165], so the environmental impact is lower [184], particularly if the suitability of camelina straw for bioenergy purposes is also included [185,186]. According to the literature, can be cultivated on poor soils and on soils with difficult conditions, even in the Mediterranean zone [187].

Concerning the economic aspects, some studies reported a breakeven point of biodiesel price lying between 0.88 € l^{-1} to 1.06 € l^{-1} in order to gain profit from the cropping [156,172] and including the cultivation costs of approximately 428.00 € ha^{-1} [188].

6.2. Camelina Mechanical Harvesting

Camelina, similarly to canola, is directly combined with traditional wheat header or swathed and then combined. Both of the methods result in similar seed yields [63]. The setting of the combine harvester is to be adapted to the species' features; e.g., speed, wind flow (fan speed), small opening screens, leak sealing, distance between the threshing cylinder, and the concave in order to prevent seed loss [64]. A certain tendency to seed shattering is reported in camelina, although not as much as in rapeseed [159].

Notwithstanding the growing attention of scientific community to such crop, very few studies dealt with mechanical harvesting analysis, setting and improvement.

Sintim et al. (2016) tested seed loss in mechanical harvesting performed with a plot combine harvester, firstly set for canola seeds, and then ongoing adjustments were carried out in order to minimize seeds loss. Seed loss was 11.70% w/w [65]. Another recent study, reported harvesting costs of 46.70 € ha^{-1} in the case of using a combine harvester equipped with wheat header, accounting for the 10% of the overall cultivation costs [66].

Needless to say, much more should be done about mechanical harvesting topic of this interesting oil crop.

7. Safflower

7.1. Safflower Main Features

Safflower (*Carthamus tinctorious* L.) is an annual oilseed crop. The main producing countries are Kazakhstan and USA, which together account for about the half of the global production [73]. Safflower oil is mainly used for biodiesel production [189] but it can be also used as a heat-stable cooking oil to fry and it is also used in cosmetics, food coatings, animal nutrition, and infant food formulations [69]. This plant has a strong central stem, a varying number of branches, and a taproot system. Each branch usually bears from one to five flower heads containing 15 to 20 seeds each [69]. The seed oil content ranges from 30% to 50%, depending on the variety and the environmental conditions [190]. Safflower is usually grown in recropping or in rotation with small grains or fallow and annual legumes. This species reaches the physiological maturity about 30 days after flowering and it is ready for the harvesting when most of the leaves have turned brown [69].

Seed shattering is a minor problem, although safflower should be harvested as soon as it is mature to minimize the danger of seed damage from excessive moisture.

Excessive rain and high humidity after physiological maturity of the seed may cause sprouting in the head. Safflower seeds are small, i.e., thousand-grain mass varies from 55.30 g to 41.30 g [191]. For a proper seed germination, both ploughing and harrowing are required. Seeding is usually performed in rows with a distance of approximately 45 cm between rows, and with a plant density of 45,000–60,000 plants per hectare. The high tolerance to drought makes the safflower suitable for cultivation in Mediterranean climate [192].

7.2. Safflower Mechanical Harvesting

Safflower harvesting is usually carried out with a combine harvester equipped with wheat header [69]. Literature lacks scientific studies regarding the harvesting of such oil crop. Indeed, only one paper dealing with this issue was found in the framework of the present review.

In particular, Pari et al. (2016) analyzed the mechanical harvesting performance of a wheat header (Figure 4) for safflower seeds collection [192].

Figure 4. Wheat header collecting safflower seeds (photo by Authors).

Table 4 provides the applied settings of the combine harvester.

Table 4. Suitable combine harvester setting for safflower [192].

Parameter	Values
Threshing speed (rpm)	800
Concave clearance (mm)	54
Fan speed (rpm)	400
Upper sieve clearance (mm)	11
Lower sieve clearance (mm)	6

The working speed was equal to 3.7 km h^{-1} and effective field capacity resulted in 1.32 ha h^{-1}. Seed loss analysis showed a value of 3.2% with 1.7% damaged seeds and 22.8% of impurities [192]. According to these results, it is possible to assert the good efficiency of a typical wheat header for the mechanical harvesting of this oil crop.

This is very interesting, because, unlike cardoon, safflower can be easily harvested with a conventional wheat header and applying fine adjustments to the combine harvester. An important feature that helps to promote the double cropping with traditional crops in both Mediterranean and Temperate zones.

8. Crambe

8.1. Crambe Main Features

Crambe abyssinica Hocst belongs to *Brassicaceae* family and it is the only cultivated species of *Crambe* genus [193]. Crambe oil can be used for the production of biodiesel [194,195] and jet fuel [12,14], but it is also suitable for non-energy applications, such as plastic films, adhesives, cosmetics, nylon, thermal insulation, corrosion inhibitors, synthetic rubber, and industrial lubricant [196–198]. After a period of decreasing attention on crambe in European Union [13], mostly due to significant yield variability [199], establishment difficulties linked to low germination energy, capsule persistence, seed dormancy [200,201], and poor interest of crambe cake for animal feeding [202]; in the last year there has been growing interest toward this species, as demonstrated by several European projects [203]. The renewed attention to this crop is related to several factors: low degree days requirement (about 1600) to reach maturity and so feasibility for double cropping [204–206], drought resistance, adaptability to various soil pH values, no seed shattering, higher dimension of the seed if compared to other oil crops [13], and plant height suitable for mechanical harvesting [207]. Moreover, crambe cultivation shows lower environmental impact than major oil crops, like, for example, canola [208].

8.2. Crambe Mechanical Harvesting

Scientific tests on mechanical harvesting of crambe were conducted by the University of Wisconsin in 1991 and University of Nebraska in 1993 [209,210] with conventional combines harvester set as for wheat harvesting. The optimal setting found is: threshing cylinder speed between 400 and 500 rpm; concave clearance of 10 mm; and, fan speed 500 rpm. The Authors also suggested to keep the frontal reel rotation speed slightly faster than the ground speed of the combine harvester in order to minimized the phenomenon of seed shattering [209,211]. Others studies also support the hypothesis to apply such strategy for crambe mechanical harvesting under the economic point of view, reporting a harvesting cost of approximately 47.00 € ha^{-1} [66,70].

Other studies, which were performed in 2012 with a 1.60 m wide plot combine harvester, aimed to provide further detailed information regarding the best combine setting. The machine was tested at three different speeds of working (3.0, 5.0, 8.2 km h^{-1}) and four rotation speeds of the threshing cylinder (400, 600, 800, and 1000 rpm). According to their findings, the lowest seeds loss, i.e., 3% w/w, was experienced when the working speed of the combine harvester is 5.04 km h^{-1} and speed rotation of the threshing cylinder is 800 rpm [71].

In light of what written above, the *status* of crambe mechanical harvesting is very similar to safflower. In fact, the only possible strategy is through using a combine harvester equipped for wheat harvesting, but set differently. Thus, the possibility to further exploit this oil crop in the future is a tangible option. Table 5 provides suitable settings of the combine harvester.

Table 5. Suitable combine harvester setting for crambe [71,209,210].

Parameter	Values
Threshing speed (rpm)	400–800
Concave clearance (mm)	10
Fan speed (rpm)	500

9. Castor

9.1. Castor Main Features

The world production of castor oil seeds (*Ricinus communis* L.) increased from 1.19 Mt in 1998 to 1.4 Mt in 2018, with a pick of 2.74 Mt in 2011 [73], highlighting a constant growing interest on this species. India is the most productive country of castor oil (more than 80% of the worldwide production) along with Mozambique, China, Brazil, Myanmar, Ethiopia, Paraguay and Vietnam [212]. Castor can grow well on marginal lands; it is resistant to drought and pests. The seeds oil content ranges from 35% to 55% w/w for high yield breed type (more than 1000 L ha^{-1} can be obtained when cultivated), has one of the highest viscosities among vegetable oils, and a molecular weight of 298 [213]. Globally, it only accounts for 0.15% of the total vegetable oil production [214] but it offers a wide spate of possible applications that stretch from pharmaceutical and cosmetic sectors to lubricants and oil-derived products [215]. The obtained bio-oil can also be used as a renewable fuel and chemical feedstock [216]. However, the content of toxic compounds as ricin in castor beans, curbs the possibility to cultivate castor plants extensively. In fact, several cases of accidental or intentional contaminations with ricin have been reported worldwide between 1981 and 2011 [217], although the scientists have never given up on studying this species.

Castor is relatively high demanding in N requirements when compared with other oil crops, as for instance soybean [209], and nitrogen availability in the soil promotes the expansion of leaves as well as the elongation of the stems [218]. Such effects trigger the development of a considerable amount of aerial biomass, which, in turn, reduces the harvest index of the crop [219].

9.2. Castor Mechanical Harvesting

Capsules are harvested when completely dried and a delay in this phase can lead to high seed loss for shattering. Castor seeds yield can reach up to 4.44 Mg ha^{-1} in the Mediterranean area [220], but the presence of green racemes bearing unripe capsules is a feature that should be reduced. At least, defoliation and fruit ripening must be artificially induced in order to have homogeneity among plants. This represents a big issue to address when it comes to harvesting stage, since more than one pass is mandatory and mechanization is barely possible when the growth is indeterminate [67,221,222]. Additionally, the use of chemicals and growth regulators are not allowed in organic farming. Therefore, harvesting is mainly performed manually with negative impact on the supply chain costs. Literature lacks the knowledge of scientific support regarding the possibility to harvest castor seeds mechanically, although unofficial tests have been conducted so far. Combine harvesters that are equipped with modified maize headers are currently used, although fine regulations of the cylinder speed, cylinder-concave clearance are strongly needed in order to reduce as much as possible seeds loss and seed damage (no data about combine harvester setting were found in literature). However, clogging may occur in the case of high quantity of aerial biomass production. The only dedicated header for castor beans harvesting is currently produced by Evogene Ltd. and Fantini s.r.l which announced in 2018 the successful results obtained: they reported a reduction of seeds loss from the current 50% to 5% in two consecutive tests that were performed in 2017 and 2018 on proprietary castor varieties [223]. Combine harvesters are usually equipped with a cutting bar that is situated in the lower part of the header, and it cuts the stem of plants at a given height. Consequently, a certain amount of aerial biomass is conveyed within the cleaning system, which, in the case of castor plants, is not negligible. On the contrary, a different approach is proposed by Zhao et al. (2019), particularly reported the possibility to harvest the capsules only, without cutting and threshing the whole plant. The innovative system seeks the use of a vibrating system instead of a cutting bar that is mounted on the header of the combine harvester [68]. Such a strategy is worthy of further investigation in order to provide scientific evidence on the reliability of such strategy. However, the upstanding residual biomass can be further collected if necessary. So far, no other scientific evidences have been found regarding the development of mechanical harvesting machines that are specifically dedicated to castor beans. However, lots of effort has been put in the selection of desirable traits, like dwarf plant type (<100 cm), early flowering and maturity (26–29 days and 120–140 days, respectively), seeds weight (70–80 g/100 seed), and high oil content (54–55%) [224].

10. Discussions and Conclusions

From the comprehensive analysis of the reviewed literature, the main aspect that stands out is the different degree of mechanical harvesting among herbaceous oil crops. Some of them share common machineries and strategies; sometimes, even specifically developed technologies are available. Others still profoundly lack this availability of a such specific machine that is able to harvest the seeds efficiently. According to the similarities and differences found, it is possible to draw some common conclusions among the reviewed species.

Sunflower and canola are currently being considered among the major herbaceous oil crops, which benefit from a well-developed technology for the seeds harvesting worldwide. Combine harvesters can be equipped with specific headers that can guarantee both high working capacity and low seeds loss. These two aspects concretely contribute to making the cultivation of them economically sustainable for farmers. In fact, the final revenue for farmers of a given crop is already jeopardized by the negative effects of unpredictable biotic factors, like climate and seasonality, and biotic factors, like pests and diseases, which can occur. Therefore, it is very important to rely on very efficient machines that accomplish the harvesting task quickly and effectively.

Significant improvements have been found in cardoon harvesting. In fact, the implementation of the header of the combine harvesters for cardoon seeds harvesting, significantly reduced the amount of seeds lost from 50% to 3%. Another problem addressed in cardoon cropping, specifically on stony soils, is the presence of stones in the field that compromise the harvesting. The development of a flexible cutting bar that can continuously adjust itself to the terrain pattern represents a valid innovation that is available to farmers. The working speed is still too low, but further research can also improve this aspect too, getting closer and closer to the combine performance known, for example, in sunflower and canola.

Differently, safflower and crambe mechanical harvesting relies on a combine harvester that is equipped with wheat header. The loss of seeds is approximately 3%, which is an acceptable value. This is a very interesting feature for farmers who cultivate cereals. In fact, they could broaden the number of crops cultivated without spending more money on new machineries. However, during the in the last years, the interest on safflower and crambe has not been constant.

Finally, it is possible to consider the mechanical harvesting of both camelina and castor not yet satisfying. They have both shown a fast growing interest in the scientific community, but, on the other hand, very little efforts have been put to the enhancement of specific mechanical harvesting. Several papers and scientific projects have been studying camelina and castor under the agronomic, genetic, and biochemical point of view, but, indeed, very little is known in the possible strategies for mechanical harvesting. Nowadays, contractors borrow the harvesting strategy from other crops (i.e., wheat header for camelina and maize one for castor), which show the major drawback to exhibit high loss of seeds. Additionally, very little is reported about seed loss, work productivity, and harvesting costs in the literature. These are the key issues for the development of efficient supply chains and future scientific research should focus on addressing this. Although the scientific community agrees on using a combine harvester for collecting the seeds, further studies on the possible regulations as well as possible modifications to apply to the machine are strongly needed and encouraged.

In oil crops cultivation, two major concerns arise: the economic sustainability (the cost of biodiesel production is currently higher than oil-derived diesel) and the competition with food crops for land use. Recently, the cultivation of industrial crops on marginal lands has gained interest throughout the Europe, and ongoing research activities focus on providing scientific evidence on such strategy. However, it is not an easy task to accomplish since the productivity of marginal lands is usually lower. Besides, if the harvesting phase is not effective, the overall strategy is not sustainable under the economic point of view. Hence, it is important to improve the harvesting machines and strategies, along with the genetic and agronomic ameliorations of the oil crops.

Author Contributions: Conceptualization, L.P., F.L., W.S.; methodology, L.P., F.L., W.S.; data curation, F.L. and W.S.; writing—original draft preparation, F.L. and W.S.; writing—review and editing, L.P., F.L., W.S.; supervision, L.P.; funding acquisition, L.P. All authors have read and agreed to the published version of the manuscript.

Funding: This research was funded by Horizon 2020 project MAGIC, grant number 727698.

Acknowledgments: In this section you can acknowledge any support given which is not covered by the author contribution or funding sections. This may include administrative and technical support, or donations in kind (e.g., materials used for experiments).

Conflicts of Interest: The authors declare no conflict of interest.

References

1. European Union. Renewable Energy Directive 2018/2001/EU. Available online: https://eur-lex.europa.eu/legal-content/EN/TXT/PDF/?uri=CELEX:32018L2001&from=EN (accessed on 10 February 2020).
2. Creutzig, F.; Ravindranath, N.H.; Berndes, G.; Bolwig, S.; Bright, R.; Cherubini, F.; Chum, H.; Corbera, E.; Delucchi, M.; Faaij, A. Bioenergy and climate change mitigation: An assessment. *Gcb Bioenergy* **2014**, *7*, 916–944. [CrossRef]
3. Suardi, A.; Saia, S.; Stefanoni, W.; Gunnarsson, C.; Sundberg, M.; Pari, L. Admixing Chaff with Straw Increased the Residues Collected without Compromising Machinery Efficiencies. *Energies* **2020**, *13*, 1766. [CrossRef]
4. Bergonzoli, S.; Suardi, A.; Rezaie, N.; Alfano, V.; Pari, L. An innovative system for Maize Cob and wheat chaff harvesting: Simultaneous grain and residues collection. *Energies* **2020**, *13*, 1256. [CrossRef]
5. Suardi, A.; Latterini, F.; Alfano, V.; Palmieri, N.; Bergonzoli, S.; Pari, L. Analysis of the Work Productivity and Costs of a Stationary Chipper Applied to the Harvesting of Olive Tree Pruning for Bio-Energy Production. *Energies* **2020**, *13*, 1359. [CrossRef]
6. Suardi, A.; Latterini, F.; Alfano, V.; Palmieri, N.; Bergonzoli, S.; Karampinis, E.; Kougioumtzis, M.A.; Grammelis, P.; Pari, L. Machine Performance and Hog Fuel Quality Evaluation in Olive Tree Pruning Harvesting Conducted Using a Towed Shredder on Flat and Hilly Fields. *Energies* **2020**, *13*, 1713. [CrossRef]
7. Cocco, D.; Deligios, P.A.; Ledda, L.; Sulas, L.; Virdis, A.; Carboni, G. LCA study of oleaginous bioenergy chains in a Mediterranean environment. *Energies* **2014**, *7*, 6258–6281. [CrossRef]
8. Dangol, N.; Shrestha, D.S.; Duffield, J.A. Life-cycle energy, GHG and cost comparison of camelina-based biodiesel and biojet fuel. *Biofuels* **2017**, *11*, 399–407. [CrossRef]
9. Graß, R.; Heuser, F.; Stülpnagel, R.; Piepho, H.P.; Wachendorf, M. Energy crop production in double-cropping systems: Results from an experiment at seven sites. *Eur. J. Agron.* **2013**, *51*, 120–129. [CrossRef]
10. Akbari, G.A.; Heshmati, S.; Soltani, E.; Amini Dehaghi, M. Influence of Seed Priming on Seed Yield, Oil Content and Fatty Acid Composition of Safflower (Carthamus tinctorius L.) Grown Under Water Deficit. *Int. J. Plant Prod.* **2019**, *14*, 245–258. [CrossRef]
11. Bušić, A.; Kundas, S.; Morzak, G.; Belskaya, H.; Mardetko, N.; Šantek, M.I.; Komes, D.; Novak, S.; Šantek, B. Recent trends in biodiesel and biogas production. *Food Technol. Biotechnol.* **2018**, *56*, 152–173. [CrossRef]
12. Carlsson, A.S. Plant oils as feedstock alternatives to petroleum—A short survey of potential oil crop platforms. *Biochimie* **2009**, *91*, 665–670. [CrossRef] [PubMed]
13. Zanetti, F.; Monti, A.; Berti, M.T. Challenges and opportunities for new industrial oilseed crops in EU-27: A review. *Ind. Crop. Prod.* **2013**, *50*, 580–595. [CrossRef]
14. Metzger, M.J.; Bunce, R.G.H.; Jongman, R.H.G.; Mücher, C.A.; Watkins, J.W. A climatic stratification of the environment of Europe. *Glob. Ecol. Biogeogr.* **2005**, *14*, 549–563. [CrossRef]

15. FAO. *Global Agriculture towards 2050: How to Feed World 2009*; FAO: Rome, Italy, 2009.

16. Pandey, V.C. Suitability of *Ricinus communis* L. cultivation for phytoremediation of fly ash disposal sites. *Ecol. Eng.* **2013**, *57*, 336–341. [CrossRef]

17. Llugany, M.; Miralles, R.; Corrales, I.; Barceló, J.; Poschenrieder, C. Cynara cardunculus a potentially useful plant for remediation of soils polluted with cadmium or arsenic. *J. Geochem. Explor.* **2012**, *123*, 122–127. [CrossRef]

18. Mauromicale, G.; Sortino, O.; Pesce, G.R.; Agnello, M.; Mauro, R.P. Suitability of cultivated and wild cardoon as a sustainable bioenergy crop for low input cultivation in low quality Mediterranean soils. *Ind. Crop. Prod.* **2014**, *57*, 82–89. [CrossRef]

19. Pavlista, A.D.; Hergert, G.W.; Margheim, J.M.; Isbell, T.A. Growth of spring camelina (Camelina sativa) under deficit irrigation in Western Nebraska. *Ind. Crop. Prod.* **2016**, *83*, 118–123. [CrossRef]

20. Schillinger, W.F. Camelina: Long-term cropping systems research in a dry Mediterranean climate. *Field Crop. Res.* **2019**, *235*, 87–94. [CrossRef]

21. Christopher, L.P.; Kumar, H.; Zambare, V.P. Enzymatic biodiesel: Challenges and opportunities. *Appl. Energy* **2014**, *119*, 497–520.

22. Acampora, A.; Croce, S.; Assirelli, A.; Del Giudice, A.; Spinelli, R.; Suardi, A.; Pari, L. Product contamination and harvesting losses from mechanized recovery of olive tree pruning residues for energy use. *Renew. Energy* **2013**, *53*, 350–353.

23. Pecenka, R.; Lenz, H.; Jekayinfa, S.O.; Hoffmann, T. Influence of tree species, harvesting method and storage on energy demand and wood chip quality when chipping poplar, willow and black locust. *Agriculture* **2020**, *10*, 116. [CrossRef]

24. Schweier, J.; Blagojević, B.; Venanzi, R.; Latterini, F.; Picchio, R. Sustainability assessment of alternative strip clear cutting operations for wood chip production in renaturalization management of pine stands. *Energies* **2019**, *12*, 3306. [CrossRef]

25. Bonfiglio, A.; Esposti, R. Analysing the economy-wide impact of the supply chains activated by a new biomass power plant. The case of cardoon in Sardinia. *Bio-Based Appl. Econ.* **2016**, *5*, 5–26. [CrossRef]

26. Cotana, F.; Cavalaglio, G.; Coccia, V.; Petrozzi, A. Energy opportunities from lignocellulosic biomass for a biorefinery case study. *Energies* **2016**, *9*, 748. [CrossRef]

27. Špokas, L.; Steponavičius, D.; Žebrauskas, G.; Čiplienė, A.; Bauša, L. Reduction in adverse environmental impacts associated with the operation of combine harvesters during the harvesting of winter oilseed rape. *J. Environ. Eng. Landsc. Manag.* **2019**, *27*, 72–81. [CrossRef]

28. Haile, T.A.; Holzapfel, C.B.; Shirtliffe, S.J. Canola genotypes and harvest methods affect seedbank addition. *Agron. J.* **2014**, *106*, 236–242. [CrossRef]

29. Fríd, M.; Dolan, A.; Celjak, I.; Filip, M.; Bartos, P. Harvest of cereals and oilseeds rape by combine harvesters New Holland CX 8090 and New Holland CR 9080. *Poljopr. Teh.* **2017**, *42*, 19–24.

30. Špokas, L.; Steponavičius, D. Fuel consumption during cereal and rape harvesting and methods of its reduction. *J. Food Agric. Environ.* **2011**, *9*, 257–263.

31. Kehayov, D.; Vezirov, C.; Atanasov, A. Some technical aspects of cut height in wheat harvest. *Agron. Res.* **2004**, *2*, 181–186.

32. Fernandez, J.; Pari, L.; García Müller, M.; Marquez, L.; Fedrizzi, M.; Curt, M.D. Strategies for the mechanical harvesting of Cynara. In Proceedings of the 15th European Biomass Conference & Exhibition, Berlin, Germany, 7–11 May 2007.

33. Picchio, R.; Proto, A.R.; Civitarese, V.; Di Marzio, N.; Latterini, F. Recent Contributions of Some Fields of the Electronics in Development of Forest Operations Technologies. *Electronics* **2019**, *8*, 1465. [CrossRef]

34. Chaplygin, M.; Bespalova, O.; Podzorova, M. Results of tests of devices for sunflower harvesting in economic conditions. *E3S Web Conf.* **2019**, *126*, 00063. [CrossRef]

35. Nalobina, O.O.; Vasylchuk, N.V.; Bundza, O.Z.; Holotiuk, M.V.; Veselovska, N.R.; Zoshchuk, N.V. A new technical solution of a header for sunflower harvesting. *INMATEH Agric. Eng.* **2019**, *58*, 129–136.

36. Csanadi, T.; Hamphoff, B. A header for efficient sunflower harvesting Deployment of special sunflower head. *VDI Berichte* **2007**, *2001*, 343.

37. Startsev, A.S.; Demin, E.E.; Danilin, A.V.; Vasilyev, O.A.; Terentyev, A.G. Results of the production test of sunflower harvesting attachment with an auger reel. *IOP Conf. Ser. Earth Environ. Sci.* **2020**, *433*, 012006. [CrossRef]

38. Shaforostov, V.D.; Makarov, S.S. The header for a breeding plot combine for sunflower harvesting. *Acta Technol. Agric.* **2019**, *22*, 60–63. [CrossRef]

39. Startsev, A.S.; Makarov, S.A.; Nesterov, E.S.; Kazakov, Y.F.; Terentyev, A.G. Comparative evaluation of the operation of a combine harvester with an additional sieve with adjustable holes for sunflower harvesting. *IOP Conf. Ser. Earth Environ. Sci.* **2020**, *433*, 012007. [CrossRef]

40. Dalmis, I.S.; Kayisoglu, B.; Bayhan, Y.; Ulger, P.; Toruk, F.; Durgut, F. Development of a chopper unit for chopping of sunflower stalk during harvesting by combine harvester. *Bulg. J. Agric. Sci.* **2013**, *19*, 1148–1154.

41. Ogilvy, S.E. The effect of timing of swathing on the quality and yield of winter oilseed rape. *Asp. Appl. Biol.* **1989**, *23*, 101–107.

42. Hobson, R.N.; Bruce, D.M. Seed loss when cutting a standing crop of oilseed rape with two types of combine harvester header. *Biosyst. Eng.* **2002**, *81*, 281–286. [CrossRef]

43. Kachel-Jakubowska, M.; Szpryngiel, M. Influence on drying condition on quality properties of rapeseed. *Int. Agrophys.* **2008**, *22*, 327–331.

44. Pari, L.; Assirelli, A.; Suardi, A.; Civitarese, V.; Del Giudice, A.; Santangelo, E. Seed losses during the harvesting of oilseed rape (*Brassica napus* L.) at on-farm scale. *J. Agric. Eng.* **2013**, *44*, 633–636. [CrossRef]

45. Pari, L.; Assirelli, A.; Suardi, A.; Santangelo, E. The yield losses during oilseed rape (*Brassica napus* L.) harvesting at on-farm scale in the Italian area. In Proceedings of the 21st European Biomass Conference and Exhibition, Copenaghen, Denmark, 3–7 June 2013; pp. 369–371.

46. Bruce, D.M.; Hobson, R.N.; Morgan, C.L.; Child, R.D. Threshability of shatter-resistant seed pods in oilseed rape. *J. Agric. Eng. Res.* **2001**, *80*, 343–350. [CrossRef]

47. Asoodar, A.M.; Izadinia, Y.; Desbiolles, J. Benefits of harvester front extension in reducing canola harvest losses. *Int. Agric. Eng. J.* **2012**, *21*, 32–37.

48. Pari, L.; Assirelli, A.; Suardi, A. Evaluation of Brassica napus and Brassica carinata losses during harvesting: Three years of experience. In Proceedings of the 18th European Biomass Conference and Exhibition Proceedings, Lyon, France, 3–7 May 2010; pp. 1790–1793.

49. Vera, C.L.; Downey, R.K.; Woods, S.M.; Raney, J.P.; McGregor, D.I.; Elliott, R.H.; Johnson, E.N. Yield and quality of canola seed as affected by stage of maturity at swathing. *Can. J. Plant Sci.* **2007**, *87*, 13–26. [CrossRef]

50. Irvine, B.; Lafond, G.P. Pushing canola instead of windrowing can be a viable alternative. *Can. J. Plant Sci.* **2010**, *90*, 145–152. [CrossRef]

51. Gan, Y.; Malhi, S.S.; Brandt, S.A.; McDonald, C.L. Assessment of seed shattering resistance and yield loss in five oilseed crops. *Can. J. Plant Sci.* **2008**, *88*, 267–270. [CrossRef]

52. Haile, T.A.; Gulden, R.H.; Shirtliffe, S.J. On-farm seed loss does not differ between windrowed and direct-harvested canola. *Can. J. Plant Sci.* **2014**, *94*, 785–789. [CrossRef]

53. Zhang, C.L.; Jun, L.I.; Zhang, M.H.; Cheng, Y.G.; Li, G.M.; Zhang, S.J. Mechanical Harvesting Effects on Seed Yield Loss, Quality Traits and Profitability of Winter Oilseed Rape (*Brassica napus* L.). *J. Integr. Agric.* **2012**, *11*, 1297–1304. [CrossRef]

54. Pari, L.; Assirelli, A.; Suardi, A.; Civitarese, V.; Del Giudice, A.; Costa, C.; Santangelo, E. The harvest of oilseed rape (*Brassica napus* L.): The effective yield losses at on-farm scale in the Italian area. *Biomass Bioenergy* **2012**, *46*, 453–458. [CrossRef]

55. Szpryngiel, M.; Wesołowski, M.; Szot, B. Economical technology of rape seed harvest. *TEKA Kom. Motoryz. Energ. Rol. Pol. Akad. Nauk Oddział Lublin.* **2004**, *4*, 19–29.

56. Szwed, G.; Lukaszuk, J. Effect of rapeseed and wheat kernel moisture on impact damage. *Int. Agrophys.* **2007**, *21*, 299–304.

57. Pari, L.; Fedrizzi, M.; Gallucci, F. Cynara cardunculus exploitation for energy applications: Development of a combine head for thesing and concurrent residues collecting and utilization. In Proceedings of the 16th European Biomass Conference & Exhibition, Valencia, Spain, 2–6 June 2008.

58. Pari, L.; Sissot, F.; Giannini, E. European Union Research Project Biocard: Mechanization Activities Results. In Proceedings of the Technology and Management to Increase the Efficiency in Sustainable Agricultural Systems, Rosario, Argentina, 1–4 September 2009.

59. Pari, L.; Alfano, V.; Acampora, A.; Del Giudice, A.; Scarfone, A.; Sanzone, E. Harvesting and Separation of Different Plant Fractions in *Cynara cardunculus* L. In *Perennial Biomass Crops for a Resource-Constrained World*; Barth, S., Murphy-Bokern, D., Kalinina, O., Taylor, G., Jones, M., Eds.; Springer International Publishing: Cham, Switzerland, 2016; pp. 261–271.

60. Pari, L.; Del Giudice, A.; Pochi, D.; Gallucci, F.; Santangelo, E. An innovative flexible head for the harvesting of cardoon (*Cynara cardunculus* L.) in stony lands. *Ind. Crop. Prod.* **2016**, *94*, 471–479. [CrossRef]

61. Pari, L.; Giudice, A.D.; Pochi, D.; Francesco, G.; Alfano, V.; Bergonzoli, S.; Santangelo, E. Use of a flexible bar in stony soil. In Proceedings of the 25th European Biomass Conference and Exhibition, Stockholm, Sweden, 12–15 June 2017; pp. 286–288.

62. Pari, L.; Alfano, V.; Mattei, P.; Santangelo, E. Pappi of cardoon (*Cynara cardunculus* L.): The use of wetting during the harvesting aimed at recovering for the biorefinery. *Ind. Crop. Prod.* **2017**, *108*, 722–728. [CrossRef]

63. Gesch, R.W.; Archer, D.W.; Berti, M.T. Dual cropping winter camelina with soybean in the northern corn belt. *Agron. J.* **2014**, *106*, 1735–1745. [CrossRef]

64. Obour, K.A. Oilseed Camelina (*Camelina sativa* L. Crantz): Production Systems, Prospects and Challenges in the USA Great Plains. *Adv. Plants Agric. Res.* **2015**, *2*. [CrossRef]

65. Sintim, H.Y.; Zheljazkov, V.D.; Obour, A.K.; Garcia y Garcia, A. Managing harvest time to control pod shattering in oilseed camelina. *Agron. J.* **2016**, *108*, 656–661. [CrossRef]

66. Stolarski, M.J.; Krzyżaniak, M.; Tworkowski, J.; Załuski, D.; Kwiatkowski, J.; Szczukowski, S. Camelina and crambe production – Energy efficiency indices depending on nitrogen fertilizer application. *Ind. Crop. Prod.* **2019**, *137*, 386–395. [CrossRef]

67. Severino, L.S.; Auld, D.L.; Baldanzi, M.; Cândido, M.J.D.; Chen, G.; Crosby, W.; Tan, D.; He, X.; Lakshmamma, P.; Lavanya, C.; et al. A review on the challenges for increased production of castor. *Agron. J.* **2012**, *104*, 853–880. [CrossRef]

68. Zhao, H.; Zhang, C.; Zhao, H. Analysis on The Research Status and Structure Characteristics of Castor Harvester. In Proceedings of the 2019 IEEE International Conference on Mechatronics and Automation (ICMA), Tianjin, China, 4–7 August 2019; pp. 415–420.

69. Berglund, D.; Riveland, N.; Bergman, J. Safflower production. *NDSU Ext. Serv. Publ.* **2019**, *870*, 4.

70. Stolarski, M.J.; Krzyżaniak, M.; Kwiatkowski, J.; Tworkowski, J.; Szczukowski, S. Energy and economic efficiency of camelina and crambe biomass production on a large-scale farm in north-eastern Poland. *Energy* **2018**, *150*, 770–780. [CrossRef]

71. Reginato, P. Colheita Mecanizada de Sementes de Crambe (*Crambe abyssinica* L.) no Cerrado sul Mato Grossense. Ph.D. Thesis, Universidade Federal da Grande Dourados, Dourados, Brazil, 2014.

72. Cruvinel, W.S.; Costa, K.A.D.P.; Da Silva, A.G.; Severiano, E.D.C.; Ribeiro, M.G. Intercropping of sunflower with Brachiaria brizantha cultivars during two sowing seasons in the interim harvest. *Semin. Agrar.* **2017**, *38*, 3173–3191. [CrossRef]

73. FAO. FAOSTAT Agriculture Data. Available online: http://www.fao.org/faostat/en/#data (accessed on 20 January 2020).

74. Dos Santos, E.R.; Barros, H.B.; Capone, A.; Ferraz, E.; Fidelis, R.R. Efeito de épocas de semeadura sobre cultivares de girassol, no Sul do Estado do Tocantins. *Rev. Ciênc. Agron.* **2012**, *43*, 199–206. [CrossRef]

75. Martinez-Force, E.; Dunford, N.T.; Salas, J.J. *2015 Sunflower: Chemistry Production, Processing and Utilization*; AOCS Press: Urbana, IL, USA, 2015.

76. Nouraein, M.; Skataric, G.; Spalevic, V.; Dudic, B.; Gregus, M. Short-term effects of tillage intensity and fertilization on sunflower yield, achene quality, and soil physicochemical properties under semi-arid conditions. *Appl. Sci.* **2019**, *9*, 5482. [CrossRef]

77. Soleymani, A.; Shayanfar, M.; Golparvar, A.R.; Shahrajabian, M.H.; Naranjani, L. Response of different sunflower cultivars to various planting densities. *Res. Crop.* **2011**, *12*, 134–138.

78. Cosentino, S.L.; Copani, V.; Patanè, C.; Mantineo, M.; D'Agosta, G.M. Agronomic, energetic and environmental aspects of biomass energy crops suitable for Italian environments. *Ital. J. Agron.* **2008**, *3*, 81–95. [CrossRef]

79. Anastasi, U.; Cammarata, M.; Sortino, O.; Abbate, V. Comportamento agronomico e composizione lipidica degli acheni di due ibridi di girasole (convenzionale e ad alto oleico) in risposta ai fattori ambientali. *Riv. Agron.* **2001**, *35*, 83–93.

80. Monotti, M.; Laureti, D. Il panorama varietale del girasole alto oleico. *Agroindustria* **2005**, *4*, 61–70.

81. Baldini, M.; Turi, M.; Giovanardi, R. Confronto tra sistemi colturali a ridotto impatto ambientale per il girasole destinato alla produzione di biodiesel. *Agroindustria* **2005**, *4*, 71–79.

82. Debaeke, P.; Bedoussac, L.; Bonnet, C.; Bret-Mestries, E.; Seassau, C.; Gavaland, A.; Raffaillac, D.; Tribouillois, H.; Véricel, G.; Justes, E. Sunflower crop: Environmental-friendly and agroecological. *Oilseeds Fats Crop. Lipids* **2017**, *24*, D304. [CrossRef]

83. Lecomte, V. Difficile de trop simplier en tournesol. *Perspect. Agric.* **2013**, *402*, 50–54.

84. Sessiz, A.; Sogut, T.; Alp, A.; Esgici, R. Tillage effects on sunflower (*Helianthus annuus* L.) emergence, yield, quality, and fuel consumption in double cropping system. *J. Cent. Eur. Agric.* **2008**, *9*, 697–710.

85. Kovacic, D.; Kralik, D.; Rupcic, S.; Jovicic, D.; Spajic, R.; Tišmac, M. Soybean straw, corn stover and sunflower stalk as possible substrates for biogas production in Croatia: A review. *Chem. Biochem. Eng. Q.* **2017**, *31*, 187–198. [CrossRef]

86. Binder, J.B.; Raines, R.T. Fermentable sugars by chemical hydrolysis of biomass. *Proc. Natl. Acad. Sci. USA* **2010**, *107*, 4516–4521. [CrossRef] [PubMed]

87. Perea-Moreno, M.A.; Manzano-Agugliaro, F.; Perea-Moreno, A.J. Sustainable energy based on sunflower seed husk boiler for residential buildings. *Sustainability* **2018**, *10*, 3407. [CrossRef]

88. Pedretti, E.F.; Del Gatto, A.; Pieri, S.; Mangoni, L.; Ilari, A.; Mancini, M.; Feliciangeli, G.; Leoni, E.; Toscano, G.; Duca, D. Experimental study to support local sunflower oil chains: Production of cold pressed oil in Central Italy. *Agriculture* **2019**, *9*, 231. [CrossRef]

89. Duca, D.; Toscano, G.; Riva, G.; Mengarelli, C.; Rossini, G.; Pizzi, A.; Del Gatto, A.; Pedretti, E.F. Quality of residues of the biodiesel chain in the energy field. *Ind. Crop. Prod.* **2015**, *75*, 91–97. [CrossRef]

90. Shie, J.-L.; Chang, C.-C.; Chang, C.-Y.; Tzeng, C.-C.; Wu, C.-Y.; Lin, K.-L.; Tseng, J.-Y.; Yuan, M.-H.; Li, H.-Y.; Kuo, C.-H. Co-pyrolysis of sunflower-oil cake with potassium carbonate and zinc oxide using plasma torch to produce bio-fuels. *Bioresour. Technol.* **2011**, *102*, 11011–11017. [CrossRef]

91. Rouilly, A.; Orliac, O.; Silvestre, F.; Rigal, L. New natural injection-moldable composite material from sunflower oil cake. *Bioresour. Technol.* **2006**, *97*, 553–561. [CrossRef]

92. Nargotra, P.; Sharma, V.; Gupta, M.; Kour, S.; Bajaj, B.K. Application of ionic liquid and alkali pretreatment for enhancing saccharification of sunflower stalk biomass for potential biofuel-ethanol production. *Bioresour. Technol.* **2018**, *267*, 560–568. [CrossRef]

93. Hesami, S.M.; Zilouei, H.; Karimi, K.; Asadinezhad, A. Enhanced biogas production from sunflower stalks using hydrothermal and organosolv pretreatment. *Ind. Crop. Prod.* **2015**, *76*, 449–455. [CrossRef]

94. Raclavska, H.; Juchelkova, D.; Roubicek, V.; Matysek, D. Energy utilisation of biowaste—Sunflower-seed hulls for co-firing with coal. *Fuel Process. Technol.* **2011**, *92*, 13–20. [CrossRef]

95. Matin, A.; Majdak, T.; Krička, T.; Grubor, M. Valorization of sunflower husk after seeds convection drying for solid fuel production. *J. Cent. Eur. Agric.* **2019**, *20*, 389–401. [CrossRef]

96. Casoni, A.I.; Bidegain, M.; Cubitto, M.A.; Curvetto, N.; Volpe, M.A. Pyrolysis of sunflower seed hulls for obtaining bio-oils. *Bioresour. Technol.* **2015**, *177*, 406–409. [CrossRef] [PubMed]

97. Xazela, N.M.; Chimonyo, M.; Muchenje, V.; Marume, U. Effect of sunflower cake supplementation on meat quality of indigenous goat genotypes of South Africa. *Meat Sci.* **2012**, *90*, 204–208.

98. John Deere Combine Setting Manual. Available online: http://manuals.deere.com/omview/OMHXE13769_19/OUO6075_00003C2_19_25JUN09_1.htm (accessed on 16 July 2020).

99. Piron, D.; Boizard, H.; Heddadj, D.; Pérès, G.; Hallaire, V.; Cluzeau, D. Indicators of earthworm bioturbation to improve visual assessment of soil structure. *Soil Tillage Res.* **2017**, *173*, 53–63. [CrossRef]

100. Venanzi, R.; Picchio, R.; Grigolato, S.; Latterini, F. Soil and forest regeneration after different extraction methods in coppice forests. *For. Ecol. Manag.* **2019**, *454*. [CrossRef]

101. Bodaghi, A.I.; Nikooy, M.; Naghdi, R.; Venanzi, R.; Latterini, F.; Tavankar, F.; Picchio, R. Ground-based extraction on salvage logging in two high forests: A productivity and cost analysis. *Forests* **2018**, *9*, 729. [CrossRef]

102. Picchio, R.; Venanzi, R.; Tavankar, F.; Luchenti, I.; Iranparast Bodaghi, A.; Latterini, F.; Nikooy, M.; Di Marzio, N.; Naghdi, R. Changes in soil parameters of forests after windstorms and timber extraction. *Eur. J. For. Res.* **2019**, *138*, 875–888. [CrossRef]

103. Picchio, R.; Latterini, F.; Mederski, P.S.; Venanzi, R.; Karaszewski, Z.; Bembenek, M.; Croce, M. Comparing accuracy of three methods based on the gis environment for determining winching areas. *Electronics* **2019**, *8*, 53. [CrossRef]

104. Botta, G.F.; Tolón-Becerra, A.; Bienvenido, F.; Rivero, D.; Laureda, D.A.; Ezquerra-Canalejo, A.; Contessotto, E.E. Sunflower (*Helianthus annuus* L.) harvest: Tractor and grain chaser traffic effects on soil compaction and crop yields. *Land Degrad. Dev.* **2018**, *29*, 4252–4261. [CrossRef]

105. Vicianová, M.; Ducsay, L.; Ryant, P.; Provazník, M.; Zapletalová, A.; Slepčan, M. Oilseed rape (*Brassica Napus* L.) nutrition by nitrogen and phosphorus and its effect on yield of seed, oil and higher fatty acids content. *Acta Univ. Agric. Silvic. Mendel. Brun.* **2020**, *68*, 129–136. [CrossRef]

106. Licata, M.; La Bella, S.; Lazzeri, L.; Matteo, R.; Leto, C.; Massaro, F.; Tuttolomondo, T. Agricultural feedstocks of two Brassica oilseed crops and energy cogeneration with pure vegetable oil for a sustainable short agro-energy chain in Sicily (Italy). *Ind. Crop. Prod.* **2018**, *117*, 140–148. [CrossRef]

107. El Sabagh, A.; Hossain, A.; Barutçular, C.; Islam, M.S.; Ratnasekera, D.; Kumar, N.; Meena, R.S.; Gharib, H.S.; Saneoka, H.; da Silva, J.A.T. Drought and salinity stress management for higher and sustainable canola (*Brassica napus* L.) production: A critical review. *Aust. J. Crop Sci.* **2019**, *13*, 88–97. [CrossRef]

108. Chew, S.C. Cold-pressed rapeseed (Brassica napus) oil: Chemistry and functionality. *Food Res. Int.* **2020**, *131*, 108997. [CrossRef] [PubMed]

109. D'Avino, L.; Dainelli, R.; Lazzeri, L.; Spugnoli, P. The role of co-products in biorefinery sustainability: Energy allocation versus substitution method in rapeseed and carinata biodiesel chains. *J. Clean. Prod.* **2015**, *94*, 108–115. [CrossRef]

110. Karimi Alavijeh, M.; Yaghmaei, S. Biochemical production of bioenergy from agricultural crops and residue in Iran. *Waste Manag.* **2016**, *52*, 375–394. [CrossRef]

111. Muszyński, S.; Sujak, A.; Stępniewski, A.; Kornarzyński, K.; Ejtel, M.; Kowal, N.; Tomczyk-Warunek, A.; Szczeniak, E.; Tomczyńska-Mleko, M.; Mleko, S. Surface tension and wetting properties of rapeseed oil to biofuel conversion by-products. *Int. Agrophys.* **2018**, *32*, 247–252. [CrossRef]

112. Palmieri, N.; Forleo, M.B.; Suardi, A.; Coaloa, D.; Pari, L. Rapeseed for energy production: Environmental impacts and cultivation methods. *Biomass Bioenergy* **2014**, *69*, 1–11. [CrossRef]

113. Gesch, R.W.; Isbell, T.A.; Oblath, E.A.; Allen, B.L.; Archer, D.W.; Brown, J.; Hatfield, J.L.; Jabro, J.D.; Kiniry, J.R.; Long, D.S.; et al. Comparison of several Brassica species in the north central U.S. for potential jet fuel feedstock. *Ind. Crop. Prod.* **2015**, *75*, 2–7. [CrossRef]

114. Yang, Y.; Zhou, X.; Tie, B.; Peng, L.; Li, H.; Wang, K.; Zeng, Q. Comparison of three types of oil crop rotation systems for effective use and remediation of heavy metal contaminated agricultural soil. *Chemosphere* **2017**, *188*, 148–156. [CrossRef]

115. Lacalle, R.G.; Gómez-Sagasti, M.T.; Artetxe, U.; Garbisu, C.; Becerril, J.M. Brassica napus has a key role in the recovery of the health of soils contaminated with metals and diesel by rhizoremediation. *Sci. Total Environ.* **2018**, *618*, 347–356. [CrossRef] [PubMed]

116. De Castro, A.M.; dos Reis Castilho, L.; Freire, D.M.G. Characterization of babassu, canola, castor seed and sunflower residual cakes for use as raw materials for fermentation processes. *Ind. Crop. Prod.* **2016**, *83*, 140–148. [CrossRef]

117. FU, D.; Jiang, L.; Mason, A.S.; Xiao, M.; Zhu, L.; Li, L.; Zhou, Q.; Shen, C.; Huang, C. Research progress and strategies for multifunctional rapeseed: A case study of China. *J. Integr. Agric.* **2016**, *15*, 1673–1684. [CrossRef]

118. Li, Y.S.; Yu, C.B.; Zhu, S.; Xie, L.H.; Hu, X.J.; Liao, X.; Liao, X.S.; Che, Z. High planting density benefits to mechanized harvest and nitrogen application rates of oilseed rape (*Brassica napus* L.). *Soil Sci. Plant Nutr.* **2014**, *60*, 384–392. [CrossRef]

119. Khan, S.; Anwar, S.; Kuai, J.; Noman, A.; Shahid, M.; Din, M.; Ali, A.; Zhou, G. Alteration in yield and oil quality traits of winter rapeseed by lodging at different planting density and nitrogen rates. *Sci. Rep.* **2018**, *8*, 1–12. [CrossRef]

120. Li, X.; Zuo, Q.; Chang, H.; Bai, G.; Kuai, J.; Zhou, G. Higher density planting benefits mechanical harvesting of rapeseed in the Yangtze River Basin of China. *Field Crop. Res.* **2018**, *218*, 97–105. [CrossRef]

121. Kuai, J.; Sun, Y.; Zuo, Q.; Huang, H.; Liao, Q.; Wu, C.; Lu, J.; Wu, J.; Zhou, G. The yield of mechanically harvested rapeseed (*Brassica napus* L.) can be increased by optimum plant density and row spacing. *Sci. Rep.* **2015**, *5*, 1–14. [CrossRef]

122. Miersch, S.; Gertz, A.; Breuer, F.; Schierholt, A.; Becker, H.C. Influence of the semi-dwarf growth type on seed yield and agronomic parameters at low and high nitrogen fertilization in winter oilseed rape. *Crop Sci.* **2016**, *56*, 1573–1585. [CrossRef]

123. Hu, Q.; Hua, W.; Yin, Y.; Zhang, X.; Liu, L.; Shi, J.; Zhao, Y.; Qin, L.; Chen, C.; Wang, H. Rapeseed research and production in China. *Crop J.* **2017**, *5*, 127–135. [CrossRef]

124. New Holland Combine TX36 Evaluation Report. Available online: https://open.alberta.ca/dataset/d62d2f3e-99bb-49c7-9adb-b31b42f61cb2/resource/ac1a629a-2b74-4f38-8586-68156e8b1f14/download/676-afmrc.pdf (accessed on 16 July 2020).

125. Gominho, J.; Curt, M.D.; Lourenço, A.; Fernández, J.; Pereira, H. *Cynara cardunculus* L. as a biomass and multi-purpose crop: A review of 30 years of research. *Biomass Bioenergy* **2018**, *109*, 257–275. [CrossRef]

126. Vasilakoglou, I.; Dhima, K. Potential of two cardoon varieties to produce biomass and oil under reduced irrigation and weed control inputs. *Biomass Bioenergy* **2014**, *63*, 177–186. [CrossRef]

127. Mauro, R.P.; Sortino, O.; Pesce, G.R.; Agnello, M.; Lombardo, S.; Pandino, G.; Mauromicale, G. Exploitability of cultivated and wild cardoon as long-term, low-input energy crops. *Ital. J. Agron.* **2015**, *10*, 44–46. [CrossRef]

128. Velez, Z.; Campinho, M.A.; Guerra, Â.R.; García, L.; Ramos, P.; Guerreiro, O.; Felício, L.; Schmitt, F.; Duarte, M. Biological characterization of *Cynara cardunculus* L. Methanolic extracts: Antioxidant, anti-proliferative, anti-migratory and anti-angiogenic activities. *Agriculture* **2012**, *2*, 472–492. [CrossRef]

129. Domínguez, M.T.; Montiel-Rozas, M.M.; Madejón, P.; Diaz, M.J.; Madejón, E. The potential of native species as bioenergy crops on trace-element contaminated Mediterranean lands. *Sci. Total Environ.* **2017**, *590–591*, 29–39. [CrossRef] [PubMed]

130. Sanz, M.; Mauri, P.V.; Curt, M.D.; Sánchez, J.; García-Müller, M.; Plaza, A.; Fernández, J. Effect of plant density on biomass yield of Cynara cardunculus. *Acta Hortic.* **2016**, *1147*, 385–391. [CrossRef]

131. Angelini, L.G.; Ceccarini, L.; o Di Nasso, N.N.; Bonari, E. Long-term evaluation of biomass production and quality of two cardoon (*Cynara cardunculus* L.) cultivars for energy use. *Biomass Bioenergy* **2009**, *33*, 810–816. [CrossRef]

132. Fernando, A.L.; Costa, J.; Barbosa, B.; Monti, A.; Rettenmaier, N. Environmental impact assessment of perennial crops cultivation on marginal soils in the Mediterranean Region. *Biomass Bioenergy* **2018**, *111*, 174–186. [CrossRef]

133. Bolohan, C.; Marin, D.I.; Mihalache, M.; Ilie, L.; Oprea, A.C. Research on *Cynara cardunculus* L. species under the conditions of southeastern Romania area. *Sci. Paper Ser. A Agron.* **2013**, *56*, 429–432.

134. Francaviglia, R.; Bruno, A.; Falcucci, M.; Farina, R.; Renzi, G.; Russo, D.E.; Sepe, L.; Neri, U. Yields and quality of *Cynara cardunculus* L. wild and cultivated cardoon genotypes. A case study from a marginal land in Central Italy. *Eur. J. Agron.* **2016**, *72*, 10–19. [CrossRef]

135. Fernández, J.; Curt, M.D.; Aguado, P.L. Industrial applications of *Cynara cardunculus* L. for energy and other uses. *Ind. Crop. Prod.* **2006**, *24*, 222–229. [CrossRef]

136. Curt, M.D.; Sánchez, G.; Fernández, J. The potential of *Cynara cardunculus* L. for seed oil production in a perennial cultivation system. *Biomass Bioenergy* **2002**, *23*, 33–46. [CrossRef]

137. Lag-Brotons, A.; Gómez, I.; Navarro-Pedreño, J.; Mayoral, A.M.; Curt, M.D. Sewage sludge compost use in bioenergy production–a case study on the effects on *Cynara cardunculus* L energy crop. *J. Clean. Prod.* **2014**, *79*, 32–40. [CrossRef]

138. Gominho, J.; Lourenço, A.; Palma, P.; Lourenço, M.E.; Curt, M.D.; Fernández, J.; Pereira, H. Large scale cultivation of *Cynara cardunculus* L. for biomass production—A case study. *Ind. Crop. Prod.* **2011**, *33*, 1–6. [CrossRef]

139. Solano, M.L.; Manzanedo, E.; Concheso, R.; Curt, M.D.; Sanz, M.; Fernández, J. Potassium fertilisation and the thermal behaviour of *Cynara cardunculus* L. *Biomass Bioenergy* **2010**, *34*, 1487–1494. [CrossRef]

140. Raccuia, S.A.; Piscioneri, I.; Sharma, N.; Melilli, M.G. Genetic variability in *Cynara cardunculus* L. domestic and wild types for grain oil production and fatty acids composition. *Biomass Bioenergy* **2011**, *35*, 3167–3173. [CrossRef]

141. Lago, C.; Herrera, I.; Sanchez, J.; Lechon, Y.; Curt, M.D. Environmental and economic assesment of Cynara cardunculus for local bioenergy production under rainfed conditions in the context of climate change impact. In Proceedings of the 27th European Biomass Conference and Exhibition, Lisbon, Portugal, 27–30 May 2019; pp. 1719–1724.

142. Sengo, I.; Gominho, J.; D'Orey, L.; Martins, M.; D'Almeida-Duarte, E.; Pereira, H.; Ferreira-Dias, S. Response surface modeling and optimization of biodiesel production from Cynara cardunculus oil. *Eur. J. Lipid Sci. Technol.* **2010**, *112*, 310–320. [CrossRef]

143. Alexandre, A.; Dias, A.M.A.; Seabra, I.J.; Portugal, A.; De Sousa, H.C.; Braga, M.E.M. Biodiesel obtained from supercritical carbon dioxide oil of *Cynara cardunculus* L. *J. Supercrit. Fluids* **2012**, *68*, 52–63. [CrossRef]

144. Mancini, M.; Lanza Volpe, M.; Gatti, B.; Malik, Y.; Moreno, A.C.; Leskovar, D.; Cravero, V. Characterization of cardoon accessions as feedstock for biodiesel production. *Fuel* **2019**, *235*, 1287–1293. [CrossRef]

145. Gil, M.; Arauzo, I.; Teruel, E.; Bartolomé, C. Milling and handling *Cynara Cardunculus* L. for use as solid biofuel: Experimental tests. *Biomass Bioenergy* **2012**, *41*, 145–156. [CrossRef]

146. Encinar, J.M.; Gonzalez, J.F.; Rodriguez, J.J.; Tejedor, A. Biodiesel fuels from vegetable oils: Transesterification of Cynara c ardunculus L. oils with ethanol. *Energy Fuels* **2002**, *16*, 443–450. [CrossRef]

147. Encinar, J.M.; Gonzalez, J.F.; Gonzalez, J. Fixed-bed pyrolysis of *Cynara cardunculus* L. Product yields and compositions. *Fuel Process. Technol.* **2000**, *68*, 209–222. [CrossRef]

148. Ballesteros, M.; Negro, M.J.; Manzanares, P.; Ballesteros, I.; Sáez, F.; Oliva, J.M. Fractionation of Cynara cardunculus (cardoon) biomass by dilute-acid pretreatment. In *Applied Biochemistry and Biotecnology*; Humana Press: Totowa, NJ, USA, 2007; pp. 239–252.

149. Kalamaras, S.D.; Kotsopoulos, T.A. Anaerobic co-digestion of cattle manure and alternative crops for the substitution of maize in South Europe. *Bioresour. Technol.* **2014**, *172*, 68–75. [CrossRef] [PubMed]

150. Gominho, J.; Lourenço, A.; Curt, M.; Fernández, J.; Pereira, H. Characterization of hairs and pappi from Cynara cardunculus capitula and their suitability for paper production. *Ind. Crop. Prod.* **2009**, *29*, 116–125. [CrossRef]

151. Coccia, V.; Cotana, F.; Cavalaglio, G.; Gelosia, M.; Petrozzi, A. Cellulose nanocrystals obtained from Cynara cardunculus and their application in the paper industry. *Sustainability* **2014**, *6*, 5252–5264. [CrossRef]

152. De Falco, B.; Incerti, G.; Amato, M.; Lanzotti, V. Artichoke: Botanical, agronomical, phytochemical, and pharmacological overview. *Phytochem. Rev.* **2015**, *14*, 993–1018. [CrossRef]

153. Pandino, G.; Lombardo, S.; Mauro, R.P.; Mauromicale, G. Variation in polyphenol profile and head morphology among clones of globe artichoke selected from a landrace. *Sci. Hortic.* **2012**, *138*, 259–265. [CrossRef]

154. Cajarville, C.; González, J.; Repetto, J.L.; Rodríguez, C.A.; Martínez, A. Nutritive value of green forage and crop by-products of Cynara cardunculus. *Ann. Zootech.* **1999**, *48*, 353–365. [CrossRef]

155. Pari, L.; Civitarese, V.; Assirelli, A.; Del Giudice, A. Il prototipo che abbatte i costi della raccolta del cardo. *Inf. Agrar.* **2009**, *29*, 8–11.

156. Natelson, R.H.; Wang, W.C.; Roberts, W.L.; Zering, K.D. Technoeconomic analysis of jet fuel production from hydrolysis, decarboxylation, and reforming of camelina oil. *Biomass Bioenergy* **2015**, *75*, 23–34. [CrossRef]

157. Larsson, M. Cultivation and processing of Linum usitatissimum and Camelina sativa in southern Scandinavia during the Roman Iron Age. *Veg. Hist. Archaeobot.* **2013**, *22*, 509–520. [CrossRef]

158. Berti, M.; Wilckens, R.; Fischer, S.; Solis, A.; Johnson, B. Seeding date influence on camelina seed yield, yield components, and oil content in Chile. *Ind. Crop. Prod.* **2011**, *34*, 1358–1365. [CrossRef]

159. Berti, M.; Gesch, R.; Eynck, C.; Anderson, J.; Cermak, S. Camelina uses, genetics, genomics, production, and management. *Ind. Crop. Prod.* **2016**, *94*, 690–710. [CrossRef]

160. Zubr, J. Qualitative variation of Camelina sativa seed from different locations. *Ind. Crop. Prod.* **2003**, *17*, 161–169. [CrossRef]

161. Vollmann, J.; Moritz, T.; Kargl, C.; Baumgartner, S.; Wagentristl, H. Agronomic evaluation of camelina genotypes selected for seed quality characteristics. *Ind. Crop. Prod.* **2007**, *26*, 270–277. [CrossRef]

162. Mupondwa, E.; Li, X.; Tabil, L.; Falk, K.; Gugel, R. Technoeconomic analysis of camelina oil extraction as feedstock for biojet fuel in the Canadian Prairies. *Biomass Bioenergy* **2016**, *95*, 221–234. [CrossRef]

163. Blackshaw, R.; Johnson, E.; Gan, Y.; May, W.; McAndrew, D.; Barthet, V.; McDonald, T.; Wispinski, D. Alternative oilseed crops for biodiesel feedstock on the Canadian prairies. *Can. J. Plant Sci.* **2011**, *91*, 889–896. [CrossRef]

164. Chen, C.; Bekkerman, A.; Afshar, R.K.; Neill, K. Intensification of dryland cropping systems for bio-feedstock production: Evaluation of agronomic and economic benefits of Camelina sativa. *Ind. Crop. Prod.* **2015**, *71*, 114–121. [CrossRef]

165. Gesch, R.W.; Archer, D.W. Double-cropping with winter camelina in the northern Corn Belt to produce fuel and food. *Ind. Crop. Prod.* **2013**, *44*, 718–725. [CrossRef]

166. Berti, M.; Samarappuli, D.; Johnson, B.L.; Gesch, R.W. Integrating winter camelina into maize and soybean cropping systems. *Ind. Crop. Prod.* **2017**, *107*, 595–601. [CrossRef]

167. Royo-Esnal, A.; Valencia-Gredilla, F. Camelina as a rotation crop for weed control in organic farming in a semiarid mediterranean climate. *Agriculture* **2018**, *8*, 156. [CrossRef]

168. Peterson, A.T.; Berti, M.T.; Samarappuli, D. Intersowing cover crops into standing soybean in the US upper midwest. *Agronomy* **2019**, *9*, 264. [CrossRef]

169. Zanetti, F.; Christou, M.; Alexopoulou, E.; Berti, M.T.; Vecchi, A.; Borghesi, A.; Monti, A. Innovative double cropping systems including camelina [*Camelina sativa* (L.) Crantz] a valuable oilseed crop for bio-based applications. In Proceedings of the European Biomass Conference and Exhibition Proceedings, Lisbon, Portugal, 27–30 May 2019; pp. 127–130.

170. Dobre, P.; Jurcoane, S.; Matei, F.; Stelica, C.; Farcas, N.; Moraru, A.C. Camelina sativa as a double crop using the minimal tillage system. *Rom. Biotech. Lett.* **2014**, *19*, 9190–9195.

171. Florin, I.; Duda, M.M. Camelina sativa: A new source of vegetal oils Camelina sativa: A new source of vegetal oils Introduction Camelina, though cultivated for over 2000 years in the area for its seeds containing. *Rom. Biotechnol. Lett.* **2011**, *16*, 6263–6270.

172. Keske, C.M.H.; Hoag, D.L.; Brandess, A.; Johnson, J.J. Is it economically feasible for farmers to grow their own fuel? A study of Camelina sativa produced in the western United States as an on-farm biofuel. *Biomass Bioenergy* **2013**, *54*, 89–99. [CrossRef]

173. Yang, J.; Caldwell, C.; Corscadden, K.; He, Q.S.; Li, J. An evaluation of biodiesel production from Camelina sativa grown in Nova Scotia. *Ind. Crop. Prod.* **2016**, *81*, 162–168. [CrossRef]

174. Tepelus, A.; Rosca, P.; Dragomir, R. Biojet from hydroconversion of camelina oil mixed with straight run gas oil. *Rev. Chim.* **2019**, *70*, 3284–3291. [CrossRef]

175. Wang, Z.; Feser, J.S.; Lei, T.; Gupta, A.K. Performance and emissions of camelina oil derived jet fuel blends under distributed combustion condition. *Fuel* **2020**, *271*, 117685. [CrossRef]

176. Pietras, M.P.; Orczewska-Dudek, S. The effect of dietary camelina sativa oil on quality of broiler chicken meat. *Ann. Anim. Sci.* **2013**, *13*, 869–882. [CrossRef]

177. Orczewska-Dudek, S.; Pietras, M. The effect of dietary Camelina sativa oil or cake in the diets of broiler chickens on growth performance, fatty acid profile, and sensory quality of meat. *Animals* **2019**, *9*, 734. [CrossRef]

178. Halmemies-Beauchet-Filleau, A.; Shingfield, K.J.; Simpura, I.; Kokkonen, T.; Jaakkola, S.; Toivonen, V.; Vanhatalo, A. Effect of incremental amounts of camelina oil on milk fatty acid composition in lactating cows fed diets based on a mixture of grass and red clover silage and concentrates containing camelina expeller. *J. Dairy Sci.* **2017**, *100*, 305–324. [CrossRef]

179. Hixson, S.M.; Parrish, C.C.; Anderson, D.M. Use of camelina oil to replace fish oil in diets for farmed salmonids and atlantic cod. *Aquaculture* **2014**, *431*, 44–52. [CrossRef]

180. Toyes-Vargas, E.A.; Parrish, C.C.; Viana, M.T.; Carreón-Palau, L.; Magallón-Servín, P.; Magallón-Barajas, F.J. Replacement of fish oil with camelina (Camelina sativa) oil in diets for juvenile tilapia (var. GIFT Oreochromis niloticus) and its effect on growth, feed utilization and muscle lipid composition. *Aquaculture* **2020**, *523*, 735177. [CrossRef]

181. Pernak, J.; Łęgosz, B.; Klejdysz, T.; Marcinkowska, K.; Rogowski, J.; Kurasiak-Popowska, D.; Stuper-Szablewska, K. Ammonium bio-ionic liquids based on camelina oil as potential novel agrochemicals. *RSC Adv.* **2018**, *8*, 28676–28683. [CrossRef]

182. Pawlowska-Olszewska, M.; Puzio, I.; Harrison, A.P.; Borkowski, L.; Tymicki, G.; Grabos, D. Supplementation with camelina oil prevents negative changes in the artery in orchidectomized rats. *J. Physiol. Pharmacol.* **2018**, *69*, 109–116. [CrossRef]

183. Omonov, T.S.; Kharraz, E.; Curtis, J.M. Camelina (Camelina Sativa) oil polyols as an alternative to Castor oil. *Ind. Crop. Prod.* **2017**, *107*, 378–385. [CrossRef]

184. Li, X.; Mupondwa, E. Life cycle assessment of camelina oil derived biodiesel and jet fuel in the Canadian Prairies. *Sci. Total Environ.* **2014**, *481*, 17–26. [CrossRef]

185. Mohammad, B.T.; Al-Shannag, M.; Alnaief, M.; Singh, L.; Singsaas, E.; Alkasrawi, M. Production of multiple biofuels from Whole Camelina Material: A renewable energy crop. *BioResources* **2019**, *13*, 4870–4883. [CrossRef]

186. Krzyżaniak, M.; Stolarski, M.J.; Graban, Ł.; Lajszner, W.; Kuriata, T. Camelina and Crambe Oil Crops for Bioeconomy—Straw Utilisation for Energy. *Energies* **2020**, *13*, 1503. [CrossRef]

187. Zanetti, F.; Gesch, R.W.; Walia, M.K.; Johnson, J.M.F.; Monti, A. Winter camelina root characteristics and yield performance under contrasting environmental conditions. *Field Crop. Res* **2020**, *252*, 107794. [CrossRef]

188. Li, X.; Mupondwa, E. Production and value-chain integration of Camelina Sativa as a dedicated bioenergy feedstock in the Canadian prairies. In Proceedings of the 24th European Biomass Conference and Exhibition, Amsterdam, The Netherlands, 6–9 June 2016; pp. 151–157.

189. Simsek, S. Effects of biodiesel obtained from Canola, safflower oils and waste oils on the engine performance and exhaust emissions. *Fuel* **2020**, *265*, 117026. [CrossRef]

190. Oba, G.C.; Goneli, A.L.D.; Masetto, T.E.; Hartmann-Filho, C.P.; Michels, K.L.L.S.; Ávila, J.P.C. Artificial drying of safflower seeds at different air temperatures: Effect on the physiological potential of freshly harvested and stored seeds. *J. Seed Sci.* **2019**, *41*, 397–406. [CrossRef]

191. Martins, E.A.S.; Goneli, A.L.D.; Gonçalves, A.A.; Hartmann Filho, C.P.; Rech, J.; Oba, G.C. Physical properties of safflower grains. Part II: Volumetric shrinkage. *Rev. Bras. Eng. Agric. Ambient.* **2017**, *21*, 350–355. [CrossRef]

192. Pari, L.; Alfano, V.; Scarfone, A.; Toscano, G. *Tecnologie Innovative per un Utilizzo Efficiente dei Co-Prodotti Agricoli*; Compagnia delle Foreste: Arezzo, Italy, 2016.

193. Desai, B.B.; Kotecha, P.M.; Salunkhe, D.K. *Seeds Handbook. Biology, Production, Processing, and Storage*; Dekker, M., Ed.; Mareel Dekker: New York, NY, USA, 2004.

194. Tavares, G.R.; Massa, T.B.; Gonçalves, J.E.; da Silva, C.; dos Santos, W.D. Assessment of ultrasound-assisted extraction of crambe seed oil for biodiesel synthesis by in situ interesterification. *Renew Energy* **2017**, *111*, 659–665. [CrossRef]

195. Costa, E.; Almeida, M.F.; Alvim-Ferraz, C.; Dias, J.M. The cycle of biodiesel production from Crambe abyssinica in Portugal. *Ind. Crop. Prod.* **2019**, *129*, 51–58. [CrossRef]

196. Falasca, S.L.; Flores, N.; Lamas, M.C.; Carballo, S.M.; Anschau, A. Crambe abyssinica: An almost unknown crop with a promissory future to produce biodiesel in Argentina. *Int. J. Hydrogen Energy* **2010**, *35*, 5808–5812. [CrossRef]

197. Zhu, L.H.; Krens, F.; Smith, M.A.; Li, X.; Qi, W.; Van Loo, E.N.; Iven, T.; Feussner, I.; Nazarenus, T.J.; Huai, D.; et al. Dedicated Industrial Oilseed Crops as Metabolic Engineering Platforms for Sustainable Industrial Feedstock Production. *Sci. Rep.* **2016**, *6*, 1–11. [CrossRef]

198. Oliveira, R.L.D.; Dias, L.A.D.S.; Corrêa, T.R.; Ferreira, P.H.S.; Silva, M.F.D. Divergence and estimates of genetic parameters in Crambe abyssinica: An oilseed plant for industrial uses. *Rev. Ceres* **2018**, *65*, 500–506. [CrossRef]

199. Meijer, W.J.M.; Mathijssen, E.; Kreuzer, A.D. Low pod numbers and inefficient use of radiation are major constraints to high productivity in Crambe crops. *Ind. Crop. Prod.* **1999**, *9*, 221–233. [CrossRef]

200. Berti, M.T.; Johnson, B.L.; Manthey, L.K. Seed physiological maturity in Cuphea. *Ind. Crop. Prod.* **2007**, *25*, 190–201. [CrossRef]

201. Gettys, L.A.; Werner, D.J. Stratification unnecessary for germination of seeds of Stokes Aster [Stokesia laevis (J. Hill) Greene]. *Proc Fla. State Hortic. Soc.* **2001**, *114*, 250–251.

202. Goos, R.J.; Johnson, B.; Bourguignon, C. Preliminary evaluation of the soil application value of crambe meal. *Commun. Soil Sci. Plant Anal.* **2009**, *40*, 3211–3224. [CrossRef]

203. Zanetti, F.; Scordia, D.; Vamerali, T.; Copani, V.; Dal Cortivo, C.; Mosca, G. Crambe abyssinica a non-food crop with potential for the Mediterranean climate: Insights on productive performances and root growth. *Ind. Crop. Prod.* **2016**, *90*, 152–160. [CrossRef]

204. Adamsen, F.J.; Coffelt, T.A. Planting date effects on flowering, seed yield, and oil content of rape and crambe cultivars. *Ind. Crop. Prod.* **2005**, *21*, 293–307. [CrossRef]

205. Soratto, R.P.; de Souza-Schlick, G.D.; Fernandes, A.M.; Souza, E.D.F.C.D. Effect of fertilization at sowing on nutrition and yield of crambe in second season. *Rev. Bras. Cienc. Solo* **2013**, *37*, 658–666. [CrossRef]

206. Wu, W. Sustainable crop rotation for improving crop productivity and environmental safety: A book review. *J. Clean. Prod.* **2018**, *176*, 555–556. [CrossRef]

207. Zorzenoni, T.O.; de Andrade, A.P.; Higashibara, L.R.; Cajamarca, F.A.; Okumura, R.S.; Prete, C.C. Sowing date and fungicide application in the agronomic performance of oleaginous brassica for the biodiesel production. *Rev. Ceres* **2019**, *66*, 257–264. [CrossRef]

208. Costa, E.; Almeida, M.F.; Alvim-Ferraz, C.; Dias, J.M. Cultivation of Crambe abyssinica non-food crop in Portugal for bioenergy purposes: Agronomic and environmental assessment. *Ind. Crop. Prod.* **2019**, *139*, 111501. [CrossRef]

209. Oplinger, E.S.; Oelke, E.A.; Kaminski, A.R.; Putnam, D.H.; Teynor, T.M.; Doll, J.D.; Kelling, K.A.; Durgan, B.R.; Noetzel, D.M. *Crambe: Alternative Field Crops Manual*; University of Wisconsin and University of Minnesota: Saint Paul, MN, USA, 1991.

210. Nelson, L.A.; Grombacher, A.; Baltensperger, D.D. Crambe Production. In *Historical Materials from University of Nebraska-Lincoln Extension*; University of Nebraska-Lincoln: Lincoln, NE, USA, 1993.

211. Santangelo, E.; Menesatti, P.; Pari, L.; Fedrizzi, M. Possibilità produttive e di meccanizzazione del crambe. *Inf. Agrar.* **1994**, *50*, 31.

212. Polito, L.; Bortolotti, M.; Battelli, M.G.; Calafato, G.; Bolognesi, A. Ricin: An ancient story for a timeless plant toxin. *Toxins* **2019**, *11*, 324. [CrossRef]

213. Khatiwora, E.; Adsul, V.B.; Kulkarni, M.M.; Deshpande, N.R.; Kashalkar, R.V. Spectroscopic determination of total phenol and flavonoid contents of ipomoea carnea. *Int. J. ChemTech Res.* **2010**, *2*, 1698–1701.

214. Scholz, V.; Silva, J. Prospects and risks of the use of castor oil as a fuel. *Biomass Bioenergy* **2008**, *32*, 95–100. [CrossRef]

215. Nautiyal, O.H. Castor Oil and Its Derivatives' With Market Growth, Commercial Perspective: Review. *Org. Med. Chem. Int. J.* **2018**, *6*, 1–13. [CrossRef]

216. Mohammed, T.H.; Lakhmiri, R.; Azmani, A.; Hassan, I.I. Bio-oil from Pyrolysis of Castor Shell. *Int. J. Basic Appl. Sci.* **2014**, *14*, 1–5.

217. Roxas Duncan, V.I. Of Beans and Beads: Ricin and Abrin in Bioterrorism and Biocrime. *J. Bioterror. Biodef.* **2011**, *8*. [CrossRef]

218. Reddy, K.R.; Matcha, S.K. Quantifying nitrogen effects on castor bean (*Ricinus communis* L.) development, growth, and photosynthesis. *Ind. Crop. Prod.* **2010**, *31*, 185–191. [CrossRef]

219. Wnuk, A.; Górny, A.G.; Bocianowski, J.; Kozak, M. Visualizing harvest index in crops. *Commun. Biometry Crop Sci.* **2013**, *8*, 48–59.

220. Alexopoulou, E.; Papatheohari, Y.; Zanetti, F.; Tsiotas, K.; Papamichael, I.; Christou, M.; Namatov, I.; Monti, A. Comparative studies on several castor (*Ricinus communis* L.) hybrids: Growth, yields, seed oil and biomass characterization. *Ind. Crop. Prod.* **2015**, *75*, 8–13. [CrossRef]

221. Baldanzi, P. 1998 Selection for Non-Branching in Castor *Ricinus communis* L. *Plant Breed.* **1998**, *117*, 392–394. [CrossRef]

222. Vandenborre, G.; Smagghe, G.; Van Damme, E.J.M. Plant lectins as defense proteins against phytophagous insects. *Phytochemistry* **2011**, *72*, 1538–1550. [CrossRef]

223. Evofuel and Fantini s.r.l Breakthrough in Mechanical Harvesting for Castor Bean. Available online: https://www.evogene.com/press_release/evofuel-fantini-s-r-l-announce-breakthrough-mechanical-harvesting-castor-bean/. (accessed on 25 February 2020).

224. Anjani, K. Castor genetic resources: A primary gene pool for exploitation. *Ind. Crop. Prod.* **2012**, *35*, 1–14. [CrossRef]

Article

Effects of Site, Genotype and Subsequent Harvest Rotation on Willow Productivity

Mariusz Jerzy Stolarski *, Michał Krzyżaniak, Dariusz Załuski, Józef Tworkowski and Stefan Szczukowski

Department of Plant Breeding and Seed Production, Faculty of Environmental Management and Agriculture, Centre for Bioeconomy and Renewable Energies, University of Warmia and Mazury in Olsztyn (UWM), Plac Łódzki 3, 10-724 Olsztyn, Poland; michal.krzyzaniak@uwm.edu.pl (M.K.); dariusz.zaluski@uwm.edu.pl (D.Z.); jozef.tworkowski@uwm.edu.pl (J.T.); stefan.szczukowski@uwm.edu.pl (S.S.)
* Correspondence: mariusz.stolarski@uwm.edu.pl; Tel.: +48-89-5234838

Received: 3 August 2020; Accepted: 14 September 2020; Published: 17 September 2020

Abstract: Perennial crops harvested in short rotations provide substantial amounts of biomass. This study determined the survival rate, biometric features and yield of fresh and dry biomass of 15 willow genotypes (including seven varieties and eight clones), cultivated at two different sites in two consecutive three-year harvest rotations. The study revealed the very high impact of the genotype (81% of the total variance) on the willow yield. The harvest rotation, along with the genotype, had a significant impact on the plant survival rate and the number of shoots per stool. Willow biomass was mainly affected by the plant height, its survival rate and shoot diameter. The significantly highest fresh (106 Mg ha^{-1}) and dry biomass yield (54.0 Mg ha^{-1}) was obtained from the Żubr variety of *S. viminalis*, which distinguished this variety from the other genotypes. The mean yield for the best three and five genotypes was 13% and 17% lower, respectively, and the mean yield for the whole experiment was 37% lower compared to the mean yield of the best variety (Żubr). Therefore, the choice of a willow genotype is of key importance for successful willow production.

Keywords: Salix; genoype × site interaction; survivability; biometric features; plant height; fresh biomass yield; dry biomass yield

1. Introduction

Perennial crops, grown on agricultural land and harvested in short rotations, can provide substantial amounts of biomass [1–4]. Until now, biomass was usually used as energy feedstock and its use for industrial purposes and in integrated biorefineries has only recently been contemplated [5–8].

There are three major categories of perennial crops grown for biomass. These are short rotation woody crops (SRWCs) such as willow, poplar, eucalyptus, black locust; herbaceous crops such as Virginia mallow, willow-leaf sunflower, cup plant; grasses such as giant miscanthus, prairie cordgrass, switchgrass and giant reed [9–11]. Although such a large diversity of perennial crops offers the advantage of providing farmers with many choices, one has to possess sufficient knowledge in this regard. Obtaining a high biomass yield throughout the period of the plantation use (which has a significant impact on the production process profitability) requires making a proper selection of species and cultivars suitable for specific weather and site conditions [12–14].

Willow has several advantages as an SRWC, including a wide range of genetic diversity, easy reproduction, tolerance to a wide range of site conditions, ability of new shoots to rapidly regrow after multiple harvests, possibility of harvests in different rotations, resistance to disadvantageous environmental conditions, e.g., morning frost, wet snow and strong winds [4,15,16]. Therefore, willow biomass production in short harvest rotations has been researched in many countries, including

Europe [17–19], Canada and the USA [20–23]. Production of willow as an SRWC can be profitable [24,25], its energy balance is positive [26,27] and it brings measurable environmental benefits [15,28–30].

The advantages of willow and benefits from its cultivation as an SRWC are very important and can encourage potential producers to decide to start willow biomass production. However, for willow production to be economically, energy-efficiently and environmentally viable, it has to provide stable and high biomass yield per unit area [31]. The main factors that determine the willow biomass yield have been analysed for years and they include: (1) the choice of a suitable species and cultivar [18,23,31,32]; (2) soil conditions [13,18]; (3) weather conditions [31,33]; (4) agrotechnical procedures, including the fertilisation type and rate [17,34]; (5) planting density and harvesting frequency [9,31]. It should be noted that all of these factors are important, and they have a combined impact on the final effect, i.e., the willow biomass yield. However, when attempting to rank these factors, the choice of the species and variety should be identified as the most important. In consequence, if a biomass producer chooses a wrong willow species or variety, he will not achieve a high biomass yield even if the plantation is set up at a very good site, with the optimum weather and agrotechnical conditions. Moreover, the biomass producer, who has a very limited influence (or sometimes none) on the site or weather conditions, enjoys the full (100%) responsibility for the choice of plant species and variety.

Therefore, because of the large diversity of willow species, varieties and genotypes fit for cultivation as SRWCs and since willow production produces higher yield (per year of plantation use) in three-year harvest rotations [4,9,31], research was conducted to determine (1) survival rate and biometric features; (2) yield of fresh and dry biomass of 15 willow genotypes (including seven varieties and eight clones), cultivated at two different sites in two consecutive three-year harvest rotations. These findings were subsequently used to: (3) quantitatively determine the relative contribution of genetic and site-related factors and their interactions in explaining the variance of willow survival rate, biometric features and biomass yield.

2. Materials and Methods

2.1. Field Experiments

Two identical single-factorial field experiments were conducted in 2013–2018 at two experimental stations of the University of Warmia and Mazury (UWM), located near the villages of Bałdy and Obory in Poland. The experiment at Bałdy (Warmińsko-Mazurskie Province, 53°35′48″ N, 20°36′12″ E) was set up on mud-muck soil developed on calcareous gyttja in loamy subsoil. Willow cuttings were planted in April 2008 at a high density of 48 thousand per ha. The cuttings were planted in two-row strips with rows spaced every 0.75 m. The strips were spaced every 0.90 m. The cuttings in each row were planted every 0.25 m. Willows were coppiced in 2008–2012 in one-year rotations because it was believed at the time that it was a potentially interesting method for willow cultivation in small farms. However, as the market situation has changed since 2013, it was decided to extend the willow harvest cycle, which is why the plants were harvested twice in two subsequent three-year rotations: the first one (I) covering the years 2013–2015, and the second (II) the years 2016–2018. An identical field experiment was set up in April 2009 at Obory (Pomorskie Province, (53°43′34″ N, 18°53′55″ E) on humic heavy alluvial soil, formed from silty clay. The planting method was the same as at the Bałdy site. The willow at Obory was also coppiced in one-year rotations in 2009–2012. Subsequently (like at Bałdy), starting with 2013, willow was harvested twice in consecutive three-year rotations (I—2013–2015, II—2016–2018). Therefore, it must be explained that although the Bałdy experiment was set up a year earlier (2008) than the Obory experiment (2009), it had no significant direct impact on the results analysis in later three-year harvest rotations (2013–2018). This was a consequence of the fact that both at Bałdy and Obory, willows had been harvested earlier every year: five and four times, respectively. Plants harvested in the first three-year rotation grew on eight- and seven-year old stools, so the root systems can be regarded as very well-developed and the plants were giving the full yield. The stools in the second three-year rotation were 10 and 11 years old.

Fifteen identical willow genotypes of nine different species and interspecies hybrids were tested at each site (Bałdy and Obory) (Table 1). The 15 willow genotypes included seven varieties registered at the Centre for Cultivar Testing in Słupia Wielka, and eight clones—all of them bred and kept at the collection of UWM. At each site (Bałdy and Obory), each genotype was planted in a 150-m^2 strip. At both sites, within each strip, three 40-m^2 plots were randomly designated to conduct biometric measurements and determine the plant density and biomass yield.

The same mineral fertilisation was applied in each experiment before the next three-year rotation was begun. Therefore, the following fertilisers were applied in April 2013 and 2016: nitrogen as ammonium nitrate (90 kg ha^{-1} N), phosphorus as triple superphosphate (13 kg ha^{-1} P) and potassium—as potassium salt (50 kg ha^{-1} K).

Table 1. Genotypes (varieties and clones) of willow tested in two identical field experiments at Bałdy and Obory.

Species	Genotype Status (Variety or Clone)	Name
Salix viminalis	variety	Start
S. viminalis	variety	Sprint
S. viminalis	variety	Turbo
S. viminalis	variety	Tur
S. viminalis	variety	Kortur
S. viminalis	variety	Oltur
S. viminalis	variety	Żubr
S. acutifolia	clone	UWM 093
S. alba	clone	UWM 095
S. dasyclados	clone	UWM 155
S. fragilis	clone	UWM 195
S. pentandra	clone	UWM 035
S. triandra	clone	UWM 198
S. viminalis × *S. amygdalina*	clone	UWM 054
S. viminalis × *S. purpurea*	clone	UWM 033

2.2. Willow Survival Rate, Biometric Features and Biomass Yield

The plant density on three plots at each site for each harvest rotation was determined for each genotype after the growing seasons ended and it was subsequently calculated per 1 ha. Moreover, the plant survival rate (%) was determined. The height (m) and diameter (mm) of shoots were measured in ten randomly selected plants on each plot. The height was measured from the ground level to the top of the tallest plant shoot. The shoot diameter was measured 0.50 m above the ground level. The number of shoots per stool was also determined in ten replicates on each plot. Only live shoots taller than 1.50 m were taken into account, whereas dry shoots were excluded. Thus, each analysed biometric feature (height, diameter, number of shoots) was measured 30 times for each genotype at each site and for each harvest rotation. This makes up a total of 1800 measurements of each feature, which constituted input data for further analyses.

The plants were harvested in winter (February 2016 and 2017) manually with a chainsaw after each three-year rotation. Immediately after harvest, the whole mass of shoots from each plot was weighed (within an accuracy of 0.01 kg) and the fresh biomass yield was calculated (Mg ha^{-1} f.m.). Representative biomass samples (approximately 3 kg) were collected during harvest from entire plants of each genotype from each plot to determine the moisture content in it. The biomass moisture content was determined at 105 °C using the oven dry method (EN ISO 18134-1:2015). Subsequently, the moisture content and the fresh biomass yield were used to calculate the dry biomass yield (Mg ha^{-1} d.m. (dry matter yield)). The biomass yields from the three-year rotations were divided by three to present the biomass yield per one year of plantation use. Moreover, a ranking of the mean dry biomass yield (Mg ha^{-1} year^{-1} d.m.) for 15 willow genotypes from two sites in two consecutive three-year rotations was developed.

2.3. Statistical Analysis

All statistical analyses were conducted with STATISTICA 13.3 software (TIBCO Software Inc., Palo Alto, CA, USA 2017). Such variables as survival rate, number of shoots per stool, shoot diameter, plant height, fresh matter yield (f.m.) and dry matter yield (d.m.) were analysed statistically by a repeated-measures ANOVA, with site and genotype as the grouping factors and the replicates were nested in the site effect. The rotation effect was the only factor used for repeated measures. The significance of the factors and their interactions were tested at the significance level of $\alpha = 0.05$. The statistics F from these analyses are shown in the table of results. The percentage share of all the analysis effects under study in the total sum square (total SS) was calculated. This measure explained the share of individual factors in the trait variance. Moreover, Tukey's honest significant difference (HSD) test with $p < 0.05$ was used to evaluate the significance of differences between the subsequently determined means and homogeneous groups. Multiple regression analysis was applied in an assessment of the relationship between fresh and dry matter yield and the morphological features of plants. The variability of the analysed features in relation to the experimental conditions was evaluated using coefficients of variation (CV%).

2.4. Soil and Weather Conditions during the Experiments

Soil examinations at Bałdy showed that the mud-muck soil was neutral, with pH 7.2 in KCl. The humus content was 16.3%. The content of P_2O_5, K_2O and Mg was: 1435 mg, 560 mg and 637 mg kg^{-1} of soil, respectively. The complete humic heavy alluvial soil at Obory was also neutral with pH 7.0 in KCl. The humus content was lower: 7.09%. The content of P_2O_5, K_2O and Mg was also lower than at Bałdy: 488 mg, 300 mg and 157 mg kg^{-1} of soil, respectively. However, despite the differences in the contents of these elements, the soil at both sites (Bałdy and Obory) was good and fertile.

The sites were about 150 km apart, which is why the weather conditions were different. The average air temperature during six growing seasons (April–October) at Obory ranged from 13.3 °C to 16.0 °C, in 2017 and 2018, respectively (Table S1). The average temperature at Obory was always higher compared to Bałdy by 0.1 °C to 1.3 °C. Therefore, the annual average air temperature during the six years of the experiment at Obory (9.0 °C) was higher by approximately 0.7 °C compared to Bałdy. July and August were always the warmest months, with average monthly temperatures reaching 21 °C. The highest temperatures at both sites during the six growing seasons of the experiment were recorded in 2018.

It was the opposite case with precipitation because the average precipitation over six years at Bałdy (598 mm) was higher by approximately 20% than at Obory (Figure S1). The average precipitation during the growing season (April–October) at Bałdy (414 mm) was approximately 18% higher than Obory. Moreover, particularly large differences in the precipitation between the study sites were recorded during the second three-year harvest rotation (2016–2018) because precipitation during all the three growing seasons was higher at Bałdy than at Obory by 80–159 mm. The largest difference was recorded during the hottest growing season of 2018, when 419 mm of precipitation was recorded at Bałdy and only 260 mm at Obory. On the other hand, the highest precipitation during the whole six-year study period was recorded during the growing season of 2017 (533 and 612 mm, at Obory and Bałdy, respectively). The annual precipitation was also the highest in 2017: 640 and 803 mm, respectively.

3. Results

3.1. Plant Survival Rate, Number of Shoots, Plant Height and Shoot Diameter

The willow survival rate, number of shoots per stool and plant height differed significantly depending on the site, genotype, harvest rotation and interactions between most of these factors (Table 2). The shoot diameter was significantly differentiated only by the genotype, harvest rotation and the genotype-site interaction. The mean plant survival rate for all genotypes, sites and harvest rotations was 42.0% after the ten years of the experiment (Figure 1). Notably, since the initial planting density was very high (48 thousand per ha), the plant density in the last three-year rotation was still high—approximately 20 thousand per ha. The willow survival rate was differentiated to the greatest extent by the genotype—61% of the total variance, followed by the harvest rotation (17%) and the genotype–site interaction (13%) (Table 2). Among the genotypes under study, the highest survival rate was observed in the hybrid *S. viminalis* × *S. amygdalina* UWM 054 (nearly 62%) and the lowest was UWM 093 of *S. acutifolia* (only 20%) (Figure 1). The mean survival rate at Obory was higher by 1.5 percentage point (pp) compared to that at Bałdy. The mean willow survival rate in the first harvest rotation was 46.4%, whereas it decreased significantly in the second rotation by 8.8 pp.

The genotype had the greatest share in the variance of the number of shoots per stool (29%), followed by the harvest rotation, site and their interaction (Table 2). Among the genotypes under study, the significantly highest mean number of shoots (5.6) per stool was found on *S. fragilis* UWM 195, and the fewest were on UWM 093 of *S. acutifolia* (Figure 2). The mean number of shoots per stool at Bałdy (4.6) was larger by one than at Obory. It was similar when this attribute in the second harvest rotation was compared to the first rotation. It should be linked to the lower plant survival rate at Bałdy than at Obory and to decreasing plant survival rate in the second harvest rotation compared to the first harvest rotation. Owing to the smaller plant density, a larger number of shoots per rootstock were able to grow and survive.

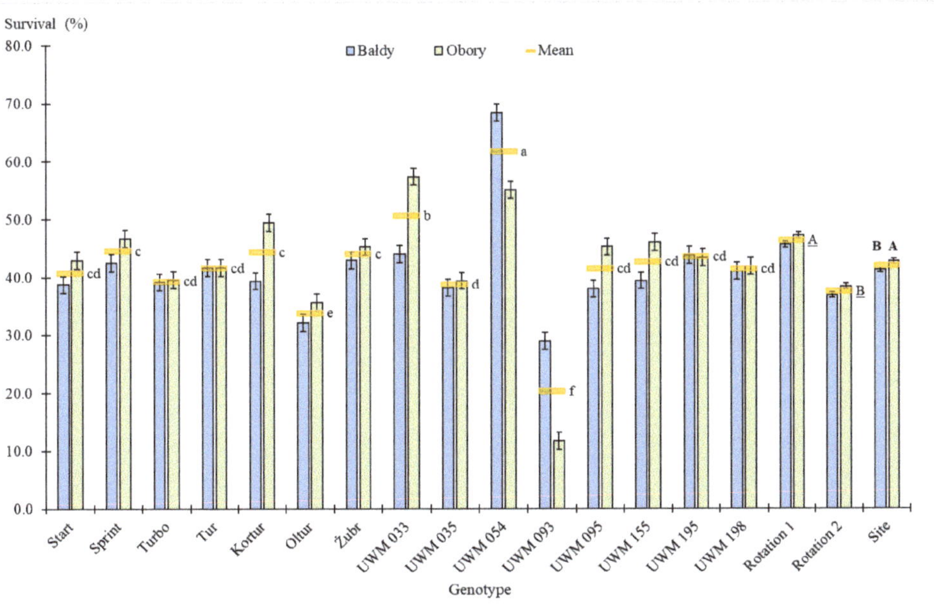

Figure 1. The survival rate of 15 willow genotypes from two sites in two consecutive three-year rotations (legend: error bars denote the standard error of mean; lower case letters a–f denote homogeneous groups from the Tukey test for mean values for genotypes regardless of the site and show differences between genotypes; underlined upper case letters A, B denote homogeneous groups for mean values for harvest rotation and show differences between rotations; bold upper case letters A, B denote homogeneous groups for mean values for sites and show differences between sites; ns denote not significant).

Table 2. Statistics F from the repeated measure variance analysis and the percentage share of effects in the total sum of squares of an attribute.

Source of Variation	df	Survival		No. of Shoots		Height		Shoot Diameter		Fresh Matter Yield		Dry Matter Yield	
		F	Share (%)	F	Share (%)	F	Share (%)	F	Share (%)	F	Share (%)	F	Share (%)
Site	1	8.1 **	0.5	68.4 ***	9.2	35.3 ***	1.6	2.3	0.5	0.9	0.0	1.8	0.1
Rep (Site)	4	1.70	0.4	3.3 *	1.8	0.7	0.1	0.3	0.2	0.4	0.0	0.4	0.0
Genotype (Gen)	14	68.4 ***	60.7	15.7 ***	29.3	131.3 ***	80.9	18.9 ***	53.3	196.5 ***	81.2	194.4 ***	81.1
Site × Gen	14	14.4 ***	12.8	5.6***	10.5	13.0 ***	8.0	5.0 ***	14.1	20.2 ***	8.3	20.0 ***	8.3
Error 1	56		3.6		7.5		2.5		11.3		1.7		1.7
Rotation (Rot)	1	425.1 ***	17.0	88.8 ***	15.5	26.2 ***	1.2	14.7 ***	3.0	136.3 ***	3.7	133.9 ***	3.7
Rot × Site	1	0.00	0.0	6.1 *	1.1	9.9 **	0.4	0.1	0.0	15.5 ***	0.4	19.5 ***	0.5
Rot × Rep (Site)	4	1.20	0.2	0.9	0.6	1.4	0.3	0.5	0.4	0.4	0.0	0.4	0.0
Rot × Gen	14	1.70	0.9	3.6 ***	8.8	3.2 ***	2.0	0.8	2.3	4.6 ***	1.8	3.8 ***	1.5
Rot × Site × Gen	14	2.8 **	1.6	2.5 **	6.1	1.0	0.6	1.1	3.3	3.4 ***	1.3	3.7 ***	1.4
Error 2	56		2.2		9.7		2.5		11.6		1.5		1.6
Total			100.0		100.0		100.0		100.0		100.0		100.0

Share (%) percentage share in the total sum of squares; * $p < 0.05$; ** $p < 0.01$; *** $p < 0.001$.

The plant height was determined to the greatest extent by the genotype (nearly 81%), followed by interaction of the site and genotype (8%) (Table 2). Among the 15 genotypes under study, plants of the Żubr variety (*S. viminalis*) were the significantly tallest (6.9 m), whereas those of the UWM 093 genotype of *S. acutifolia* were the shortest (4.1 m) (Figure 3). Plants at Bałdy were taller by 0.2 m on average compared to those growing at Obory and the plants were taller in the second rotation.

An assessment of the impact of the studied factors on the shoot diameter showed similar dependence for the plant height, as this attribute is determined the most strongly by the genotype (53%), followed by the interaction of site and genotype (14%) (Table 2). Shoots of the Żubr variety (*S. viminalis*) were also the thickest: mean 35.5 mm. On the other hand, shoots in the UWM 093 genotype and the Start variety were the thinnest (Figure 4). It is also noteworthy that individual variance in the objects under study had a greater share compared to the other shoot diameter attributes and the number of shoots per stool.

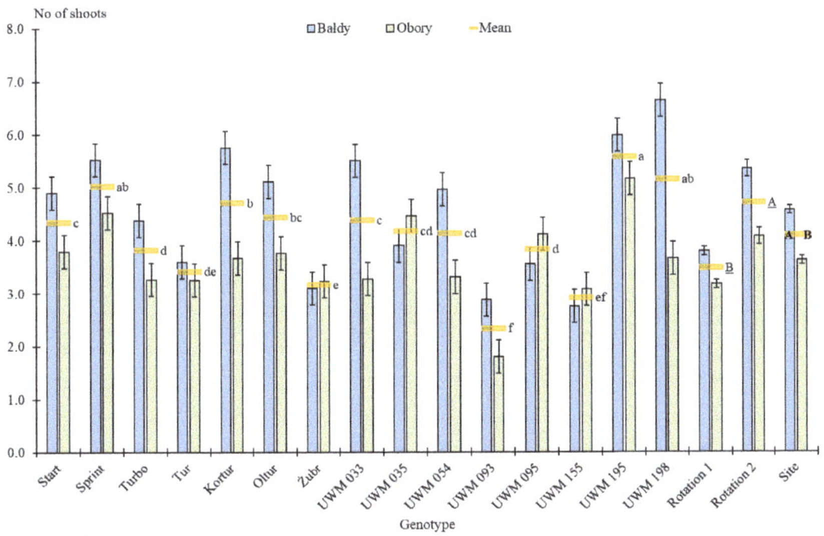

Figure 2. The number of shoots per stool of 15 willow genotypes from two sites in two consecutive three-year rotations (legend: see Figure 1).

3.2. Biomass Yield

Both the fresh and dry biomass yield were differentiated by the genotype and harvest rotation and by their interaction. However, no site impact on the biomass yield was observed (Table 2). The biomass yield was differentiated to the greatest extent by the genotype (81% of the total variance), followed by the interaction of genotype and site (8%) and harvest rotation (4%). Among the genotypes under study, the significantly highest fresh biomass yield (106 Mg ha^{-1}) was obtained from the Żubr variety of *S. viminalis*, which distinguished this variety against the other genotypes (Figure 5). On the other hand, because of the lowest survival rate and the poorest biometric features, the fresh biomass yield for the UWM 093 genotype of *S. acutifolia* was very low (8 Mg ha^{-1}), which practically eliminates this genotype from further research devoted to willow yield assessment. Furthermore, the mean yield of the other 13 genotypes was lower from 17% up to 54%, for *S. fragilis* UWM 195 and Start, respectively, when compared with the Żubr variety. The fresh biomass yield in the first three-year harvest rotation, when the plants were harvested from the 7-year-old stool, was higher by over 9 Mg ha^{-1} compared to the second rotation yield.

When the biomass moisture content was taken into account, the relationships for dry biomass yield were similar to those for the fresh biomass yield. The significantly highest mean dry biomass yield (54.0 Mg ha^{-1}) was obtained from the Żubr variety of species *S. viminalis* (Figure 6). On the other hand, the poorest UWM 093 genotype of species *S. acutifolia* yielded only 4 Mg ha^{-1}. The other 13 genotypes gave a lower yield, from 17% up to 56%, for UWM 195—an *S. fragilis* genotype—and UWM 155 of *S. dasyclados*, respectively. The mean dry biomass yield in the first three-year harvest rotation (36.6 Mg ha^{-1}) was higher by 4.8 Mg ha^{-1} compared to the second rotation yield.

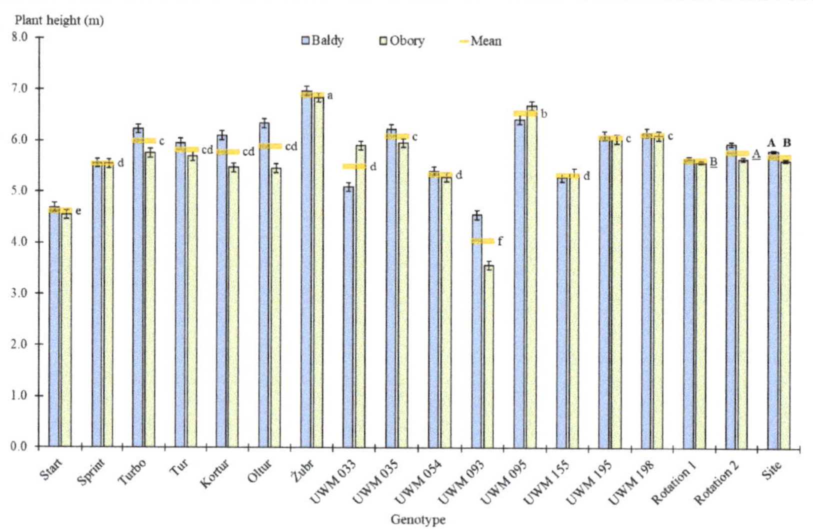

Figure 3. The plant height of 15 willow genotypes from two sites in two consecutive three-year rotations (legend: see Figure 1).

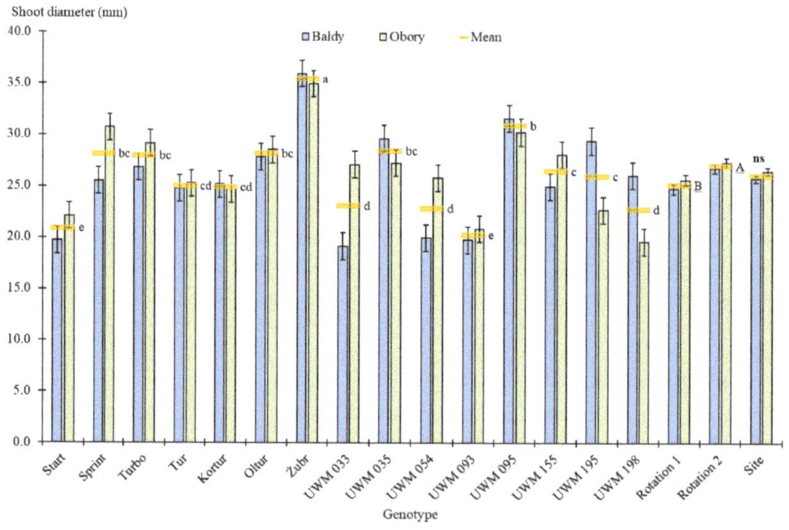

Figure 4. The shoot diameter of 15 willow genotypes from two sites in two consecutive three-year rotations (legend: see Figure 1).

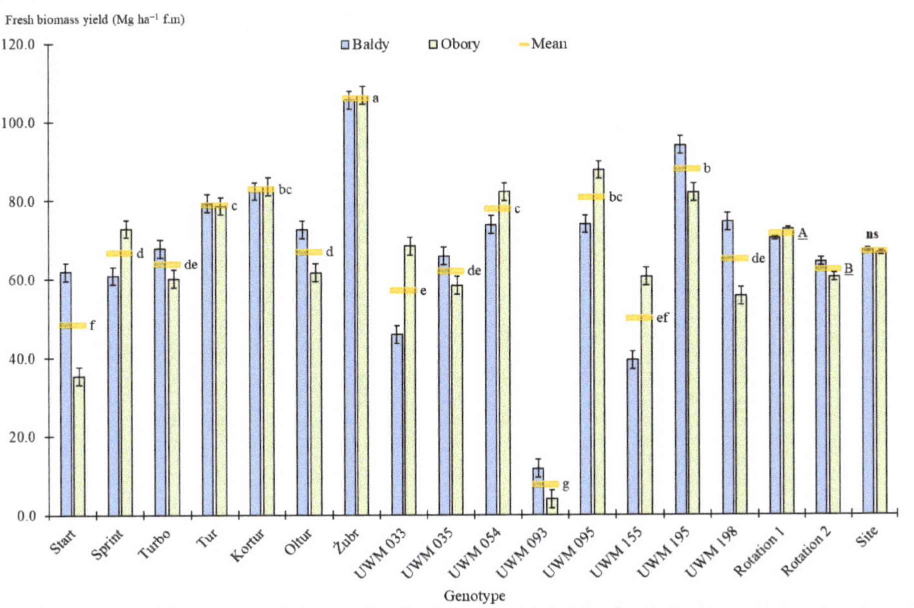

Figure 5. Fresh biomass yield of 15 willow genotypes from two sites in two consecutive three-year rotations (legend: see Figure 1).

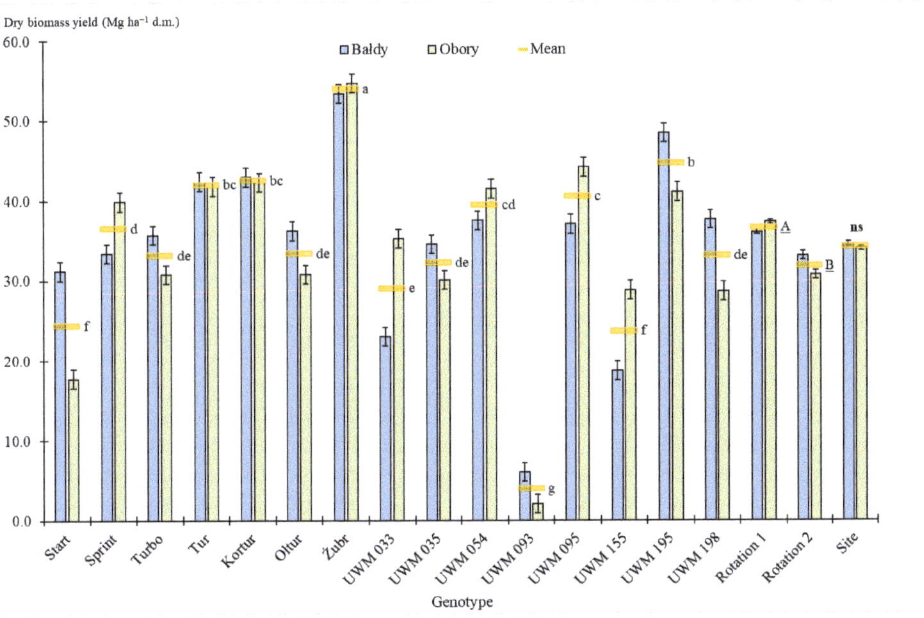

Figure 6. Dry biomass yield of 15 willow genotypes from two sites in two consecutive three-year rotations (legend: see Figure 1).

Both the fresh and dry biomass yields were significantly affected by the plant survival rate and their biometric features. The Pearson correlation coefficients (Table 3) showed significant inter-relations between the fresh or dry biomass yield and the plant height, shoot diameter and the plant survival rate. However, for the latter feature, the authors considered the number of plants, which survived until the subsequent 3-year harvest rotation rather than the initial planting density.

The greatest correlation strength was observed for the plant height (up to 0.76). It is this feature that was the principal determinant of the biomass yield, although the number of shoots did not correlate significantly with the biomass yield. Multiple regression models (Table 4) have confirmed these relations and the R^2 coefficients not only expand interpretations by providing information on a good fit of the regression models to experimental data, but also these three significant characteristics alone explained 75% of the biomass yield variance.

Table 3. Pearson correlation coefficients between the fresh and dry biomass yield and the yield structure elements: survival rate, number of shoots, height and diameter.

Variable	Survival	No. of Shoots	Height	Shoot Diameter
Fresh matter yield	0.57 **	0.15	0.76 **	0.55 **
Dry matter yield	0.56 **	0.17	0.75 **	0.54 **

$** p < 0.01.$

Table 4. Multiple regression analysis in an assessment of the relationship between fresh or dry matter yield and biometric features.

Parameter	Parameter Value	*p*-Value
Fresh matter yield		
Intercept	**−98.95**	**<0.0001**
Survival	**2.18**	**<0.0001**
No. of shoots	0.85	0.4826
Height	**15.12**	**<0.0001**
Shoot diameter	**1.21**	**0.0186**
R^2	0.76	
$R^2_{adj.}$	0.74	
Dry matter yield		
Intercept	**−51.27**	**<0.0001**
Survival	**1.11**	**<0.0001**
No. of shoots	0.58	0.3615
Height	**7.64**	**0.0001**
Shoot diameter	**0.64**	**0.0180**
R^2	0.75	
$R^2_{adj.}$	0.73	

Significant parameters are shown in bold.

The coefficients of variation analyses showed the lowest variation (13.3%) in the total approach for height and the highest (38.3%) for the number of shoots (Table 5). The same pattern was observed when this variation was broken down into the principal components: site, genotype and rotation. The biomass yield variation was 36%. Therefore, since biomass yield depends on several uncontrollable biotic and abiotic factors, it can be claimed that these results are good and relatively stable. Moreover, taking into account the effect of the principal components on the variation of the characteristics, the lowest variation was determined by the genetic factor (genotype). The genotype variation ranged from 5.9% (height) to 30.7% (number of shoots) and the effect of the other two principal factors (site and rotation) had a similar impact on individual variables.

It is very important in an assessment of the willow biomass yield to present the dry biomass yield per one year of plantation use. Such yield in the whole experiment presented in this paper amounted to 11.4 Mg ha^{-1} year^{-1} d.m., regardless of the genotype, site or harvest rotation (Figure 7). Moreover, seven out of the 15 genotypes under study gave a higher yield than the mean from the whole experiment. However, the differences between genotypes, sites and rotations were large because the highest-yielding Żubr variety gave a mean yield of 18 Mg ha^{-1} year^{-1} d.m., and 20.3 Mg ha^{-1} year^{-1} d.m. in the first harvest rotation at Obory (Table 6). The mean yield for the best three and five genotypes was lower by 13% and 17%, respectively, and the mean yield of the whole experiment was lower by 37% compared to the mean yield of the best variety (Żubr). It is also noteworthy that all of the genotypes in the second harvest rotation at each site gave a lower yield than in the first rotation. The yield decrease in the second rotation at Obory was much larger (mean 18%) compared to Bałdy (mean 8%). Moreover, only in the first rotation was the mean dry biomass yield higher at both sites than the mean for the whole experiment.

Table 5. Percent coefficients of variation (CV%) for site, genotype, harvest rotation and total for the willow characteristics under study.

Variable	Site	Genotype	Rotation	Total
Survival	25.42	17.43	23.59	25.50
No. of shoots	34.95	30.73	33.10	38.31
Height	13.24	5.90	13.27	13.32
Shoot diameter	20.25	14.25	20.01	20.24
Fresh matter yield	35.89	18.61	35.19	35.82
Dry matter yield	36.27	18.97	35.59	36.19

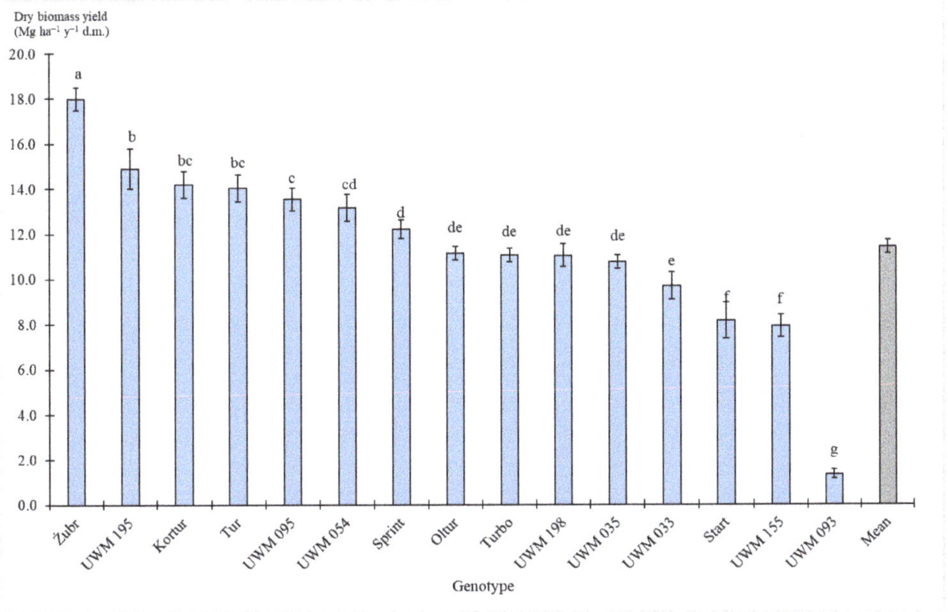

Figure 7. Ranking of the mean dry biomass yield (Mg ha^{-1} year^{-1} d.m.) for 15 willow genotypes from two sites in two consecutive three-year rotations (error bars denote the standard error of mean; letters a–f denote homogeneous groups).

Table 6. First and second three-year rotation dry biomass yields (Mg ha^{-1} year^{-1} d.m.) and changes in yield for the top 1, 3, 5 and all 15 willow genotypes at two sites in northern Poland.

Genotype		Bałdy	Obory	Mean for Two Sites
Top 1	1st rotation yield	18.0	20.3	19.2
	2nd rotation yield	17.6	16.1	16.8
	yield difference	−0.4	−4.3	−2.3
	% of yield decrease	2.1	20.9	12.1
	mean yield from two rotations	17.8	18.2	18.0
	% of mean yield compared to the highest-yielding genotype	100.0	100.0	100.0
Top 3	1st rotation yield	16.7	17.6	17.1
	2nd rotation yield	15.4	13.1	14.3
	yield difference	−1.3	−4.4	−2.9
	% of yield decrease	7.8	25.3	16.8
	mean yield from two rotations	16.1	15.3	15.7
	% of mean yield compared to the highest-yielding genotype	90.4	84.2	87.3
Top 5	1st rotation yield	15.6	17.0	16.3
	2nd rotation yield	14.3	12.9	13.6
	yield difference	−1.3	−4.1	−2.7
	% of yield decrease	8.5	24.0	16.5
	mean yield from two rotations	14.9	14.9	14.9
	% of mean yield compared to the highest-yielding genotype	84.0	81.9	83.0
Mean for all 15 genotypes	1st rotation yield	12.0	12.4	12.2
	2nd rotation yield	11.0	10.2	10.6
	yield difference	−1.0	−2.2	−1.6
	% of yield decrease	8.2	17.7	13.0
	mean yield from two rotations	11.5	11.3	11.4
	% of mean yield compared to the highest-yielding genotype	64.7	62.1	63.4

4. Discussion

In the present study, the willow plant survival rate decreased naturally with increasing plantation age. It was also found in other research by the authors that the willow survival rate over 12 consecutive years was significantly differentiated by the genotype, harvest rotation and the initial planting density [31]. The willow survival rate in that research decreased rapidly in the first and second harvest rotation (88% and 58%). It was more stable in the third and fourth rotation, with a slightly decreasing trend (51% and 47%). The highest survival rate in the four three-year harvest rotations was determined for the UWM 095 genotype in the first and fourth rotation (91% and 53%, respectively).

The mean survival rate of 30 willow genotypes in research conducted in Canada was also highly varied [35]. The mean willow survival rate after the second three-year rotation in that research was 60% (ranging from 3% to 94%) and it was lower by 25% compared to the first rotation. Furthermore, the survival rate of 12 willow genotypes in the first three-year harvest rotation in a study conducted in the USA ranged from 63% to 100%, and from 60% to 97% in the subsequent rotation [36]. In another study, the mean survival rate of seven willow genotypes after the first three-year rotation was 75%, but the inter-genotype differences were large, ranging from 44% to 85% [37].

In the present study, the number of shoots per rootstock ranged widely from 1.8 to 6.6, whereas in research on 25 willow plantations in Denmark, the number of shoots also ranged widely—from 1.4 to 9.9 shoots per stool depending on the cultivar, plantation age and harvest rotation [38]. The number of three-year-old willow shoots per stool ranged from 2.1 to 3.1. A similar number of shoots per stool (2.7) for seven willow genotypes in a three-year harvest rotation were found in another study [37]. An even smaller mean number of shoots per stool (1.4–2.3) were found in the willow harvested in the first five-year harvest rotation in Spain [39].

In the present study, the Żubr variety was particularly distinctive since it produced the tallest plants with the largest shoot diameters. Moreover, this plant's survivability was slightly higher, although the number of shoots was lower than the average for all the 15 genotypes. In other studies, three-year-old plants of the Żubr variety (formerly UWM 006) were also the tallest (7.28 m) and the thickest (48.6 mm) [37]. Willows were found to be shorter in the research conducted in Canada, where the mean height for 30 genotypes was found to be 2.55 m, with the best three genotypes being 3.76 m tall [35]. The shoot diameter for 30 genotypes in Canada measured only 11.9 mm and 15 mm for the best three genotypes, which was markedly lower than in the present study.

Willow harvest in three-year rotations can give a higher dry biomass yield compared to the yield in shorter one- or two-year rotations [4,31]. Moreover, the willow biomass yield in the first three-year rotation is generally lower compared to subsequent three-year harvest rotations [13,21,36,40,41]. The yield increase in the second and subsequent harvest rotations was particularly manifest in plantations that were not coppiced in the first growing year to increase the number of shoots per stool. However, in some cases, the biomass yield during the first three-year harvest rotation was much higher than in the subsequent harvest rotations [9,31,42,43].

It must be stressed that the current study dealt with two consecutive three-year harvest rotations and one cannot (and must not) regard them as the first and the second three-year harvest rotation in a direct meaning of the word. This is a consequence of the fact that plants in the experiment were obtained earlier in one-year harvest rotations, and a change of the harvest rotation to three years took place in the fifth and the sixth year of the plantation use, when the plants' root systems were strong (which is also explained in the Materials and Methods section). Plants harvested in the first three-year rotation grew on eight-and seven-year-old stools, so the root systems can be regarded as very well-developed and the plants were giving the full yield. Nevertheless, the study showed that the duration of a willow harvest rotation can be changed. This knowledge is also important to potential investors, as it gives them, in a sense, some flexibility in choosing the willow harvest rotation.

The mean biomass yield from two three-year harvest rotations in this study for most of the 15 genotypes under study (except one) ranged from 8 to 18 Mg ha^{-1} year^{-1} d.m., with the mean yield being 11.4 Mg ha^{-1} year^{-1} d.m. However, because of the plantation (and the plants' root systems) age, the biomass yield obtained in the current experiment should be compared with the findings of other experiments, starting with the second three-year harvest rotation. For this reason, literature data were used and Figure 8 shows a comparison of the willow biomass yield mainly in the second and the third harvest rotation [9,13,21,31,35,36,38,40,41,44–52], although the harvest rotation was shorter or longer by one year in some studies. Since the studies were conducted in different countries, the site conditions (soil and weather) were clearly different. Moreover, the plantations were set up with various clones and various agrotechnical and plantation management procedures were applied. Nevertheless, the findings of this study concerning the mean dry biomass yield in the three-year harvest rotation were on a medium or high level compared to the other studies cited in Figure 8. Moreover, the current results were similar to those of a study conducted at five different sites in the USA (mean range 10.0–11.6 Mg ha^{-1} year^{-1} d.m.) [9].

In the present study, the best-yielding variety—Żubr—gave a higher biomass yield at both sites and harvest rotations (range: 16–20 Mg ha^{-1} year^{-1} d.m.) compared to the levels achieved for the best genotypes in the USA (range 11–14 Mg ha^{-1} year^{-1} d.m.) [9]. A large increase in the biomass yield in the second three-year harvest rotation was also achieved in Canada, mean 13.3 Mg ha^{-1} year^{-1} d.m. [44]. Similar mean yield as in Canada was obtained in the authors' earlier research of five willow genotypes grown at four planting densities and harvested in four three-year rotations [31]. The highest mean yield (14.5 Mg ha^{-1} year^{-1} d.m.) was achieved in the cited experiment from the UWM 095 genotype, i.e., it was higher only by one Mg ha^{-1} year^{-1} d.m. compared to the mean value for this genotype achieved in the authors' research. The other genotypes gave lower yields of 2.3%, 5.1%, 8.2% and 24.7%, respectively, compared to the UWM 095 genotype [31]. However, one must add that the willow yield in four consecutive three-year harvest rotations, depending on the experimental factors as determined in the cited research, varied strongly and ranged from 4.8 to 23.2 Mg ha^{-1} year^{-1} d.m.

Considerable diversity (from 3.5 to 13.6 Mg ha^{-1} year^{-1} d.m.) was also shown in other research conducted in the USA using 18 willow genotypes at two sites in a three-year harvest rotation [23]. Furthermore, in a study conducted in Sweden, the mean yield for two willow cultivars grown at five sites with no fertilisation amounted to 5.9 Mg ha^{-1} year^{-1} d.m. [17]. The highest yield (13.2 Mg ha^{-1} year^{-1}) was achieved with intensive fertilisation. The willow yield obtained in Denmark also ranged widely from 2.4 to 15.1 Mg ha^{-1} year^{-1} d.m. [38]. Moreover, as in other studies, the yield in the second rotation was higher than in the first one.

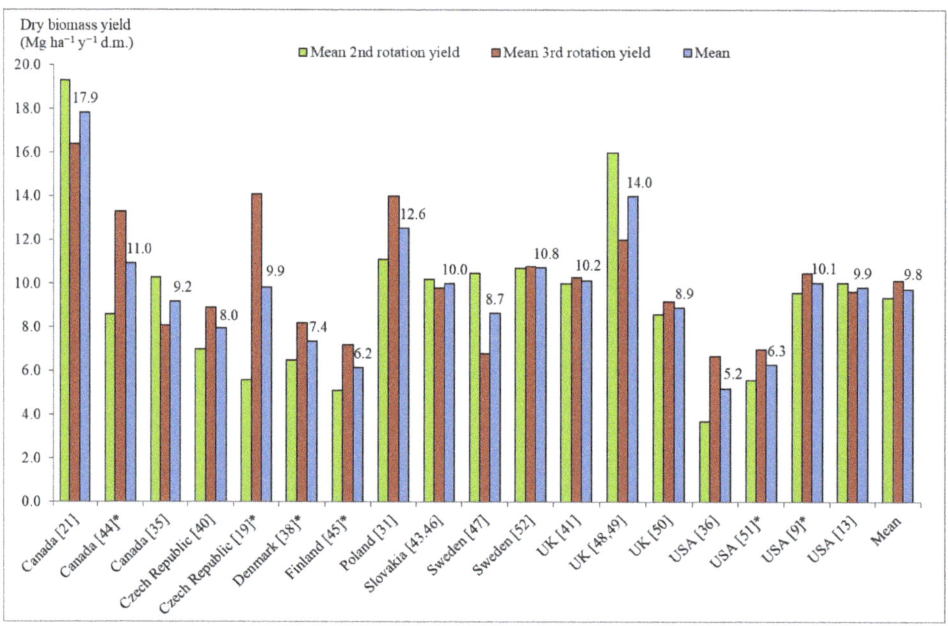

Figure 8. A comparison based on literature data of willow biomass yield (Mg ha^{-1} year^{-1} d.m.) in the second and the third harvest rotation, although the harvest rotation was shorter or longer by one year in some studies (* data from the first and second three-year willow harvest rotation).

5. Conclusions

The current study analysed 15 genotypes grown at two different sites and harvested in two consecutive three-year harvest rotations and demonstrated the very high impact of the genotype (81%) on the yield of willow grown as SRWC. The harvest rotation, along with the genotype, had a significant impact on the plant survival rate and the number of shoots per stool. The results showed that the willow biomass was mainly affected by the plant height, survival rate and shoot diameter. The differences in the biomass yield between the genotypes under study were very large, and the Żubr variety of species *S. viminalis* gave a particularly high yield. The mean yield for the best three and five genotypes was 13% and 17% lower, respectively, and the mean yield for the whole experiment was 37% lower compared to the mean yield of the best variety (Żubr). Therefore, the choice of a willow genotype is of key importance to success in SRWC willow production, since single-genotype monoculture on a commercial plantation may be a significant source of future problems with disease development or pest infestations. Therefore, further research is needed in subsequent harvest rotations to verify the yielding stability and to assess the economic, energy and environmental viability of biomass production based on the highest-yielding genotypes. Such knowledge will improve production and increase interest among producers and consumers of this renewable biomaterial.

Supplementary Materials: The following are available online at http://www.mdpi.com/2077-0472/10/9/412/s1, Figure S1: Monthly precipitation (mm) in the years of the experiment at two sites Bałdy (a) and Obory (b) and mean value for years 2013–2018., Table S1: Mean monthly air temperature (°C) in the years of the experiment at two sites (Bałdy and Obory) and average temperatures in the growing seasons (IV-X) and throughout the whole year.

Author Contributions: Conceptualisation, M.J.S., S.S., J.T.; Data curation, M.J.S., M.K.; Formal analysis, M.J.S., M.K., D.Z.; Funding acquisition, M.J.S., S.S.; Investigation, M.J.S., S.S. and M.K.; Methodology, M.J.S., S.S., J.T. and M.K.; Project administration, M.J.S.; Validation, M.J.S., M.K., and D.Z.; Visualisation, M.J.S., D.Z.; Writing—original draft, M.J.S.; Writing—review and editing, M.J.S., M.K., D.Z., J.T. and S.S. All authors have read and agreed to the published version of the manuscript.

Funding: This paper is the result of a long-term study carried out at the University of Warmia and Mazury in Olsztyn, Faculty of Environmental Management and Agriculture, Department of Plant Breeding and Seed Production, topic number 20.610.008–300 and it was co-financed by the National (Polish) Centre for Research and Development (NCBiR), entitled "Environment, agriculture and forestry", project: BIOproducts from lignocellulosic biomass derived from MArginal land to fill the Gap In Current national bioeconomy, No. BIOSTRATEG3/344253/2/NCBR/2017.

Acknowledgments: We would also like to thank the staff of the Department of Plant Breeding and Seed Production for their technical support during the experiments.

Conflicts of Interest: The authors declare no conflict of interest.

References

1. Tharakan, P.J.; Volk, T.A.; Abrahamson, L.P.; White, E.H. Energy feedstock characteristics of willow and hybrid poplar clones at harvest age. *Biomass Bioenergy* **2003**, *25*, 571–580. [CrossRef]
2. US. Department of Energy. *US Billion-Ton Update: Biomass Supply for a Bioenergy and Bioproducts Industry*; U.S. Department of Energy: Oak Ridge, TN, USA, 2011; p. 227.
3. Nordborg, M.; Berndes, G.; Dimitriou, I.; Henriksson, A.; Mola-Yudego, B.; Rosenqvist, H. Energy analysis of willow production for bioenergy in Sweden. *Renew. Sustain. Energy Rev.* **2018**, *93*, 473–482. [CrossRef]
4. Stolarski, M.J.; Niksa, D.; Krzyżaniak, M.; Tworkowski, J.; Szczukowski, S. Willow productivity from small- and large-scale experimental plantations in Poland from 2000 to 2017. *Renew. Sustain. Energy Rev.* **2019**, *101*, 461–475. [CrossRef]
5. Gonçalves, F.A.; Sanjinez-Argandona, E.J.; Fonseca, G.G. Cellulosic ethanol and its co-products from different substrates, pretreatments, microorganisms and bioprocesses: A review. *Nat. Sci.* **2013**, *5*, 624–630. [CrossRef]
6. Monedero, E.; Portero, H.; Lapuerta, M. Pellet blends of poplar and pine sawdust: Effects of material composition, additive, moisture content and compression die on pellet quality. *Fuel Process. Technol.* **2015**, *132*, 15–23. [CrossRef]
7. Stolarski, M.J.; Krzyżaniak, M.; Łuczyński, M.; Załuski, D.; Szczukowski, S.; Tworkowski, J.; Gołaszewski, J. Lignocellulosic biomass from short rotation woody crops as a feedstock for second-generation bioethanol production. *Ind. Crops Prod.* **2015**, *75*, 66–75. [CrossRef]
8. Tyśkiewicz, K.; Konkol, M.; Kowalski, R.; Rój, E.; Warmiński, K.; Krzyżaniak, M.; Gil, Ł.; Stolarski, M.J. Characterization of bioactive compounds in the biomass of black locust, poplar and willow. *Trees* **2019**, *33*, 1235–1263. [CrossRef]
9. Sleight, N.J.; Volk, T.A.; Johnson, G.A.; Eisenbies, M.H.; Shi, S.; Fabio, E.S.; Pooler, P.S. Change in yield between first and second rotations in willow (Salix spp.) biomass crops is strongly related to the level of first rotation yield. *BioEnergy Res.* **2016**, *9*, 270–287. [CrossRef]

10. PocienÈ, L.; KadŽIulienÈ, Ž. Biomass yield and fibre components in reed canary grass and tall fescue grown as feedstock for combustion. *Zemdirb. Agric.* **2016**, *103*, 297–304. [CrossRef]

11. Stolarski, M.J.; Śnieg, M.; Krzyżaniak, M.; Tworkowski, J.; Szczukowski, S.; Graban, Ł.; Lajszner, W. Short rotation coppices, grasses and other herbaceous crops: Biomass properties versus 26 genotypes and harvest time. *Ind. Crops Prod.* **2018**, *119*, 22–32. [CrossRef]

12. Monti, A.; Zanetti, F.; Scordia, D.; Testa, G.; Cosentino, S.L. What to harvest when? Autumn, winter, annual and biennial harvesting of giant reed, miscanthus and switchgrass in northern and southern Mediterranean area. *Ind. Crops Prod.* **2015**, *75*, 129–134. [CrossRef]

13. Sleight, N.J.; Volk, T.A. Recently bred willow (Salix spp.) biomass crops show stable yield trends over three rotations at two sites. *BioEnergy Res.* **2016**, *9*, 782–797. [CrossRef]

14. Stolarski, M.J.; Śnieg, M.; Krzyżaniak, M.; Tworkowski, J.; Szczukowski, S. Short rotation coppices, grasses and other herbaceous crops: Productivity and yield energy value versus 26 genotypes. *Biomass Bioenergy* **2018**, *119*, 109–120. [CrossRef]

15. Volk, T.A.; Verwijst, T.; Tharakan, P.J.; Abrahamson, L.P.; White, E.H. Growing fuel: A sustainability assessment of willow biomass crops. *Front. Ecol. Environ.* **2004**, *2*, 411–418. [CrossRef]

16. Djomo, S.N.; Kasmioui, O.E.; Ceulemans, R. Energy and greenhouse gas balance of bioenergy production from poplar and willow: A review. *GCB Bioenergy* **2011**, *3*, 181–197. [CrossRef]

17. Aronsson, P.; Rosenqvist, H.; Dimitriou, I. Impact of nitrogen fertilization to short-rotation willow coppice plantations grown in sweden on yield and economy. *BioEnergy Res.* **2014**, *7*, 993–1001. [CrossRef]

18. Larsen, S.U.; Jørgensen, U.; Lærke, P.E. Willow yield is highly dependent on clone and site. *BioEnergy Res.* **2014**, *7*, 1280–1292. [CrossRef]

19. Weger, J.; Hutla, P.; Bubeník, J. Yield and fuel characteristics of willows tested for biomass production on agricultural soil. *Res. Agric. Eng.* **2016**, *62*, 155–161. [CrossRef]

20. Labrecque, M.; Teodorescu, T.I. Field performance and biomass production of 12 willow and poplar clones in short-rotation coppice in southern Quebec (Canada). *Biomass Bioenergy* **2005**, *29*, 1–9. [CrossRef]

21. Guidi Nissim, W.; Pitre, F.E.; Teodorescu, T.I.; Labrecque, M. Long-term biomass productivity of willow bioenergy plantations maintained in southern Quebec, Canada. *Biomass Bioenergy* **2013**, *56*, 361–369. [CrossRef]

22. Volk, T.A.; Abrahamson, L.P.; Nowak, C.A.; Smart, L.B.; Tharakan, P.J.; White, E.H. The development of short-rotation willow in the northeastern United States for bioenergy and bioproducts, agroforestry and phytoremediation. *Biomass Bioenergy* **2006**, *30*, 715–727. [CrossRef]

23. Serapiglia, M.; Cameron, K.; Stipanovic, A.; Abrahamson, L.; Volk, T.; Smart, L. Yield and woody biomass traits of novel shrub willow hybrids at two contrasting sites. *BioEnergy Res.* **2012**, *6*, 1–14. [CrossRef]

24. Stolarski, M.J.; Rosenqvist, H.; Krzyżaniak, M.; Szczukowski, S.; Tworkowski, J.; Gołaszewski, J.; Olba-Zięty, E. Economic comparison of growing different willow cultivars. *Biomass Bioenergy* **2015**, *81*, 210–215. [CrossRef]

25. Styles, D.; Thorne, F.; Jones, M.B. Energy crops in Ireland: An economic comparison of willow and Miscanthus production with conventional farming systems. *Biomass Bioenergy* **2008**, *32*, 407–421. [CrossRef]

26. Stolarski, M.J.; Krzyżaniak, M.; Tworkowski, J.; Szczukowski, S.; Niksa, D. Analysis of the energy efficiency of short rotation woody crops biomass as affected by different methods of soil enrichment. *Energy* **2016**, *113*, 748–761. [CrossRef]

27. Heller, M.C.; Keoleian, G.A.; Volk, T.A. Life cycle assessment of a willow bioenergy cropping system. *Biomass Bioenergy* **2003**, *25*, 147–165. [CrossRef]

28. Krzyżaniak, M.; Stolarski, M.J.; Szczukowski, S.; Tworkowski, J. Life cycle assessment of new willow cultivars grown as feedstock for integrated biorefineries. *BioEnergy Res.* **2016**, *9*, 224–238. [CrossRef]

29. Londo, M.; Dekker, J.; ter Keurs, W. Willow short-rotation coppice for energy and breeding birds: An exploration of potentials in relation to management. *Biomass Bioenergy* **2005**, *28*, 281–293. [CrossRef]

30. Feledyn-Szewczyk, B.; Matyka, M.; Staniak, M. Comparison of the effect of perennial energy crops and agricultural crops on weed flora diversity. *Agronomy* **2019**, *9*, 695. [CrossRef]

31. Stolarski, M.J.; Szczukowski, S.; Tworkowski, J.; Krzyżaniak, M.; Załuski, D. Willow production during 12 consecutive years—The effects of harvest rotation, planting density and cultivar on biomass yield. *GCB Bioenergy* **2019**, *11*, 635–656. [CrossRef]

32. Tharakan, P.J.; Volk, T.A.; Nowak, C.A.; Abrahamson, L.P. Morphological traits of 30 willow clones and their relationship to biomass production. *Can. J. For. Res.* **2005**, *35*, 421–431. [CrossRef]

33. Weih, M. Intensive short rotation forestry in boreal climates: Present and future perspectives. *Can. J. For. Res.* **2004**, *34*, 1369–1378. [CrossRef]

34. Sevel, L.; Nord-Larsen, T.; Ingerslev, M.; Jørgensen, U.; Raulund-Rasmussen, K. Fertilization of SRC willow, I: Biomass production response. *BioEnergy Res.* **2014**, *7*, 319–328. [CrossRef]

35. Amichev, B.Y.; Volk, T.A.; Hangs, R.D.; Bélanger, N.; Vujanovic, V.; Van Rees, K.C.J. Growth, survival, and yields of 30 short-rotation willow cultivars on the Canadian Prairies: 2nd rotation implications. *New For.* **2018**, *49*, 649–665. [CrossRef]

36. Wang, Z.; MacFarlane, D.W. Evaluating the biomass production of coppiced willow and poplar clones in Michigan, USA, over multiple rotations and different growing conditions. *Biomass Bioenergy* **2012**, *46*, 380–388. [CrossRef]

37. Stolarski, M.J.; Krzyżaniak, M.; Szczukowski, S.; Tworkowski, J.; Śnieg, M. Willow productivity on a commercial plantation in triennial harvest cycle. *Bulg. J. Agric. Sci.* **2016**, *22*, 65–72.

38. Nord-Larsen, T.; Sevel, L.; Raulund-Rasmussen, K. Commercially grown short rotation coppice willow in Denmark: Biomass production and factors affecting production. *BioEnergy Res.* **2015**, *8*, 325–339. [CrossRef]

39. Castaño-Díaz, M.; Barrio-Anta, M.; Afif-Khouri, E.; Cámara-Obregón, A. Willow short rotation coppice trial in a former mining area in northern Spain: Effects of clone, fertilization and planting density on yield after five years. *Forests* **2018**, *9*, 154. [CrossRef]

40. Weger, J.; Havlíčková, K.; Bubeník, J. Results of testing of native willows and poplars for short rotation coppice after three harvests. *Asp. Appl. Biol.* **2011**, *112*, 335–340.

41. Lindegaard, K.N.; Carter, M.M.; McCracken, A.; Shield, I.; MacAlpine, W.; Jones, M.H.; Valentine, J.; Larsson, S. Comparative trials of elite Swedish and UK biomass willow varieties 2001–2010. *Asp. Appl. Biol.* **2011**, *112*, 57–66.

42. Bullard, M.J.; Mustill, S.J.; Nixon, P.M.I.; McMillan, S.D.; Britt, C.P. The effect on yield of initial planting density of short rotation coppice. *Asp. Appl. Biol.* **2001**, *65*, 167–171.

43. Otepka, P.; Habán, M. Biomass yield of basket willow (*Salix viminalis* L.) cultivated as energy plant in a long-term experiment. *Acta Fytotech. Zootech.* **2006**, *9*, 70–74.

44. Labrecque, M.; Teodorescu, T.I. High biomass yield achieved by Salix clones in SRIC following two 3-year coppice rotations on abandoned farmland in southern Quebec, Canada. *Biomass Bioenergy* **2003**, *25*, 135–146. [CrossRef]

45. Hytönen, J. Ten-year biomass production and stand structure of Salix 'aquatica' energy forest plantation in Southern Finland. *Biomass Bioenergy* **1995**, *8*, 63–71. [CrossRef]

46. Otepka, P.; Habán, M.; Habánová, M. Cultivation of fast-growing woody plant basket willow (*Salix viminalis* L.) and their bioremedial abilities while fertilized with wood ash. *Res. J. Agric. Sci.* **2011**, *43*, 218–222.

47. Nordh, N.E. Long Term Changes in Stand Structure and Biomass Production in Short Rotation Willow Coppice. Ph.D. Thesis, Swedish University of Agricultural Sciences, Uppsala, Sweden, 2005.

48. McElroy, G.H.; Dawson, W.M. Biomass from short-rotation coppice willow on marginal land. *Biomass* **1986**, *10*, 225–240. [CrossRef]

49. McCracken, A.R.; Dawson, W.M. Short rotation coppice willow in Northern Ireland since 1973: Development of the use of mixtures in the control of foliar rust (*Melampsora* spp.). *Eur. J. For. Pathol.* **1998**, *28*, 241–250. [CrossRef]

50. Lindegaard, K.N.; Parfitt, R.I.; Donaldson, G.; Hunter, T.; Dawson, W.M.; Forbes, E.G.A.; Carter, M.M.; Whinney, C.C.; Whinney, J.E.; Larsson, S. Comparative trials of elite Swedish and UK biomass willow varieties. *Asp. Appl. Biol.* **2001**, *65*, 183–192.

51. Volk, T.A.; Abrahamson, L.P.; Cameron, K.D.; Castellano, P.; Corbin, T.; Fabio, E.; Johnson, G.; Kuzovkina-Eischen, Y.; Labrecque, M.; Miller, R. Yields of willow biomass crops across a range of sites in North America. *Asp. Appl. Biol.* **2011**, *112*, 67–74.

52. Larsson, S. Commercial varieties from the Swedish willow breeding programme. *Asp. Appl. Biol.* **2001**, *65*, 193–198.

 *agriculture*

Article

The Influence of Three Years of Supplemental Nitrogen on Above- and Belowground Biomass Partitioning in a Decade-Old *Miscanthus × giganteus* in the Lower Silesian Voivodeship (Poland)

Izabela Gołąb-Bogacz [1], Waldemar Helios [2], Andrzej Kotecki [2], Marcin Kozak [2] and Anna Jama-Rodzeńska [2,*]

[1] Bugaj Sp. z o.o., Bugaj Zakrzewski 5, 97-512 Kodrąb, Poland; iza.golab@o2.pl
[2] Institute of Agroecology and Plant Production, Wroclaw University of Environmental and Life Sciences, Pl. Grunwaldzki 24A, 50-363 Wrocław, Poland; waldemar.helios@upwr.edu.pl (W.H.); andrzej.kotecki@upwr.edu.pl (A.K.); marcin.kozak@upwr.edu.pl (M.K.)
* Correspondence: anna.jama@upwr.edu.pl; Tel.: +48-713201627

Received: 3 July 2020; Accepted: 23 September 2020; Published: 13 October 2020

Abstract: Because of the different opinions regarding nitrogen (N) requirements for *Miscanthus × giganteus* biomass production, we conducted an experiment with a set dose of nitrogen. The objective of this study was to examine the effects of nitrogen fertilization on the biomass yield, water content, and morphological features of rhizomes and aboveground plant parts in various terms during a growing season over the course of three years (2014–2016) in Lower Silesia (Wroclaw, Poland). The nitrogen fertilization (dose 60 kg/ha and control) significantly affected the number of shoots ($p = 0.0018$), the water concentration of rhizomes ($p = 0.0004$) and stems ($p = 0.0218$), the dry matter yield of leaves ($p = 0.0000$), and the nitrogen uptake ($p = 0.0000$). Nitrogen fertilization significantly affected the nitrogen uptake in all plant parts ($p = 0.0000$). Although low levels of nitrogen appeared to be important in maintaining the maximum growth potentials of mature *Miscanthus × giganteus*, the small reductions in the above- and belowground biomass production are unlikely to outweigh the environmental costs of applying nitrogen. More studies should use the protocols for the above- and belowground yield determination described in this paper in order to create site- and year-specific fertilizer regimes that are optimized for quality and yield for autumn (green) and spring (delayed) harvests.

Keywords: *Miscanthus*; nitrogen fertilization; rhizomes; stem; leaves

1. Introduction

New technologies, excessive fossil fuel combustion, and future fossil fuel depletion will contribute to permanent changes in the natural environment. One of the most pivotal environmental problems is climate change, which is caused by the anthropogenic heating of the atmosphere as a result of rising greenhouse gas concentrations [1–5]. To overcome this difficulty, we must increase the use of renewable energy sources. Renewable energy sources play an increasingly essential role in the energy policy of European countries [6]. Among all renewable energy sources, plant biomass deserves special attention. Fast growing bioenergy crops are characterized by a great potential to provide raw material for renewable energy. *Miscanthus* has been proposed as a biomass energy crop in Europe [7,8], and its use could increase in the near future, as it is one of the most productive plants among bioenergy crops [9–13]. Additionally, biomass combustion is regarded to be more beneficial for the environment than fossil fuel combustion [14–16].

The success of this bioenergy crop is also determined by its low environmental requirements—for instance, its low nitrogen and water requirements, the mechanization of its planting and harvesting, and the resistance of the plants to diseases and pests [13,14,16,17]. Because of its low nutrient requirements, *Miscanthus* can be successfully cultivated on sandy and high organic matter soils with a wide pH range. Additionally, it is being successfully grown in unused marginal areas and has a tolerance to various abiotic stresses, including excessive salinity, low humidity, or the presence of heavy metals [7,18,19]. According to Galatsidas et al. (2018) [20], the total area of marginal land that is appropriate for *Miscanthus* cultivation in Europe is thought to be as high as 11.11 million ha.

For the successful development of *Miscanthus* production, it is necessary to consider the end specific uses and precise information on the effective management of nitrogen fertilization for different soil types under various climatic and growth conditions [14,21]. Although nitrogen is the main element that determines the efficiency of biomass production, it can have negative environmental effects such as water eutrophication and increased carbon dioxide emissions [22,23].

The literature varies regarding the nitrogen requirements for *Miscanthus* × *giganteus* biomass production [14,24–27], because the nitrogen applications of *Miscanthus* × *giganteus* are characterized by variable productivity results. The N requirements of *Miscanthus* × *giganteus* are low compared to those of other bioenergy crops [16,28,29]. According to Cadoux et al. (2011, 2012) [30,31], these low nutrient requirements are caused by various factors, including a high nutrient use efficiency and the nutrient recycling accumulated in the rhizomes. However, there is a serious debate about the exact need for N fertilizer in a given crop and whether N fertilizer should be required at all. The translocation of nitrogen to rhizomes during the late vegetation period is a major factor in the high efficiency of nitrogen utilization [17]. There are divergent results regarding the requirements of Mischanthus ×*giganteus* for N fertilization. The findings are divided on this matter; some studies have shown that the yield increases after the application of N fertilizer [14,25–27,32], while some state the contrary [13,33–38].

There are many European studies that provide estimates of the belowground biomass for *M. giganteus* at a single point in time [29,36–39]; however, there have been few previous studies that determined the dynamics of the rhizome yield which were not based on regular sampling through the growing season [40,41].

The organ of wintering in the *Mischanthus* is the rhizome, an underground part that grows horizontally that is important for nutrient storage and accumulation. Most research on the yield and biometric traits of *Mischanthus* is concentrated on the aboveground parts of the plants [9,14,29,42]. The main aspects of experimental research are mainly focused on the environmental impact of *Mischanthus*, the different terms of harvesting, the different genotypes of *Mischanthus*, and its chemical composition during multiannual study periods. Thus far, the elemental composition and resistance to frost and salinity have been examined in the rhizomes; however, there is a lack of information on the water content in the rhizomes during the whole growing season [43–46]. A new aspect of our research is the determination of the changes in the rhizome water content during the entire vegetation period (May–December) on a 10-year-old plant.

The objective of this study was to examine the effects of nitrogen fertilization on the number, height, and diameter of leaves on a shoot, as well as the water concentration, dry mass yield, and nitrogen uptake of *Miscanthus* × *giganteus*. The growth rate of the aboveground and belowground biomass of *Miscanthus* × *giganteus* (Greef et Deu) was evaluated in the conditions of southwest Poland, with and without nitrogen fertilization. Additionally, research was undertaken to determine the influence of nitrogen fertilization on the dynamics of the water content changes in the rhizomes during the whole vegetation period.

2. Materials and Methods

2.1. Study Site and Fertilization Treatments of Miscanthus × Giganteus

An investigation of *Miscanthus* × *giganteus* and nitrogen fertilization was conducted after 10 years of establishing crops (2014–2016) at the Experimental Station of Wroclaw University of Environmental and Life Sciences, Pawlowice (geographical location 17°7′ E and 51°08′ N in the Lower Silesian Voivodship, Wrocław, Poland). Pawlowice is characterized by a vegetation period (March–November) that lasts 223–230 days, with an average temperature during the growing season of 14.5 °C and an annual rainfall ranging between 500 and 600 mm (around 350 mm during the growing season). The soil conditions were defined as alluvial soil, very light on loose sand, and sandy gravel (V grade) (soil classification used in Poland). These soils are weak with a low humus level, and are poor in organic matter. The fifth class of soil quality (6 classes of soil quality: I class—the best arable land; VI—the weakest arable soil) comprises weak arable soils [47].

Plowing was carried out in 2003 at the depth of 20–25 cm, followed by rotary harrowing before planting. Miscanthus rhizomes (10 cm long with 3–6 nodes) were planted in a row spaced 75 cm apart and another row spaced 48 cm apart (on 1 ha–27,777 rhizomes). *Miscanthus* × *giganteus* was planted in 2004. Plantation was fertilized annually from the year 2004 to 2013 at the beginning of the growing season using the following doses: 40 kg ha^{-1} of N ammonium nitrate 32%, 17.5 kg ha^{-1} of P 40% enriched superphosphate, and 50 kg ha^{-1} of K potassium salt. The plots were separated by a distance of 1.0 m, and all measurements (non-destructive and destructive) were taken at least 0.2 m from the edge of the plot in the years 2014–2016. The dimension of the plot was around 20 m^2. Nitrogen treatments of 0 and 60 kg/ha were applied in March/April during each of the 3 years (17/3/2014, 18/3/2015, 17/4/2016) after pulling out the bedding. Fertilization was annually (from 2014 to 2016) applied during the field experiment, where the following doses were used: 17.5 kg ha^{-1} of P 40% enriched superphosphate, 50 kg ha^{-1} of K potassium salt. After fertilization, the mulch was placed in its original position.

Fertilization was applied via a hand broadcast at the beginning of the vegetation period.

No significant pests and weeds were found in the *Miscanthus* cultivation during the experiment, so the use of herbicides was not necessary.

2.2. Plant Growth Measurement

Miscanthus sampling started from the 30th day of the vegetation period and every 30 days until the end of vegetation period (June, July, August, September, October, November, and December) in the years 2014–2016. At each date of sampling, a plant sample of the aboveground part of the plant and rhizomes was sampled from an area of 0.25 m^2. The fresh mass of the rhizomes and the aboveground part was determined. Additionally, 10 randomly selected shoots were sampled from each replication to perform measurements on plant material—the height of the upper leaf, the diameter measured 10 cm from the soil surface, and the number of leaves per one stem. All the measurements (except the number of shoots) were made on 10 shoots per plot. The number of shoots was counted from a unit of 0.25 m^2 from each replication. Both white and yellow rhizomes were sampled.

Terminal (from outer rows) plants from the external rows were not included in the analysis because of the so-called edge effect. After the end of the vegetation period, *Miscanthus* was harvested at 10–15 cm using a circular saw. Harvested crops were weighed and the percentage of dry matter was determined. The dry biomass weight was determined by drying samples (specific weight, 500 g) to 60 °C for up to 48 h, then drying them at 105 °C for 4 h. Further, the harvested crops were weighed and the fresh mass yield was determined. The dry biomass weight was determined by drying samples (specific weight, 500 g) to 60 °C for up to 48 h, then drying them at 105 °C for 4 h. On this basis, the dry biomass yield per 1 m^2 in a given year was calculated.

Water concentration was calculated according to the Formula (1):

$$\text{Water concentration (\%)} = (100 \times (FM - DM))/FM. \tag{1}$$

FM—fresh mass.
DM—dry mass.

2.3. Soil and Weather Conditions

Tables 1 and 2 summarize the soil conditions for the *Miscanthus* plantation in this trial. Soil samples were twice taken (April, July) during the vegetation period and after its end (November) each year. These dates were presented as annual mean values. Soil samples were taken from the experimental field at a 0–20 cm soil depth and were thoroughly mixed to make a representative composite soil sample. The analysis was comprised of pH, humus, C, N, P, K, S, and micronutrients. Analyses were performed according to the following methods: the soil reaction (pH/KCl (potassium chloride)) was found using the potentiometric method; the total organic carbon was found using Tiurin's method [48]; the total nitrogen (classical distillation) content was found using the Kjehdal method both in soil and plant material [48]; the available forms of potassium and phosphorus were found using the Egner–Rhiem method; magnesium was found using the Schachtschabel method [49]; the total carbon content (TOC) was found via oxidimetric titration [50]; sulfur in the extract was found using the Johnson–Nishita procedure [51]; humic substances (HS) were found using the short fractionation method [52]; and the total contents of Fe, Mn, Zn, and Cu were found using an atomic absorption spectrophotometer (ASA) after mineralization with a concentrated mixture of acids using atomic-absorbent flame spectrophotometry Varian spectra AA 200 [52].

Table 1. The content of organic matter and soil abundance in macronutrients for a depth of 0–20 cm in 2014–2016.

Year	pH 1 N KCl	C g kg^{-1}	Humus g kg^{-1}	N g kg^{-1}	C:N	P mg kg^{-1}	K mg kg^{-1}	Mg mg kg^{-1}	S mg kg^{-1}
2014	5.0	5.82	10.00	0.58	10.53	119.6	114.0	24.3	188.0
2015	5.0	5.86	10.05	0.60	10.60	119.6	115.3	27.3	192.6
2016	4.8	5.86	10.05	0.59	10.63	119.7	112.6	26.0	190.0

Table 2. Soil abundance in the micronutrients at the depth of 0–20 cm in 2014–2016.

Year	Fe mg kg^{-1}	Mn mg kg^{-1}	Zn mg kg^{-1}	Cu mg kg^{-1}
2014	428	93.4	82.3	1.82
2015	461	97.1	79.4	1.69
2016	463	95.2	78.5	1.78

The soil's carbon stock was typical for light alluvial soils, and the C: N ratio was on average 10.6:1, which indicates the appropriate process of the organic decomposition (Table 1). In the experimental years, the soil reaction ranged from 4.8 to 5.0 (acidic), which was favorable for *Miscanthus* cultivation, and the arable layer's richness in nutrients was as follows: P—very high; K—medium; Mg—low; S—medium; Fe—low; Mn—medium; Zn—high; and Cu—low (Tables 1 and 2). The assessment of the soil's nutrient content was determined by limit numbers to assess the content of elements developed by the Polish Institute of Soil and Plant Cultivation in Puławy [47].

Monthly data on the temperature and precipitation in the years 2014–2016 are presented in Table 3. The temperatures in the years 2014–2016 oscillated between ±9 °C in IV through to an average of ±17 °C from V to VIII. During the experimental years, the thermal conditions were favorable for the development of *Miscanthus*, with mild winters characterized by positive temperatures. The highest temperatures were recorded in 2015, while the lowest were in 2016 (Table 3).

The optimal amount of rainfall for *Miscanthus* × *giganteus* depends on many factors, including the air temperature, soil type, and groundwater level; however, 600 mm was sufficient for the development of *Miscanthus* [14,26]. The year with the lowest rainfall was 2015. Despite the lack of rainfall, there were no reduction in the yield. The highest rainfall during the growing season was recorded in 2016 (Table 3).

Table 3. Weather conditions during 2014–2016 with a 30-year average for Wroclaw, Lower Silesia (Poland).

Month	Temperature [°C]				Precipitation [mm]			
	2014	2015	2016	Average 1981–2010	2014	2015	2016	Average 1981–2010
I	0.0	2.3	−1.2	−0.8	35.8	46.0	33.4	31.9
II	3.7	1.5	3.8	0.3	1.2	15.6	56.2	26.7
III	7.0	5.4	4.3	3.8	40.1	39.5	55.9	31.7
IV	10.6	8.9	8.7	8.9	55.2	15.8	46.4	30.5
V	13.3	13.5	15.3	14.4	101.4	21.0	5.3	51.3
VI	16.6	16.6	18.6	17.1	40.2	73.3	44.6	59.5
VII	21.2	20.3	19.5	19.3	52.9	55.6	114.3	78.9
VIII	17.3	22.7	17.9	18.3	75.0	5.6	27.1	61.7
IX	15.5	15.1	16.4	13.6	72.2	23.2	44.7	45.3
X	10.7	8.4	8.5	9.1	59.4	20.0	83.8	32.3
XI	6.6	6.2	3.4	3.9	15.5	52.4	36.3	36.6
XII	2.3	5.4	1.2	0.2	17.5	24.0	36.1	37.4
Average annual air temperature and total precipitation	10.4	10.5	9.7	9.0	566.4	392.0	584.1	523.8

2.4. Statistical Analysis

The experiment was conducted with a randomized block design in four replications to test the effects of N fertilization on the morphological traits and yield of *Mischanthus*. The analysis of variance (ANOVA) and the mixed model with repeated measurements was used. The doses of nitrogen fertilizers were assumed to be a fixed factor, while the years were random. The results of the biometric measurements of the *Mischanthus* were analyzed via ANOVA in the Statistica program (13.1 StatSoft, Kraków, Poland).

3. Results

3.1. Effect of Nitrogen Fertilization on Morphological Features of Miscanthus × Giganteus

Nitrogen fertilization had a significant influence on the number of leaves on the shoot ($p = 0.0018$) during the field experiment (Table 4). Both the number of shoots and the height of the plants increased significantly until the end of vegetation period (Figures 1 and 2). Without N fertilization, the shoots reached 3.34 m in height, whereas the height of plants after an application of 60 kg ha^{-1} N was 3.31 m. The highest increases in height of shoots on unfertilized plots were found between June and July, while in fertilized plants they was found between July and August. The greatest increase in shoot diameter was found at the beginning of the vegetation period (Figure 3). A fast increase in the number of leaves on the shoot was observed in September. Between September and November, the differences were insignificant (Figure 4). The number of leaves on both fertilized and unfertilized shoots increased until November. After this period, it decreased.

Table 4. Morphological features of *Miscanthus* × *giganteus* (average for years 2014–2016).

Dose kg ha^{-1} N	Number of Days Starting from Beginning of Vegetation Period	Number of Shoots per 1 m^2	Height of Plants (m)	Diameter of Shoots (mm)	Number of Leaves on Shoot
0	June	52	0.18	8.5	3.1
	July	59	1.23	10.3	5.9
	August	64	2.14	9.7	8.7
	September	64	2.33	10.5	10.7
	October	66	2.98	9.6	11.3
	November	74	3.07	10.0	11.7
	December	72	3.34	10.8	8.9
60	June	53	0.21	9.1	3.8
	July	64	1.04	9.9	5.9
	August	66	2.24	10.1	8.8
	September	76	2.71	10.8	11.3
	October	78	3.09	10.2	12.1
	November	72	3.13	10.5	12.4
	December	78	3.31	10.8	11.2
	p value	0.2884	0.0001	0.4553	0.0322
Averages for Factors and Years					
0	-	64	2.18	9.9	8.6
60	-	70	2.25	10.2	9.3
	p value	0.0018	0.7020	0.1004	0.1484
2014		67	2.28	10.0	8.5
2015	-	66	2.20	10.0	8.6
2016		68	2.16	10.2	9.9
	p value	0.4112	0.8354	0.4200	0.4040

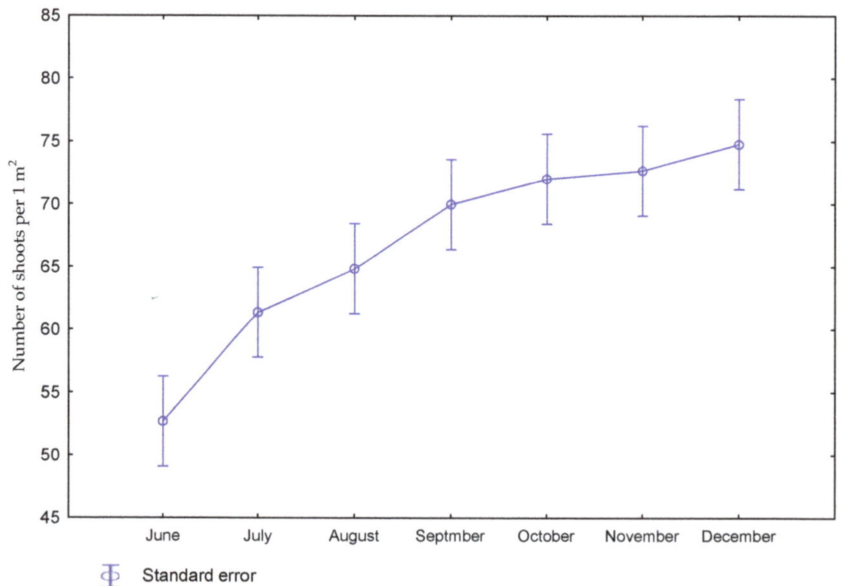

Figure 1. The number of shoots during the vegetation period in the years 2014–2016 (average for years).

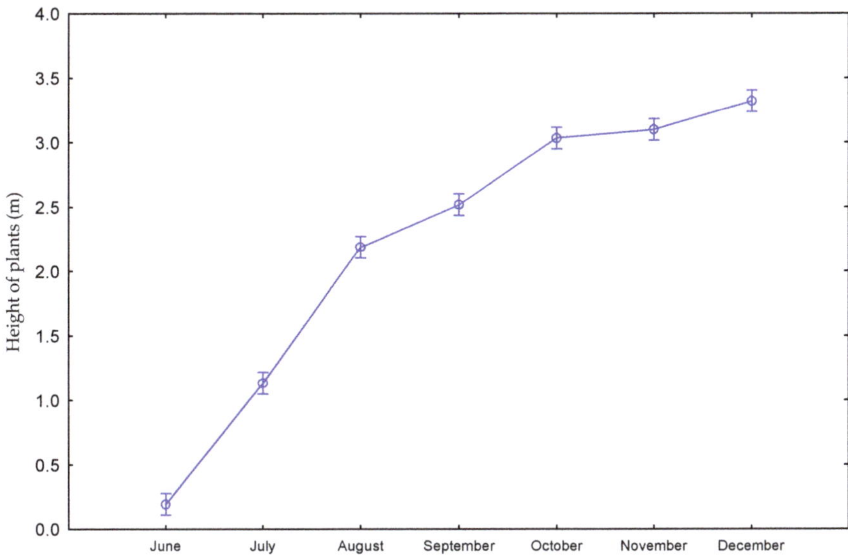

Figure 2. The height of plants during the vegetation period in the years 2014–2016 (average for years).

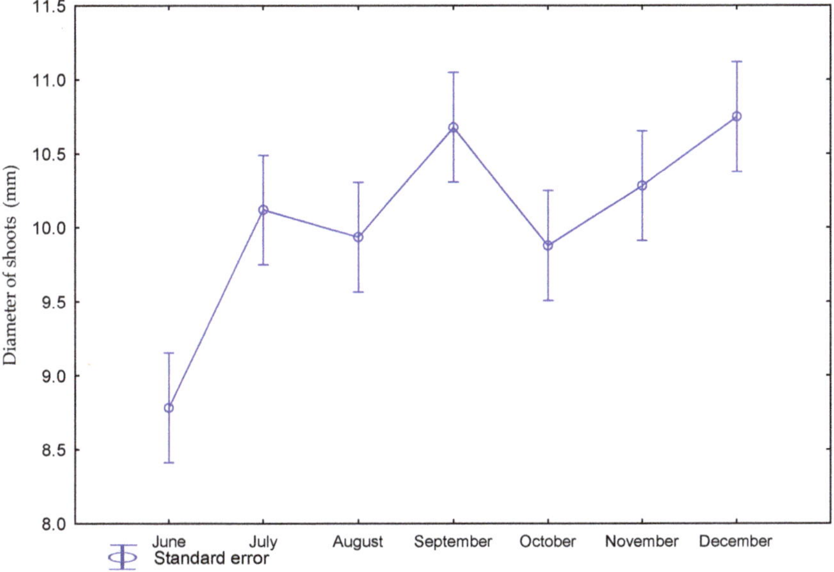

Figure 3. Diameter of shoots during the vegetation period in the years 2014–2016 (average for years).

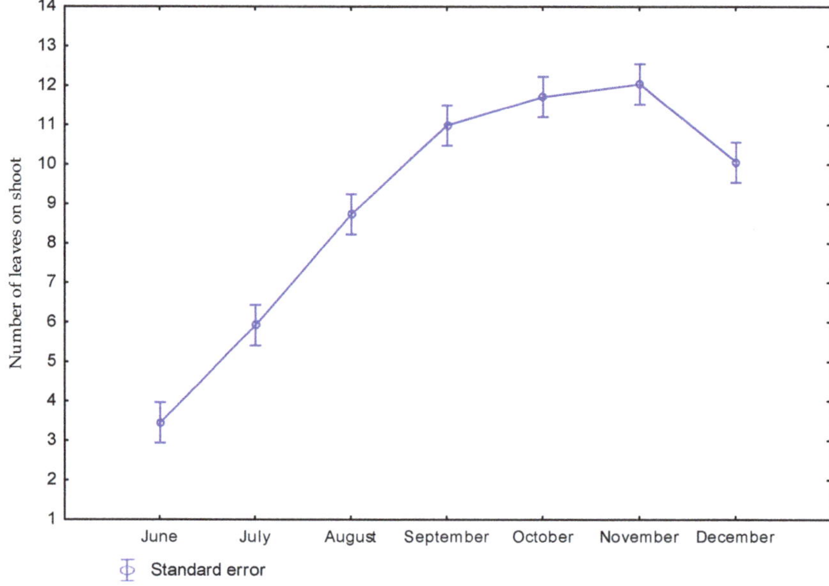

Figure 4. Number of leaves on the shoots during the vegetation period in the years 2014–2016 (average for years).

3.2. Effect of Nitrogen Fertilization on Water Concentration of Miscanthus × Giganteus

The water concentration was characterized with differences between the examined parts of plants. The rhizomes, stems, and leaves were characterized by a higher water concentration at the beginning of the growing season (Table 5, Figures 5 and 6). On fertilized and unfertilized plots, the water content in the leaves ($p = 0.0260$) and stems ($p = 0.0015$) decreased until the end of the vegetation period. For rhizomes, the water content decreased until October and then increased at about 7 g in the unfertilized plot and 31 g in the fertilized plot. There was a significantly higher water concentration found in the rhizomes ($p = 0.004$) and stems ($p = 0.0218$) fertilized with nitrogen. The water concentration was significantly different during the experimental years. The highest content of water was observed in the rhizomes ($p = 0.0000$), stems ($p = 0.0022$), leaves ($p = 0.0000$), and whole aboveground parts of plants ($p = 0.0025$) in the third year of the study (Table 5). A greater water content in the aboveground part of plants was observed until November (Figure 6).

Table 5. Water concentration in the fresh mass of *Miscanthus × giganteus* (g kg^{-1}) (average for years 2014–2016).

Dose kg ha^{-1} N	Number of Days Starting from Beginning of Vegetation Period	Rhizomes	Stems	Leaves	Aboveground Part of Plant
	June	722	-	-	882
	July	689	870	879	875
	August	709	772	777	775
0	September	684	697	715	702
	October	663	691	702	694
	November	663	662	698	672
	December	670	622	679	635

Table 5. *Cont.*

Dose kg ha^{-1} N	Number of Days Starting from Beginning of Vegetation Period	Rhizomes	Stems	Leaves	Aboveground Part of Plant
	June	744	-	-	883
	July	707	867	853	862
	August	712	827	810	820
60	September	713	764	781	769
	October	673	720	740	726
	November	683	676	707	685
	December	704	661	701	672
	p value	0.4958	0.0015	0.0260	0.00120
Average for Factors and Years					
0		686	719	742	748
60		705	752	765	774
	p value	0.0004	0.0218	0.0669	0.0693
2014		673	714	722	738
2015		693	722	735	750
2016		721	771	803	795
	p value	0.0000	0.0022	0.0000	0.0025

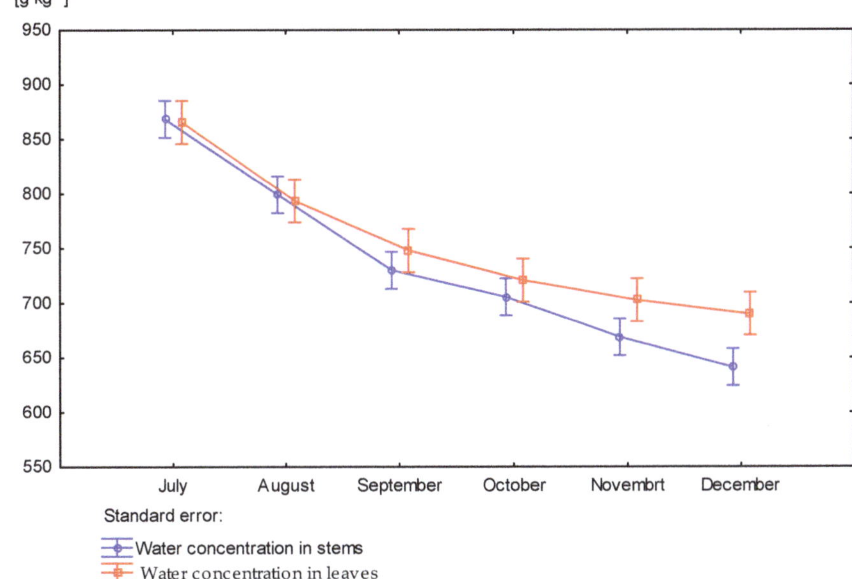

Figure 5. Water concentration in leaves and stems during the vegetation period of the years 2014–2016 (average for years).

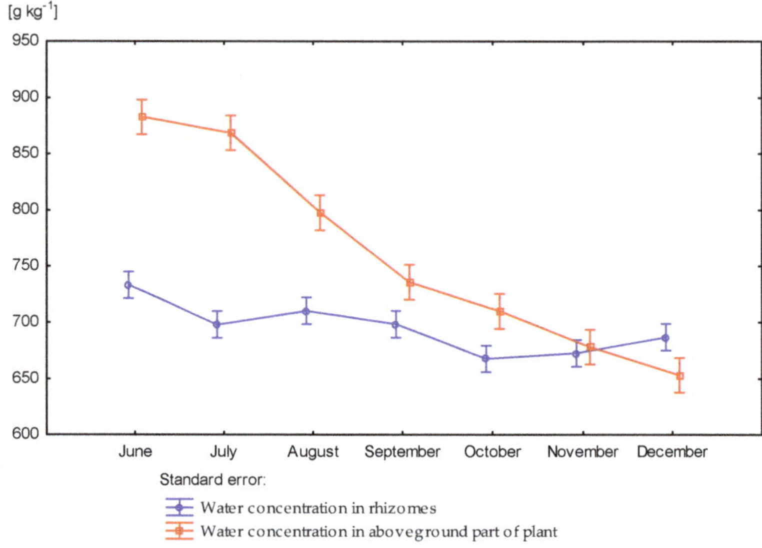

Figure 6. Water concentration in the rhizomes and aboveground part of plants during the vegetation period of the years 2014–2016 (average for years).

3.3. Effect of Nitrogen Fertilization on Dry Matter Yield of Miscanthus × Giganteus

Nitrogen fertilization significantly contributed to an increase in the dry matter yield of leaves ($p = 0.0000$). The nitrogen fertilization and lack of fertilization of biomass sampling was characterized by an increasing tendency in the dry mass of rhizomes and aboveground parts of plants. The dry mass of the stems grew faster than that of the leaves over the whole vegetation period (Figure 7). The highest yield growth dynamics of the whole plant was observed between August and September (Table 6, Figure 8).

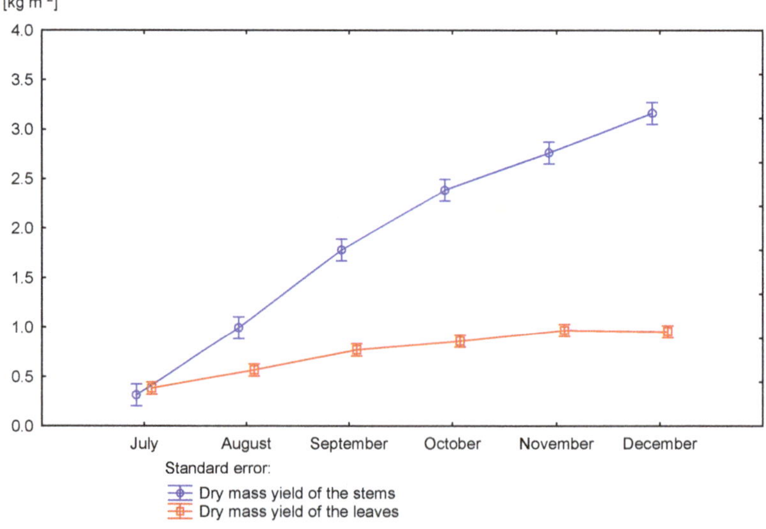

Figure 7. Dry mass yield of the leaves and stems during the vegetation period in the years 2014–2016 (average for years).

Table 6. The yield of the dry mass of *Miscanthus* × *giganteus* (kg m^{-2}) (average for years 2014–2016).

Dose kg ha^{-1} N	Number of Days Starting from Beginning of Vegetation Period	Rhizomes	Aboveground Part			Rhizomes and Aboveground Part
			Stems	Leaves	All Together	
	June	0.52	-	-	0.34	0.86
	July	0.65	0.28	0.32	0.60	1.25
	August	0.77	0.94	0.45	1.39	2.16
0	September	1.15	1.67	0.63	2.30	3.45
	October	1.34	2.22	0.78	3.00	4.34
	November	1.67	2.81	0.89	3.70	5.37
	December	1.73	3.08	0.84	3.92	5.65
	June	0.73	-	-	0.44	1.17
	July	0.77	0.35	0.45	0.80	1.57
	August	0.97	1.06	0.68	1.74	2.71
60	September	1.22	1.90	0.92	2.82	4.04
	October	1.64	2.55	0.94	3.49	5.13
	November	1.64	2.72	1.06	3.78	5.42
	December	1.83	3.25	1.07	4.32	6.15
	p value	0.0125	0.1223	0.1393	0.0153	0.0056
Average for Factors and Years						
0		1.12	1.83	0.65	2.18	3.29
60		1.26	1.97	0.85	2.48	3.74
	p value	0.0524	0.4310	0.0000	0.1586	0.1181
2014		1.20	2.00	0.75	2.41	3.61
2015		1.15	1.91	0.71	2.30	3.45
2016		1.21	1.79	0.79	2.27	3.48
	p value	0.7318	0.6005	0.3165	0.8522	0.8881

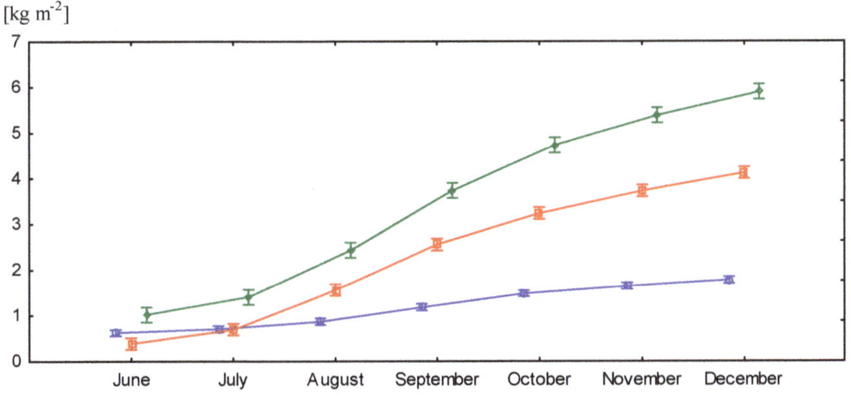

Figure 8. Dry mass yield of the rhizomes and the aboveground part of the plant during the vegetation period of the years 2014–2016 (average for years).

The dry mass of aboveground parts of plants (p = 0.0153) and rhizomes (p = 0.0125) in 30-day intervals significantly differentiated from June to November, in which we obtained the highest values in December (Table 6, Figure 8). In July, the dry matter of leaves was slightly greater than that of the stems, and from this month the increase in the dry matter of stems was greater than that of the leaves. The period between June and July and the November and December vegetation days, constituted 29%

of the entire vegetation period. During this time, a more than 18% increase in the dry weight of the rhizomes and aboveground parts was observed.

3.4. Nitrogen Uptake by Miscanthus × Giganteus

Nitrogen fertilization caused a significant increase in the nitrogen uptake in all the examined parts of plants ($p = 0.0000$). For the control object, the nitrogen uptake by rhizomes decreased until July, whereas in fertilized plots it decreased until August ($p = 0.0118$) (Table 7). The highest uptake of nitrogen in rhizomes was found in December, while in whole plants it was found in November. Therefore, it can be presumed that rhizomes can be a nitrogen reserve for shoots. In the initial vegetation period, the nitrogen uptake in leaves was higher than that in stems. The accumulation of nitrogen in stems was found to be higher than in leaves starting in August (Figure 9). The highest nitrogen uptake was found in the case of whole plants, with an increasing tendency from July to September, where the differences became insignificant (Figure 10). The fastest increase in the N uptake by rhizomes was observed from October to November (Figure 10). In the case of the aboveground parts of plants, the nitrogen uptake increased from June to September and then decreased (Figure 10).

Table 7. Nitrogen uptake of *Miscanthus × giganteus* (kg m^{-2}) (average for years 2014–2016).

Dose kg ha^{-1} N	Number of Days from the Start of the Growing Season	Rhizomes	Aboveground Part of Plants			Rhizomes and Aboveground Part of Plants
			Stems	Leaves	Together	
	June	5.35	-	-	4.75	10.10
	July	3.25	3.16	4.34	7.50	10.75
	August	3.65	8.24	4.79	13.03	16.68
0	September	5.29	10.24	5.99	16.24	21.53
	October	6.62	8.50	6.04	14.54	21.16
	November	10.19	8.61	6.66	15.27	25.46
	December	10.68	7.97	3.97	11.95	22.63
	June	9.24	-	-	7.60	16.84
	July	5.79	5.08	6.98	12.05	17.84
	August	5.40	8.63	9.52	18.15	23.55
60	September	6.91	14.36	11.01	25.37	32.28
	October	7.72	15.71	8.78	24.51	32.23
	November	14.07	12.82	7.19	20.01	34.08
	December	14.94	12.40	5.74	18.15	33.09
	p value	0.0118	0.0000	0.000	0.0000	0.0000
Means for Factors and Years						
0		6.43	7.79	5.30	11.90	18.33
60		9.15	11.50	8.20	17.98	27.13
	p value	0.0000	0.0000	0.0000	0.0000	0.0000
	2014	8.18	10.07	6.53	15.06	23.24
	2015	7.48	10.10	7.05	15.55	23.03
	2016	7.71	8.77	6.67	14.19	21.89
	p value	0.6315	0.1925	0.5895	0.5205	0.6493

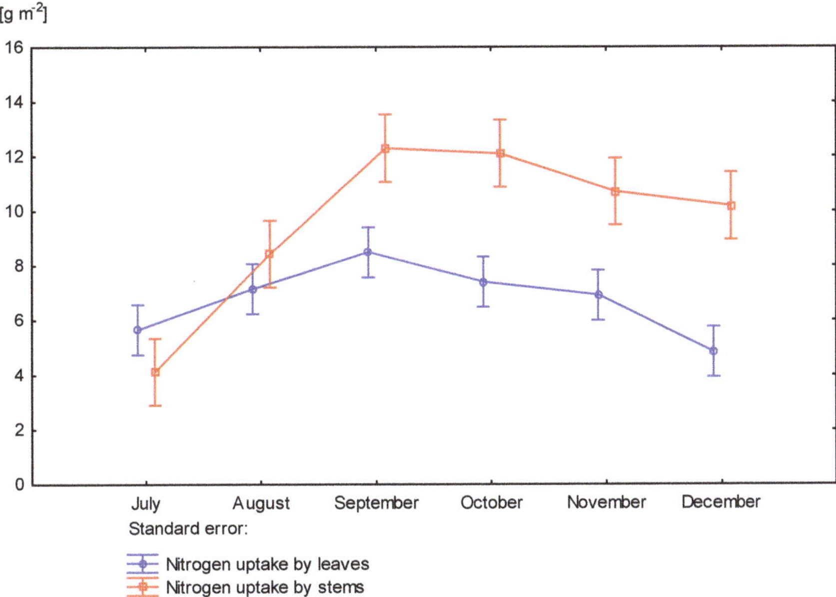

Figure 9. Nitrogen uptake by leaves and stems during the vegetation period in the years 2014–2016 (average for years).

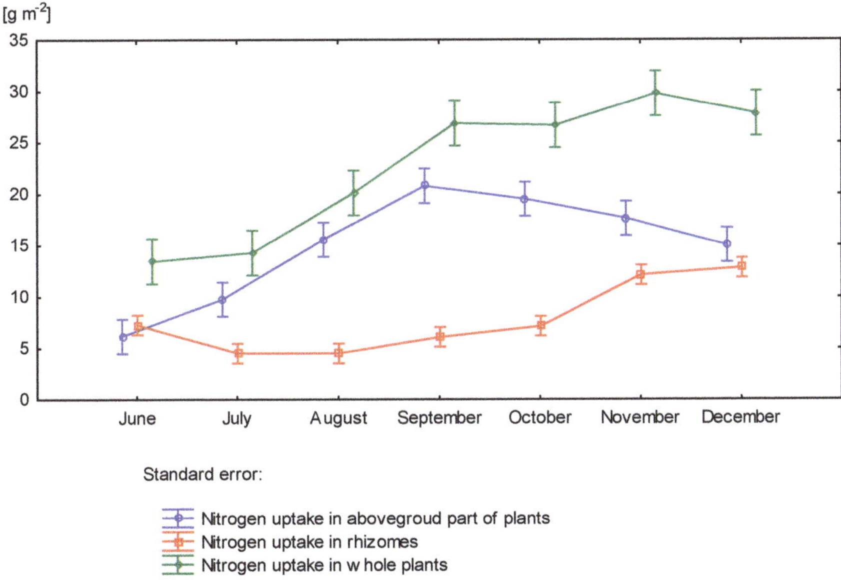

Figure 10. Nitrogen uptake by the whole plants (average for years).

4. Discussion

Nitrogen fertilization is important for biomass production and its components. The results provided statistical evidence to prove that the number of shoots responded positively to N fertilization. Other studies have also shown an increase in the number of shoots after applying N [53–55]. The water concentration in rhizomes and stems, the yield of dry mass leaves, and the nitrogen uptake was dependent on the level of nitrogen fertilization. Higher water content promoted metabolic processes and faster dry mass accumulation [56]. Therefore, research has been undertaken to determine the influence of nitrogen fertilization on the dynamics of the water content changes in rhizomes during the whole vegetation period. According to Drazic et al. (2017) [25], the number of stems per rhizome depended strongly on the soil type and was in strong positive correlation with the yield in all years. In our own research, the number of shoots were not significantly different during the experimental years.

In our research, the application of nitrogen stimulated the number of shoots. The plant height was also increased by N fertilization in various terms of harvesting. The plant height increased after the application of N, which was also reported by Cosentino et al. (2007) [54] and Finnan and Burke (2014) [39].

There have been conflicting results concerning the yield response of *Miscanthus* × *giganteus* to nitrogen fertilization and its yield components. Our positive responses to nitrogen fertilization were in agreement with Arundale et al. 2014 [57]. Moreover, Greef, J.M. (1995) [35] and Lee and Boe (2005) [26] obtained similar results when applying a 60 kg ha^{-1} N dose as appropriate for proper rhizome development and *Miscanthus* × *giganteus* yield increase. In the research of Dierking et al. (2017) [17], a dose of 75 kg ha^{-1} N contributed to the increase in the *Miscanthus* biomass yield, and this amount was applied annually. In the research of Lee and Boe (2005) [26], the dry matter yield visibly increased when the nitrogen fertilization increased up to 60 kg ha^{-1} N. However, increasing the nitrogen dose further did not contribute to an increase in the *Miscanthus* yields. The Miscanthus dry matter yields obtained in this research were 2.55 and 2.49 kg m^{-2} for 60 and 120 kg ha^{-1} N, while in the control plant it was 1.3 kg m^{-2}. Schwarz et al., 1994 [34], conducted an experiment involving nitrogen fertilization that did not have a significant impact on the *Miscanthus* yield. In their second year of cultivation, they obtained a yield of 0.8 kg m^{-2}, and in the third year they obtained 2.2 kg m^{-2}. Moreover, many other studies have shown that nitrogen fertilization is not required to obtain high yields of *Miscanthus* × *giganteus* biomass [58]. Christian et al. (2008) [33] did not find any answer to the applied N in 14 consecutive harvests. This result is supported by other studies that showed no response to N fertilization. However, some experiments have been concerned with soils featuring a large N content [13,21,25,34]. No reaction to nitrogen was found during the first two years after planting. Maughan et al. 2012 [21] reported a small positive reaction in a dose of 100 kg ha^{-1} N of fertilizer. According to Kering et al. (2012) [13], Himken et al. (1997) [58], and Miquez et al. (2008) [21], *Miscanthus* yields are not dependent on the level of nitrogen fertilization, as they determined 2.5–3.0 kg m^{-2} of D.M. and even 3.8 kg m^{-2} of D.M. In our research, the dry matter yield with the nitrogen fertilization of all examined plants was insignificantly higher compared to the control. Only the leaf yields of D.M. depended on nitrogen fertilization.

The ambiguous response to nitrogen fertilization results from several reasons:

1. Most research on *Miscanthus* productivity has been conducted in Europe (different soils, different spatial diversity, and topographic diversity);
2. The studies carried out are generally short-term;
3. The soil type and soil texture [21,37,59];
4. The potential share of nitrogen reserves in rhizomes and soil nitrogen increases the uncertainty of the *Miscanthus* nitrogen requirements [29].

Precipitation is the most important factor that directly and indirectly affects the biomass yield of *Miscanthus* × *giganteus*. Plant biomass production reacts positively to annual rainfall [60], and the seasonal distribution of rainfall is a key factor that determines the formation of perennial grasses and biomass yield [26,60]. In this experiment, the precipitation was variable during the 3-year study period, with much less precipitation than 2015. In our research, the most favorable year with a high and evenly distributed precipitation was in 2016; however, this did not translate into dry matter yields but rather translated to the water content in all the examined plant parts. According to Heaton et al. (2004) [46], the biomass yield may be affected by rainfall during the growing season from April to September.

The nitrogen uptake was significantly affected by the analyzed factors—nitrogen fertilization and the term of harvesting. According to Roncucci et al. (2014) [14], the time of harvest is the most relevant factor in influencing the miscanthus nutrient uptakes. Late harvesting (W) led to a reduction in the nitrogen uptake of about 80% in the aboveground biomass. This nitrogen uptake is observed to be lower than the literature data. In 10 years of research in the UK, Christian et al. (2008) [33] reported that the N is 76 and 6 kg ha^{-1} N. According to Roncucci et al. (2014) [14], N fertilization affected the nutrient uptake mainly in autumn, with no differences in winter. These results are in agreement with those of Himken et al. (1997) [58], who observed a higher N uptake with higher N fertilization rates in November, which is confirmed by our results. Nitrogen fertilization in the fertilizer treatments significantly affected the nitrogen uptake by all plant parts, which is confirmed by Strullu et al. (2011) [30].

Slightly higher results relating to the nitrogen uptake under various N doses in the harvest biomass of giant miscanthus were found in Christian et al. (2008) [33]. In Beale et al. (1997) [29], the rhizome nitrogen uptake decreased until July and then increased until December. Similar conclusions were presented in our research.

5. Conclusions

Nitrogen fertilization did not contribute to the increase in all the examined yield components. The proposed dose caused an increase in all the components of features and the dry matter yield. However, the differences were mostly insignificant. Only the dry mass of leaves increased significantly in the experiment. The water content in the rhizomes and stems increased under nitrogen fertilization. Therefore, we can assume that rhizomes, because of their significant nitrogen uptake, can constitute a nitrogen reserve for elements in the initial growth and development stages of plants. The results coming from our 3-year field experiment suggest that N fertilization is unnecessary for sustainable biomass production.

Author Contributions: Conceptualization, I.G.-B., W.H., and A.K.; data curation, M.K.; formal analysis, W.H. and A.J.-R.; investigation, I.G.-B., W.H., and M.K.; methodology, I.G.-B. and A.K.; project administration, I.G.-B.; resources, A.K. and A.J.-R.; software, M.K.; supervision, M.K.; visualization, M.K. and A.J.-R.; writing—original draft, I.G.-B. and W.H.; writing—review and editing, A.K. and A.J.-R. All authors have read and agreed to the published version of the manuscript.

Funding: This research received no external funding.

References

1. Baxter, X.C.; Darvell, L.I.; Jones, J.M.; Barraclough, T.; Yates, N.E.; Shield, I. Miscanthus combustion properties and variations with Miscanthus agronomy. *Fuel* **2014**, *117*, 851–869. [CrossRef]
2. Karl, T.R.; Trenberth, K.E. Modern global climate change. *Science* **2003**, *302*, 1719–1723. [CrossRef] [PubMed]
3. Easterling, D.R.; Meehl, G.A.; Parmesan, C.; Changnon, S.A.; Karl, T.R.; Mearns, L.O. Climate extremes: Observations, modeling, and impacts. *Science* **2000**, *289*, 2068–2074. [CrossRef] [PubMed]
4. Capellán-Pérez, I.; Mediavilla, M.; de Castro, C.; Carpintero, Ó.; Miguel, L.J. Fossil fuel depletion and socio-economic scenarios: An integrated approach. *Energy* **2014**, *77*, 641–666.
5. Shafiee, S.; Topal, E. When will fossil fuel reserves be diminished? *Energy Policy* **2009**, *37*, 181–189. [CrossRef]
6. Hager, H.A.; Sinasac, S.E.; Gedalof, Z.; Newman, J.A. Predicting potential global distributions of two miscanthus grasses: Implications for horticulture, biofuel production, and biological invasions. *PLoS ONE* **2014**, *9*, e100032. [CrossRef]
7. Hastings, A.; Clifton-brown, J.; Wattenbach, M.; Mitchell, C.P.; Stampfl, P.; Smith, P. Future energy potential of Miscanthus in Europe. *GCB Bioenergy* **2009**, *30*, 1058. [CrossRef]

8.   Lewandowski, I.; Clifton-Brown, J.; Trindade, L.M.; Van Der Linden, G.C.; Schwarz, K.U.; Müller-Sämann, K.; Anisimov, A.; Chen, C.L.; Dolstra, O.; Donnison, I.S.; et al. Progress on optimizing miscanthus biomass production for the european bioeconomy: Results of the EU FP7 Project OPTIMISC. *Front. Plant Sci.* **2016**, *7*, 1620. [CrossRef]

9.   Lewandowski, I.; Clifton-Brown, J.; Scurlock, J.; Huisman, W. Miscanthus: European experience with a novel energy crop. *Biomass Bioenergy* **2000**, *19*, 209–227. [CrossRef]

10.   Von Cossel, M.; Lewandowski, I.; Elbersen, B.; Staritsky, I.; Van Eupen, M.; Iqbal, Y.; Mantel, S.; Scordia, D.; Testa, G.; Cosentino, S.L.; et al. Marginal agricultural land low-input systems for biomass production. *Energies* **2019**, *12*, 3123. [CrossRef]

11.   Jenkins, B.; Baxter, L.; Miles, T. Combustion properties of biomass. *Fuel Process. Technol.* **1998**, *54*, 17–46. [CrossRef]

12.   Sahoo, K.; Milewski, A.; Mani, S.; Hoghooghi, N.; Panda, S. Assessment of miscanthus yield potential from strip-mined lands (SML) and its impacts on stream water quality. *Water* **2019**, *11*, 546. [CrossRef]

13.   Kering, M.K.; Butler, T.J.; Biermacher, J.T.; Guretzky, J.A. Biomass yield and nutrient removal rates of perennial grasses under nitrogen fertilization. *BioEnergy Res.* **2011**, *5*, 61–70. [CrossRef]

14.   Roncucci, N.; Di Nasso, N.N.O.; Tozzini, C.; Bonari, E.; Ragaglini, G. *Miscanthus × giganteus* nutrient concentrations and uptakes in autumn and winter harvests as influenced by soil texture, irrigation and nitrogen fertilization in the Mediterranean. *GCB Bioenergy* **2015**, *7*, 1009–1018. [CrossRef]

15.   Heaton, E.A.; Dohleman, F.G.; Long, S.P. Meeting US biofuel goals with less land: The potential of Miscanthus. *Glob. Chang. Biol.* **2008**, *14*, 2000–2014. [CrossRef]

16.   Heaton, E.A.; Dohleman, F.G.; Miguez, A.F.; Juvik, J.A.; Lozovaya, V.; Widholm, J.; Zabotina, O.A.; McIsaac, G.F.; David, M.B.; Voigt, T.B.; et al. Miscanthus. A promising biomass crop. *Adv. Bot. Res.* **2010**, *56*, 75–137. [CrossRef]

17.   Dierking, R.M.; Allen, D.J.; Cunningham, S.M.; Brouder, S.M.; Volenec, J.J. Nitrogen reserve pools in two *Miscanthus × giganteus* genotypes under contrasting N managements. *Front. Plant Sci.* **2017**, *8*, 1618. [CrossRef]

18.   Stavridou, E.; Hastings, A.; Webster, R.; Robson, P. The impact of soil salinity on the yield, composition and physiology of the bioenergy grass *Miscanthus × giganteus*. *GCB Bioenergy* **2017**, *9*, 92–104. [CrossRef]

19.   Stefanovska, T.; Pidlisnyuk, V.; Lewis, E.; Gorbatenko, A. Herbivorous insects diversity at miscanthus × giganteus in Ukraine. *Agriculture* **2017**, *63*, 23–32. [CrossRef]

20.   Galatsidas, S.; Gounaris, N.; Vlachaki, D.; Dimitriadis, E.; Kiourtsis, F.; Keramitzis, D.; Gerwin, W.; Repmann, F.; Rettenmaier, N.; Reinhardt, G.; et al. Revealing bioenergy potentials: Mapping marginal lands in Europe—The seemla approach. In Proceedings of the European Biomass Conference and Exhibition, Copenhagen, Denmark, 14–17 May 2018.

21.   Maughan, M.; Bollero, G.; Lee, D.K.; Darmody, R.; Bonos, S.; Cortese, L.; Murphy, J.; Gaussoin, R.; Sousek, M.; Williams, D.; et al. Miscanthus × giganteus productivity: The effects of management in different environments. *GCB Bioenergy* **2012**, *4*, 253–265. [CrossRef]

22.   Zub, H.W.; Brancourt-Hulmel, M. Agronomic and physiological performances of different species ofMiscanthus, a major energy crop. A review. *Agron. Sustain. Dev.* **2010**, *30*, 201–214. [CrossRef]

23.   Styles, D.; Jones, M.B. Energy crops in Ireland: Quantifying the potential life-cycle greenhouse gas reductions of energy-crop electricity. *Biomass Bioenergy* **2007**, *31*, 759–772. [CrossRef]

24.   Di Nasso, N.N.O.; Roncucci, N.; Triana, F.; Tozzini, C.; Bonari, E. Seasonal nutrient dynamics and biomass quality of giant reed (*Arundo donax* L.) and miscanthus (Miscanthus × giganteus Greef et Deuter) as energy crops. *Ital. J. Agron.* **2011**, *6*, 152–158. [CrossRef]

25.   Drazic, G.; Milovanovic, J.; Stefanovic, S.; Petric, I. Potential of Miscanthus × Giganteus for heavy metals removing from industrial deposol. *Acta Reg. Environ.* **2017**, *2*, 56–58.

26.   Lee, D.K.; Boe, A. Biomass production of switchgrass in central south Dakota. *Crop. Sci.* **2005**, *45*, 2583. [CrossRef]

27.   Dohleman, F.G.; Heaton, E.A.; Arundale, R.A.; Long, S.P. Seasonal dynamics of above- and below-ground biomass and nitrogen partitioning in *Miscanthus × giganteus* and *Panicum virgatum* across three growing seasons. *GCB Bioenergy* **2012**, *4*, 534–544. [CrossRef]

28.   Clifton-Brown, J.C.; Lewandowski, I. Water use efficiency and biomass partitioning of three different miscanthus genotypes with limited and unlimited water supply. *Ann. Bot.* **2000**, *86*, 191–200. [CrossRef]

29. Beale, C.; Long, S. Seasonal dynamics of nutrient accumulation and partitioning in the perennial C4-grasses Miscanthus × giganteus and Spartina cynosuroides. *Biomass Bioenergy* **1997**, *12*, 419–428. [CrossRef]

30. Strullu, L.; Cadoux, S.; Preudhomme, M.; Jeuffroy, M.H.; Beaudoin, N. Biomass production and nitrogen accumulation and remobilisation by *Miscanthus × giganteus* as influenced by nitrogen stocks in belowground organs. *Field Crop. Res.* **2011**, *121*, 381–391. [CrossRef]

31. Cadoux, S.; Riche, A.B.; Yates, N.E.; Machet, J.M. Nutrient requirements of *Miscanthus × giganteus*: Conclusions from a review of published studies. *Biomass Bioenergy* **2012**, *38*, 14–22. [CrossRef]

32. Arundale, R.A.; Dohleman, F.G.; Heaton, E.A.; McGrath, J.M.; Voigt, T.B.; Long, S.P. Yields of *Miscanthus × giganteus* and panicum virgatum decline with stand age in the Midwestern USA. *GCB Bioenergy* **2014**, *6*, 71–87. [CrossRef]

33. Christian, D.; Riche, A.; Yates, N. Growth, yield and mineral content of *Miscanthus × giganteus* grown as a biofuel for 14 successive harvests. *Ind. Crop. Prod.* **2008**, *28*, 320–327. [CrossRef]

34. Schwarz, H.; Liebhard, P.; Ehrendorfer, K.; Ruckenbauer, P. The effect of fertilization on yield and quality of Miscanthus sinensis 'Giganteus'. *Ind. Crop. Prod.* **1994**, *2*, 153–159. [CrossRef]

35. Greef, J.M. *Etablierung und Biomassebildung von Miscanthus × Giganteus*; Cuvillier Verlag: Gottingen, Germany, 1995; pp. 1–162.

36. Kahle, P.; Beuch, S.; Boelcke, B.; Leinweber, P.; Schulten, H.R. Cropping of Miscanthus in Central Europe: Biomass production and influence on nutrients and soil organic matter. *Eur. J. Agron.* **2001**, *15*, 171–184. [CrossRef]

37. Lee, M.-S.; Wycislo, A.; Guo, J.; Lee, D.K.; Voigt, T. Nitrogen fertilization effects on biomass production and yield components of *Miscanthus × giganteus*. *Front. Plant Sci.* **2017**, *8*, 1–9. [CrossRef]

38. Yost, M.A.; Randall, B.K.; Kitchen, N.R.; Heaton, E.A.; Myers, R.L. Yield potential and nitrogen requirements of miscanthus × giganteus on eroded soil. *Agron. J.* **2017**, *109*, 684–695. [CrossRef]

39. Finnan, J.; Burke, B. Nitrogen dynamics in a mature Miscanthus × giganteus crop fertilized with nitrogen over a five year period. *Irish J. Agric. Food Res.* **2014**, *53*, 171–188.

40. Wildová, R.; Wild, J.; Herben, T. Fine-scale dynamics of rhizomes in a grassland community. *Ecography* **2007**, *30*, 264–276. [CrossRef]

41. Mann, J.J.; Barney, J.N.; Kyser, G.B.; DiTOMASO, J.M. Root system dynamics of *Miscanthus × giganteus* and panicum virgatum in response to rainfed and irrigated conditions in California. *BioEnergy Res.* **2013**, *6*, 678–687. [CrossRef]

42. Ameen, A.; Liu, J.; Han, L.P.; Xie, G.H. Effects of nitrogen rate and harvest time on biomass yield and nutrient cycling of switchgrass and soil nitrogen balance in a semiarid sandy wasteland. *Ind. Crop. Prod.* **2019**, *136*, 1–10. [CrossRef]

43. Peixoto, M.; Friesen, P.C.; Sage, R.F. Winter cold-tolerance thresholds in field-grown Miscanthus hybrid rhizomes. *J. Exp. Bot.* **2015**, *66*, 4415–4425. [CrossRef]

44. Clifton-Brown, J.C.; Lewandowski, I. Overwintering problems of newly established Miscanthus plantations can be overcome by identifying genotypes with improved rhizome cold tolerance. *New Phytol.* **2000**, *148*, 287–294. [CrossRef]

45. Ahmadi, S.H.; Agharezaee, M.; Kamgar-Haghighi, A.A.; Sepaskhah, A.R. Comparing canopy temperature and leaf water potential as irrigation scheduling criteria of potato in water-saving irrigation strategies. *Int. J. Plant Prod.* **2017**, *11*, 333–347.

46. Heaton, E. A quantitative review comparing the yields of two candidate C4 perennial biomass crops in relation to nitrogen, temperature and water. *Biomass Bioenergy* **2004**, *27*, 21–30. [CrossRef]

47. IUNG-PIB. Available online: http://www.iung.pulawy.pl/ (accessed on 13 September 2020).

48. Avramidis, P.; Nikolaou, K.; Bekiari, V. Total organic carbon and total nitrogen in sediments and soils: A comparison of the wet oxidation—Titration method with the combustion-infrared method. *Agric. Agric. Sci. Procedia* **2015**, *4*, 425–430. [CrossRef]

49. Kondratowicz-Maciejewska, K.; Kobierski, M. Content of available magnesium, phosphorus and potassium forms in soil exposed to varied crop rotation and fertilisation. *J. Elementology* **2011**, *16*, 543–553. [CrossRef]

50. Slepetiene, A.; Slepetys, J. Status of humus in soil under various long-term tillage systems. *Geoderma* **2005**, *127*, 207–215. [CrossRef]

51. Kilmer, V.J.; Nearpass, D.C. The determination of available sulfur in soils. *Soil Sci. Soc. Am. J.* **1960**, *24*, 337–340. [CrossRef]

52. De Rooij, G.H. Methods of soil analysis. *Vadose Zone J.* **2004**, *24*, 337–340.
53. Finnan, J.; Burke, B. Nitrogen fertilization of Miscanthus × giganteus: Effects on nitrogen uptake, growth, yield and emissions from biomass combustion. *Nutr. Cycl. Agroecosyst.* **2016**, *106*, 249–256. [CrossRef]
54. Cosentino, S.; Patanè, C.; Sanzone, E.; Copani, V.; Foti, S. Effects of soil water content and nitrogen supply on the productivity of Miscanthus×giganteus Greef et Deu. in a Mediterranean environment. *Ind. Crop. Prod.* **2007**, *25*, 75–88. [CrossRef]
55. Wiesler, F.; Dickmann, J.; Horst, W.J. Effects of nitrogen supply on growth and nitrogen uptake byMiscanthus sinensis during establishment. *J. Plant Nutr. Soil Sci.* **1997**, *160*, 25–31. [CrossRef]
56. Chen, Z.; Tao, X.; Khan, A.; Tan, D.K.Y.; Luo, H. Biomass Accumulation, photosynthetic traits and root development of cotton as affected by irrigation and nitrogen-fertilization. *Front. Plant Sci.* **2018**, *9*, 1–14. [CrossRef] [PubMed]
57. Anderson, E.; Arundale, R.; Maughan, M.; Oladeinde, A.; Wycislo, A.; Voigt, T. Growth and agronomy of Miscanthus × giganteus for biomass production. *Biofuels* **2011**, *2*, 71–87. [CrossRef]
58. Himken, M.; Lammel, J.; Neukirchen, D.; Czypionka-Krause, U.; Olfs, H.W. Cultivation of Miscanthus under West European conditions: Seasonal changes in dry matter production, nutrient uptake and remobilization. *Plant Soil* **1997**, *189*, 117–126. [CrossRef]
59. Miguez, F.E.; Villamil, M.; Long, S.P.; Bollero, G.A. Meta-analysis of the effects of management factors on Miscanthus×giganteus growth and biomass production. *Agric. For. Meteorol.* **2008**, *148*, 1280–1292. [CrossRef]
60. Paruelo, J.; Lauenroth, W.K.; Burke, I.C.; Sala, O.E. Grassland precipitation-use efficiency varies across a resource gradient. *Ecosystems* **1999**, *2*, 64–68. [CrossRef]

Article

The Eucalyptus Firewood: Understanding Consumers' Behaviour and Motivations

Nadia Palmieri *, Alessandro Suardi, Francesco Latteriniand Luigi Pari

CREA Research Centre for Engineering and Agro-Food Processing, Via della Pascolare, 16, Monterotondo, 00015 Rome, Italy; alessandro.suardi@crea.gov.it (A.S.); francesco.latterini@crea.gov.it (F.L.); luigi.pari@crea.gov.it (L.P.)
* Correspondence: nadia.palmieri@crea.gov.it; Tel.: +39-06-9067-5219

Received: 1 October 2020; Accepted: 28 October 2020; Published: 30 October 2020

Abstract: Italy is one of the world's major importers of firewood, despite the large amount of Italian eucalyptus plantations that could satisfy part of the country's internal demand. The demand is critical for farmers to understand developing market dynamics and people's willingness to buy a product is related to several parameters, including different supply methods. This study aimed to analyse the willingness to consume domestic eucalyptus firewood, and the related motivations of consumers considering the preferred supply method. Data was collected through a web-survey and analysed applying a multilevel regression. In general, the sample showed that attention is paid to both the type of wood and its origin, and that there is a preference for loose firewood as a supply method. Our findings suggest that factors such as age, experience, and familiarity with a product, the supply method, attitude towards novelty, provenience, and energetic density of firewood have an important role in shaping individual inclination towards consuming domestic eucalyptus firewood. This implies that the owners of eucalyptus plantations should target mostly young and detail-oriented consumers, and should also try to clearly give information regarding the origin of the product and its technical characteristics.

Keywords: consumer choices; eucalyptus; firewood; Italy; multilevel logistic regression model; willingness to consume

---

1. Introduction

Eucalyptus (*Eucalyptus* spp.) forests and agro-forestry plants cover about 20 million hectares in the world. These species show faster biomass growth if compared to other species [1] and can be used to obtain pulp, paper, and firewood. In regards to the environment, the management of eucalyptus plantations can be considered more sustainable than other energy crops [2]. Due to its fast growth and modesty, eucalyptus can contribute to biodiversity conservation [1,3]. Moreover, it shows an important role in climate change mitigation [1] due to its high capacity of carbon sequestration during growth [4,5].

In Italy, agro-forestry plants mainly aim towards the bioenergy production of eucalyptus (*Eucalyptus* spp.), poplar (*Populus* spp.), and black locust (*Robinia pseudoacacia* L.), which cover more than 100,000 ha [6]. Considering this, eucalyptus could provide a biomass that can fulfill about 72% of Italian demand [7–9]. Notwithstanding this large availability of wooden biomass, Italy is one of the major global importers of wood for energy purpose [10], used particularly for domestic heating [11,12]. The major part of imported fuelwood originates from the Balkans' area, and this is mainly due to the lower cost of labour and of raw materials in these countries [11,13].

Considering the growing importance of Eucalyptus plantations throughout the world, many researches have focused on both environmental impacts related to eucalyptus management [1,4,14–16] and cultivation methodology [10]. However, lower attention has been put on the economic aspects of cultivation, i.e., investigation of the supply chain analysis [2,17,18] as well as of the demand of such species' wood.

Consumer choice is a key variable for farmers to understand developing market dynamics [19], moreover people's willingness to buy a given product is related to several parameters [20]. In fact, consumer attitude is an important aspect to analyse because it allows us to study the acceptance of a particular good by consumers [21]. According to this, it could be important to investigate the willingness to consume domestic eucalyptus fuelwood, given that eucalyptus firewood shows similar heating value to other species, such as oak [22], and thus it could be an interesting alternative.

In this framework, the study aimed to understand people's willingness to consume domestic eucalyptus firewood and their motivations, considering also people's preferred supply method. This last parameter is indeed important in consumer behaviour [23], even if, to the best of our knowledge, current literature studies that consider this wood supply chain are lacking.

This paper is structured as follows: Section 2 describes the materials and methods used. The results and discussion are presented in Section 3 and the conclusions are handled in Section 4.

2. Materials and Methods

2.1. Data Collection, Sample and Questionnaire

Data was collected through a web-based survey performed during the period January–April 2020, from an initial sample of 300 Italian people. In particular, consumers were recruited through invitations to participate in the online survey (performed by Google drive) via social networks (Instagram, Twitter, and Facebook). Moreover, snowball sampling recruitment was also adopted, using the interpersonal relations of the authors (via email) to reach a larger number of participants [24]. For these reasons, the sample was not representative of the Italian population, which happens in many studies about consumer behaviour where a convenience sample was used [21,25–30]. Subsequently, 18 respondents were excluded from the survey because they stated not to be domestic firewood consumers. The final sample was consequently made up of 282 consumers. Before starting the survey, an interview with 80 consumers was carried out in order to understand if the investigated topic was understandable through our questionnaire.

The questionnaire was made up of three sections: (1) Consumers' behaviour towards firewood; (2) consumers' willingness to consume domestic eucalyptus firewood and related motivations; and (3) socio-demographic features of respondents.

The first two sections applied a five-point Likert scale (1 = totally disagree; 2 = disagree; 3 = indifferent; 4 = agree, and 5 = totally agree), with the exception of some questions in Section 2 in which categorical variables (i.e., *Forn*, *Will*, *Cons*, and *Familiarity*) have been used [23]. Furthermore, the Cronbach Alpha coefficient for each item group was calculated to assess the reliability of the scale, which showed a good level (from 0.70 to 0.90).

In the first section of the questionnaire we analysed the consumers' attitudes about firewood species, ethical aspects of the choice, geographic provenience (*Prov*) (i.e., if firewood comes from tropical countries or Mediterranean ones), and the origin of firewood (*Origin*) (if firewood comes from an agro-forestry plant or natural stand) [29].

The second part of the questionnaire investigated respondents' willingness to consume domestic eucalyptus fuelwood (*Will*) with a binary choice (Yes vs. No), their willingness to pay per 100 kg of biomass (*Price*), and the amount of eucalyptus firewood which the consumer would be willing to consume yearly (*Will_q*). It is important to underline that, if respondents were not willing to consume it, the quantity of eucalyptus firewood was considered zero.

The questionnaire also requested to indicate the consumer's familiarity (*Familiarity*) with eucalyptus fuelwood and if they have consumed it in the past (*Cons*) [30], and in both cases the question was asked as a binary choice (Yes vs. No). It is important to underline that familiarity was investigated using the question: "Have you ever heard about eucalyptus as firewood alternative?"

Moreover, another important aspect considered in the survey is consumers' motivation to use eucalyptus firewood. These aspects were investigated by asking questions related to curiosity (*Curiosity*), to technical characteristics (*Energetic*), as well as to environmental aspects [29]. In particular, the question about curiosity was "You are willing to consume domestic eucalyptus firewood for curiosity (How much do you agree with the following statements? Express your judgment by putting a tick from 1 to 5. 1 = totally disagree. 5 = totally agree)", while questions on technical characteristics were "You are willing to consume domestic eucalyptus firewood if it had better combustion behaviour (wood burning duration) than other firewood species (How much do you agree with the following statements? Express your judgment by putting a tick from 1 to 5. 1 = totally disagree. 5 = totally agree)" [23]. Moreover, it is important to highlight that the questions concerning the environmental aspects of firewood were asked to respondents but these questions were subsequently excluded because, using the stepwise procedure, they did not fit in the applied model. Finally, respondents had to select their preferred supply method (*Forn*), i.e., loose firewood, firewood arranged in pallets, or firewood in 10–15 kg bags. Other questions about where consumers usually buy firewood (i.e., *Woodman*, *Retail*, *Internet*, etc.) were asked to respondents but these questions were also excluded because they did not fit in the applied model.

In the last section of the questionnaire, data about socio-demographic features of the sample, such as age (*Age*), gender, area of residence, and education level were collected [31–34].

2.2. Statistical Analysis

A logistic regression model describes the relationships between the willingness to consume a particular product and the motivations of consumers, without including the variability among predictors of different levels (in case of data with nested structure) [35]. Given the nested structure of the studied sample, this study utilized multilevel logistic regression model to understand the consumers' willingness to consume domestic eucalyptus firewood and their motivations, considering the people's preferred supply method.

The multilevel logistic regression model consists of an extension of the regression model in which data are arranged in groups and coefficients can differ among the various groups [19].

In particular, several steps were followed to lead the analysis [19,35]. In fact, in the first step, it is necessary to understand if the dataset show a nested structure calculating the intra-class correlation (ICC) coefficient [36]. After such a check, analysis can be carried out, calculating, and comparing: The simplest two level model, an intermediate model, and the full multilevel logistic regression model [35]. This procedure allowed us to have a final model accounting both for the effects of the lower-level predictor variables and for higher level ones [35].

In particular, following some authors [19,35] and as above mentioned, to examine the existence of a nested structure of the data, an intra-class correlation (ICC) coefficient was calculated [19,35,37]. The ICC coefficient estimates the heterogeneity of the dependent variables among groups i.e., people's preferred supply method. Values of ICC range from 0 to 1, in which 0 indicates that probability does not vary among groups while 1 means the result probability only differentiate between groups. In our case, the ICC coefficient was 0.1432, therefore the calculated ICC of the total dataset means 14% of the difference in the probability of willingness to consume domestic eucalyptus fuelwood was related to the difference in people's preferred supply method. Therefore, to build up the single-level logistic regression model would not be appropriate to describe the relationship between the probability of willingness to consume domestic eucalyptus fuelwood and the motivations of consumers without considering the preferred supply method [38].

Successively, the following steps were applied:

- *The first step: Building up of the simplest two level model.*

The simplest two level model represents the model in which intercepts casually vary among groups [19]. In particular, the simplest two level model was described as:

$$Pr(Will = Yes|x) = \gamma0 + u0j + rij \tag{1}$$

where *Will* refers to the probability that people are willing to consume domestic eucalyptus firewood, $\gamma0$ is the fixed intercept, *u0j* represents the random intercept, and *rij* represent the error. In other words, $\gamma0$ represents the overall average probability that people are willing to consume domestic eucalyptus firewood of the total dataset, while *u0j* represents the variety in the average probability that people are willing to consume domestic eucalyptus firewood. Equation (1) shows two sources of errors in its random part (*u0j* + *rij*) i.e., the between-groups variance (σ_1) and the within-group variance (σ_2). The parametric estimation results for the empty model are given in Table 1.

Table 1. Parametric estimation results for the simplest two level model Equation (1).

AIC	BIC	LogLik			
300.3	307.2	−148.1			
Random effects		$\sigma_1 = 0.38$	$\sigma_2 = 0.62$		
Fixed effects		Value	Standard Error	z value	*p*-value
Intercept		0.51	0.1739	2.965	<0.001

Note: The AIC (Akaike information criterion) and the BIC (Bayesian information criterion) are the well-known model fit indices.

Source: Our elaboration.

- *The second step: Building of the intermediate model.*

The result of the first phase (i.e., ICC estimation) allowed the identification of the two-level model as the most suitable for the analysis. In particular, Equation (1) represents the model in which intercepts casually vary among groups while a general model considers fixed predictors, both at an individual and group level [19]. Using a stepwise procedure [24], several models were identified and the one which showed the lowest AIC was selected. The AIC calculates the likelihood of a model for future estimations and in particular, a smaller AIC means that the corresponding model shows a better prediction performance [19,35].

Moreover, the likelihood ratio test (LRT) was applied to compare the simplest two level model and intermediate ones and according to this, the best model was the intermediate one (Tables 2 and 3).

The intermediate model was described as:

$$Pr(Will = Yes|x) = \gamma0 + \gamma1\,Prov + \gamma2\,Origin + \gamma3\,Familiarity + \gamma4Cons + \gamma5Will\_q + \\ \gamma6Curiosity + \gamma7Energetic + \gamma8Age + u0j + rij. \tag{2}$$

Table 2. Likelihood ratio test (LRT): The simplest two level model and intermediate model.

Model	df	AIC	BIC	LogLik	LRT	*p*-Value
Simplest two level model	2	300.3	307.2	−148.1		
Intermediate model	11	187.6	225.5	−82.8	130.7	<0.0001

Note: The AIC (Akaike information criterion) and the BIC (Bayesian information criterion) are the well-known model fit indices.

Source: Our elaboration.

Table 3. Parametric estimation results for the intermediate model Equation (2).

AIC	BIC	LogLik			
187.6	225.5	−82.8			
Random effects		$\sigma_1 = 0.37$	$\sigma_2 = 0.60$		
Fixed effects		Value	Standard Error	z value	*p*-value
Intercept		−2.06	1.29	−1.59	n.s.
Prov		0.99	0.27	3.56	<0.0001
Origin		−0.88	0.32	−2.73	<0.001
Familiarity		1.26	0.50	2.51	<0.01
Cons		1.81	0.57	3.18	<0.001
Will_q		0.05	0.01	3.32	<0.0001
Curiosity		0.57	0.15	3.61	<0.0001
Energetic		0.71	0.21	3.25	<0.001
Age		−0.05	0.01	−2.85	<0.001

Note: n.s. means that variable is not significant. The AIC (Akaike information criterion) and the BIC (Bayesian information criterion) are the well-known model fit indices.

Source: Our elaboration.

- *The third step: Building up the full model.*

In this case, the final step consisted of building a full model to account for the direct effect of the lower-lever predictor variables and higher-level predictor ones, including the effect of the interaction terms and random intercept effect. The final model was described as follow:

$$Pr(Will = Yes|x) = \gamma 0 + \gamma 1 Prov + \gamma 2\,Origin + \gamma 3 Familiarity + \gamma 4 Cons + \gamma 5 Will\_q + \gamma 6 Curiosity + \gamma 7 Energetic + \gamma 8 Age + \gamma 9 Forn + u0j + rij. \tag{3}$$

In order to understand if the full model was more suitable than the intermediate one, the ANOVA test was applied. In particular, the ANOVA test (Table 4) between the intermediate model Equation (2) and the full model Equation (3) showed that the willingness to consume domestic eucalyptus firewood is significantly influenced by all variables including the supply method (*Forn*) (i.e., loose firewood, firewood arranged in pallets, or firewood in 10–15 kg bags). It can be seen that the AIC values of the full model is smaller than the intermediate model one. This means that adding the supply method (*Forn*) in the model improves the quality of the regression.

Table 4. Likelihood ratio test between the intermediate model Equation (2) and the full model Equation (3).

Model	df	AIC	BIC	LogLik	LRT	*p*-Value
Intermediate model	11	187.6	225.5	−82.8		
Full Model	21	177.5	249.8	−67.7	30.008	<0.0001

Note: The AIC (Akaike information criterion) and the BIC (Bayesian information criterion) are the well-known model fit indices.

Source: Our elaboration.

Finally, the odds' ratio was calculated to show the probability increase/decrease of the willingness to consume domestic eucalyptus firewood when the considered variable increases or decreases.

The analysis was performed using R version 3.6.2 [39].

3. Results and Discussion

3.1. Descriptive Statistics

Within the sample of 282 individuals, 62% were males, average age resulted in about 43 years, 59% of respondents presented a low education level (i.e., primary or secondary school), and 57% of the sample lives in small towns.

Table 5 shows the variables used in the full model, their average values, and standard deviation while Table 6 shows the Pearson correlations among explanatory variables indicating low correlations index among variables used in the full model.

Table 5. Variables used in the full multilevel logistic regression model (n = 282).

No	Label Variables	Description	Mean Value (M) and Standard Deviation (SD)
1	*Will*	Willingness to consume domestic eucalyptus firewood (Yes = 1; No = 0)	M: 0.64; SD: 0.36
		1st level variables	
2	*Prov*	The degree to which consumers pay attention to provenience of firewood (i.e., if firewood comes from tropical countries or Mediterranean ones)	M: 3.53; SD: 1.35
3	*Origin*	The degree to which consumers pay attention to origin of firewood (i.e., if firewood comes from an agro-forestry plant or natural woodland)	M: 3.65; SD: 1.35
4	*Familiarity*	The familiarity with eucalyptus firewood (Yes = 1; No = 0)	M: 0.44; SD: 0.56
5	*Cons*	Consume eucalyptus firewood in the past (Yes = 1; No = 0)	M: 0.17; SD: 0.38
6	*Will_q*	The quantity of eucalyptus firewood that they were willingness to consume (quintals)	M: 14.53; SD: 31.89
7	*Curiosity*	The degree to which consumers are willing to consume eucalyptus firewood for curiosity	M: 3.00; SD: 1.63
8	*Energetic*	The degree to which consumers are willing to consume eucalyptus firewood for its energetic characteristics	M: 3.85; SD: 1.15
9	*Age*	Consumer age	M: 42.95; SD:12.21
		2nd level variable	
10	*Forn*	The eucalyptus firewood supply methods, i.e., firewood in 10–15 kg bags (1), loose firewood (2), and (3) firewood arranged in pallets	M: 1.93; SD: 0.64

Source: Our elaboration on data survey.

Table 6. Pearson correlations between explanatory variables of the full multilevel logistic regression model.

	Prov	*Origin*	*Familiarity*	*Cons*	*Will*	*Will_q*	*Forn*	*Curiosity*	*Energetic*	*Age*
Prov	1.00									
Origin	0.40	1.00								
Familiarity	0.23	0.27	1.00							
Cons	0.17	0.20	0.19	1.00						
Will	0.14	0.01 *	0.15 *	0.03 *	1.00					
Will_q	0.01 *	0.05	0.08 *	0.08	0.23	1.00				
Forn	0.05 **	0.08	0.07 *	0.04	0.10 **	0.12 **	1.00			
Curiosity	0.16	0.24	0.18	0.32	0.13	0.12	0.01	1.00		
Energetic	0.39	0.34	0.09	0.17	0.03	0.11 *	0.06 **	0.04 *	1.00	
Age	0.03	0.08 **	0.15	0.11	0.04	0.01	0.05	0.21	0.09 **	1.00

* p-value < 0.05; ** p-value < 0.01.

In the first parameter of Table 5 (*Will*), 64% of the sample was willing to consume domestic eucalyptus firewood and the theoretical average annual consumption was about 1.5 Mg·yr^{-1}. This finding may indicate that people are becoming more receptive towards eucalyptus, considering it as a suitable firewood alternative.

Moreover, generally, the sample showed attention both to provenience and to the origin of fuelwood and preferred loose firewood as the supply method.

On average, the sample could be willing to consume domestic eucalyptus firewood for its energetic characteristics, even if the respondents did not show curiosity about it. In addition, as a general trend, the sample had neither familiarity (*Familiarity*) or consumed eucalyptus firewood in the past (*Pass*).

3.2. The Full Multilevel Logistic Regression Model

The result of the full multilevel logistic regression model is given in Table 7.

Table 7. Parametric estimation results for the full model Equation (3).

AIC	BIC	LogLik			
177.5	249.8	−67.7			
Random effects		$\sigma_1 = 0.23$	$\sigma_2 = 0.48$		
Fixed effects		Value	Standard Error	z value	*p*-value	Odds' Ratio
Intercept		−5.73	6.08	−0.94	n.s	-
		1st level variables				
Prov		2.44	1.44	1.68	<0.05	3.03
Origin		−2.74	1.73	−1.58	n.s.	-
Familiarity		7.75	2.34	3.31	<0.0001	5.9
Cons		9.18	2.72	3.37	<0.0001	5.5
Will_q		−0.04	0.09	−0.49	n.s.	-
Curiosity		1.34	0.74	1.80	<0.05	1.3
Energetic		3.12	1.08	2.88	<0.001	2.1
Age		−0.16	0.08	−1.89	<0.05	−0.1
		2nd level variable				
Forn		1.38	2.76	0.50	<0.0001	2.2

Note: n.s. means that variable is not significant. The AIC (Akaike information criterion) and the BIC (Bayesian information criterion) are the well-known model fit indices.

Source: Authors' elaboration.

The parameters *Familiarity*, *Cons*, *Energetic*, and *Forn* resulted to be the most significant in the full model, followed by *Prov*, *Curiosity*, and *Age*. All variables showed a positive sign except *Age*.

Unfortunately, there are not many studies focused on the willingness to consume domestic eucalyptus firewood which could help us to evaluate the findings of the present work, which represent a novelty in this field.

Our findings show that consumers who pay attention to the supply method (*Forn*), i.e., loose firewood, firewood arranged in pallets, or firewood in 10–15 kg bags, are 2.2 times more likely to consume domestic eucalyptus firewood than other consumers.

The knowledge of the existence of eucalyptus (*Familiarity*) as a firewood alternative resulted in an important factor in the consumers' willingness, indeed, those who reported to know it, asserted to be more willing to consume it in the future. In particular, consumers that showed familiarity with eucalyptus firewood are 5.9 times more likely to consume it than other people. Our findings are in line with current literature [40] where familiarity is a key attribute that influences consumer behaviour and, consequently, decision making [41]. In fact, familiarity showed a significant influence on the acceptability of a given product [42]. Moreover, increasing familiarity, through further information, about new products could improve their accepting rate [43] and this aspect resulted to be a key concerning firewood [44]. Johnson et al., (1984) suggested that familiarity involves a process of looking for information and processing both new and already existing goods [45]. In addition, Fischer et al., (2008) showed that the more information the consumers gather, the more products they are willing to purchase [46]. This is an important aspect, taking into consideration that firewood consumption is linked with information received [23], and in general the familiarity with a given product, through available information, could increase the likelihood of consuming it in the future [47].

More recently, Seo et al., (2013) highlighted that familiarity with a product is linked to both level of information [41] and, in accordance with Ryynänen et al., (2018), to previous experience [41,48]. Our findings confirmed this assertion, considering that previous experience with eucalyptus firewood showed an important role in the consumers' willingness. Indeed, respondents who reported previous experiences with eucalyptus firewood (*Cons*) were more willing to consume it again. In particular, people who had previous experiences with eucalyptus firewood are 5.5 times more likely to consume it than other people. Our results are in line with literature about consumer behaviour, where previous experience with an innovative product enhances consumer acceptance [49] and people with previous experiences resulted in having a higher probability to willingly consume it again in future [25].

In our case, the geographic provenience of firewood (*Prov*) (i.e., if firewood comes from tropical countries or Mediterranean ones) also seemed to be one of the reasons that could push people to consume eucalyptus firewood. Indeed, respondents who pay attention to the geographic provenience of firewood are 3.03 times more likely to consume it than other consumers. These findings are also in line with previous literature [50] where the provenience of wood was a key factor that influenced the consumer behaviour. Moreover, according to Paletto et al., (2017) the geographic provenience of wood is one of the most relevant factors in the enhancement strategies for local fuelwood [29]. The consumer relies on the image of place of production (negative or positive) as a quality standard [24]. In this framework, attention towards geographic provenience of firewood could imply benefits for Italian local economies [29], especially in Southern Italy where eucalyptus is usually located [2].

Moreover, curiosity towards a new product also demonstrated a certain importance in the respondents' choice. In particular, those who showed a positive attitude about new products (*Curiosity*) resulted to be 1.3 times more willing to consume domestic eucalyptus firewood. Similar results were reported by Palmieri et al., (2020), which showed that willingness to consume a new product is linked to consumers' curiosity towards new products [25].

It is well known that the combustion of wood in fireplaces involves a very low level of technology, with an equally low energy conversion efficiency. For this reason, it is not surprising that the consumer described by the sample showed a marked interest in technical aspects related to the quality of the wood to be burned. In fact, energetic density of firewood (*Energetic*) was another factor that had an important influence on respondents' choice. Technical aspects such as energetic density could shape the respondents' behaviour. In particular, people who pay attention to the energetic aspect of firewood are 2.1 times more likely to consume it than other people. Vásquez Lavin et al., (2020) also reported that technical aspects such as heating value and moisture of firewood were considered very important for what concern consumer's choice [51]. Similar results were confirmed by other studies [23] where energetic density of firewood played an important role in consumer choices. This aspect could be very interesting, taking into account that eucalyptus firewood has similar heating value to other fuelwood species, for example oak [22].

Another significant variable in our model was the consumer's age. Indeed, respondents' age resulted to be an important driver and was negatively associated with willingness to consume domestic eucalyptus firewood. According to our results, younger people showed a higher willingness to consume eucalyptus firewood than older ones. In fact, young people are 0.1 times more likely to consume it than older people. Similar results were found by some authors [25], where age exerted a negative effect on the probability to willingly consume a given new product. On the other hand, this aspect cannot be generalized because other authors reported that respondents' age had no significant influence on preference [50].

4. Conclusions

Italy is one of the world's major importers of firewood and medium rotation eucalyptus plantations could represent a source of biomass for the firewood market. Although, excluding some areas of Southern Italy, the eucalyptus is a wood species not extensively used as firewood. People's willingness to consume a product depend on several parameters including the different supply methods of a product.

This study aimed to understand consumer willingness to consume domestic eucalyptus firewood and their motivations by considering the preferred supply method.

Although the sample of this research cannot be considered representative of the entire Italian population, as happens in many studies about consumer behaviour, the results obtained gave interesting hints to understand the process of consumer decision-making. In fact, further studies are necessary to better understand the Italian consumers' propensity towards eucalyptus fuelwood acceptance, in terms of their individual preferences, attitudes, or concerns.

This research suggested that factors such as age, previous experience, familiarity, supply method, attitude towards new product, provenience, and technical characteristics of firewood, i.e., energetic density, play an important role in shaping individual behaviour concerning eucalyptus firewood consumption. In summary, the owners of eucalyptus plantations should address themselves towards informed young people, curious, who already know eucalyptus as fuelwood species, and who pay attention to the supply method, provenience, and technical characteristics of firewood (i.e., heating value). Our results are interesting, taking into consideration that eucalyptus plantations are less environmentally impactful than other crops, therefore developing a eucalyptus fuelwood value chain could reduce the environmental impacts linked to firewood production.

Even if the conclusions of the present work cannot be over-generalised, the present findings could open new spaces for domestic eucalyptus firewood, considering that the quantity and quality of information could shape the probability that people would be willing to consume it. In conclusion, a growing eucalyptus demand as sustainable firewood alternative would be an interesting opportunity for farmers to enter the sector and target themselves to specific market niches.

Author Contributions: N.P. contributed to the study design, data collection, data analysis, writing, and revising of the whole manuscript. A.S. and F.L. contributed to the study design and revising of the whole manuscript. L.P. contributed to the study supervision, project administration, and funding acquisition. All authors have read and agreed to the published version of the manuscript.

Funding: This research was funded by the European Union's Horizon 2020 Research and Innovation Programme under grant agreement No. 744821 and the APC was funded by Becool Project. This research was partly carried out within both the Becool Project funded by the European Union's Horizon 2020 Research and Innovation Programme under grant agreement No. 744821; and the AGROENER project (D.D. n. 26329, 1 April 2016) funded by the Italian Ministry of Agriculture (MiPAAF). The ideas expressed do not represent either those of the European Commission or the Italian Ministry of Agriculture (MiPAAF).

Conflicts of Interest: The authors declare no conflict of interest.

References

1. Bayle, G.K. Ecological and social impacts of eucalyptus tree plantation on the environment. *J. Biodivers. Conserv. Bioresour. Manag.* **2019**, *5*, 93–104. [CrossRef]
2. Sgroi, F.; Di Trapani, A.M.; Foderà, M.; Testa, R.; Tudisca, S. Economic assessment of Eucalyptus (spp.) for biomass production as alternative crop in Southern Italy. *Renew. Sustain. Energy Rev.* **2015**, *44*, 614–619. [CrossRef]
3. Boothroyd-Roberts, K.; Gagnon, D.; Truax, B. Hybrid poplar plantations are suitable habitat for reintroduced forest herbs with conservation status. *SpringerPlus* **2013**, *2*, 507. [CrossRef]
4. Burrows, W.H.; Henry, B.K.; Back, P.V.; Hoffmann, M.B.; Tait, L.J.; Anderson, E.R.; Menke, N.; Danaher, T.; Carter, J.O.; McKeon, G.M. Growth and carbon stock change in eucalypt woodlands in northeast Australia: Ecological and greenhouse sink implications. *Glob. Chang. Biol.* **2002**, *8*, 769–784. [CrossRef]

5. Du, H.; Zeng, F.; Peng, W.; Wang, K.; Zhang, H.; Liu, L.; Song, T. Carbon storage in a Eucalyptus plantation chronosequence in Southern China. *Forests* **2015**, *6*, 1763–1778. [CrossRef]

6. Pettenella, D.; Masiero, M. *Una Nuova Economia del Legno-Arredo tra Industria, Energia e Cambiamento Climatico*; Gargiulo, T.Z.R., Ed.; FrancoAngeli: Monza, Italy, 2007.

7. Deidda, A.; Buffa, F.; Linaldeddu, B.T.; Pinna, C.; Scanu, B.; Deiana, V.; Satta, A.; Franceschini, A.; Floris, I. Emerging pests and diseases threaten eucalyptus camaldulensis plantations in Sardinia, Italy. *iForest Biogeosci. For.* **2016**, *9*, 883–891. [CrossRef]

8. Pari, L.; Bergonzoli, S.; Suardi, A.; Scarfone, A.; Alfano, V.; Mattei, P.; Lazar, S. Produttività dell'eucalipto Un impianto quinquennale in Italia centrale. *Sherwood-Foreste ed Alberi Oggi* **2019**, *241*, 70–72.

9. Mughini, G.; Rosso, L. Selezioni di cloni di eucalitto per la destinazione da biomassa. *Sherwood-Foreste ed Alberi Oggi* **2017**, 45–52.

10. Mughini, G. Suggestions for sustainable Eucalyptus clonal cultivation in Mediterranean climate areas of central and southern Italy. *For. Riv. Selvic. Ed. Ecol. For.* **2016**, *13*, 47–58. [CrossRef]

11. Pierobon, F.; Zanetti, M. Life cycle environmental impact of firewood production—A case study in Italy. *Appl. Energy* **2015**, *150*, 185–195. [CrossRef]

12. Caserini, S.; Fraccaroli, A.; Monguzzi, A.; Moretti, M.; Angelino, E. *Stima dei Consumi di Legna da Ardere per Riscaldamento ed Uso Domestico in Italia [Estimation of the Domestical Firewood Consumption in Italy]*; Agenzia Nazionale per la Protezione dell'Ambiente e per i Servizi Tecnici (APAT), Agenzia Regionale per la Protezione dell'Ambiente (ARPA): Milano, Italy, 2008; p. 60.

13. Mairota, P.; Manetti, M.C.; Amorini, E.; Pelleri, F.; Terradura, M.; Frattegiani, M.; Savini, P.; Grohmann, F.; Mori, P.; Terzuolo, P.G.; et al. Opportunities for coppice management at the landscape level: The Italian experience. *iForest Biogeosci. For.* **2016**, *9*, 775–782. [CrossRef]

14. Medeiros, G.; Florindo, T.; Talamini, E.; Fett Neto, A.; Ruviaro, C. Optimising Tree Plantation Land Use in Brazil by Analysing Trade-Offs between Economic and Environmental Factors Using Multi-Objective Programming. *Forests* **2020**, *11*, 723. [CrossRef]

15. Garcia-Gonzalo, J.; Pais, C.; Bachmatiuk, J.; Barreiro, S.; Weintraub, A. A progressive hedging approach to solve harvest scheduling problem under climate change. *Forests* **2020**, *11*, 224. [CrossRef]

16. Luwesi, C.N.; Obando, J.A.; Shisanya, C.A. The Impact of a Warming Micro-Climate on Muooni Farmers of Kenya. *Agriculture* **2017**, *7*, 20. [CrossRef]

17. Simioni, F.J.; de Almeida Buschinelli, C.C.; Moreira, J.M.M.Á.P.; dos Passos, B.M.; Girotto, S.B.F.T. Forest biomass chain of production: Challenges of small-scale forest production in southern Brazil. *J. Clean. Prod.* **2018**, *174*, 889–898. [CrossRef]

18. Cuong, T.; Chinh, T.T.Q.; Zhang, Y.; Xie, Y. Economic Performance of Forest Plantations in Vietnam: Eucalyptus, Acacia mangium, and Manglietia conifera. *Forests* **2020**, *11*, 284. [CrossRef]

19. Mastronardi, L.; Romagnoli, L.; Mazzocchi, G.; Giaccio, V.; Marino, D. Understanding consumer's motivations and behaviour in alternative food networks. *Br. Food J.* **2019**. [CrossRef]

20. Botti, S.; McGill, A.L. The locus of choice: Personal causality and satisfaction with hedonic and utilitarian decisions. *J. Consum. Res.* **2011**, *37*, 1065–1078. [CrossRef]

21. Palmieri, N.; Perito, M.A.; Lupi, C. Consumer Acceptance of Cultured Meat: Some Hints from Italy. *Br. Food J.* **2020**. [CrossRef]

22. Mughini, G.; Gras, M.; Salvati, L.; Filippelli, S.; Tanchis, U. Velino and Viglio: Two eucalypt hybrid clones for Italy. *Sherwood-Foreste ed Alberi Oggi* **2012**, *187*, 41–45.

23. Palmieri, N.; Suardi, A.; Pari, L. Italian consumers' willingness to pay for eucalyptus firewood. *Sustainability* **2020**, *12*, 2629. [CrossRef]

24. Palmieri, N.; Perito, M.A. Consumers' Willingness To Consume Sustainable and Local Wine in Italy. *Ital. J. Food Sci.* **2020**, *32*, 222–233.

25. Palmieri, N.; Perito, M.A.; Macrì, M.C.; Lupi, C. Exploring consumers' willingness to eat insects in Italy. *Br. Food J.* **2019**, *121*, 2937–2950. [CrossRef]

26. Aguilar, F.X.; Cai, Z. Conjoint effect of environmental labeling, disclosure of forest of origin and price on consumer preferences for wood products in the US and UK. *Ecol. Econ.* **2010**, *70*, 308–316. [CrossRef]

27. Wöhler, M.; Andersen, J.S.; Becker, G.; Persson, H.; Reichert, G.; Schön, C.; Schmidl, C.; Jaeger, D.; Pelz, S.K. Investigation of real life operation of biomass room heating appliances—Results of a European survey. *Appl. Energy* **2016**, *169*, 240–249. [CrossRef]

28. Adapa, S. Factors influencing consumption and anti-consumption of recycled water: Evidence from Australia. *J. Clean. Prod.* **2018**, *201*, 624–635. [CrossRef]

29. Paletto, A.; Notaro, S.; Pastorella, F.; Giacovelli, G.; Giovannelli, S.; Turco, R. Certificazione forestale in Calabria: Attitudini, preferenze e disponibilità a pagare delle imprese di seconda trasformazione del legno. *For. Silvic. For. Ecol.* **2017**, *14*, 107. [CrossRef]

30. Jensen, K.L.; Jakus, P.M.; English, B.C.; Menard, J. Consumers' willingness to pay for eco-certified wood products. *J. Agric. Appl. Econ.* **2004**, *36*, 617–626. [CrossRef]

31. Altamore, L.; Ingrassia, M.; Columba, P.; Chironi, S.; Bacarella, S. Italian Consumers' Preferences for Pasta and Consumption Trends: Tradition or Innovation? *J. Int. Food Agribus. Mark.* **2020**, *32*, 337–360. [CrossRef]

32. Ceschi, S.; Canavari, M.; Castellini, A. Consumer's Preference and Willingness to Pay for Apple Attributes: A Choice Experiment in Large Retail Outlets in Bologna (Italy) Consumer's Preference and Willingness to Pay for Apple Attributes: A Choice Experiment in Large Retail Outlets in Bol. *J. Int. Food Agribus. Mark.* **2018**, *30*, 305–322. [CrossRef]

33. Ingrassia, M.; Sgroi, F.; Tudisca, S.; Chironi, S. Study of Consumer Preferences in Regard to the Blonde Orange Cv. Washington Navel "Arancia Di Ribera PDO". *J. Food Prod. Mark.* **2017**, *23*, 799–816. [CrossRef]

34. Marinda, P.A. Child–mother nutrition and health status in rural Kenya: The role of intra-household resource allocation and education. *Int. J. Consum. Stud.* **2006**, *30*, 327–336. [CrossRef]

35. Shi, S.; Li, H.; Ding, X.; Gao, X. Effects of household features on residential window opening behaviors: A multilevel logistic regression study. *Build. Environ.* **2020**, *170*, 106610. [CrossRef]

36. Ruscone, M.N. *Utilizzo dell'ICC Come Indicatore dell'Esistenza di Una Struttura Gerarchica*; Università Cattolica del Sacro Cuore, Dipartimento di Scienze Statistiche: Milano, Italy, 2012; pp. 1–28.

37. Iezzi, E. Modelli multilivello a scelta binaria: I vantaggi di una loro applicazione ai dati sanitari. *Politiche Sanit.* **2009**, *10*, 69–78.

38. Bacci, S.; Chiandotto, B. Mobilità Dei Laureati per Motivi di Lavoro: UN' Analisi Multilivello. *Stat. Appl.* **2007**, *19*, 5–40.

39. R Development Core Team. *R: A Language and Environment for Statistical Computing*; R Foundation for Statistical Computing: Vienna, Austria, 2019.

40. Hiser, J.; Nayga, R.M.; Capps, O. An exploratory analysis of familiarity and willingness to use online food shopping services in a local area of Texas. *J. Food Distrib. Res.* **1999**, *30*, 78–90.

41. Seo, S.; Kim, O.Y.; Oh, S.; Yun, N. Influence of informational and experiential familiarity on image of local foods. *Int. J. Hosp. Manag.* **2013**, *34*, 295–308. [CrossRef]

42. Tan, H.S.G.; van den Berg, E.; Stieger, M. The influence of product preparation, familiarity and individual traits on the consumer acceptance of insects as food. *Food Qual. Prefer.* **2016**, *52*, 222–231. [CrossRef]

43. Jang, S.C.S.; Kim, D.H. Enhancing ethnic food acceptance and reducing perceived risk: The effects of personality traits, cultural familiarity, and menu framing. *Int. J. Hosp. Manag.* **2015**, *47*, 85–95. [CrossRef]

44. Van Kempen, L.; Muradian, R.; Sandóval, C.; Castañeda, J.P. Too poor to be green consumers? A field experiment on revealed preferences for firewood in rural Guatemala. *Ecol. Econ.* **2009**, *68*, 2160–2167. [CrossRef]

45. Johnson, E.J.; Russo, J.E. Product familiarity and learning new information. *J. Consum. Res.* **1984**, *11*, 542–550. [CrossRef]

46. Fischer, A.R.H.; De Vries, P.W. Everyday behaviour and everyday risk: An approach to study people's responses to frequently encountered food related health risks. *Health. Risk Soc.* **2008**, *10*, 385–397. [CrossRef]

47. Iannuzzi, E.; Sisto, R.; Nigro, C. The willingness to consume insect-based food: An empirical research on italian consumers. *Agric. Econ.* **2019**, *65*, 454–462. [CrossRef]

48. Ryynänen, T.; Heinonen, V. From nostalgia for the recent past and beyond: The temporal frames of recalled consumption experiences. *Int. J. Consum. Stud.* **2018**, *42*, 186–194. [CrossRef]

49. Verbeke, W. Profiling consumers who are ready to adopt insects as a meat substitute in a Western society. *Food Qual. Prefer.* **2015**, *39*, 147–155. [CrossRef]

50. Aguilar, F.X.; Vlosky, R.P. Consumer willingness to pay price premiums for environmentally certified wood products in the US. *For. Policy Econ.* **2007**, *9*, 1100–1112. [CrossRef]

51. Vásquez Lavin, F.; Barrientos, M.; Castillo, Á.; Herrera, I.; Ponce Oliva, R.D. Firewood certification programs: Key attributes and policy implications. *Energy Policy* **2020**, *137*, 111160. [CrossRef]

Publisher's Note: MDPI stays neutral with regard to jurisdictional claims in published maps and institutional affiliations.

Article

Biomass Characteristics and Energy Yields of Tobacco (*Nicotiana tabacum* L.) Cultivated in Eastern Poland

Adam Kleofas Berbeć *and Mariusz Matyka

Department of Systems and Economics of Crop Production, Institute of Soil Science and Plant Cultivation-State Research Institute, Czartoryskich 8, 24-100 Puławy, Poland; mmatyka@iung.pulawy.pl
* Correspondence: aberbec@iung.pulawy.pl; Tel.: +48-81-4786-824

Received: 14 October 2020; Accepted: 15 November 2020; Published: 17 November 2020

Abstract: The present pilot study examined the potential of tobacco (*Nicotiana tabacum* L.) as an energy source. The fresh matter of whole tobacco plants, the yield of dry matter of stems and leaves, as well as the higher heating value and methane production potential from tobacco biomass were determined. The yield of tobacco leaves was on average 4.69 Mg ha^{-1} (dry matter) and 76.90 GJ ha^{-1} yr^{-1} (biomass energy yield). Tobacco stems yielded on average 8.55 Mg ha^{-1} and 150.69 GJ ha^{-1} yr^{-1}, while yields of whole tobacco crops were (on average) 13.24 Mg ha^{-1} and 227.59 GJ ha^{-1} yr^{-1}. Methane potential of tobacco plants was (on average) 248 Nm^3 Mg^{-1} VS (volatile solids). The tobacco plants tested in the study could be used as energy crops as their dry matter and energy yields are similar to those of the most popular energy crops being currently used in biomass production in Poland and the European Union. Nevertheless, further studies to choose the *Nicotiana* species and varieties most suitable for energy production and to assess the cost-effectiveness of tobacco biomass production are needed.

Keywords: tobacco biomass; energy yield; higher heating value; biogas potential; *Nicotiana tabacum*

1. Introduction

Most agricultural land is used for food production purposes as food production is the most important role of agriculture. A portion of agricultural land has always been devoted to non-food products that are used as industrial biomaterials and bioenergy sources [1]. One of the plants that can be used for various non-food purposes is tobacco (*Nicotiana tabacum* L.). Leaves, the main yield of tobacco, contain about 1–4% fatty acids per dry weight, but that can be raised to almost 7% by metabolic engineering (expression of other plant species genes), which makes tobacco a good production platform for biofuel [2]. Tobacco leaves have non-structural sugar content comparable to switchgrass (*Panicum virgatum* L.) and miscanthus (*Miscanthus* L.) while their lignin content is low [3]. Moreover, tobacco seeds contain about 40% oil, which can be used as a diesel fuel in turbocharged, indirect injection engines [4]. The yield of tobacco seeds is, however, insignificant (about 400–600 kg/ha according to the authors' knowledge). Tobacco also has a high polysaccharide content. Stems are built of 60% polysaccharides, which could possibly make tobacco a significant source of bioethanol [5]. Other energy carriers, such as biomass and biogas, could also be obtained from tobacco plants.

Currently, tobacco is cultivated in approximately 130 countries, in fields covering almost 3.4 mln hectares, with more than 6 million tons of tobacco harvested each year [6]. More than 40% of the world's tobacco is produced by China. All the largest producers are located outside the European Union (Table 1). Poland is currently the second largest tobacco producer in the European Union, surpassed only by Italy. Domestic plantations produce about 33,000 metric tons (Mg) of tobacco leaves [7], which is only half of the Polish tobacco industry's needs [8]. Tobacco is currently cultivated in Poland by approximately 11,500 farmers on between 13 to 17 thousand hectares [7,9]. The exact area of cultivation and yields are probably strongly underestimated. This is because the statistics of production of tobacco were limited in previous years [10]. According to the Central Statistical Office [7], the average yield of tobacco leaves is about 2.5 Mg per hectare, but according to the authors' knowledge, farmers often get yields as much as two times that amount.

Table 1. The world's largest producers of tobacco (leaves).

Area	Production (Thousands of Mg)	Area Harvested (Thousands of ha)	Average Yield (Mg ha^{-1})
China	2242.2	1003.7	2.2
Brazil	762.3	356.5	2.1
India	749.9	417.7	1.8
USA	241.9	117.9	2.0
Indonesia	181.1	203.0	0.9
Pakistan	106.7	46.3	2.3
Malawi	95.4	86.1	1.1
Argentina	104.1	54.7	1.9
Zambia	115.9	65.7	1.8
Italy	59.3	17.2	3.4
Poland	33.2	16.4	2.0
UE-27 *	178.3	77.9	2.3
World (total)	6094.9	3368.9	1.8

Source: Food and Agriculture Organisation Data (FAOSTAT)(2018); * according to FAOSTAT (2018) tobacco was cultivated in 15 European Union (EU) countries: Austria, Belgium, Bulgaria, Croatia, Cyprus, France, Greece, Germany, Hungary, Italy, Poland, Portugal, Romania, Slovakia and Spain.

The most common types of tobacco cultivated in Poland are flue-cured tobacco (Virginia type) varieties (65–75% of plantations) and Burley type varieties (25–35% of plantations) [10]. Tobacco leaves are the main yield of tobacco, while the other parts of the plants are waste and of no-value for the industry. The leaves are often damaged (especially during summertime) by intensive rain and hail. Tobacco can also be damaged by fungal (e.g., *Botrytis cinerea, Sclerotinia sclerotiorum*) and viral diseases (e.g., TMV, tobacco mosaic virus) [11]. Damaged leaves are worthless for the tobacco industry, and diseased crops have to be liquidated immediately to avoid the spread of diseases in the following years. According to the Central Statistical Office [7], about 20% of polish tobacco plantations are insured for weather-caused damage. This means that about 80% of farms that cultivate tobacco have a high risk of loss of their entire income if adverse weather conditions occur. To minimize the negative impact of weather-caused damage to tobacco leaves, alternative ways of management of tobacco yield should be established. One of those alternative ways of management could be the use of damaged plants for energy purposes as a renewable energy source.

Currently, at least 6 million tons of tobacco leaves are produced worldwide [6]. After harvesting, stems of a height of approximately 2.5 m are left on the fields. It is estimated that about 60% of tobacco biomass is unsuitable for industry and is unnecessary waste. Those wastes are most often burned or left on the field's surface. When managed properly, they can contribute to biogas, bioethanol, and biodiesel production [12]. Those stems have to be managed properly, to avoid phyto-sanitary issues (spread of diseases); however, due to the number of plants and their size, management is often too costly for farmers, and thus residues of plants are often crushed and plowed into the soil. The total amount of waste tobacco biomass must be significant, but there are no exact figure in the literature on this topic. However, some authors estimate the total amount of Virginia dried stalks at 6–10 Mg per hectare [13].

The development of the global economy and rapid growth of consumption increases the needs for conventional energy. According to various scenarios, the share of conventional fuels will be reduced as the result of depletion of resources and the related increase in energy prices [14,15]. The adopted European Union (EU) climate-energy package requires the market share of energy consumption from renewables to reach 20% with a 10% share in liquid biofuels by 2020, while the newest ambitions of the EU is to reach 32% renewable energy by 2030. In addition, a 20% reduction of greenhouse gases emissions (40% by 2030) and an increase in energy efficiency are expected [16]. Implementation of these objectives will increase the demand for agricultural substrates intended for use as an energy source [17,18]. Currently, the development of biomass renewable energy is dependent on the availability of appropriate biomass material, which, in most cases, is corn silage. Due to the multifaceted environmental-safety conditions, the growth of the area of this species may be subject to some restrictions. This is why there is a need to search for alternative biomass materials, which should be tested to confirm their suitability as a renewable source of energy [19]. Management of food-crop residues for energy purposes (e.g., cereal straw) and a search for new biomass sources seem to be two of the ways to meet those needs. The production of energy from tobacco waste biomass (stems or damaged tobacco plants) can increase farmers income while preserving the low environmental costs of the biomass production. The production of energy from wastes could also contribute to the goal of sustainable development and fits into the goal of a bioeconomy.

The aim of this study was to evaluate potential biomass yields of tobacco under typical Polish production plantation conditions. Moreover, the calorific value and biogas yield of *Nicotiana tabacum* were estimated.

2. Materials and Methods

The present study was carried out as pilot study in 2016 on three tobacco plantations located near Puławy, Poland. Virginia (flue cured) and Burley types of tobacco were chosen to analyze their biomass yields and potential as energy crops. The tobacco plants were collected both from production plantations (two locations: Virginia A—51°23′11.3″ N 21°53′55.4″ E and Virginia B—51°22′45.2″ N 22°06′19.5″ E). The third plantation was set in the Institute of Soil Science and Plant Cultivation's State Research Institute's (IUNG's) experimental plantation in Osiny (51°27′59.98″ N 21°39′44.28″ E). The third field was planted with Virginia (Virginia C) and Burley (Burley) types of tobacco and established as a split-plot field experiment. The production plantations chosen to be part of the experiment were both of about 0.5 ha, planted with the same Virginia flue-cured type of tobacco, of the same variety (HYV 23). The experimental crop planted at the IUNG experimental station was 200 m^2. The experimental crop was planted with Virginia VRG 10TL (Virginia C) and Burley TN10 (Burley) varieties. The fertilization schemes and other properties of the tobacco plantations are given in Table 2. Plants of tobacco were collected in early October to estimate yields and for further research. The leaves were separated from stems and then dried and weighed. Dried samples of tobacco were burned in a calorimeter (Precyzja-Bit KL-12Mn calorimeter) in order to assess the higher heating value. The biomass energy yield of tobacco was calculated on a basis of dry matter yield and the corresponding heating value.

Table 2. Soil properties and fertilization of tested tobacco plantations.

Tobacco Plantation	Virginia A	Virginia B	Virginia C	Burley
Variety of tobacco	HYV 23	HYV 23	VRG 10TL	TN10
Soil classification	Podzoluvisol S—Sand	Loess soil	Pseudo Podzoluvisol LS- loamy sand (on sandy loam)	
Complex of agricultural suitability of soils: from 1 (best) to 9 (worst)	6	2	5	
Soil pH	5.0	6.0	5.9	
Soil P_2O_5 content (mg 100g^{-1})	16.2	18.1	16.8	
Soil K_2O content (mg 100g^{-1})	15.8	19.1	17.0	
Soil Mg content (mg 100g^{-1})	5.9	7.6	7.2	
Mineral fertilization N-P_2O_5-K_2O (kg ha^{-1})	130-50-130	50-45-100	100-43-103	
Organic fertilization N-P_2O_5-K_2O (kg ha^{-1})	0	100-33-38	0	
Density of plants (per hectare)	22,000	20,000	23,000	
Irrigation	Yes	No	No	

Methane potential of tobacco silage was estimated on the basis of three out of the four plantations of tobacco: Virginia A, Virginia B, and Burley. Moreover, estimation of methane potential of maize (*Zea mays* L.) was performed (maize was cultivated in the same research station as tobacco, but was a part of an independent experiment). This was to compare tobacco methane potential to maize methane potential, as maize is currently most widely used substrate in agricultural biogas plants. Both tobacco and maize silage was made from whole plants harvested in October. Plants were chopped and ensilaged for 2 months. Total solids (TS), volatile solids (VS), and ash were determined in accordance to the Polish norms PN-EN 12,880 and PN-EN 12,779 using the gravimetric method after drying at 105 °C and 550 °C, respectively. The potential methane (CH_4) yield from tobacco and maize silage was evaluated by using The Automatic Methane Potential Test System II (AMPTS II) manufactured by Bioprocess Control. Bacterial inoculum was collected from an active, agricultural biogas plant, which used mostly corn silage to produce biogas. The initial ratio of VS content in inoculum to VS content in substrate was 2:1. A single reactor had a volume of 500 mL and it was filled with 400 mL of inoculum-substrate mixture. Twelve reactors were used in the experiment. The substrate in the reactors was mixed continuously for 30 s with 30 s interval, at a speed of 70 RPM. The fermentation temperature was set to 37 °C. The process was performed in three independent replications for about 35 days until no significant methane volume increases were found. The values of the biogas volume obtained were converted into standard conditions (1013 mbar, 273 K). Statistical analysis of data for different plantations was completed on the basis of one-way ANOVA (Analysis of variance) with a post-hoc Tukey HSD (Honestly Significant Difference) test. Significance of difference of other biomass parameters (e.g., average higher heating value (HHV) of stems vs. average HHV of leaves) was assessed with the Mann–Whitney test or *t*-test (depending on data distribution parameters). The aim of the pilot study was to assess the potential range of biomass yields, biomass energy, and methane potential of tobacco. The plants from all plantations were picked up randomly, but production plantations (Virginia A and Virginia B) themselves were not designed as experiments and no further measures were implemented to prevent an error variation (e.g., resulting from variability of soil). In contrast, Virginia C and Burley plantations were established as field experiments with a split-plot design (tobacco types as independent variable) with four replications to minimize the impact of soil variability.

3. Results and Discussion

3.1. Biomass Energy Potential

After harvesting tobacco leaves, plant residues are left on the field. The biomass of those residues (mostly stems) is large, but there is hardly any data about it in the literature. The present study on tobacco biomass yield was carried out in different conditions (different types and varieties of tobacco, different soil conditions and agro-techniques in the fields) in order to gather general knowledge of the tobacco biomass yield potential and the residue biomass potential in varied conditions. The fresh matter of tobacco (whole crop) varied significantly between plantations from 34.8 (Virginia C) to 62.3 Mg ha^{-1} (Virginia A) with an average of 53.2 Mg ha^{-1}. Total average dry biomass yield of tobacco varied significantly between locations from 9.4 Mg ha^{-1} (Virginia C) to 15.8 Mg ha^{-1} (Virginia B), with an average of 13.2 Mg ha^{-1}. The tobacco leaves dry matter differed between locations from 3.4 Mg of dry matter per hectare (flue cured tobacco planted in the IUNG's experimental station) to 5.6 Mg ha^{-1} in one of the production fields (Virginia A). An average yield of leaves for all plantations was 4.7 Mg ha^{-1}. Yield of stems varied between locations of tobacco. The lowest yield of stems was observed again for flue cured tobacco planted at the IUNG's experimental station (6.0 Mg of dry matter per hectare) while flue cured tobacco planted in the second plantation field (Virginia B) yielded the most (10.8 Mg ha^{-1}), while the average yield of stems for all plantation was 8.6 Mg ha^{-1} (Table 3).

Table 3. Yields and properties of the tested tobacco biomass.

Plantation and Type of Tobacco	Virginia A	Virginia B	Virginia C	Burley	Average
Yield of fresh matter (FM), (leaves + stems) (Mg ha^{-1})	62.3 [a] *	55.7 [a]	34.8 [b]	59.8 [a]	53.2
Total dry matter (DM) in FM (%)	23.0 [a]	28.4 [b]	27.1 [b]	22.4 [a]	25.2
Volatile Solids (VS) content in FM (%)	20.5 [a]	21.1 [a]	16.2 [b]	15.9 [b]	18.4
Yield of leaves (Mg ha^{-1}) DM	5.6 [a]	5.0 [a]	3.5 [b]	4.7 [a]	4.7
Yield of stems (Mg ha^{-1}) DM	8.7[a]	10.8 [b]	6.0 [c]	8.7 [a]	8.6
Leaf/stem ratio	0.6	0.5	0.6	0.5	0.6
Yield (leaves + stems) (t·ha^{-1}) DM	14.3 [ab]	15.8 [a]	9.4 [c]	13.4 [b]	13.2
Higher heating value of leaves (MJ kg^{-1})	17.2 [a]	16.4 [a]	16.2 [a]	15.6 [a]	16.3
Higher heating value of stems (MJ kg^{-1})	18.4 [a]	17.6 [ab]	17.4 [ab]	17.0 [b]	17.6
Higher heating value (leaves + stems) (MJ kg^{-1})	18.0 [a]	17.0 [ab]	16.8 [ab]	16.3 [b]	17.0
Biomass energy yield (GJ ha^{-1} yr^{-1}) of leaves	96.3 [a]	81.8 [ab]	56.5 [c]	73.0 [bc]	76.9
Biomass energy yield (GJ ha^{-1} yr^{-1}) of stems	160.3 [a]	191.0 [b]	103.8 [c]	147.7 [a]	150.7
Biomass energy yield (GJ ha^{-1} yr^{-1}) of leaves + stems	256.6 [ab]	272.8 [a]	160.3 [c]	220.7 [b]	227.6
Methane potential (Nm^3 Mg^{-1} VS)	298 [b]	220 [a]	nd	226 [a]	248
Yield of methane (Nm^3 ha^{-1} yr^{-1})	3802 [b]	2590 [a]	nd	2149 [a]	2847

Source: own study; * different lowercase letters indicate significant differences according to one-way ANOVA with a post-hoc Tukey HSD test at $p < 0.05$.

Bilalis et al. (2009) [20] reported that the dry matter of tobacco leaves ranged from 1.9 Mg ha^{-1} to 5.0 Mg ha^{-1} and was strongly influenced by irrigation and fertilization. In the present study, tobacco yields varied among locations. According to the authors' knowledge, there are only a few studies about dry matter yields of stems or whole tobacco plants. Radojičić et al. (2014) [13] estimated the yields of Virginia tobacco stems at about 6–10 Mg per hectare. Dry matter yields are expected to be highly variable and dependent on soil and weather conditions, fertilization, and agrotechnical treatments, in particular loosening the soil. Szewczuk (2009) [21] found that Virginia type tobacco industrial yields (leaves) can vary (in Poland) from 2.5 Mg ha^{-1} to 3.7 Mg ha^{-1}, and they are dependent on fertilization scheme. According to Pawełek (2000) [22], industrial yields of Virginia tobacco can be even greater and could reach from 2.7 Mg to 5.0 Mg per hectare. Annual yields of other widely used biomass plants are strongly differentiated depending on genotype and site conditions. According to McKendry (2002) [23], poplar (*Populus* L.) and willow (*Salix* L.) yields could reach 10 to 15 Mg of dry matter per hectare per year (which is very similar to the tobacco biomass yields presented in this study), while switchgrass yields are estimated at 8 Mg ha^{-1}, and yields of miscanthus biomass could reach up to 12–30 Mg ha^{-1} (Table 4). Yields of industrial hemp (*Cannabis sativa* L.), according to Prade et al. (2011) [24], reach 9.9–14.4 Mg ha^{-1}, depending on harvest period (Table 4). Concerning maize dry matter (DM) (whole aboveground biomass), Jurekova et al. [25] found variation of yields at 8 and 32 Mg ha^{-1} while Stolarski et al. (2020) [26] found that yields of willow in Poland can range from 2 to 18 Mg ha^{-1} yr^{-1} (11 Mg ha^{-1} yr^{-1} on average) (Table 4). Niemczyk et al. [27] showed that in the conditions found in Poland DM yields of poplar range from 2 to 8 Mg ha^{-1} yr^{-1} and are lower than those in countries located in southern parts of Europe. Yields of DM of tobacco and other energy sources are given in Table 4. Sheen (1983) [28] cultivated tobacco for biomass yield under high plant density conditions (about 766,000 per hectare). Sheen found that tobacco fresh biomass, harvested once a year in June, can reach from 44 to 70 Mg ha^{-1} depending on tobacco cultivar. Moreover, Wildman (1979) [29] found that tobacco fresh biomass yields from a stand of 370,000 plants and multiple harvests could reach up to 150 Mg ha^{-1} yr^{-1}. In the present study, the density of plants was not that high (up to 23,000 plants per hectare) and reached on average 53.2 Mg ha^{-1} (Table 3). Thus, tobacco could be an important source of biomass, especially when cultivated for energy purposes, with the introduction of new management techniques (higher density of plants, different fertilization schemes, and multiple harvests during vegetation).

Table 4. Yields, higher heating value (HHV), and energy yield of selected energy crops.

Crop	Crop Yield DM (Mg ha^{-1} yr^{-1})	HHV (MJ kg^{-1})	Biomass Energy Yield (GJ ha^{-1} yr^{-1})
Tobacco	9–16	16.3–18.0	160–273
Wheat (*Triticum aestivum* L.)	14 [1] 7–13 [3]	12.3 [1]	123 [1] 128–231 [3]
Poplar	10–15 [1] 2–8 [6]	17.3 [1]	173–259 [1] 33–230 [8]
Willow	10–15 [1] 2–18 [5]	18.7 [1]	187–280 [1] 203–210 [7]
Switchgrass	8 [1]	17.4 [1]	139 [1]
Miscanthus	12–30 [1] 11–29 [3]	18.5 [1]	222–555 [1] 186–486 [3]
Industrial hemp	10–14 [2]	17.5–19.1[2]	246–296 [2]
Maize	8–32 [3]	17.5 [4]	142–545 [3]

Sources: own study and: [1]—McKendry (2002) [23], [2]—Prade et al. (2011) [24], [3]—Jurekova et al. (2015) [25], [4]—Jóvér et al. (2018) [30], [5]—Stolarski et al. (2020) [26]. [6]—Niemczyk et al. (2016) [27], [7]—Stolarski et al. (2014) [31], [8]—Stolarski et al. (2020) [30].

The average HHV of tobacco leaves was significantly lower (16.3 MJ kg^{-1}) than the average HHV of its stems (17.6 MJ kg^{-1}) (*t*-test, *p*-value < 0.001). The lowest HHV for whole tobacco plants was observed for the Burley type of tobacco (16.3 MJ kg^{-1}), while the highest at 18.0 MJ kg^{-1} was for flue cured tobacco cultivated on sandy soil (Virginia A). The average HHV for whole tobacco plants was 17.0 MJ kg^{-1} (Table 3). There are not many accurate reports about the energy value of tobacco plants in the literature. Some authors [13,32] have estimated (on the basis of chemical analysis of tobacco stems) the average higher heating value of Virginia (flue cured) stems at around 18.2 MJ kg^{-1}, which is similar to the HHV of Virginia stems presented in this study. In addition, Indahsari (2018) [33] found tobacco stems' HHV to be 18.2 MJ kg^{-1}. Moreover, Mijailovic et al. (2014) [32] found Burley stems' HHV to be 18.3 MJ kg^{-1}, which is a higher value than the average HHV of Burley stems measured in the present study (17.0 MJ kg^{-1}). The difference in Burley's HHV value is probably due to rather poor nutrition and growth rate of burley plants at the IUNG's experimental station. Nevertheless, the average HHV for all tested tobacco varieties was at a rather good level, comparable with other energy plants. According to McKendry (2002) [23] the HHVs of poplar and switchgrass are at a comparable level (17.3 MJ kg^{-1} and 17.4MJ kg^{-1}, respectively) to tobacco's HHV presented in this study (17.03 MJ ha^{-1}), while the HHVs of willow and miscanthus are higher (18.7 MJ kg^{-1} and 18.5 MJ kg^{-1}, respectively) (Table 4). Prade et al. (2011) [24] found that industrial hemp's HHV can reach from 17.5 to 19.1 MJ kg^{-1}, depending on the time of harvest (Table 4).

Energy yield of whole tobacco plants reached values from 160.3 GJ ha^{-1} to 272.8 GJ ha^{-1} per year. On average, tobacco energy yield was estimated at 227.6 GJ ha^{-1} (Table 3). This means that one hectare of tobacco plants had the same energy potential as 9.5 tons of coal (24 MJ kg^{-1}). Plant residue management (mostly stems) is an important issue in tobacco plantations. The amount of that waste biomass is often large. The present study showed that tobacco stems are about 65% of the tobacco dry biomass yield (average leaf/stem ratio of 0.55). In addition, energy yields of stems were about two times higher than energy yields of leaves, which was due to the significantly higher yields of stems (Mann–Whitney test, *p*-value < 0.001) and also the significantly higher heating value of stems (*t*-test, *p*-value < 0.001). According to Sheen (1983) [28], tobacco plants planted in more dense stands (up to 670,000 plants per hectare) produce more fresh biomass of leaves than that of stems (leaf/stem ratio from 1.77 to 3.02). The differences were probably due to density of stands, genotypes of plants, and harvesting time. This study showed that the energy yields of stems were approximately two times higher (150.7 GJ ha^{-1} yr^{-1}) than the energy yields of leaves (76.9 GJ ha^{-1} yr^{-1}) (Table 3). This can be valuable information for farmers who want to use tobacco residues as energy biomass. Yields of energy of other plants are strongly differentiated. For example, McKendry (2002) [23] found energy yields of wheat (123 GJ ha^{-1} yr^{-1}) and switchgrass (139 GJ ha^{-1} yr^{-1}) to be lower than the energy yields of tobacco estimated in the present study. Moreover, the author found that willow (187–280 GJ ha^{-1} yr^{-1}) and poplar (173–259 GJ ha^{-1} yr^{-1}) yields of energy are at a similar level as the tobacco energy yields presented in this study, while miscanthus energy yields can outperform other crops (222–555 GJ ha^{-1} yr^{-1}). Jurekova et al. (2015) [25] found miscanthus and maize energy yields to be at a similar level (186–486 GJ ha^{-1} yr^{-1} and 142–545 GJ ha^{-1} yr^{-1}, respectively). According to Prade et al. (2011) [24], energy yields of industrial hemp can reach up to 246–296 GJ ha^{-1} yr^{-1} (Table 4). Stolarski et al. (2020) [30] found poplar's energy yield (cultivated in Poland) to be able to reach up to 230 GJ ha^{-1} yr^{-1} for the best-performing clones while the average energy yield for all genotypes and rotations of poplar was at a level of 115 GJ ha^{-1} yr^{-1}. This is lower than the energy yields for Virginia A and Virginia B presented in this study.

3.2. Methane Potential of Tobacco Silage

According to Weiland (2010) [34], methane yield of different plant substrates is strongly affected by the chemical composition of the crop. Generally, the methane potential of plants is lower in older plants, which are strongly lignified, which causes slower anaerobic decomposition of a substrate. Biogas production from different agricultural substrates can vary from 231 Nm^{3}CH$_4$ t^{-1} VS (volatile solids) for sunflower to 400 Nm^{3}CH$_4$ t^{-1} DMO (organic dry matter) for sugar and fodder beet (Table 5).

Table 5. Gross crop yield (fresh) and methane potential of different crops.

Crop	Crop yield (t FM ha^{-1})	Nm^3 CH$_4$ t^{-1} VS (MIN)	Nm^3 CH$_4$ t^{-1} VS (MAX)
Tobacco [1]	35–62	220	298
Maize [1]	36	346	437
Corn-cob-mix (CCM) [2]	10–15	350	360
Fodder beet [2]	80–120	398	424
Grass [2]	22–31	286	324
Maize [2]	40–60	291	338
Red clover [2]	17–25	297	347
Rye grain [2]	4–7	297	413
Sorghum [2]	40–80	286	319
Sugar beet [2]	40–70	387	408
Sunflower [2]	31–42	231	297
Triticale [2]	28–33	319	335
Wheat [2]	30–50	351	378
Wheat grain [2]	6–10	371	398

Sources: [1] own study, [2] Weiland (2010) [34].

Some authors have cultivated energy crops in similar conditions as those presented in this study. Matyka and Księżak (2013) [19] cultivated canary grass (*Phalaris arundinacea* L.) in the same experimental farm that cultivated the Virginia C and Burley tobacco presented in this study. Authors reported that ensiled canary grass can produce up to 273 Nm^{3}CH$_4$ t^{-1} VS. Kacprzak et al. (2012) [35] reported that, on average, 389 Nm^{3}CH$_4$ t^{-1} VS can be produced from ensiled sorghum, while maize silage can produce 566 Nm^{3}CH$_4$ t^{-1} VS. Kacprzak et al. (2013) [35] also tested canary grass and maize cultivated in the Osiny Experimental Farm for its biogas potential. They found that ensiled canary grass produced, on average, 187 Nm^3 t^{-1} DM (dry matter) of CH$_4$ while maize silage produced 208 Nm^3 t^{-1} DM of CH$_4$. Oleszek et al. (2014) [36] found that biogas yield from canary grass, also cultivated in the Osiny Experimental Farm, can reach up to 406 Nm^3 t^{-1} VS. This shows how varied are the outputs of studies on methane potential of different crops. In contrast, the tobacco silage tested in this study seems to be a rather poor substrate for methane production. Methane potential for tested tobacco varied between 220 to 298 Nm^{3}CH$_4$ t^{-1} VS (248 Nm^{3}CH$_4$ t^{-1} VS on average) (Table 3), and was lower than the methane potential of maize, which was tested during the same analysis as tobacco silage (Figures 1 and 2). Li et al. (2019) [37] tested four varieties of tobacco stalks for their methane potential and found that they can reach up to 132 Nm^3CH_4 t^{-1}, which is less than half of the maximum methane potential of tobacco shown in the present study. The differences are probably due to the substrate; while Li et al. tested only tobacco stems (which are strongly lignified), the present study shows the methane potential of the whole tobacco plant. According to Weiland (2010) [34], the optimal dry matter contents of plant substrates for ensilage should be 25% to 35%. The general quality of silage made of a substrate of TS (total solids) content below 25% can be poor and has high leachate formation, which results in slow fermentation and low methane yields. It is possible that tobacco harvested earlier, when plants are less lignified, would have better methane potential, but this still needs to be tested. The production of methane (Nm^3 $CH_4 \cdot t^{-1}$ VS) from tobacco was at a level of 65% of what can be produced from maize (methane potential of maize was tested in the same conditions and at the same time as methane potential of tobacco,) (Table 5, Figure 2). Moreover, the calculated average yield of methane per hectare of tobacco (2847 Nm^3 ha^{-1} yr^{-1} (Tabel 3)) was only half of calculated annual yields of methane of maize (5900 Nm^3 $ha^{-1}yr^{-1}$, not showed in a table). Daily methane production from tobacco was as intensive, or even more rapid, as methane production from maize in the first four days of production. In the following days, methane potential of tobacco declined rapidly, while the decrease in methane production from maize was slower (Figure 1). Low methane production potential of tobacco, comparable to sunflower methane production in Weiland's (2010) [34] study, probably was due to a rather low dry matter content of the tobacco substrate. Moreover, the tobacco plants of the present study at the time of harvesting (October) were in large part lignified, which probably resulted in slow anaerobic decomposition of silage. This might also explain why methane production from tobacco was less efficient (compared to maize) especially at the later stages of methane production (Figure 1). Tobacco can contain more than a dozen (up to 20%) carbohydrates [38,39]. Probably, as a large part of those carbohydrates are easily digestible simple soluble sugars, the methane fermentation process of tobacco was most efficient at the initial stage of fermentation. With the depletion of simple substances, fermentation slowed down significantly (Figure 1).

Tobacco samples in the present study were taken from plantations cultivated as typical production plantations (leaves as main yield), with low plant density (20,000–23,000 plants per hectare). Tobacco, however, probably could be a higher biomass yielder, especially when cultivated for energy purposes in more dense stands (according to Sheen (1983) [28] even up to 670,000 plants per hectare). Moreover, tobacco plants have high biomass growth with increased nitrogen fertilization. This is, however, linked with deterioration of leaves' quality for the industry and thus use of N fertilizers on production plantations of tobacco is usually low. Tobacco plants often create additional stems, which are unwanted by farmers as they result in lower yield and quality of leaves. This natural ability of tobacco to create extra stems could, however, be useful when cultivating tobacco for energy purposes. This process could also be stimulated by cutting in early growth phases (according to authors' observations, tobacco plants can easily regrow when the main stem is cut off; side shoots in such conditions appear in large numbers and easily grow on the plant). Tobacco in the present study was cultivated in soils of different quality (Table 2), but even in the soil of the poorest quality (Virginia A), yields of biomass were acceptable (Table 3). While tobacco is cultivated on production plantations as an annual plant (planted as seedlings each year) it is actually a perennial plant that can regrow (from stem and/or roots) after being cut down. It cannot be excluded that other species of tobacco could be cultivated as perennials, or even mowed (harvested) 2–3 times per year in moderate climate conditions.

The profitability and risks (lack of established knowledge) linked with production of tobacco entirely for energy purposes compared to production of leaves for the industry would probably be an issue for Polish farmers. Economic efficiency of tobacco cultivation is often at a significantly higher level than economic efficiency of energy crops. According to Stolarski et al. (2014) [31], the direct surplus of the plantation of willow in Poland can reach up to 1860 PLN ha^{-1} $year^{-1}$, while the direct surplus of tobacco production is at a level of 10,000–20,000 PLN ha^{-1} $year^{-1}$ (assuming the 100% share of labor in cultivation and harvesting of tobacco). Currently, due to the ending of support for tobacco cultivation (subsidies), the expected economic efficiency will drop significantly. It is therefore highly probable that in the coming years, due to the decline in tobacco production profitability, a transition from cultivation of tobacco for leaves to energy purposes would become more attractive for farmers. Tobacco cultivation requires up to 4000 man-hours per hectare [40]. A multi-stage collection of leaves and their preparation for drying consumes most of this time. In the case of production for energy purposes, a large part of this work could be saved and allocated elsewhere on the farm.

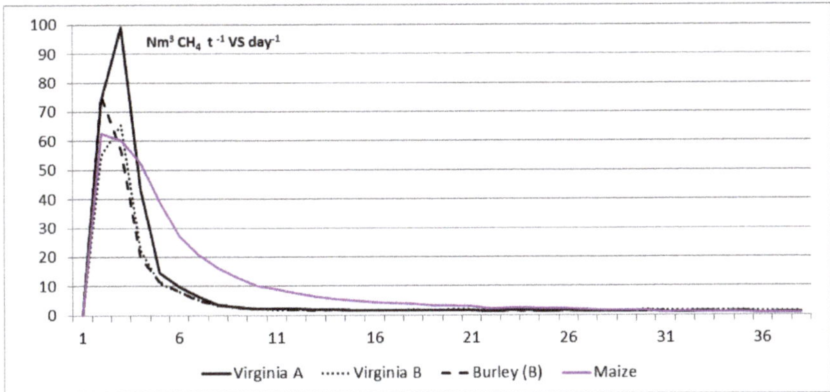

Figure 1. Daily rate of methane production from different substrates.

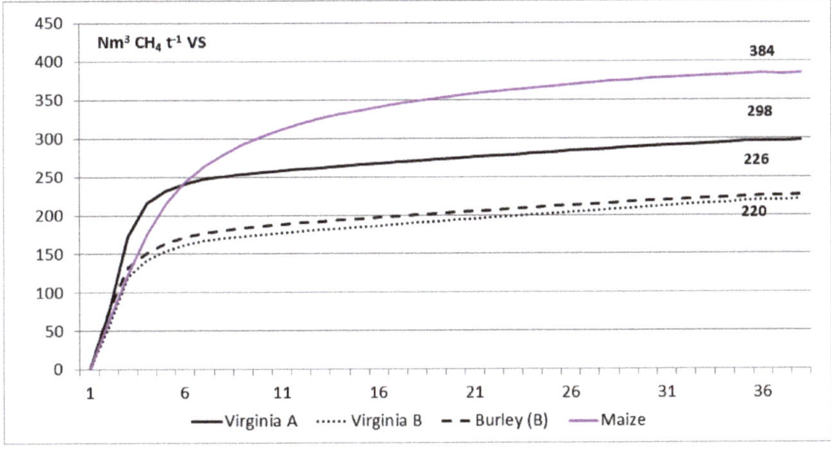

Figure 2. Methane potential from different tobacco plantations and from maize.

The present pilot study covered just one year of vegetation. It is expected that different years with different weather conditions would affect the results to a greater or lesser extent. Nevertheless, this study showed that tobacco biomass can match the energy potential of other commercial energy crops. Residues (or damaged plants) of tobacco can be a valuable source of energy that could be used directly on farms (e.g., as pellet biomass). Moreover, it seems that with the right choice of varieties and species of tobacco and the adjustment of its management (planting density, harvest cycles) it would be possible to cultivate this species for energy purposes. Therefore, further studies to choose *Nicotiana* species and varieties most suitable for energy production as well as to assess the cost-effectiveness of tobacco biomass production are needed.

4. Conclusions

The present study shows that tobacco biomass and energy yields can match other commercial energy crops in Poland. Its energy yields varied between 160 and 273 GJ ha^{-1} yr^{-1}, which is comparable to poplar and willow energy yields. Tobacco, however, could be a valuable biomass energy crop, especially when cultivated deliberately to obtain biomass. This study showed that tobacco can be cultivated on soils of poor quality and high yields can be still maintained. The tested tobacco (harvested in October) seemed to be a rather poor source of substrate for biogas production, which was probably due to harvesting time and the high lignification rate of tobacco tissues. Yield of methane per hectare of tobacco was 2847 Nm^3 ha^{-1} yr^{-1}.

Author Contributions: Conceptualization, A.K.B.; methodology, A.K.B. and M.M.; software, A.K.B.; investigation, A.K.B.; resources, A.K.B.; data curation, A.K.B.; writing—original draft preparation, A.K.B.; writing—review and editing, M.M. and A.K.B.; visualization, A.K.B.; supervision, M.M. All authors have read and agreed to the published version of the manuscript.

Funding: This paper is the result of a study carried out at the Institute of Soil Science and Plant Cultivation, State Research Institute in Pulawy, and it was co-financed by the National (Polish) Centre for Research and Development (NCBiR), project: BIOproducts from lignocellulosic biomass derived from MArginal land to fill the Gap In Current national bioeconomy, No. BIOSTRATEG3/344253/2/NCBR/2017. This research acknowledges financial support from the Widening Program ERA Chair: project BioEcon (H2020), contract number: 669062.

Conflicts of Interest: The authors declare no conflict of interest. The funders had no role in the design of the study; in the collection, analyses, or interpretation of data; in the writing of the manuscript, or in the decision to publish the results.

References

1. Diamantidis, N.D.; Koukis, E.G. Agricultural crops and residues as feedstock for non-food products in Western Europe. *Ind. Crop. Prod.* **2000**, *11*, 97–106. [CrossRef]
2. Andrianov, V.; Borisjuk, N.; Pogrebnyak, N.; Brinker, A.; Dixon, J.; Spitsin, S.; Flynn, J.; Matyszczuk, P.; Andryszak, K.; Laurelli, M.; et al. Tobacco as a production platform for biofuel: Overexpression of Arabidopsis DGAT and LEC2 genes increases accumulation and shifts the composition of lipids in green biomass. *Plant Biotechnol. J.* **2010**, *8*, 277–287. [CrossRef] [PubMed]
3. Barla, F.G.; Kumar, S. Tobacco biomass as a source of advanced biofuels. *Biofuels* **2016**, *10*, 335–346. [CrossRef]
4. Usta, N. Use of tobacco seed oil methyl ester in a turbocharged indirect injection diesel engine. *Biomass Bioenergy* **2005**, *28*, 77–86. [CrossRef]
5. Guo, G.; Zhang, J.; Li, D.; Cai, B.; Zhang, X. Enhanced saccharification and ethanolic fermentation of tobacco stalks by chemical pretreatment. *J. Biotechnol.* **2010**, *150*, 173–174. [CrossRef]
6. FAOSTAT. Food and Agriculture Organization of the United Nations, Database on Crops Statistics. 2018. Available online: http://www.fao.org/faostat/en/#data/QC (accessed on 8 May 2020).
7. GUS (Main Statistical Office of Poland). *Rocznik Statystyczny Rolnictwa*; Zakład Wydawnictw Statystycznych: Warsaw, Poland, 2018; ISSN 2080-8798.
8. Gwiazdowski, R.; Barna, A.; Wepa, M.; Marchewka, R. *Skutki Wdrożenia Dyrektywy Tytoniowej: Raport Centrum im. Adama Smitha o Ekonomicznych Skutkach Wdrożenia Rewizji Dyrektywy 2001/37/WE Parlamentu Europejskiego i Rady Europy z 5 czerwca 2001 Roku*; Centrum im. Adama Smitha: Warsaw, Poland, 2013; ISBN 978-83-86885-65-7. (In Polish)
9. Barna, A.; Marchewka, R.; Wepa, M. *Raport—Uwarunkowania Prawne oraz Polityka Podatkowa Państwa Wobec Wyrobów Tytoniowych a Rynek Pracy w Polskiej branży Tytoniu*; Związek Przedsiębiorców i Pracodawców: Warsaw, Poland, 2013; p. 72. (In Polish)
10. Berbeć, A.; Madej, A. Obecna sytuacja i perspektywy uprawy tytoniu w Polsce na tle świata i Unii Europejskiej (in polish). *Studia Rap. IUNG PIB* **2012**, *31*, 51–67.
11. Doroszewska, T.; Berbeć, A.; Czarnecka, D.; Kawka, M. *Choroby i Szkodniki Tytoniu*; IUNG-PIB: Puławy, Poland, 2013; ISBN 978-83-7562-154-9. (In Polish)
12. Budzianowska, A.; Budzianowski, J. Tytoń źródłem biopaliw. *Przegl Lek* **2012**, *69*, 1149–1152. (In Polish)
13. Radojičić, V.; Ećim-Djurić, O.S.M.; Djulančić, N.; Kulić, G. Possibilities of Virginia Tobacco Stalks Utilization. *Tutun Tob.* **2014**, *64*, 71–76.

14. Cornelissen, S.; Koper, M.; Deng, Y.Y. The role of bioenergy in a fully sustainable global energy system. *Biomass Bioenergy* **2012**, *41*, 21–33. [CrossRef]
15. Tsavkelova, E.A.; Netrusow, A.I. Biogas production from cellulose-containing substrates. A review. *Appl. Biochem. Microbiol.* **2012**, *48*, 421–433. [CrossRef]
16. EU Climate Strategies and Targets: 2030 Climate & Energy Framework. Available online: https://ec.europa.eu/clima/policies/strategies/2030_en (accessed on 12 October 2020).
17. Matyka, M. Production and economic aspects of cultivation of perennial plants for energy purposes. *Monogr. Rozpr. Nauk. IUNG PIB* **2013**, *35*, 1–98.
18. Oniszk-Popławska, A.; Matyka, M.; Dagny-Ryńska, E. Evaluation of a long-term potential for the development of agricultural biogas plants: A case study for the Lubelskie Province, Poland. *Renew. Sustain. Energy Rev.* **2014**, *36*, 329–349. [CrossRef]
19. Matyka, M.; Księżak, J. Potencjalne możliwości produkcji biogazu z mozgi trzcinowatej. *Probl. Inżynierii Rol.* **2013**, *80*, 79–86.
20. Bilalis, D.; Karkanis, A.; Efthimiadou, A.; Konstantas, A.; Triantafyllidis, V. Effects of irrigation system and green manure on yield and nicotine content of Virginia (flue-cured) Organic tobacco (Nicotiana tabaccum), under Mediterranean conditions. *Ind. Crop. Prod.* **2009**, *29*, 388–394. [CrossRef]
21. Szewczuk, C. Wpływ dolistnego dokarmiania tytoniu na plony i jakość liści. *Ann. Univ. Mariae Curie-Skłodowska Lub. Sect. E* **2009**, *64*, 46–51. (In Polish)
22. Pawełek, E. Zbiór i suszenie Virginii w regionie leżajskim. *Przegląd Tyton.* **2000**, *9*, 6–9. (In Polish)
23. McKendry, P. Energy production from biomass (part 1): Overview of biomass. *Bioresour. Technol.* **2002**, *83*, 37–46. [CrossRef]
24. Prade, T.; Svensson, S.E.; Andersson, A.; Mattsson, J.E. Biomass and energy yield of industrial hemp grown for biogas and solid fuel. *Biomass Bioenergy* **2011**, *35*, 3040–3049. [CrossRef]
25. Jureková, Z.; Húska, D.; Kotrla, M.; Prčík, M.; Hauptvogl, M. Comparison of Energy Sources Grown on Agricultural Land. *Acta Reg. Environ.* **2015**, *12*, 38–43. [CrossRef]
26. Stolarski, M.J.; Krzyżaniak, M.; Załuski, D.; Tworkowski, J.; Szczukowski, S. Effects of Site, Genotype and Subsequent Harvest Rotation on Willow Productivity. *Agriculture* **2020**, *10*, 412. [CrossRef]
27. Niemczyk, M.; Wojda, T.; Kantorowicz, W. Przydatność hodowlana wybranych odmian topoli w plantacjach energetycznych o krótkim cyklu produkcji. *Sylwan* **2016**, *160*, 292–298.
28. Sheen, S.J. Biomass and chemical composition of tobacco plants under high density growth. *Beitr Tabakforsch. Int.* **1983**, *12*, 35–42. [CrossRef]
29. Wildman, S.G. Tobacco, a potential food crop. Crops and Soil Magazine, January 1979, pp. 7–9. *Beitr Tabakforsch. Int.* **1979**, *12*, 35–42.
30. Stolarski, M.J.; Warmiński, K.; Krzyżaniak, M. Energy Value of Yield and Biomass Quality of Poplar Grown in Two Consecutive 4-Year Harvest Rotations in the North-East of Poland. *Energies* **2020**, *13*, 1495. [CrossRef]
31. Stolarski, M.J.; Krzyżaniak, M.; Szczukowski, S.; Tworkowski, J. Efektywność energetyczna produkcji biomasy wierzby w jednorocznym i trzyletnim cyklu zbioru. *Fragm. Agron* **2014**, *31*, 88–95.
32. Mijailovic, I.; Radojicic, V.; Ecim-Djuric, O.; Stefanovic, G.; Kulic, G. Energy potential of tobacco stalks in briquettes and pellets production. *J. Environ. Prot. Ecol.* **2014**, *15*, 1034–1041.
33. Indahsari, O.P. Briquettes from Tobacco Stems as the New Alternative Energy. *J. Kim. Terap. Indones. (Indones. J. Appl. Chem.)* **2018**, *19*, 73–80. [CrossRef]
34. Weiland, P. Biogas production: Current state and perspectives. *Appl. Microbiol. Biotechnol.* **2010**, *85*, 849–860. [CrossRef]
35. Kacprzak, A.; Matyka, M.; Krzystek, L.; Ledakowicz, S. Wpływ stopnia nawożenia i czynników atmosferycznych na wydajność produkcji biogazu z wybranych roślin energetycznych. *Inż. Ap. Chem.* **2012**, *4*, 141–142. (In Polish)
36. Oleszek, M.; Król, A.; Tys, J.; Matyka, M.; Kulik, M. Comparison of biogas production from wild and cultivated varieties of reed canary grass. *Bioresour. Technol.* **2014**, *156*, 303–306. [CrossRef]
37. Li, L.; Wang, R.; Jiang, Z.; Li, W.; Liu, G.; Chen, C. Anaerobic digestion of tobacco stalk: Biomethane production performance and kinetic analysis. *Environ. Sci. Pollut. Res.* **2019**, *26*, 14250–14258. [CrossRef] [PubMed]

38. Leffingwell, J. Basic chemical constituents of tobacco leaf and differences among tobacco types. Reprinted from Tobacco. In *Production, Chemistry, and Technology*; Layten, D., Davis, M.T., Eds.; Blackwell Science: Hoboken, NJ, USA, 1999; pp. 265–284, Chapter 8.
39. Miś, T. Ocena odmian tytoniu uprawianych w rejonie województwa podkarpackiego (in polish). *Biul. Inst. Hod. Aklim. Roślin* **2003**, *228*, 363–371.
40. Wijatyk, B. Rynek tytoniu. In *Strategiczne Opcje dla Polskiego Sektora Agrobiznesu w Świetle Analiz Ekonomicznych*; Majewski, E., Dalton, G., Eds.; SGGW: Warszawa, Poland, 2000; p. 408. (In Polish)

Publisher's Note: MDPI stays neutral with regard to jurisdictional claims in published maps and institutional affiliations.

 *agriculture*

Article

Planting Density Effects on Grow Rate, Biometric Parameters, and Biomass Calorific Value of Selected Trees Cultivated as SRC

Adam Kleofas Berbeć * and Mariusz Matyka

Department of Systems and Economics of Crop Production, Institute of Soil Science and Plant Cultivation-State Research Institute, Czartoryskich 8, 24-100 Puławy, Poland; mmatyka@iung.pulawy.pl
* Correspondence: aberbec@iung.pulawy.pl; Tel.: +48-81-4786-824

Received: 7 October 2020; Accepted: 25 November 2020; Published: 26 November 2020

Abstract: Agricultural land is mostly devoted to food production. Production of biomass is limited, as it competes for land with basic food production. To reduce land loss for growing food, biomass can be grown on marginal lands that are not usable for food production. The density of plantings have to be optimized to maximize yield potential. The presented study compares yield parameters end energy potential of six species of biomass plants (poplar, Siberian elm, black alder, white birch, boxelder maple, silver maple) cultivated in 18 planting densities from 3448 to 51,282 plants per hectare as short rotation coppice (SRC). Biomass yield parameters depended on both cultivated species and planting density. Green mass, dry mass, and shoot diameter was dropping with the increasing planting density for most tested species. Calculated yield of dry mass was dropping with increasing planting density for black alder, increasing for Siberian elm and boxelder maple. White birch and silver maple yields were optimal at moderate planting densities (25,000–30,000). White birch and boxelder maple had the highest average higher heating value (HHV). The optimal density of plantings should be chosen to best suit both the needs of cultivated species and to optimize the most important parameters of produced biomass.

Keywords: energy crops; planting density; calorific value; SRC

1. Introduction

Although the main role of agriculture is food production, a part of agricultural land has always been devoted to non-food products, mainly within the framework of emerging technologies. Such uses include the production of bioenergy and various biomaterials [1]. The application of biomass for energy generation and for industry is of great importance and brings benefits to: (i) energy independence, (ii) environmental protection, (iii) economy, and (iv) society [2,3]. Non-food crops cultivation should not compete with food and fodder cultivation on high-yielding, fertile soils of good agricultural quality. On the other hand, the number of available food and fodder species able to produce a satisfactory yield on light, sandy soils of rather poor agricultural quality is not wide [4,5]. The cultivation of tree species using the short rotation coppice technique of lignocellulose biomass production had spread following the first oil crisis. Plants cultivated as short rotation coppices (SRCs) are characterized by a high growth rate, adequate sprouting of the stool bed, and an adaptation to sub-optimal environmental conditions [6]. Plantations for woody biomass production can be adapted depending on planting density and rotation length. Available experimental results indicate that a decrease in stem circumference is a commonly observed response to increasing planting density in poplar. However, studies also showed fast growing hardwoods tree height can increase, decrease, or remain unchanged with increasing planting density [7,8].

Agricultural biodiversity is of great importance nowadays. Plants of different genotype (species, varieties) including those cultivated as SRC, differ in habitat requirements for optimal growth and development and create different habitats for wildlife. Production conditions such as soil quality, water availability, harvest cycle, and technology (i.e., planting density) have an impact on biomass production, but the strength and direction of this impact could vary between species.

The goal of this study was to the determine the grow rate, biometric parameters, and biomass caloric value of six trees species cultivated as SRC depending on planting density.

2. Materials and Methods

2.1. Site Characteristics and Experimental Design

The experiment was conducted in the experimental farm of Institute of Soil Science and Plant Cultivation—State Research Institute in Osiny, Poland (N: 51°28′16.37″, E: 22°3′5.11″). The experiment was established in spring 2010 on heavy black soil with a heavy clay granulometric composition. The following trees were included in the experiment:

- Poplar (*Populs* L.), AF2 variety;
- Siberian elm (*Ulmus pumila*);
- Black alder (*Alnus glutinosa*);
- White birch (*Betula pubecsens*);
- Boxelder maple (*Acer negundo*);
- Silver maple (*Acer saccharinum* L.).

Trees were planted in April 2010 at 18 ranges of density in a "Nelder wheel design" [9] (see Figure 1). Eighteen concentric rings of trees were planted at radii ranging from 1.5 to 11 m as indicated in Table 1. An additional outer ring was planted as a guard ring to minimize the edge effect. The center of the circle was planted with a small ring to also form a guard. Each of the experimental rings contained 24 trees (six tested species in four replications) planted et equal distances around the circumference, giving a range from 3448 to 51,282 trees per hectare in equivalent planting density (Table 1). Those 24 trees planted as a concentric ring of increasing diameter formed 24 rows of experiment. Each tested species was represented in four rows (replications). Two rows of the same species were always adjacent to each other, while the other two were on the opposite side of the circle—creating an experimental arrangement in the form of a mirror reflection (see Figure 1).

Table 1. Planting distances and tree densities.

Ring No.	Distance to the Center [m]	Distance to Previous Ring [m]	Density [Plants ha^{-1}]
1	1.0	Guard ring	
2	1.5	0.5	51,282
3	2.0	0.5	48,462
4	2.5	0.5	30,769
5	3.0	0.5	25,477
6	3.5	0.5	21,739
7	4.0	0.5	19,231
8	4.5	0.5	16,949
9	5.0	0.5	15,384
10	5.5	0.5	14,286
11	6.0	0.5	12,739
12	6.5	0.5	11,764
13	7.0	0.5	10,929
14	7.5	0.5	10,204
15	8.0	0.5	9569
16	8.5	0.5	9009
17	9.0	0.5	8510
18	9.5	0.5	8064
19	10.0	Guard ring	
20	11.0	1.0	3448
21	12.0	Guard ring	

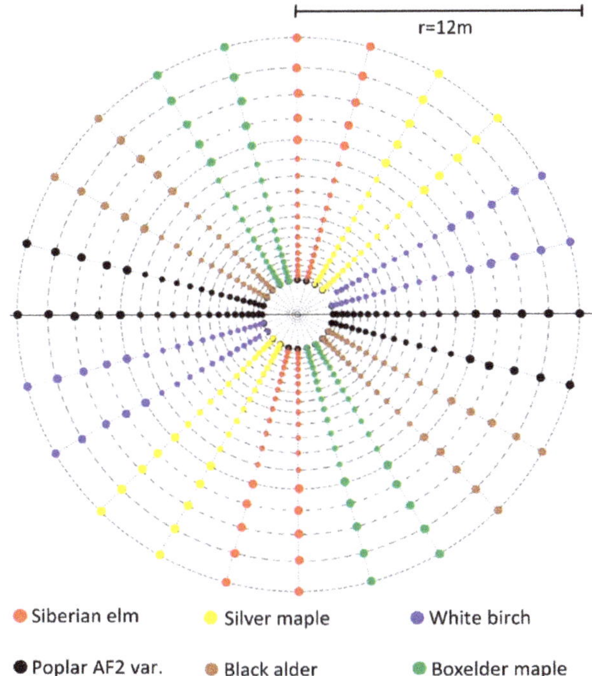

Figure 1. Distribution of species in the experiment design according to Nelder (1962) design (dots of different color represents trees of different species).

2.2. Biomass Analyses

Trees were harvested after 7 years of growth in February 2017. The harvest was made by hand. Shoot diameter at a height of 15 cm was determined using an electronic caliper with an accuracy of 0.1 mm. Each plant was cut at a height of 5 cm above the ground level. Plant height was determined from the cut place to the top of the plant with the accuracy of 1 cm. Yield of green mass was determined by weighing whole plants immediately after harvesting (green mass included limbs and/or barks). Whole plants were chipped and carefully mixed. Representative biomass samples (in 5 replications) were taken to determine the share of dry mass. In the laboratory, biomass moisture was determined by a drying method at a temperature of 80 °C for a period of 14 days. The dry mass yield was determined from the ratio of the green mass yield and its moisture content. Dried samples of trees were burned in calorimeter (KL-12Mn calorimeter, Precyzja-Bit, Bydgoszcz, Poland) in order to assess the higher heating value—HHV (or gross calorific value) of the biomass.

2.3. Statistical Analyses

Statistical analysis of results was performed using Statistica 10.0 software (StatSoft Polska, Krakow, Poland)). Few clear extreme outliers (observations located outside upper or lower quartile) were removed from the dataset according to the software manual. The level of significance for analysis was set to $p = 0.05$.

Subsequently, the normality of the distribution was tested using the Shapiro–Wilk and Kolmogorow–Smirnow tests. The vast majority of the examined features were characterized by a non-normal distribution. Data transformation attempts have not changed the data distribution. Therefore, nonparametric Kruskal–Wallis tests for comparison of many groups of independent variables were used to assess the significance of differences. Because of data distribution, all average values presented in the study are medians.

The green mass, the share of dry mass, dry mass, and the biometric features of the examined trees, depending on the plant density, were characterized by using the trend equation. The criterion for selecting the trend equation was the highest value of the determination coefficient.

In addition, for each tree species correlation coefficient for the tested parameters relationship was calculated.

Woody plants intended for industry are, by definition, grown in long cycles, and therefore, their growth in subsequent years was not studied. Annual yields were not assessed as experimental design was not adapted to it (it would result in a complete failure of the basic methodological assumptions due to invasive nature of such measurements (annual harvesting)). However, potential annual biomass of dry mass was calculated Y_{dm} (t ha^{-1} year^{-1}). To calculate the potential annual biomass yield of dry mass at given (chosen) density, the measured average dry mass of a single plant (P_{dm}) (kg) was multiplied by actual density of plantings (plants per hectare) (D_{act}). The obtained result (t ha $^{-1}$) was divided by the number of years of cultivation, which was 7.

Potential annual biomass yield of dry mass:

$$Y_{dm} = \frac{P_{dm}D_{act}}{7}$$

3. Results and Discussion

3.1. Green Mass

The highest green mass of plants (GM) were observed in poplar AF2. On average, a single plant of poplar weighted 29.1 kg (median). Boxelder maple green mass was on average at a level of 9.3 kg per plant. Black alder, white birch, silver maple, and Siberian elm had a GM on a similar level (Table 2). Nevertheless, the lowest GM was recorded for silver maple, and it was only 2.6 kg per plant. GM was positively correlated with plant height and shoot diameter for all tested plants. For Polar AF2 and Siberian elm, a negative correlation between the green mass of the plants and the share of dry mass was noted. The GM was dropping significantly with the increasing planting density for all tested species. The strength of response of species to the increasing stand density showed some differences. The strongest negative response was observed in black alder, while white birch and silver maple were the least sensitive species to increasing planting density. In the case of other species, the strength of dependence was at a similar level (Figure 2, Table 3). Wilkinson et al. [10] showed that yield of green matter of 1-year-old and 3-year-old willow (*Salix* ssp.) was increasing significantly with increasing plant density (density from 10,000 to 25,000 of plants per hectare).

Table 2. Median values of tested plants and their biometric features.

Specification	Green Mass [kg]	Share of Dry Mass [%]	Dry Mass [kg]	Plant Height [m]	Shoot Diameter [mm]	Higher Heating Value [J g^{-1}]	Potential Yield of Dry Mass [t ha^{-1} yr^{-1}]
Poplar AF2 var.	29.1 a *	42.8 a	13.1 a	10.6 a	113.5 a	17,908 a	15.8 a
Siberian elm	2.8 b	58.5 b	1.7 b	4.7 d	41.8 b	18,664 bc	5.2 b
Black alder	5.3 bc	48.2 c	2.5 bc	6.6 bc	60.4 c	18,366 ab	3.9 b
White birch	5.5 bc	52.4 c	2.9 bc	7.0 b	59.4 bc	19,509 c	2.2 b
Boxelder maple	9.3 c	50.8 c	4.7 c	6.8 b	63.8 c	19,158 c	9.8 c
Silver maple	2.6 b	51.8 c	1.4 b	5.2 cd	39.2 b	18,726 bc	4.7 b

* Data marked with the same letter do not differ significantly between species ($\alpha = 0.05$).

Table 3. Correlation matrix for the studied features.

Plant	Specification	Green Mass [kg Plant $^{-1}$]	Share of Dry Mass [%]	Dry Mass [kg Plant $^{-1}$]	Plant Height [m]	Shoot Diameter [mm]	Higher Heating Value [J g^{-1}]	Potential Yield of Dry Mass [t ha^{-1} yr^{-1}]
Poplar AF2 var.	Plant density [plants ha^{-1}]	**−0.619** *	0.152	**−0.651**	−0.384	−0.367	**−0.485**	0.025
	Green mass [kg]		**−0.542**	**0.993**	**0.822**	**0.815**	0.435	0.377
	Share of dry matter [%]			−0.454	**−0.650**	−0.389	−0.350	−0.131
	Dry mass [kg plant $^{-1}$]				**0.797**	**0.796**	0.428	0.399
	Plant height [m]					**0.676**	0.443	**0.525**
	Shoot diameter [mm]						0.346	0.130
	Higher heating value [J g^{-1}]							0.086
Siberian elm	Plant density [plants ha^{-1}]	**−0.660**	0.457	**−0.663**	−0.402	**−0.628**	0.248	0.414
	Green mass [kg]		**−0.504**	**1.000**	**0.774**	**0.923**	−0.065	0.309
	Share of dry matter [%]			**−0.479**	−0.376	−0.439	0.356	−0.087
	Dry mass [kg plants $^{-1}$]				**0.777**	**0.925**	−0.061	0.306
	Plant height [m]					**0.793**	−0.023	**0.518**
	Shoot diameter [mm]						−0.124	0.234
	Higher heating value [J g^{-1}]							0.235
Black alder	Plant density [plants ha^{-1}]	**−0.833**	0.409	**−0.804**	**−0.775**	**−0.874**	0.115	**−0.539**
	Green mass [kg]		−0.357	**0.929**	**0.616**	**0.884**	−0.166	0.437
	Share of dry matter [%]			−0.025	−0.257	−0.293	−0.317	−0.257
	Dry mass [kg plant $^{-1}$]				**0.603**	**0.850**	−0.304	0.437
	Plant height [m]					**0.841**	−0.129	0.346
	Shoot diameter [mm]						−0.129	0.280
	Higher heating value [J g^{-1}]							−0.149

Table 3. *Cont.*

Plant	Specification	Green Mass [kg Plant^{-1}]	Share of Dry Mass [%]	Dry Mass [kg Plant^{-1}]	Plant Height [m]	Shoot Diameter [mm]	Higher Heating Value [J g^{-1}]	Potential Yield of Dry Mass [t ha^{-1} yr^{-1}]
White birch	Plant density [plants ha^{-1}]	**−0.589**	0.301	**−0.589**	−0.188	**−0.639**	−0.095	−0.175
	Green mass [kg]		−0.278	**0.998**	0.598	**0.679**	−0.139	**0.728**
	Share of dry matter [%]			−0.227	−0.457	−0.175	**−0.520**	−0.185
	Dry mass [kg plant $^{-1}$]				0.588	**0.684**	−0.170	**0.727**
	Plant height [m]					**0.563**	0.062	0.434
	Shoot diameter [mm]						0.169	0.233
	Higher heating value [J g^{-1}]							−0.288
Boxelder maple	Plant density [plants ha^{-1}]	**−0.684**	**−0.486**	**−0.670**	**−0.642**	**−0.706**	0.321	**0.493**
	Green mass [kg]		0.185	**1.000**	0.587	**0.956**	−0.388	−0.068
	Share of dry matter [%]			0.245	0.119	0.229	−0.109	−0.360
	Dry mass [kg plant $^{-1}$]				0.580	**0.957**	−0.395	−0.095
	Plant height [m]					**0.630**	−0.163	0.121
	Shoot diameter [mm]						−0.313	−0.130
	Higher heating value [J g^{-1}]							0.196
Silver maple	Plant density [plants ha^{-1}]	**−0.590**	−0.209	**−0.601**	−0.124	**−0.621**	−0.064	0.378
	Green mass [kg]		0.041	**0.994**	0.567	**0.865**	−0.112	0.326
	Share of dry matter [%]			0.144	−0.450	−0.140	−0.172	−0.281
	Dry mass [kg plant $^{-1}$]				0.518	**0.839**	−0.121	0.295
	Plant height [m]					**0.668**	−0.087	**0.549**
	Shoot diameter [mm]						−0.013	0.197
	Higher heating value [J g^{-1}]							**−0.539**

* Significant correlations are in bold ($\alpha = 0.05$).

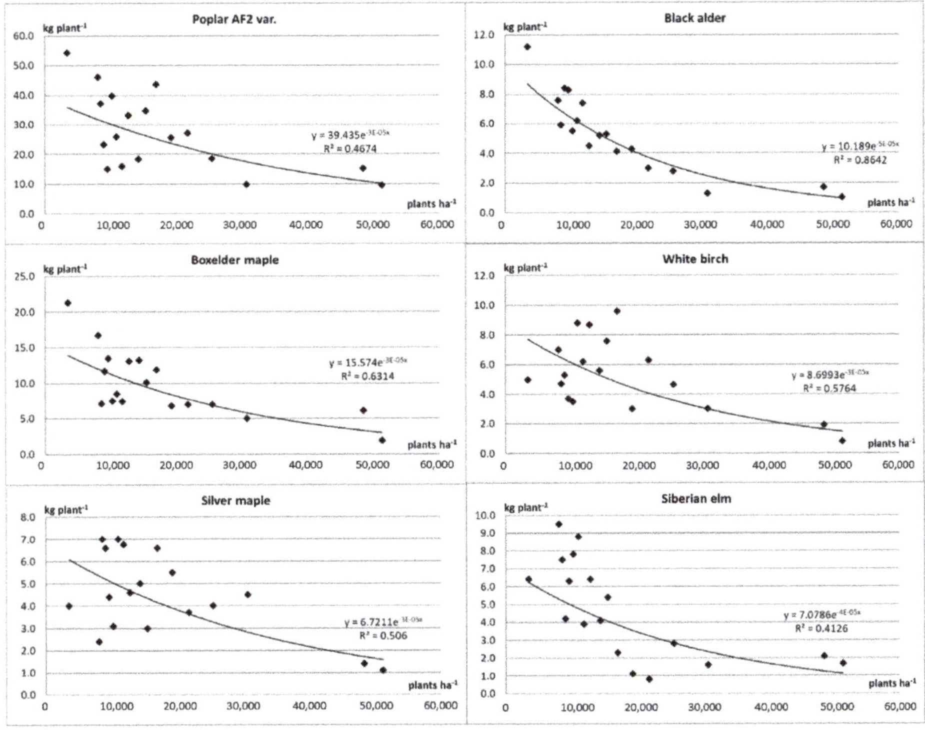

Figure 2. Relationship between green mass of plants (kg plant^{-1}) and planting density (plants ha^{-1}).

3.2. Dry Mass

Very similar relationships were found for the dry mass (DM) of plants as for green matter. This shows that the dry mass/green mass ratio (or in other words—moisture content) is at constant level at different plating densities and, therefore, have little or no effect on yields. The highest dry mass (DM) of plants were observed in poplar AF2. On average, a plant of poplar weighted 13.1 kg (median). Boxelder maple green mass was on average at a level of 4.7 kg per plant. Black alder, white birch, silver maple, and Siberian elm had a dry mass of plant on a similar level (Table 2). Nevertheless, the lowest dry mass of plants was recorded for silver maple, and it was only 1.4 kg per plant. Walle et al. [11] compared the dry mass of 4-year-old SRC of poplar (*Populus trichocarpa* × *deltoids*), birch (*Betula pendula* Roth), and maple (*Acer pseudoplatanus* L.) cultivated with a density of 20,000, 6667, and 6667 plants per hectare, respectively. The authors found that the average dry mass of plants grown under these conditions was 831, 2007, and 738 g, respectively, which was a noticeably different result than in presented study; however, the growing conditions were also different to the plants tested by Walle et al. [11], which were also about 3 years younger than in the presented study. DM in the presented study was positively correlated with plant height and shoot diameter for all tested plants. For Siberian elm, a negative correlation between the dry mass of the plants and the share of dry mass was noted. The dry mass of plants dropped significantly with the increasing planting density for all tested species. The strength of response of species to the increasing stand density showed some differences. The strongest negative response was observed for black alder, while white birch and silver maple were the least sensitive species to increasing plant density. In the case of other species, the strength of dependence was at a similar level (Figure 3, Table 3). Other authors investigated the response of dry mass of plants to increasing density and found out that it was also dropping for willow (*Salix* ssp.) [12] and for oak (*Quercus robur*) [13].

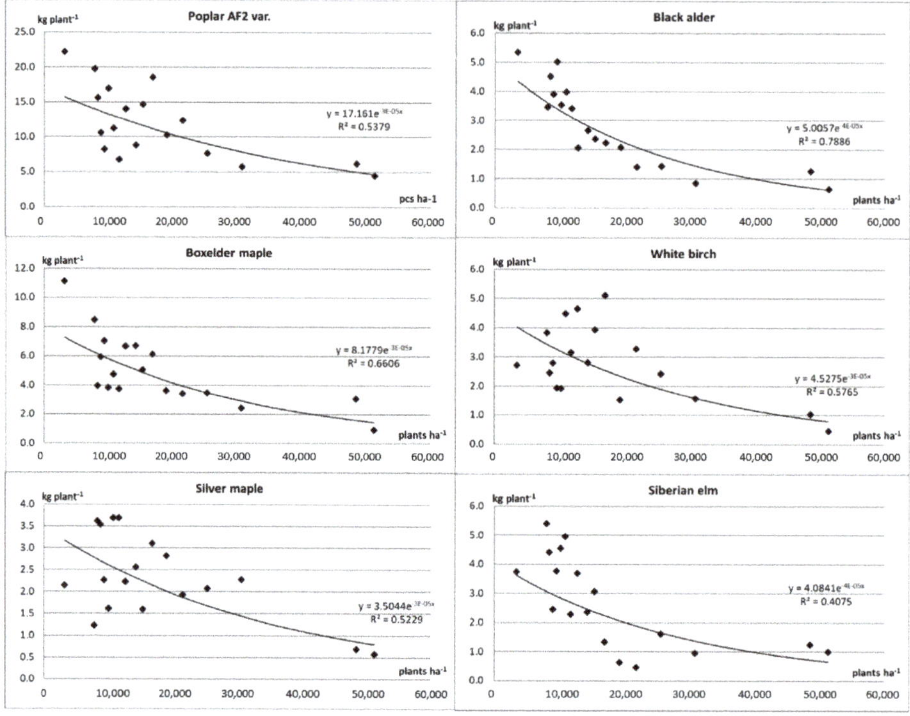

Figure 3. Relationship between dry mass of plant biomass (kg plant^{-1}) and planting density (plants ha^{-1}).

3.3. Share of Dry Mass

The highest dry to green mass (DM to GM) ratio was observed for Siberian elm (58.5%), while the lowest was observed for poplar AF2 (42.8%). White birch, silver maple, boxelder maple, and black alder DM to GM ratio was on a similar level (52.4, 51.8, 50.8, and 48.2%, respectively) (Table 2). DM to GM ratio varied depending on the stand density, but also on the tested species. Share of dry mass was negatively correlated with green mass of plants for three tested species (poplar AF2, Siberian elm and boxelder maple). Share of dry mass also negatively correlated with plant height for poplar AF2 and with higher heating value for white birch (Table 3). Poplar AF2 and silver maple showed no reaction of DM to GM ratio to increasing density of plants, while Siberian elm, black alder, and white birch showed moderately positive increase in DM to GM ratio with increasing plant density. Boxelder maple showed negative response of DM to GM ratio to increasing plant density (Figure 4). Wilkinson et al. [10] found dry mass of 1-year-old and 3-year-old willow (Salix ssp.) not dependent on density of plantings (similar to poplar AF2 and silver maple reaction in presented study). In addition, Stolarski et al. [14] and Kulig et al. [15] found no effect of planting density on the fresh–dry matter ratio of willow. This was also confirmed by Elfeel and Elmagboul [16] for other woody species—*leucaena leucocephala*. On the other hand, Achinelli et al. [17] found that dry to fresh matter content ratio in willow was higher in more dense stands (similar effect to Siberian elm, black alder, and white birch in the discussed study).

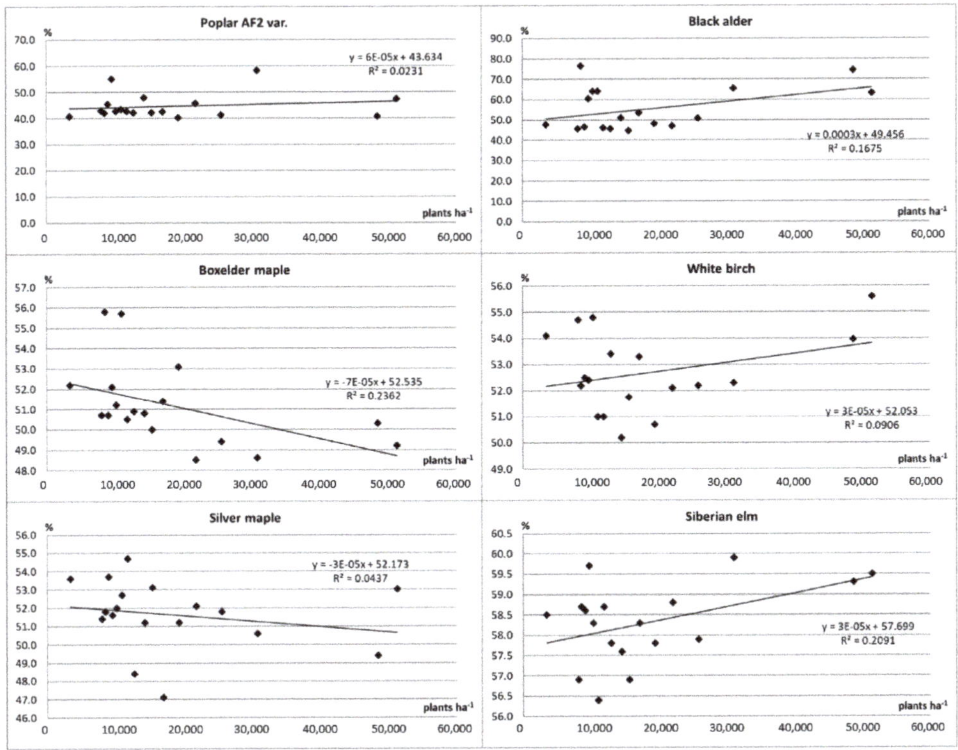

Figure 4. Relationship between dry mass share (%) and planting density (plants ha^{-1}).

3.4. Potential Yield of Dry Mass

Poplar AF2 had the highest calculated potential of dry mass yield (after 7 years) of about 15 t ha^{-1} for all tested planting densities. Boxelder maple was able to match yielding potential that AF2 poplar only at planting density of about 40,000 plants per hectare. Silver maple reached about 50% of this potential (about 7.5–8 t ha^{-1}), while other tested species reached about 5 t ha^{-1} of dry mass yield. (Figure 3). The calculated annual yield of dry mass was positively correlated with plant height for poplar AF2, Siberian elm, and silver maple. There was also a positive correlation with plant density for boxelder maple and negative for black alder. In the case of white birch, a strong positive correlation was found with green and dry mass of plants (Table 3). Despite the fact that the dry mass of individual trees decreased with increasing density, the calculated annual yield of dry matter of Siberian elm, boxelder maple, and silver maple was increasing with increasing density of planting (Figure 5). The same was also observed by Geyer, Argent, and Walawender [18] for 7-year-old Siberian elm. Authors found that annual yields of dry matter of Siberian elm were at a level of 4.7 t ha^{-1} with stand density of 1400 plants per hectare, while they increased significantly to 9.8 t ha^{-1} with stand density of 7000 plants per hectare. In the presented study, calculated annual dry mass yields of Siberian elm varied between 1.8 t ha^{-1} for the lowest density and 7.4 t ha^{-1} for the highest density. Geyer and Walawender [19] reported that annual dry mass yield of 7-year-old silver maple was increasing from 5.3 t ha^{-1} at 1400 plants per hectare to 11.2 t ha^{-1} at 7000 plants per hectare. In addition, other authors found that for some species such as willow (*Salix* L.) [10] and black locust (*Robinia pseudoacacia*) [20] annual yield of dry mass was increasing with increasing planting density. Niemczyk et al. [21] found that annual yields of a 7-year-old poplar can reach up to 8 t ha^{-1}. Truax et al. [22] also found a positive correlation of planting density and yield of 8-year-old hybrid poplar. Stolarski et al. [23] found that annual dry mass yield of willow planted with densities from 12,000 to 96,000 was increasing from 12,000 to 24,000 (optimal density) and decreasing with increasing density from 24,000 to 96,000. Similar reaction was found in presented study silver maple (optimal planting density of about 25,000–30,000 plants per hectare).

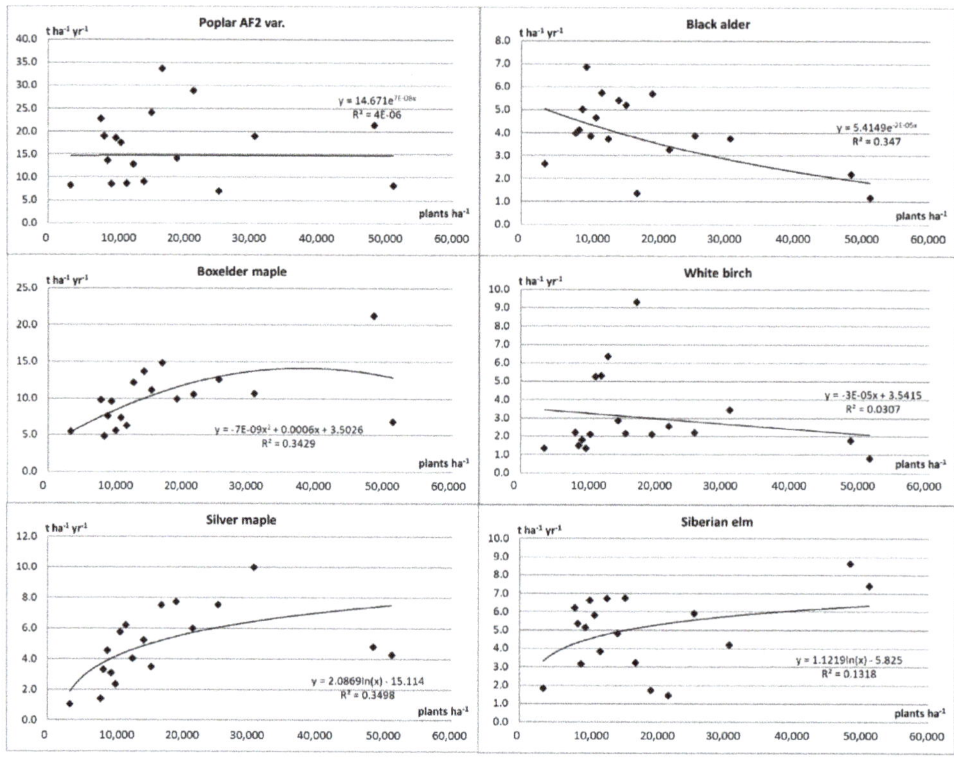

Figure 5. Relationship between potential yield of dry mass per year (t ha^{-1} r^{-1}) and planting density (plants ha^{-1}).

3.5. Height of Plants

Plants of poplar AF2 were, on average, the highest of all tested species (10.6 m). Black alder, boxelder maple, and white birch's height was on a similar level of 6.6–7.0 m. Silver maple and Siberian elm had, on average, the lowest plants of both 5.2 and 4.7 m, respectively (Table 2). Siberian elm plants in Geyer, Argent, and Walawender [18] were, on average, higher (6.3 m) than in presented study (4.7 m). Geyer, Barden, and Preece [24] found that 6-year-old silver maple clones of dense stands (no accurate data available) were 7.3 m high. Plant height for all tested species was positively correlated with their green and dry mass and shoot diameter. In addition, for poplar AF2, plant height was negatively correlated with share of dry mass, and, for black alder, with plant density (Table 3). Height of tested plants varied on planting density. In most cases, the decrease in plant height with increasing density of planting started from the very begging—from about 3500 plants per hectare (poplar AF2, black alder, boxelder maple, and Siberian elm) (Figure 6). Geyer, Argent, and Walawender [18] also found this relationship for the 7-year-old Siberian elm, whose height was, on average, 6.4 m for the stand density of 1400 plants per hectare and was dropping to 6.2 m for the density of 7000 plants per hectare. A similar relationship was presented by Perez et al. [25] for a 3-year-old Siberian elm. In addition, it was confirmed by Geyer and Walawender for silver maple [19] and black locust [20]. On the other hand, Toillon et al. [8] found that height of poplar increases with increased planting density in favorable site conditions (as an effect of increased competition for light), while in less favorable conditions, height of plants remained unaffected by increasing planting density. Benomar et al. [7] showed that the relationship between the density and height of plants is also strongly influenced by the genotype.

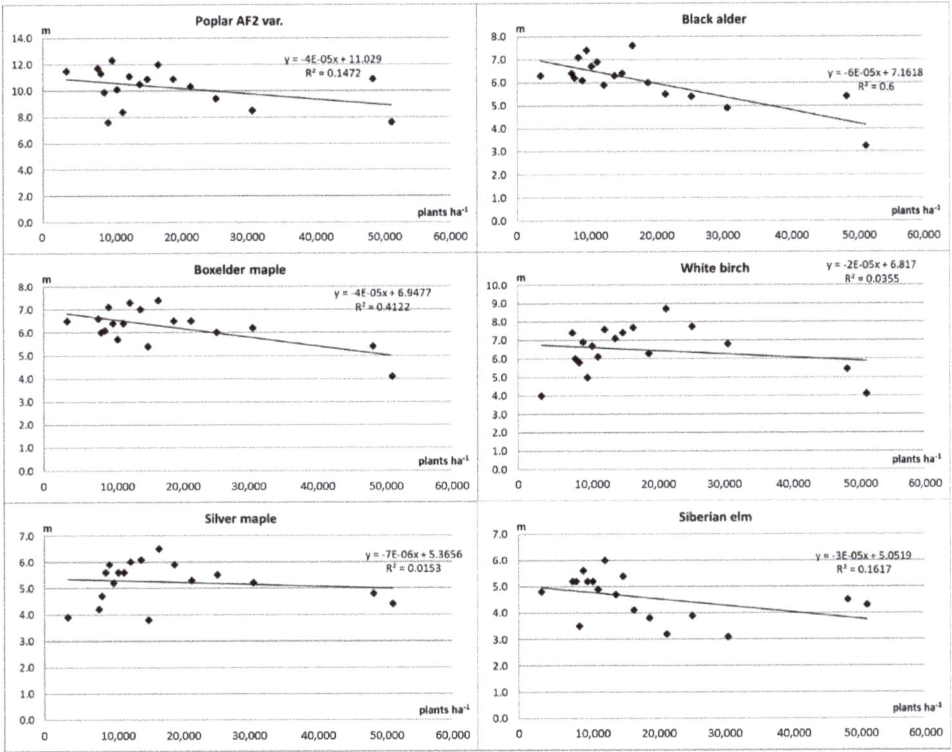

Figure 6. Relationship between plant height (m) and planting density (plants ha^{-1}).

3.6. Shoots Diameter

Plants of poplar AF2 were, on average, of the greatest shoot diameter (113.5 mm). Boxelder maple, black alder, and white birch's shoot diameter was on similar level (63.8, 604, 59.4 mm, respectively). Siberian elm and silver maple had the lowest shoot diameter of 41.8 mm both (Table 2). Geyer, Argent, and Walawender [18] found 7-year-old Siberian elm to have, on average, stems of diameter of 109 mm at 10 cm. Walle et al. [11] found that shoots diameter (at 30 cm) of 4-year-old poplar (*Populus trichocarpa* × *deltoids*) (336 mm) and maple (*Acer pseudoplatanus* L.) (310 mm) were lower than shoot diameter of birch (*Betula pendula* Roth) (493 mm). These authors conducted research on other clones/species of trees in different habitat conditions than in presented study; however, differences in results show the importance of the selection of appropriate species/cultivars and proper density of planting to make the best use of habitat conditions ensuring high biomass accumulation.

Shoot diameter for all tested species was positively correlated with their green and dry mass and shoot diameter. For all tested species, a negative correlation with the planting density was found (Table 3). Shoot diameter of all tested plant species were decreasing with increasing planting density, but the strength of this reaction varied between species (Figure 7). The decrease in stem diameter is a common response to increasing planting density [18,26–28], which can also be modified by species genotype (clone) [29].

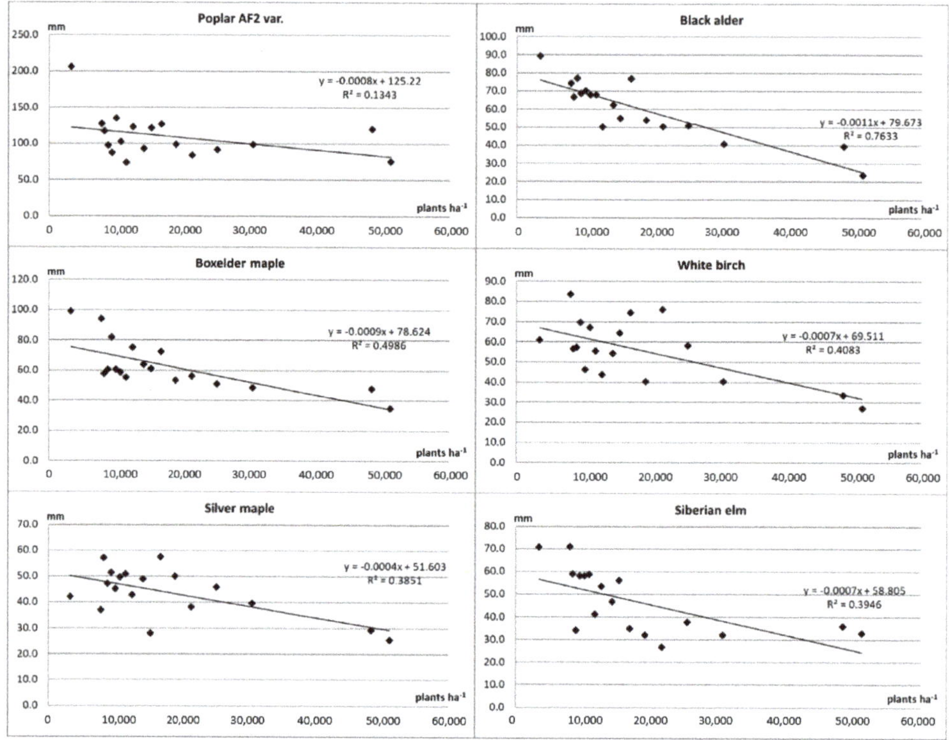

Figure 7. Relationship between plants shoot diameter (mm) and planting density (plants ha^{-1}).

3.7. Higher Heating Value

Tested energy sources differed significantly in higher heating value (HHV). White birch and boxelder maple had the highest average HHV value (19,509 and 19,158 J g^{-1} o, respectively). The lowest HHV was observed for poplar AF2 (17,908 J g^{-1} o). Dry mass of Siberian elm, black alder, and silver maple had a similar HHV value of around 18,500 J g^{-1} o (Table 2). Geyer, Argent, and Walawender [18] found that HHV of 7-year-old Siberian was between 18,900 and 20,200 J g^{-1} (19,700 J g^{-1} on average), which is a higher HHV value than in the presented. Study stand density had no effect on most species' HHV. A statistically significant negative correlation of HHV with the plant density was found only for poplar AF2 and white birch (Table 3, Figure 8). In the case of other species, no relationship between HHV and planting density was found. Literature study shows that HHV is highly variable and depends on both genetic factors (species, cultivars) [15,30,31] and cultivation conditions (soil quality, management methods) [30,32] or even plant age [33].

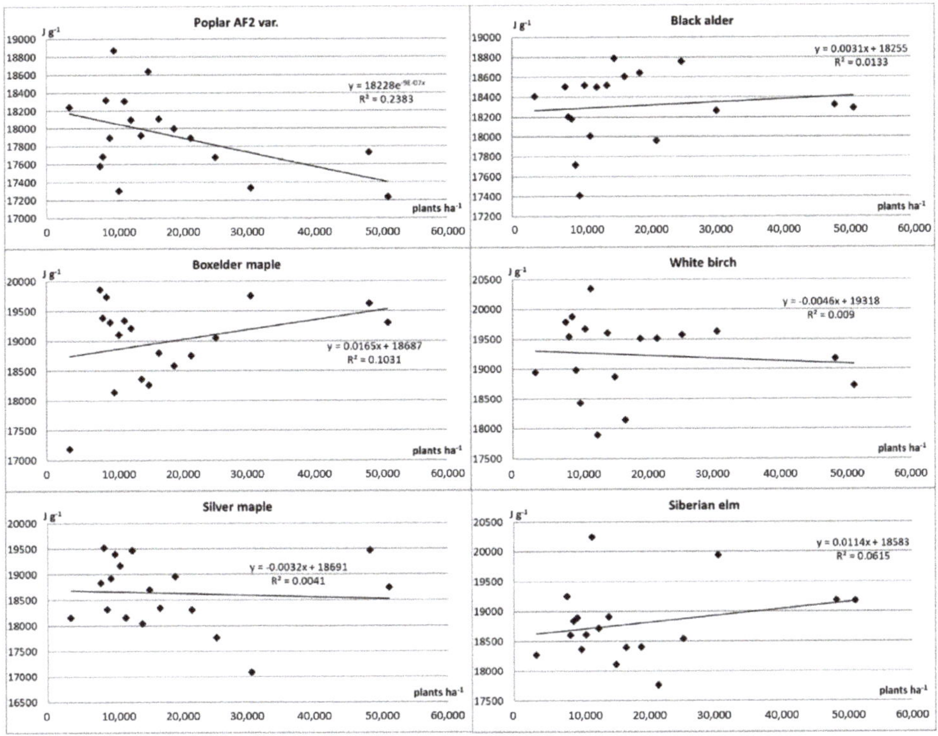

Figure 8. Relation of higher heating value (J g^{-1}) and planting density (plants ha^{-1}).

3.8. Survival Rate

The highest survival rate after 7 years was observed for boxelder maple (100%), silver maple, and Siberian elm (both 99%). Black alder survival rate was also at moderately good level (81%), while poplar AF2 and white birch survival rate was the lowest (69% and 54%, respectively). Survival rate in the presented study was dropping with increasing planting densities (for whole experimental design), but some species were unaffected by increasing density (boxelder and silver maple, Siberian elm), while white birch, poplar AF2, and black alder survival rates were dropping with increasing densities (Figure 9). According to Trnka et al. (2008) [34], survival rate of 6-year-old poplar at dense stand (10,000 plants per hectare) varied between 37% and 73% depending on the genotype. Geyer et al. (1987) [18] found that survival rate of 7-year-old Siberian elm was "almost perfect" and did not vary on planting densities from 1400 to 7000 plants per hectare (stands of lower densities than in presented study). Moreover, Geyer (2006) [35] found that survival rate decreases substantially for most tested species at spacing distances less than 1 m (more than 10,000 plants per hectare), while 2 m spacing (2500 plants per hectare) was found optimal for high biomass production and high longevity of plantation. Authors also found that silver maple survival after five years of cultivation was at a level of 97%.

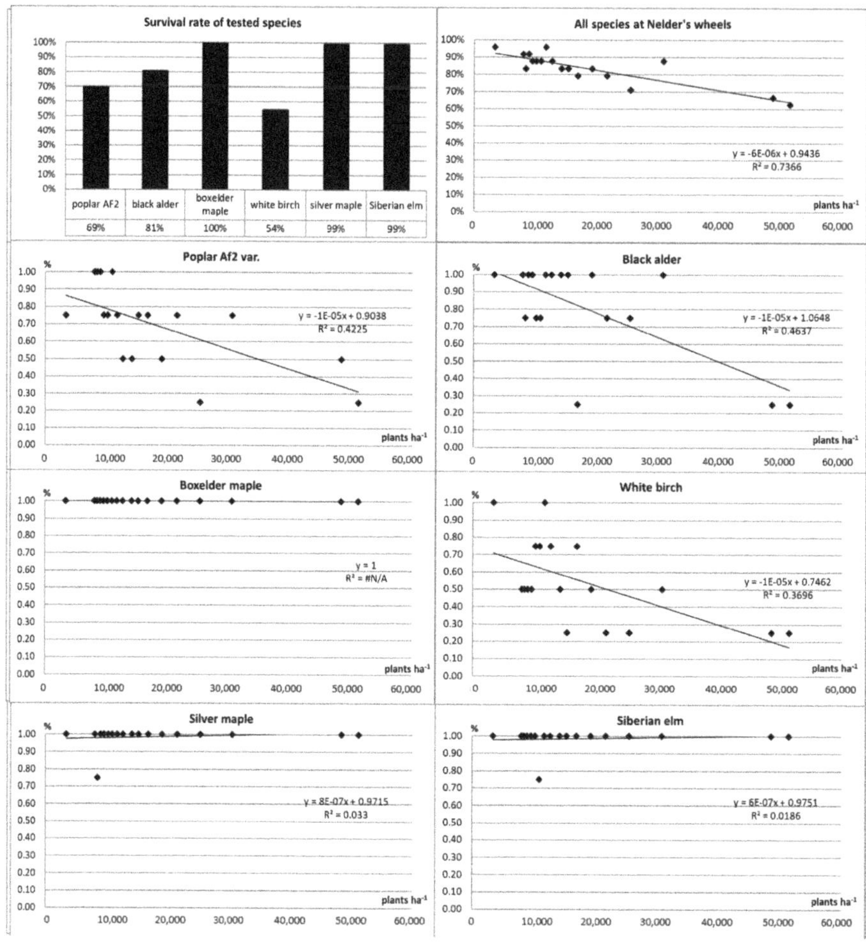

Figure 9. Survival rate after 7 years for different species and different planting densities.

4. Conclusions

Biomass yield parameters of six tested SRC plants strongly depended on both genotype (species) and planting density. The strength and direction of reaction of plants to increasing planting density varied between species. The green mass, dry mass, and shoot diameter of plants was dropping with the increasing planting density for almost all tested species (insignificant negative correlation of shoot diameter and planting density of poplar AF2). At the same time, function curve of calculated potential yield of dry mass and planting density was dropping with increasing planting density for black alder, increasing for Siberian elm and boxelder maple, and was concave for white birch and silver maple (with optimal planting density of about 25,000–30,000 plants per hectare). Planting density seemed to have no effect on calculated potential yield of dry mass for AF2 poplar.

White birch and boxelder maple had the highest average higher heating value (HHV) (19,509 and 19,158 J g^{-1}, respectively). The lowest HHV was observed for poplar AF2 (17,908 J g^{-1}). Dry mass of all other species had similar HHV value of around 18,500 J g^{-1}. A tendency towards reduced higher heating value of plants in more dense stands was observed, but it was confirmed only for AF2 poplar.

Presented study showed that, in the study conditions, poplar AF2 was the most promising SRC plant, with calculated potential dry mass yield of about 15 t ha^{-1} at wide range of planting densities, with boxelder maple being the only species able to match AF2 poplar's yielding performance (at 40,000 plants per hectare). The study showed the importance of testing and selection of best-performing species, to maximize biomass accumulation at local site conditions. The optimal density of plantings should be chosen to best suit the needs of cultivated species but also to meet the needs of the industry by optimizing the most important (for the industry) parameters of produced biomass. Cultivation of renewable resources, such as energy crops, must be optimized to site-specific conditions. This includes cultivation on poor soils or marginal soils to minimize their competitiveness against food crops. Optimization of energy crop yields can contribute to the goals of bioeconomy and sustainable development.

Author Contributions: Conceptualization, M.M.; methodology, M.M.; software, A.K.B. and M.M.; investigation, M.M. and A.K.B.; data curation, A.K.B. and M.M.; writing—original draft preparation, A.K.B.; writing—review and editing, M.M. and A.K.B.; visualization, M.M. and A.K.B. All authors have read and agreed to the published version of the manuscript.

Funding: This paper is the result of a long-term study carried out at the Institute of Soil Science and Plant Cultivation, State Research Institute in Pulawy, and it was co-financed by the National (Polish) Centre for Research and Development (NCBiR), project: BIOproducts from lignocellulosic biomass derived from MArginal land to fill the Gap in Current national bioeconomy, No. BIOSTRATEG3/344253/2/NCBR/2017. This research acknowledges financial support from the Widening Program ERA Chair: project BioEcon (H2020), contract number: 669062.

Conflicts of Interest: The authors declare no conflict of interest.

References

1. Diamantidis, N.D.; Koukis, E.G. Agricultural crops and residues as feedstock for non-food products in Western Europe. *Ind. Crops Prod.* **2000**, *11*, 97–106. [CrossRef]
2. El Kasmioui, O.; Ceulemans, R. Financial analysis of the cultivation of poplar and willow for bioenergy. *Biomass Bioenergy* **2012**, *43*, 52–64. [CrossRef]
3. Stolarski, M.J.; Śnieg, M.; Krzyżaniak, M.; Tworkowski, J.; Szczukowski, S. Short rotation coppices, grasses and other herbaceous crops: Productivity and yield energy value versus 26 genotypes (2018). *Biomass Bioenergy* **2018**, *119*, 109–120. [CrossRef]
4. Nabel, M.; Barbosa, D.B.P.; Horsch, D.; Jablonowski, N.D. Energy crop (Sida hermaphrodita) fertilization using digestate under marginal conditions: A dose-response experiment. *Energy Procedia* **2014**, *54*, 127–133. [CrossRef]
5. Matyka, M.; Kuś, J. Influence of soil quality for yield and biometric features of Sida hermaphrodita L. Rusby. *Pol. J. Environ. Stud.* **2018**, *27*, 2669–2675. [CrossRef]
6. Navarro, A.; Facciotto, G.; Campi, P.; Mastrorilli, M. Physiological adaptations of five poplar genotypes grown under SRC in the semi-arid Mediterranean environment. *Trees* **2014**, *28*, 983–994. [CrossRef]
7. Benomar, L.; DesRochers, A.; Larocque, G. The effects of spacing on growth, morphology and biomass production and allocation in two hybrid poplar clones growing in the boreal region of Canada. *Trees* **2012**, *26*, 939–949. [CrossRef]
8. Toillon, J.; Fichot, R.; Dallé, E.; Berthelot, A.; Brignolas, F.; Marron, N. Planting density affects growth and water-use efficiency depending on site in Populus deltoids x P. nigra. *For. Ecol. Manag.* **2013**, *304*, 345–354. [CrossRef]
9. Nelder, J.A. New kinds of systematic design for spacing experiments. *Biometrics* **1962**, *18*, 283–307. [CrossRef]
10. Wilkinson, J.M.; Evans, E.J.; Bilsborrow, P.E.; Wright, C.; Hewison, W.O.; Pilbeam, D.J. Yield of willow cultivars at different planting densities in a commercial short rotation coppice in the north of England. *Biomass Bioenergy* **2007**, *31*, 469–474. [CrossRef]
11. Walle, I.V.; Van Camp, N.; Van de Casteele, L.; Verheyen, K.; Lemeur, R. Short-rotation forestry of birch, maple, poplar and willow in Flanders (Belgium) I—Biomass production after 4 years of tree growth. *Biomass Bioenergy* **2007**, *31*, 267–275. [CrossRef]
12. Bullard, M.J.; Mustill, S.J.; McMillan, S.D.; Nixon, P.M.; Carver, P.; Britt, C.P. Yield improvements through modification of planting density and harvest frequency in short rotation coppice *Salix* spp.—1. Yield response in two morphologically diverse varieties. *Biomass Bioenergy* **2002**, *22*, 15–25. [CrossRef]

13. Dahlhausen, J.; Uhl, E.; Heym, M.; Biber, P.; Ventura, M.; Panzacchi, P.; Tonon, G.; Horváth, T.; Pretzsch, H. Stand density sensitive biomass functions for young oak trees at four different European sites. *Trees* **2017**, *31*, 1811–1826. [CrossRef]

14. Stolarski, M.J.; Szczukowski, S.; Tworkowski, J.; Wróblewska, H.; Krzyżaniak, M. Short rotation willow coppice biomass as an industrial and energy feedstock. *Ind. Crops Prod.* **2011**, *33*, 217–223. [CrossRef]

15. Kulig, B.; Gacek, E.; Wojciechowski, R.; Oleksy, A.; Kołodziejczyk, M.; Szewczyk, W.; Klimek-Kopyra, A. Biomass yield and energy efficiency of willow depending on cultivar, harvesting frequency and planting density. *Plant. Soil Environ.* **2019**, *65*, 377–386. [CrossRef]

16. Elfeel, A.A.; Elmagboul, A.H. Effect of planting density on eucaena leucocephala forage and Woody stems production under arid dry climate. *Int. J. Agric. Res.* **2016**, *2*, 7–11.

17. Achinelli, F.G.; Doffo, G.; Barotto, A.J.; Luquez, V.; Monteoliva, S. Effects of irrigation, plantation density and clonal composition on woody biomass quality for bioenergy in a short rotation culture system with willows (*Salix* spp.). *Rev. Árvore* **2018**, *42*, 1–8. [CrossRef]

18. Geyer, W.A.; Argent, R.M.; Walawender, W.P. Biomass properties and gasification behavior of 7-year-old Siberian elm. *Wood Fiber Sci.* **1987**, *19*, 176–182.

19. Geyer, W.A.; Walawender, W.P. Biomass properties and gasification behavior of young silver maple trees. *Wood Fiber Sci.* **1997**, *29*, 85–90.

20. Geyer, W.A.; Walawender, W.P. Biomass properties and gasification behavior of young black locust. *Wood Fiber Sci.* **1994**, *26*, 354–359.

21. Niemczyk, M.; Kaliszewski, A.; Jewiarz, M.; Wróbel, M.; Mudryk, K. Productivity and biomass characteristics of selected poplar (*Populus* spp.) cultivars under the climatic conditions of northern Poland. *Biomass Bioenergy* **2018**, *111*, 46–51. [CrossRef]

22. Truax, B.; Fortier, J.; Gagnon, D.; Lambert, F. Planting density and site effects on stem dimensions, stand productivity, biomass partitioning, carbon stocks and soil nutrient supply in hybrid poplar plantations. *Forests* **2018**, *9*, 293. [CrossRef]

23. Stolarski, M.J.; Szczukowski, S.; Tworkowski, J.; Krzyżaniak, M.; Załuski, D. Willow production during 12 consecutive years—The effects of harvest rotation, planting density and cultivar on biomass yield. *Glob. Chang. Biol. Bioenergy* **2019**, *11*, 635–656. [CrossRef]

24. Geyer, W.A.; Barden, C.; Preece, J.E. Biomass and wood properties of young silver maple clones. *Wood Fiber Sci.* **2008**, *40*, 23–28.

25. Perez, I.; Perez, J.; Carrasco, J.; Ciria, P. Siberian elm responses to different culture conditions under short rotation forestry in Mediterranean areas. *Turk. J. Agric. For.* **2014**, *38*, 652–662. [CrossRef]

26. Di Matteo, G.; Sperandio, G.; Verani, S. Field performance of poplar for bioenergy in southern Europe after two coppicing rotations: Effects of clone and planting density. *iForest Biogeosci. For.* **2012**, *5*, 224. [CrossRef]

27. Hummel, S. Height, diameter and crown dimensions of Cordia alliodora associated with tree density. *For. Ecol. Manag.* **2000**, *127*, 31–40. [CrossRef]

28. Mehtätalo, L.; de-Miguel, S.; Gregoire, T.G. Modeling height-diameter curves for prediction. *Can. J. For. Res.* **2015**, *45*, 826–837. [CrossRef]

29. Castaño-Díaz, M.; Barrio-Anta, M.; Afif-Khouri, E.; Cámara-Obregón, A. Willow short rotation coppice trial in a former mining area in northern Spain: Effects of clone, fertilization and planting density on yield after five years. *Forests* **2018**, *9*, 154. [CrossRef]

30. Sabatti, M.; Fabbrini, F.; Harfouche, A.; Beritognolo, I.; Mareschi, L.; Carlini, M.; Paris, P.; Scarascia-Mugnozza, G. Evaluation of biomass production potential and heating value of hybrid poplar genotypes in a short-rotation culture in Italy. *Ind. Crops Prod.* **2014**, *61*, 62–73. [CrossRef]

31. Navarro, A.; Stellacci, A.M.; Campi, P.; Vitti, C.; Modugno, F.; Mastrorilli, M. Feasibility of SRC species for growing in Mediterranean conditions. *Bioenergy Res.* **2016**, *9*, 208–223. [CrossRef]

32. Nebeská, D.; Trögl, J.; Žofková, D.; Voslařová, A.; Štojdl, J.; Pidlisnyuk, V. Calorific values of miscanthus x giganteus biomass cultivated under suboptimal conditions in marginal soils. *Stud. Oecol.* **2019**, *13*, 61–67. [CrossRef]

33. Kumar, R.; Pandey, K.K.; Chandrashekar, N.; Mohan, S. Study of age and height wise variability on calorific value and other fuel properties of Eucalyptus hybrid, Acacia auriculaeformis and Casuarina equisetifolia. *Biomass Bioenergy* **2011**, *35*, 1339–1344. [CrossRef]

34. Trnka, M.; Trnka, M.; Fialová, J.; Koutecky, V.; Fajman, M.; Zalud, Z.; Hejduk, S. Biomass production and survival rates of selected poplar clones grown under a short-rotation system on arable land. *Plant Soil Environ.* **2008**, *54*, 78–88. [CrossRef]
35. Geyer, W.A. Biomass production in the central great plains USA under various coppice regimes. *Biomass Bioenergy* **2006**, *30*, 778–783. [CrossRef]

Publisher's Note: MDPI stays neutral with regard to jurisdictional claims in published maps and institutional affiliations.

Article

The Energy and Environmental Potential of Waste from the Processing of Hulled Wheat Species

Jaroslav Bernas [1,*], Petr Konvalina [1], Daniela Vasilica Burghila [2], Razvan Ionut Teodorescu [2] and Daniel Bucur [3]

[1]   Faculty of Agriculture, University of South Bohemia, Studentska 1668,
       37005 Ceske Budejovice, Czech Republic; konvalina@zf.jcu.cz
[2]   Department of Environment and Land Improvements, Faculty of Land Improvements and
       Environmental Engineering, University of Agronomical Sciences and Veterinary Medicine of Bucharest,
       59 Marasti Blvd., 011464 Bucharest, Romania; burghila.daniela@edu.gov.ro (D.V.B.);
       razvan.teodorescu@usamv.ro (R.I.T.)
[3]   Department of Pedotechnics, Faculty of Agriculture, University of Agricultural Sciences and Veterinary
       Medicine of Iasi, 3 Mihail Sadoveanu, 700490 Iasi, Romania; dbucur@uaiasi.ro
*   Correspondence: bernas@zf.jcu.cz; Tel.: +420-389-032-401

Received: 12 October 2020; Accepted: 27 November 2020; Published: 1 December 2020

Abstract: Organic farmers farming on arable land have often had, in addition to the cultivation of common species of cultivated crops (such as wheat, rye, triticale or potatoes), interest in the cultivation of marginal crops such as hulled wheat species (Einkorn, Emmer and Spelt wheat). The production of marginal cereals has seen significant developments in the European Union related to the development of the organic farming sector. Just the average annual organic production of spelt in the Czech Republic reached more than 9000 tons in 2018. The cultivation of these cereals requires post-harvest treatment in the special method of dehulling. The waste emerging after dehulling of spikelet (i.e., chaff) accounts for about 30% of the total amount of harvest and can be used as an alternative fuel material. When considering the energy utilization of this waste, it is also necessary to obtain information on the energy quality of the material, as well as environmental aspects linked to their life cycle. For evaluating the energy parameters, the higher and lower heating value, based on the elemental (CHNS) analysis, was determined. The environmental aspects were determinate according to the Life Cycle Assessment (LCA) methodology where the system boundary includes all the processes from cradle to farm gate, and the mass unit was chosen. The SimaPro v9.1.0.11 software and ReCiPe Midpoint (H) within the characterization model was used for the data expression. The results predict the energy potential of chaff about 50–90 TJ per year. The results of this study show that in some selected impact categories, 1 kg of chaff, as a potential fuel, represents a higher load on the environment than 1 kg of lignite, respectively potential energy gain (1 GJ) from the materials.

Keywords: hulled wheat species; energy; life cycle assessment

1. Introduction

The production of marginal cereals has seen significant developments in the European Union related to the development of the organic farming sector. The typical marginal wheat species in the Czech Republic are Einkorn, Emmer, and especially Spelt wheat [1,2]. They can be defined as the cultural hulled wheat species, which replace, expand, and supplement the existing range of cereals and contribute to broadening the spectrum of crop production [3]. These marginal cereals have usually lower harvest index but have less input intensity requirements. Thanks to this aspect, these grains are particularly suitable for organic farming systems [3–5]. The benefits of introducing these marginal cereal species include extending the food spectrum, maintaining the production capacity of the soil, and the efficient use of marginal and less-favored areas [6,7]. However, the disadvantages are low yields (low harvest index) and uneven ripening, which causes large losses during harvesting [5,8]. The processing of these hulled wheat species generates a relatively large amount of waste-chaff [9], which can be used in various ways, for example, can be composted [10], used as litter [9] or as an additive to building materials [11], or could be directly put back to the agricultural land to help to maintain the soil fertility [2,5]. Due to the high content of mycotoxins, however, it is not recommended to use chaff from hulled wheat as litter and it is preferable to use it for energy production and to burn it [9], optimally in the form of pellets [12,13]. The energy use of such residual agricultural biomass has great potential not only across the EU [14]. The potential for energy utilization, the energy parameters of chaff and the removal rate, including pelleting issues, have been summarized for Spelt and Emmer in some studies [9,10,13,15]. However, the crop residue removal for biofuel production can have a significant impact on crop productivity, soil health, and greenhouse gas emissions [16].

In addition to Spelt and Emmer wheat, this manuscript also evaluates the energy parameters of Einkorn chaff. The energy use of chaff of these marginal cereal species is often perceived as an environmentally friendly source of energy, because it is energy from biomass. However, it has to be considered in terms of inputs into the cultivation process. This work aims to point out environmental aspects related to hulled wheat chaff and quantitative and qualitative parameters of individual wheat species using the Life Cycle Assessment (LCA) method.

2. Materials and Methods

2.1. Field Trails

Data for evaluation was based on field trials (the University of South Bohemia, Faculty of Agriculture, location: Zvíkov, GPS 48.9758531N, 14.6245594E; production region: cereal production region; altitude: 490 m; year temperature: 7.2 °C; year rainfall: 634 mm; type of soil: brown soil; sort of soil: loamy soil—medium) realized in the regime of organic farming. The field trails are used to assess yield potential, environmental, and economic aspects of cultivation. The fieldwork methodology was chosen based on commonly used organic farming technologies. Selected cultivation practices are typical for the conditions of the Czech Republic [2]. The design of randomized field trails in three replicates with an average area of an experimental plot of 10 m^2 was used. The growing period related to this study was from September 2017 to August 2018. Fodder pea—winter type was the preceding crop. Organic fertilizers were applied to the soil before sowing (6.6 ton ha^{-1} of manure/solid cattle; organic production) and the soil was loosened with a mid-deep ploughing to a 14–18 cm depth and levelled with a cultivator within the framework of pre-sowing preparatory works. Sowing of Spelt (winter type) was carried out in October 2017. Sowing of Emmer and Einkorn was carried out in April 2018 (spring types). The sowing depth was 3–4 cm. The amount of seed was 180 kg per ha. The harvest was performed during the July and August 2018. After harvest were taken samples from each replication and homogenized (hammer mill PSY MP 20/MP 40). Individual operations are shown in Table 1.

Table 1. Input data inventory.

Input	Unit	Investigated Crop		
		Spelt [W]	Emmer	Einkorn
Manure (solid cattle)	ton ha^{-1}	6.6	6.6	6.6
Solid manure loading and spreading	ton ha^{-1}	6.6	6.6	6.6
Tillage, ploughing	ha	1	1	1
Tillage, cultivating, chiselling	ha	1	1	1
Sowing	ha	1	1	1
Seeds, organic	kg ha^{-1}	180	180	180

Table 1. *Cont.*

Input	Unit	Investigated Crop		
Tillage, rolling	ha	1	-	-
Tillage, harrowing	ha	1	1	1
Combine harvesting	ha	1	1	1
Transport, tractor and trailer	tkm	50	50	50
Electricity for processing	kWh ton^{-1}	0.3	0.3	0.3

Transport included in the process as a flat rate 50 tkm. It is calculated with the same weight for all crops (max 8 tons per load); [W] = Winter Spelt wheat.

2.2. Analysis of Phytomass

For the purposes of this study, the elemental composition of chaff that remained after the dehulling of grains of Spelt, Emmer, and Einkorn was determined. The design of randomized field trails in three replicates for Spelt, Emmer, and Einkorn was used. After harvest, samples of chaff were taken from each replication and homogenized. For the elemental composition of chaff, two homogenized samples were used from each hulled wheat species. The CHNS analysis (the elemental composition of chaff) was carried out using the elemental analyzer (Vario EL CUBE). The method of direct jet injection of oxygen and combustion in the high furnace temperatures of up to 1200 °C with a complete conversion of the sample to measuring gas was used. The higher heating value (HHV) was calculated using the Mendeleev's equation (Equation (1)) [17], as well as lower heating value (LHV) from the equation (Equation (2)) [18], where Qv is the heat of combustion in kcal kg^{-1} [18]. Based on the observed elementary composition and empirical formulas, the HHV and LHV of the chaff were determined.

$$Q_s{}^r = [81 \times C + 300 \times H - 26 \times (O - S)] \times 4.186 \; (kJ \; kg^{-1}), \tag{1}$$

where: * $Q_s{}^r$ = HHV [kJ kg^{-1}]; C = carbon in the sample (%); H = hydrogen in the sample (%); O = oxygen in the sample (%); S = sulphur in the sample (%); 4.186 = conversion factor from kcal kg^{-1} to kJ kg^{-1}

$$Q_u = Q_v - 5.85 \; (W + 8.94 \times H) \times 4.186 \; (kJ \; kg^{-1}), \tag{2}$$

where: * Q_v = HHV in kcal kg^{-1}; Q_u = LHV in kcal kg^{-1}; W = moisture (%) in the sample (average amount of moisture in the sample of chaff was determined according to Beloborodko et al. [19] and Žandeckis et al. [20]); H = hydrogen in the sample; 4.186 = conversion factor from kcal kg^{-1} to kJ kg^{-1}.

2.3. Environmental Aspects

A life cycle assessment method was used for environmental load quantification. This method is defined by the international standards of ČSN EN ISO 14 040 [21] and ČSN EN ISO 14 044 [22]. The system boundaries are set within the chaff from the cradle to the farm gate and within the lignite from the cradle to the gate (from mining to raw material ready for use). The results of this study are related to the 19 impact categories (characterization model). SimaPro v9.1.0.11 software and ReCiPe Midpoint (H) V1.13/Europe Recipe H., an integrated method, were used for environmental load quantification. One GJ of energy (from potential energy profit) was used as the defined unit/functional units (FU). The technological processes of growing the hulled wheat species were set up based on primary data (field trials carried out on plots at the University of South Bohemia) and secondary data (data gained from the Ecoinvent v3.6 database [23] and commonly used cultivation practices of organic farming [2]). Mass allocation approach was used (grain/straw/chaff). Data geographically related to central Europe was used. The primary data was collected between 2018 and 2019. The data selected for modelling is based on the average of commonly applied organic farming technologies [2,24]. Agrotechnical operations from seedbed preparation, the number of seeds, the use of agrotechnological operations for plant protection, treatment and application of organic fertilizers to harvesting, transporting of the harvested grain and processing the grain were included into the model system.

The usage of mineral and organic nitrogenous fertilizers and lime application results in the release of so-called direct and indirect emissions of N_2O, CO_2, NH_3, NO_3^- and NO_x. The following were taken into account in the monitoring of field and agricultural emissions: liming, NH_3 and NOx volatilization, and nitrogen loss from leaching and surface outflow. The emission load was determined following the IPCC (Intergovernmental Panel on Climate Change) methodology called Tier 1 [25,26], and with Nemecek and Kägi [27] and the national greenhouse gas inventory report of the Czech Republic (the agricultural section) [28]. Emissions of phosphorus due to leaching and run-off were estimated following recommendations from Nemecek and Kägi [27].

The individual steps determined by the methodology are shown in Figure 1.

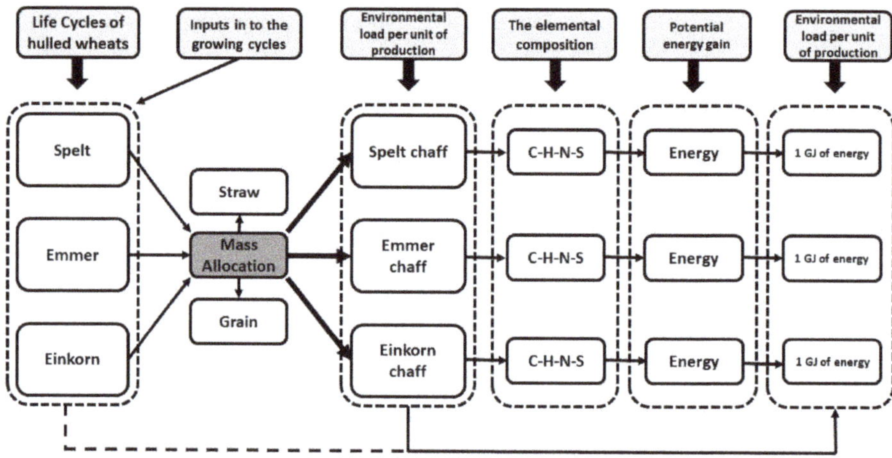

Figure 1. Workflow scheme.

3. Results and Discussion

3.1. Field Trial Results

The basic data source were field trials established according to the methodology plan. The data obtained from field trials are included in Table 2.

Table 2. Field trials results.

	Yield Range (t ha^{-1})	Grain Average Yield (t ha^{-1})	Straw Average Yield (t ha^{-1})	Chaff Rate (%)	Grain Net Yield (t ha^{-1})	Chaff Yield (t ha^{-1})
Spelt	2.80–3.27	2.96	4.75	33.23	1.98	0.99
Emmer	1.77–2.90	2.40	3.73	23.82	1.83	0.57
Einkorn	1.17–1.84	1.65	3.13	26.16	1.21	0.43

The results are based on one-year field trials under the organic farming system. The study aimed to obtain samples of waste material (chaff) and information about its quantity independence on yield level. The largest amount of chaff is produced during the peeling of Spelt wheat (average 1.45 t ha^{-1}), reps. the removal rate was 33.23%. This corresponds to the results reported in the study by Weiss and Glasner [13], which reported the removal rate of about 33%, and according to Wiwart et al. [9], of about 30%–35%. In the comparison of selected hulled wheat species, spelt wheat is also the most represented in the Czech Republic (3400 ha) [29]. This represents only 0.35% of the sowing areas of all winter cereals in the Czech Republic [29]. The lowest removal rate was recorded for Emmer wheat (23.82%), but the lowest chaff yield was for Einkorn (0.43 t ha^{-1} on average), given the lowest grain yield per hectare (1.65 t ha^{-1} on average).

3.2. Elemental Composition and Statistical Evaluation

For the study, elementary analysis of representative samples was carried out according to the methodology (Section 2.2). The results of elemental analysis are an essential source of information for the determination of HHV and LHV. The results of this analysis are included in Table 3.

Table 3. Elemental composition of chaff.

Sample	Sample Weight (mg)	N Ø%	N S.D.	C Ø%	C S.D.	H Ø%	H S.D.	S Ø%	S S.D.
Emmer	4.8545	0.6025	0.000707	41.5810	0.098995	6.1008	0.002121	0.0896	0.016971
Spelt	5.4315	1.1434	0.072832	42.3710	0.098995	6.4031	0.002121	0.0915	0.023335
Einkorn	5.1870	0.9230	0.028991	41.0150	0.007071	6.3185	0.002828	0.0697	0.054447

The average values of the homogenized samples; S.D. = standard deviation; Ø% = average percentage.

Based on the results of the elementary analysis, statistical evaluation was carried out. The ANOVA and Tukey HSD test were used. Individual samples within the percentage content of C, N, H, and S were shown to be statistically demonstrably different from each other at a significance level of $p = 0.05$ (Table 4).

Table 4. Statistical evaluation—variance analysis ANOVA.

C	$F_{2.3} = 141; p = 0.001075$
N	$F_{2.3} = 72.268; p = 0.002900$
H	$F_{2.3} = 38.5; p = 0.007272$
S	$F_{2.3} = 49.794; p = 0.005001$

p = level of significance ($p = 0.05$).

The following post-hoc Tukey HSD test (Table 5) showed that all crops differed within percentage content of C and N. However, within the percentage content of H, the samples of Einkorn and Spelt chaff did not differ from each other and within the percentage content of S, the samples of Einkorn and Emmer chaff did not differ from each other.

Table 5. Statistical evaluation—post-hoc Tukey HSD test.

Category	C	N	H	S
Emmer × Einkorn	$p = 0.004837$	$p = 0.002753$	$p = 0.007136$	no difference
Einkorn × Spelt	$p = 0.001136$	$p = 0.033646$	no difference	$p = 0.006019$
Emmer × Spelt	$p = 0.012420$	$p = 0.011883$	$p = 0.017988$	$p = 0.007922$

p = level of significance ($p = 0.05$).

The ash content of the sample was derived from the Sheng and Azevedo study [30] and the percentage content of oxygen was determined based on the difference. Moisture in the chaff sample was derived from the study by Beloborodko et al. [19] and Zandeckis et al. [20].

3.3. HHV, LHV and Potential Energy Profit Ha

Based on the data obtained from the elementary analysis, the HHV and LHV were determined. The resulting values are included in Table 6.

Table 6. Energy parameters of hulled wheat chaff.

	HHV (MJ kg^{-1})	LHV (MJ kg^{-1})	Energy Potential (GJ ha^{-1}) E	Energy Potential for CZ (TJ rok^{-1})
Spelt	17.74	15.90	13.21–26.41	50–90
Emmer	16.92	15.14	6.49–10.46	-
Einkorn	16.99	15.16	4.76–7.53	-

E = The energy potential is related to the yield range obtained in the field trials; HHV = higher heating value; LHV = Lower heating value; CZ = Czech Republic.

In the frame of the study, the HHV and LHV of Spelt chaff (17.74 MJ kg^{-1} respectively 15.90 MJ kg^{-1}), Emmer chaff (16.92 MJ kg^{-1} resp. 15.14 MJ kg^{-1}), and Einkorn chaff (16.99 MJ kg^{-1}, resp. 15.16 MJ kg^{-1}) were determined. The HHV and LHV of Spelt and Emmer chaff did not differ significantly from those reported in some earlier studies [9,10,13,15]. For example, according to the study by Wiwart et al. [9] the energy values (HHV and LHV) of Spelt and Emmer chaff were also determined. The chaff of Spelt and Emmer are generally defined by the higher HHV (18.75 MJ kg^{-1} resp. 18.31 MJ kg^{-1}), higher LHV (16.74 MJ kg^{-1} resp. 16.35 MJ kg^{-1}), significantly lower ash content (3.79% resp. 6.16%), and also lower content of the volatile matter (70.3% resp. 74.9%) in comparison with wheat and barley straw. Despite the relatively high Sulphur content (0.148%), the Emmer chaff has significant energy potential. Considering the LHV of 15.1 MJ kg^{-1} and the removal rate of 0.33, winter wheat and Spelt chaff has a theoretical potential of 191 PJa^{-1} in the EU [13]. For the Czech Republic only, the energy potential of Spelt is about 50–90 TJ year^{-1}.

3.4. Environmental Impact Assessment and Economy Aspects

For the study, an evaluation of the environmental load related to individual hulled wheat species and waste (chaff) resulting from their processing was compared with the traditional non-renewable fuel type lignite. The inputs to the growing cycle are part of the Field Trails methodology. The results are generated within the characterization model (Table 7).

Table 7. Environmental load per 1 GJ of potential energy profit.

Impact Category	Unit	Spelt	Emmer	Einkorn	Lignite [EI]
Climate change	kg CO$_2$ eq	1.35×10^1	1.75×10^1	2.25×10^1	1.60
Ozone depletion	kg CFC-11 eq	6.25×10^{-7}	7.95×10^{-7}	1.02×10^{-6}	1.23×10^{-7}
Terrestrial acidification	kg SO$_2$ eq	1.21×10^{-1}	1.58×10^{-1}	2.03×10^{-1}	5.22×10^{-3}
Freshwater eutrophication	kg P eq	1.69×10^{-3}	2.15×10^{-3}	2.76×10^{-3}	2.35×10^{-1}
Marine eutrophication	kg N eq	3.57×10^{-2}	4.72×10^{-2}	6.06×10^{-2}	4.92×10^{-2}
Human toxicity	kg 1,4-DB eq	1.30	1.61	2.07	1.32
Photochemical oxidant formation	kg NMVOC	5.04×10^{-2}	6.40×10^{-2}	8.22×10^{-2}	4.48×10^{-2}
Particulate matter formation	kg PM10 eq	2.92×10^{-2}	3.76×10^{-2}	4.82×10^{-2}	2.16×10^{-3}
Terrestrial ecotoxicity	kg 1,4-DB eq	1.17×10^{-3}	1.52×10^{-3}	1.95×10^{-3}	7.64×10^{-5}
Freshwater ecotoxicity	kg 1,4-DB eq	1.61×10^{-1}	2.00×10^{-1}	2.56×10^{-1}	3.28
Marine ecotoxicity	kg 1,4-DB eq	1.44×10^{-1}	1.79×10^{-1}	2.29×10^{-1}	3.13×10^{-1}
Ionising radiation	kBq U235 eq	2.85×10^{-1}	3.59×10^{-1}	4.60×10^{-1}	4.80×10^{-1}
Agricultural land occupation	m^{2}a	8.46×10^1	1.12×10^2	1.44×10^2	1.17×10^{-1}
Urban land occupation	m^{2}a	1.06×10^{-1}	1.30×10^{-1}	1.67×10^{-1}	1.19×10^{-1}
Natural land transformation	m^2	1.36×10^{-3}	1.73×10^{-3}	2.22×10^{-3}	1.64×10^{-3}
Water depletion	m^3	7.52×10^{-2}	9.85×10^{-2}	1.26×10^{-1}	6.82×10^{-2}
Metal depletion	kg Fe eq	8.77×10^{-1}	1.06	1.36	8.21×10^{-2}
Fossil depletion	kg oil eq	1.38	1.75	2.24	2.30×10^1

[EI] = source from Ecoinvent library (3.6 v); ReCiPe Midpoint (H) method, Characterization model, Results are expressed per GJ of potential energy profit; eq = equivalent; CFC-11 = Trichlorofluoromethane; 1,4-DB = 1,4-dichlorobenzene; NMVOC = Non-methane volatile organic compound; PM10 = Particulate matter <10 μm; U235 = Uranium235; m^{2}a = Potentially disappeared fraction (PDF)*m^2*year/m^2.

Due to the different material properties (such as combustion rate in the incineration unit), the resulting values are recalculated to potential energy gain (1GJ) compared to 1 kg of lignite with LHV of 9.9 MJ kg^{-1} (resp. 1GJ of potential energy gain). The most significant environmental savings compared to lignite can be found in the impact categories of freshwater eutrophication (k P eq), human toxicity (kg 1,4-DB eq), freshwater ecotoxicity (kg 1,4-DB eq), marine ecotoxicity (kg 1,4-DB eq) and fossil depletion (kg oil eq). On the other hand, the potential gain of 1 GJ from lignite was associated with lower environmental impacts for the other selected impact categories. The most important difference was determined among the impact categories of climate change (kg CO_2 eq), terrestrial acidification (kg SO_2 eq), photochemical oxidant formation (kg NMVOC), particulate matter formation (kg PM10 eq), terrestrial ecotoxicity (kg 1,4-DB eq), and metal depletion (kg Fe eq). In general terms, the highest environmental load was associated with the Einkorn chaff (within all impact categories). It is caused by the grain yield per hectare and lower LHV compared to Spelt chaff.

The evaluation results are influenced by the selected allocation approach—in this case, mass allocation (Chart 1). In terms of mass allocation, the environmental load associated with Spelt, Emmer, and Einkorn chaff was 12.8%, 9.3%, resp. 9.0% of the total environmental load connected with the growing cycle. On the other hand, the environmental load associated with the production of Spelt, Emmer, and Einkorn chaff would be, according to the economy allocation, 5.8%, 4.6% and 3.8%, respectively, of the total environmental load linked to the growing cycle (Table 8). However, price relations are highly volatile, and the economy has no direct impact on yield relations and therefore the economy allocation is not considered appropriate in this assessment. A comparison of allocation approaches and market price data is also included in Table 8.

Table 8. Allocation approach and market price.

	Product	Field Production (t ha^{-1})	Mass Allocation (%)	Market Price (Eur ton^{-1}) without VAT	Economy Allocation (%)
Spelt	Grain	1.98	25.6	400	78.5
	Straw	4.75	61.5	80	15.7
	Chaff	0.99	12.8	30	5.8
Emmer	Grain	1.83	29.9	560	83.5
	Straw	3.73	60.8	80	11.9
	Chaff	0.57	9.3	30	4.6
Einkorn	Grain	1.21	25.4	680	86.1
	Straw	3.13	65.6	80	10.1
	Chaff	0.43	9.0	30	3.8

VAT = value-added tax.

From the values given in Table 8, the price for 1GJ of potential energy gain and the amount of material needed to obtain the same amount of energy can be predicted (Table 9).

Table 9. Price relations.

	LHV (MJ kg^{-1})	Market Price (Eur ton^{-1}) without VAT	Price (Eur per GJ) (Potential)	Amount of Material to Obtain the Same Amount of Energy (kg)
Spelt chaff	15.90	30	1.88	1
Emmer chaff	15.14	30	1.98	1.05
Einkorn chaff	15.16	30	1.98	1.05
Lignite	9.9	140	14.14	1.61

VAT = value-added tax; HHV = Higher heating value; LHV = Lower heating value.

The results of the environmental impact assessment show that the use of waste material (chaff) arising after processing hulled wheat species for energy purposes does not necessarily mean lower environmental load. This is due to inputs into the growing cycles, yield level, chosen technological processes, and selected allocation approach. The advantages of the lignite are the relatively high yield per area unit, easier logistics and generally better fuel properties, and currently also the price and availability. However, it is a non-renewable energy source that is generally considered to be very problematic, especially concerning climate change, air quality impacts, landscape, water quality and other environmental categories. Biomass is generally considered to be a renewable source of energy [31], but from the LCA methodology point of view, this is not the case even when organic farming production is involved. The results of this study show that in some selected impact categories (e.g., climate change, terrestrial acidification, photochemical oxidant formation, particulate matter formation, terrestrial ecotoxicity, and metal depletion), 1 kg of chaff, as a potential fuel, represents a higher load on the environment than 1 kg of lignite, resp. potential energy gain (1 GJ) from the materials.

4. Conclusions

In a comparison of the monitored wheat species, the largest amount of the chaff is generated after processing of Spelt wheat (33.23% removal rate) with an average HHV value of 17.74 MJ kg^{-1} and LHV 15.9 MJ kg^{-1}. Compared to that, in the case of Einkorn and Emmer 26.16% resp. 23.82% of chaff with HHV 16.99 MJ kg^{-1} resp. 16.92 MJ kg^{-1} and LHV 15.16 MJ kg^{-1} resp. 15.14 MJ kg^{-1} can be expected. Based on the yields obtained in field trials, a potential energy gain of 26.41 GJ ha^{-1} for spelt wheat, 19.84 GJ ha^{-1} for Einkorn, and 18.03 GJ ha^{-1} for Emmer wheat can be predicted. Only Spelt wheat is grown in the Czech Republic at around 3400 ha per year, and the energy potential of chaff at 50–90 TJ year^{-1} concerning the yield can be estimated. This can be expressed by the lignite equivalent (with LHV of 9.9 MJ kg^{-1}) corresponding to a very rough estimate of 103.3–206.5 boxcars/wagons of lignite. Concerning the environmental aspects, hulled wheat chaff is an interesting alternative energy source, ideally in the region of cultivation and processing. Regarding the assessment of the environmental aspects, it is also necessary to choose an appropriate allocation approach. According to study results, when using the mass allocation principle, the share of the total environmental load associated with the production of chaff is 9.0%–12.8%, but when using the economy allocation principle, it is only 3.8%–5.8%. An appropriate allocation approach can improve the quality of data and their interpretation. The results also show that hulled wheat chaff can be a cheaper source of energy compared to lignite.

Author Contributions: Conceptualization, J.B.; methodology, J.B. and P.K.; software, J.B.; validation, J.B., P.K., D.V.B., R.I.T. and D.B.; formal analysis, J.B.; investigation, J.B.; resources, J.B.; data curation, J.B.; writing—original draft preparation, J.B.; writing—review and editing, J.B.; visualization, J.B.; supervision, J.B. All authors have read and agreed to the published version of the manuscript.

Funding: This research received no external funding

Conflicts of Interest: The authors declare no conflict of interest.

References

1. Grausgruber, H.; Janovská, D.; Káš, M.; Konvalina, P.; Moudrý, J.; Peterka, J.; Štěrba, Z. *Cultivation and Use of Marginal Cereals and Pseudocereals in Organic Farming*, 1st ed.; South Bohemia University, Faculty of Agriculture: České Budějovice, Czech Republic, 2012; ISBN 978-80-87510-24-7.
2. Konvalina, P. *Selected Crops Growing in Organic Farming*, 1st ed.; South Bohemia University, Faculty of Agriculture: České Budějovice, Czech Republic, 2014; p. 284, ISBN 978-80-87510-32-2.
3. Longin, C.F.H.; Ziegler, J.; Schweiggert, R.; Koehler, P.; Carle, R.; Würschum, T. Comparative study of hulled (einkorn, emmer, and spelt) and naked wheats (durum and bread wheat): Agronomic performance and quality traits. *Crop Sci.* **2016**, *56*, 302–311. [CrossRef]
4. Zaharieva, M.; Ayana, N.G.; Al Hakimi, A.; Misra, S.C.; Monneveux, P. Cultivated emmer wheat (*Triticum diccocum* Schrank), an old crop with a promising future: A review. *Genet. Resour. Crop Evol.* **2010**, *57*, 937–962. [CrossRef]
5. Moudry, J. *Alternativní Plodiny*, 1st ed.; Profi Press sro: Praha, Česká Republika, 2011; ISBN 978-80-86726-40-3.
6. Castagna, R.; Borghi, B.; Di Fonzo, N.; Heum, M.; Salamini, F. Yield and related traits of einkorn (*T. monococcum* spp. *monococcum*) in different environments. *Eur. J. Agron.* **1995**, *4*, 371–378. [CrossRef]
7. Moudrý, J.; Strašil, Z. *Alternative Crops*, 1st ed.; South Bohemia University: České Budějovice, Czech Republic, 1996; ISBN 80-7040-198-2.

8. D'Antuono, L.F.; Galletti, G.C.; Bocchini, P. Fiber quality of emmer (*Triticum dicoccum* Schubler) and einkorn wheat (*T. monococcum* L.) landraces as determined by analytical pyrolysis. *J. Sci. Food Agric.* **1998**, *78*, 213–219. [CrossRef]

9. Wiwart, M.; Bytner, M.; Graban, Ł.; Lajszner, W.; Suchowilska, E. Spelt (*Triticum spelta*) and Emmer (*T. dicoccon*) Chaff Used as a Renewable Source of Energy. *BioResources* **2017**, *12*, 3744–3750. [CrossRef]

10. Jovičić, N.; Matin, A.; Kalambura, S. The energy potential of spelt biomass. *Krmiva: Časopis o Hranidbi Životinja Proizvodnji i Tehnologiji Krme* **2015**, *57*, 23–28.

11. Bucur, R.D.; Barbuta, M.; Konvalina, P.; Serbanoiu, A.A.; Bernas, J. Studies for understanding effects of additions on the strength of cement concrete. In *IOP Conference Series: Materials Science and Engineering*; IOP Publishing: Bristol, UK, 2017; Volume 246, p. 012036.

12. Brlek, T.; Bodroža Solarov, M.; Vukmirovic, D.; Čolović, R.; Vuckovic, J.; Levic, J. Utilization of spelt wheat hull as a renewable energy source by pelleting. *Bulg. J. Agric. Sci.* **2012**, *18*, 752–758.

13. Weiß, B.D.; Glasner, C. Evaluation of the process steps of pretreatment, pellet production and combustion for an energetic utilization of wheat chaff. *Front. Environ. Sci.* **2018**, *6*, 36. [CrossRef]

14. Bentsen, N.S.; Felby, C.; Thorsen, B.J. Agricultural residue production and potentials for energy and materials services. *Prog. Energy Combust. Sci.* **2014**, *40*, 59–73. [CrossRef]

15. Kiš, D.; Jovičić, N.; Matin, A.; Kalambura, S.; Vila, S.; Guerac, S. Energy value of agricultural spelt residue (*Triticum spelta* L.)—Forgotten cultures. *Tech. Gaz.* **2017**, *24*, 369–373.

16. Battaglia, M.; Thomason, W.; Fike, J.H.; Evanylo, G.K.; von Cossel, M.; Babur, E.; Iqbal, Y.; Diatta, A.A. The broad impacts of corn stover and wheat straw removal for biofuel production on crop productivity, soil health and greenhouse gases emissions—A review. *GCB Bioenergy* **2020**, 1–13. [CrossRef]

17. Štindl, P.; Kolář, L.; Kužel, S. Higher heating value of biomass and its calculation from elementary composition. In *Agroregion 2006–Zvyšování Konkurenceschopnosti V Zemědělství (Půda–Základ Konkurenceschopnosti Zemědělství)*, 1st ed.; Jihočeská univerzita v Ceských Budějovicích: České Budějovice, Česká Republika, 2006; pp. 136–140.

18. Hubáček, J.; Kessler, F.; Ludmila, J.; Tejnický, B. *Chemie Uhlí*, 1st ed.; Státní nakladatelství technické literatury: Praha, Česká Republika, 1962; p. 472.

19. Beloborodko, A.; Klavina, K.; Romagnoli, F.; Kenga, K.; Rosa, M.; Blumberga, D. Study on availability of herbaceous resources for production of solid biomass fuels in Latvia. *Agron. Res.* **2013**, *11*, 283–294.

20. Žandeckis, A.; Romagnoli, F.; Beloborodko, A.; Kirsanovs, V.; Blumberga, D.; Menind, A.; Hovi, M. Briquettes from mixtures of herbaceous biomass and wood: Biofuel investigation and combustion tests. *Chem. Eng. Trans.* **2014**, *42*, 67–72.

21. ISO. *ISO 14040—Environmental Management–Life Cycle Assessment–Principles and Framework*; International Organization for Standardization: Geneva, Switzerland, 2006.

22. ISO. *ISO 14044—Environmental Management–Life Cycle Assessment–Requirements and Guidelines*; International Organization for Standardization: Geneva, Switzerland, 2006.

23. Wernet, G.; Bauer, C.; Steubing, B.; Reinhard, J.; Moreno-Ruiz, E.; Weidema, B. The ecoinvent database version 3 (part I): Overview and methodology. *Int. J. Life Cycle Assess.* **2016**, *21*, 1218–1230. [CrossRef]

24. Konvalina, P. *Cultivation Marginal Cereals and Pseudocereals in Organic Farming*, 1st ed.; Faculty of Agriculture, South Bohemia University: České Budějovice, Czech Republic, 2008; ISBN 978807394116.

25. De Klein, C.; Novoa, R.S.; Ogle, S.; Smith, K.A.; Rochette, P.; Wirth, T.C.; McConkey, B.G.; Mosier, A.; Rypdal, K.; Walsh, M.; et al. N_2O emissions from managed soils, and CO_2 emissions from lime and urea application. In *IPCC Guidelines for National Greenhouse Gas Inventories, Prepared by the National Greenhouse Gas Inventories Programme*; IPCC: Geneva, Switzerland, 2006; Volume 4, pp. 1–54.

26. IPCC. IPCC. IPCC guidelines for national greenhouse gas inventories. In *The National Greenhouse Gas Inventories Programme*; Eggleston, H.S., Buendia, L., Miwa, K., Ngara, T., Tanabe, K., Eds.; Institute for Global Environmental Strategies (IGES): Kanagawa, Japan, 2006.

27. Nemecek, T.; Kägi, T. Life Cycle Inventories of Swiss and European Agricultural Production Systems. In *Agroscope Reckenholz-Taenikon Research Station ART*, 1st ed.; Final Report Ecoinvent V2.0 No. 15a; Swiss Centre for Life Cycle Inventories: Zürich, Switzerland; Dübendorf, Switzerland, 2007; p. 360.

28. Exnerova, Z.; Beranova, J. Agriculture (CRF sector 3). In *National Greenhouse Gas Inventory Report of the Czech Republic (Reported Inventories 1990–2015)*, 1st ed.; Krtkova, E., Ed.; Czech Hydrometeorological Institute: Prague, Czech Republic, 2017; pp. 225–252.

29. CZSO (The Czech Statistical Office) Prague: Integrated operational program, European Union. Available online: https://www.czso.cz/csu/czso/home (accessed on 1 April 2020).
30. Sheng, C.; Azevedo, J.L.T. Estimating the higher heating value of biomass fuels from basic analysis data. *Biomass Bioenergy* **2005**, *28*, 499–507. [CrossRef]
31. Demirbaş, A. Combustion characteristics of different biomass fuels. *Prog. Energy Combust. Sci.* **2004**, *30*, 219–230. [CrossRef]

Publisher's Note: MDPI stays neutral with regard to jurisdictional claims in published maps and institutional affiliations.

Article

Growth Potential of Yellow Mealworm Reared on Industrial Residues

Anna Bordiean , Michał Krzyżaniak *, Mariusz J. Stolarski and Dumitru Peni

Department of Plant Breeding and Seed Production, Faculty of Environmental Management and Agriculture, University of Warmia and Mazury in Olsztyn, Plac Łódzki 3, 10-724 Olsztyn, Poland; anna.bordiean@uwm.edu.pl (A.B.); mariusz.stolarski@uwm.edu.pl (M.J.S.); dumitru.peni@uwm.edu.pl (D.P.)
* Correspondence: michal.krzyzaniak@uwm.edu.pl

Received: 9 November 2020; Accepted: 1 December 2020; Published: 3 December 2020

Abstract: Since the world's population will continue to grow in the next decades, the problem of providing people with food will deepen. One-third of the food production volume is wasted while nearly one in ten people in the world suffer from hunger. To reduce the negative impact of human activity on the environment and meet the needs of the population, alternative sources of protein are proposed. Yellow mealworm larvae can be used as a source of food and animal feed. Therefore, this study aimed to compare the growth performance, feed conversion ratio (FCR) and efficiency of ingested feed (ECI) by yellow mealworm larvae fed 13 different diets containing chicken feed (CF), rapeseed meal (RM), wheat bran (WB) and willowleaf sunflower (WS) residues after the process of supercritical CO_2 extraction. The mean dry individual bodyweight for all diets used in the experiment was 31.44 mg dry matter (d.m.) Mealworms fed diet mixes that contained WB demonstrated the highest dry individual larval weight (from 40.9 to 47.9 mg d.m.). A significantly lower dry individual larval weight was found for mealworms fed solely WS residues (3.9 mg d.m.). The FCR ranged from 1.57 to 2.08, for pure CF and pure WS diet, respectively. The ECI of yellow mealworm larvae varied significantly (mean value 20.1%) and depended on the diet. Moreover, the ECI of mealworm was significantly the lowest and amounted to 5.9% for the pure WS diet. The industrial residues investigated in this study can be successfully used for mealworm farming, excluding pure willowleaf sunflower residues.

Keywords: *Tenebrio molitor*; edible insects; larval development; feed conversion ratio; agricultural and industrial residues; lignocellulosic biomass; bioconversion

---

1. Introduction

Since the world's population is predicted to reach 10 billion in the next three decades [1], it poses a problem for humanity and a challenge [2] for researchers and scientists. Nowadays, one-third of food is wasted [3,4] and nearly one in ten people suffer from hunger and food insecurity [5]. Another problem related to food quality, i.e., obesity, results from the overconsumption of some products, especially animal products [6]. Thus, there is enormous pressure on the environment to meet the demand for food [7]. About 26% of all arable land are pastures designated for animal farming, while 33% of arable land is used for feed crops. Thus, out of the total arable land, about 70% is used for animal husbandry [4]. The consumption of meat and other proteins from livestock, especially cattle, is inefficient and involves excessive expenses and activities to the detriment of the environment [8]. It was estimated that if cattle farming was reduced at least 50%, it would offer abundant environmental benefits and reduce the area of land used for the cultivation of maize used for feed [9]. An increase in the demand for poultry, eggs [10], fish and seafood [11] is expected. Consequently, new feed sources will be sought to improve growth rates and energy efficiency by replacing soymeal and fishmeal [12].

Food loss occurs throughout the whole supply chain, causing various economic, environmental, and social issues [13]. Frequently, such losses are caused by the end consumer, especially living in developed countries [14,15]. Annually, 1.3 billion tonnes of food are wasted [16]. In Europe (EU27) the food lost is estimated at approximately 89 million tonnes, or 179 kg per capita, excluding the losses from agricultural production [17].

To reduce the negative impact of food production operations on the environment while meeting the demand of the population, new alternative sources of protein and foods are proposed [9]. Insects are known to feature high bioconversion ratios of residues from agriculture and food waste [18]. In this context, insects can be a source of protein that can contribute to more efficient protein conversion. The consumption of insects, also called entomophagy, has been known since ancient times [19]. There are currently over 2000 known insects in different development stages that are consumed around the world [20]. The relatively recent use of insects as animal feed, components for food and feed as well as protein on a larger scale has begun in Europe, especially following the recommendation of the European Food Safety Authority (EFSA) in 2015 [21]. In some western countries, stores/markets and restaurants already offer insects for human consumption [22], while the use of insects as feed for small pets, poultry and fishery and other livestock is becoming more popular [23–27]. According to recent studies, the yellow mealworm (*Tenebrio molitor* L.) is one of the most reared insects in Europe [28–30]. This species can convert many substrates originating from the agricultural, baking and brewing industries [31,32]. This insect has been studied broadly to confirm its nutritional values and resistance to harmful compounds (mycotoxins, pesticides, heavy metals, etc.) [28]. It is highly likely that in the near future yellow mealworm will be used on a large scale as food (e.g., animal protein), either whole or only certain compounds (chitin, fatty acids, amino acids, proteins) or other beneficial components. Recent Regulation 2283/2015 and Regulation 2017/893 of the European Commission authorise the production of animal protein from insects (including yellow mealworm) for food and feed purposed in the European Union [33,34].

This paper presents preliminary results of a study which aimed to compare the growth performance, feed conversion ratio, efficiency of ingested feed by mealworm larvae fed with different types of agricultural and industrial residues.

2. Materials and Methods

2.1. Experimental Insects

The mealworm *Tenebrio molitor* (Linnaeus, Coleoptera: Tenebrionidae) larvae were obtained from a commercial supplier (CRICKETSFARM, Motycz-Józefin, Poland) specializing in growing insects in Poland. The insects were maintained under laboratory conditions (relative humidity: 55–60%, air temperature: 28 °C, photoperiod: 12 h). The purchased mealworm larvae were reared to obtain a colony of matured individuals. The mealworms were fed ad libitum with chicken feed milled at 3 mm and supplied with fresh carrots three times a week. The insects were kept in plastic containers (35 × 23 × 13 cm) with 3 mm aeration holes in the sides, closed with a lid. Hygrometers were installed to enable the verification of the parameters in the boxes. Three weeks later, the first pupas appeared. A total of 6–7 days later after first pupation a new adult generation was obtained. The adults were placed in the "adult box" that was built with the mosquito mesh on the bottom that allowed the eggs to fall into a collection box. The pupas/beetles were collected until it was sufficient to obtain a high number of eggs in a short collecting period (3–4 days). Hatching was estimated to last 7–10 days after egg collection. Newly hatched larvae were fed ad libitum on the chicken feed, supplied with fresh carrot slices for four weeks. The chicken feed was composed of corn, wheat gluten feed, wheat, soybean meal (with GMO), calcium carbonate, vegetable oil, sodium chloride and rapeseed meal. This procedure allowed the larvae to grow to a size enabling their easy and safe collection for the main experiment (feed conversion).

2.2. Diet Preparation, Larval Growth and Measurements during the Experiment

Before the beginning of the experiment, some of the cheapest types of agricultural/industrial residues were selected. Diets were obtained from commercial companies from Poland (except for willowleaf sunflower pellets, which were obtained from another experiment conducted in the Department of Plant Breeding and Seed Production of the University Warmia and Mazury in Olsztyn, Poland). The experimental diet mixes were composed of (1) chicken feed used as a control feed (CF), (2) rapeseed meal (RM), (3) wheat bran (WB) and (4) pelleted willowleaf sunflower (WS) biomass originating from residues after supercritical CO_2 extraction of active compounds.

All feeds were comminuted to obtain 3 mm particles. The obtained meal was sieved through a 300 μm sieve to remove the smallest particles with sizes similar to the larva faeces. Subsequently, the nutritional value of all the feeds was determined, including dry matter, ash, nitrogen/protein content and crude fat (Table 1). Experimental feeds were mixed in different weight proportions with chicken feed: only residual feed (0%), 25%, 50%, and 75% CF (Table 2). In this way, 13 types of feed mixes were obtained. The diets were stored at −20 °C until the start of the experiment.

Table 1. Main components of chicken feed (control) and industrial residues used for different diets.

Substrate	Ash (% d.m.)	Fibre Content (% d.m.)	Crude Fat (% d.m.)	Protein (% d.m.)
Chicken feed (CF)	12.89	3.23	2.38	20.11
Rapeseed meal (RM)	7.39	11.12	1.78	35.52
Wheat bran (WB)	5.82	8.68	2.9	18.63
Willowleaf sunflower (WS)	9.49	22.64	0.76	10.56

Table 2. Composition of experimental diets for the feed conversion experiment.

No.	Diet	Residual Feed Proportion (%)	Chicken Feed Proportion (%)
1.	CF (control)	-	100
2.	RM 100	100	-
3.	RM 75/CF 25	75	25
4.	RM 50/CF 50	50	50
5.	RM 25/CF 75	25	75
6.	WB 100	100	-
7.	WB 75/CF 25	75	25
8.	WB 50/CF 50	50	50
9.	WB 25/CF 75	25	75
10.	WS 100	100	-
11.	WS 75/CF 25	75	25
12.	WS 50/CF 50	50	50
13.	WS 25/CF 75	25	75

CF (chicken feed), RM (rapeseed meal), WB (wheat bran), WS (willowleaf sunflower).

The experimental trial began after hatching and after 4 weeks of undisturbed growth of mealworm, which was fed ad libitum with chicken feed. Fifty small larvae were placed in a plastic container ("diet box") (22 × 13 × 5 cm) with aeration holes on the sides, covered by a lid. Each diet box was provided with 3.1 g of experimental feed (weekly) and 2 g of carrots (twice a week). Each diet was tested in three replications. To maintain the same humidity in boxes, carrots were changed twice a week, except in the first week when it was replaced only once. Weekly larval development parameters were monitored by counting the larval survival rate and mealworm weight. The feed and carrot amounts were measured and the non-consumed feed and carrot remains were dried to calculate the amount of consumed feed and carrots. When the first pupae were observed, all larvae were counted per container and left for fasting for 24 h before harvesting. The next day, the larvae were weighed and sacrificed by freezing at −20 °C.

2.3. Feed Conversion Efficiency Experiment

Determinations of live larvae weight gain (LWG) as well as individual weights (IW) (mg fresh matter (f.m.)) were carried out weekly. For this purpose, 20 randomly chosen larvae from each box were weighed and placed back in the container for further experiments.

To determine the feed consumption, the feed and carrots not consumed during each week were weighed and dried at 105 °C until a constant mass was achieved. Feed consumption in the dry and wet form (FC) was calculated by subtracting the remaining feed mass from the feed provided at the start of each experimental week. A similar procedure was used for carrots twice a week. From week 7, when the feed and carrot consumption increased in some diet boxes, the decision was made to increase the feed and carrot amount from 3.1 to 4.5 g for feed and from 2 to 3 g and week later to 4 g of carrot. Frass was removed by sieving through 300 μm openings (from week 4), while 400 and 500 μm openings were used for larvae fed with wheat bran diets due to larger faeces particles.

Feed conversion efficiency was calculated weekly based on fresh feed conversion ratio (FCR) and the FCR and efficiency of conversion of ingested feed (ECI) were recorded at the end of the experiment. The ECI was [35] calculated as:

$$ECI = (final\ weight/weight\ of\ ingested\ food) \times 100\ [\%] \tag{1}$$

and expressed for the dry form. The feed conversion ratio (FCR) was calculated by dividing the total mean individual consumption by the total mean individual weight gain [36] and expressed for the fresh form:

$$FCR = weight\ of\ ingested\ food/weight\ gained \tag{2}$$

2.4. Statistical Analysis

Individual fresh larval weight and larval weight gain were analysed with a one-way ANOVA for multiple comparisons and Wilk's multidimensional test for multiple comparisons at $\alpha = 0.05$. Data on final survival rate, final larval dry weight, FCR and ECI were analysed with a one-way ANOVA. Homogeneous groups for all features were determined using Tukey's (HSD) multiple test at $\alpha = 0.05$. All statistical analyses were performed with the Statistica 13 software package (TIBCO Software Inc., Palo Alto, CA, USA).

3. Results and Discussion

The multidimensional test for multiple comparisons (Table 3) presents *p*-values of the individual weights and weight gains depending on the week and the interaction of week and feed. It was found that both traits (dependent variables) significantly differed and these differences were determined by the factor and the interactions between factors. Moreover, the results of one-way ANOVA (Table 4) for final larval dry weight, FCR and ECI differed significantly. However, the type of feed had no significant effect on the survival rate of yellow mealworm larvae.

Table 3. Results of the multidimensional test (Wilk's) for multiple comparisons.

Source of Variation	Individual Weight			Weight Gain		
	df	F	*p*	df	F	*p*
Week	8	640.6	<0.001	7	721.3	<0.001
Week×Feed	98	3.3	<0.001	84	3.7	<0.001

Table 4. Results of a one-way ANOVA.

Source of Variation	Survival	Dry Weight	FCR	ECI
df			12	
F	0.80	41.9	39.7	53.0
p	0.65	<0.001	<0.001	<0.001

A nutritive control feed (CF) was chosen for the first weeks after the hatching of mealworms, which provided them with enough elements to grow until week 4. This feed was also safely used and verified before the experiment. In other experiments, however, wheat bran was used as the control feed [37,38]. In the present study, at the beginning of the experiment, the mean weight of mealworms was 0.88 mg f.m. (Figure 1). At the end of week 1 of the experiment, mealworm weight almost doubled for mealworms reared on wheat bran (2.05 mg f.m.). The lowest weight was found for larvae grown on willowleaf sunflower (WS 100) (1.27 mg f.m.). During the experimental period, mealworms doubled their mass weekly, except for those fed with WS 100. Overall, the mealworms fed diets mixes with wheat brans had the highest body weight during the rearing experiment compared with the mealworms fed other diets. Thus, at the last measurement (week 9) mealworms weight was 134.4., 125.9, 124.6 and 114.2 mg f.m. for WB 100, WB 75, WB 50, WB 25, respectively. A slightly lower fresh weight (114.7 mg per larvae) was found in another study for mealworms fed with wheat bran and cabbage leaves as a water source [39]. The second-highest final weight (75 mg f.m.) in the penultimate week of measurements was found for mealworms fed the RM 25/CF 75 diet, followed by the control feed (67.7 mg f.m.). However, during the last week of the experiment, the weight of mealworms fed control feed (111 mg f.m.) outperformed the weight of mealworms (99.2 mg f.m.) fed RM 25/CF 75 diet. This indicates that the growing mealworms rely on significant amounts of CF rather than on RM in the diet. An almost similar weight of mealworms was noted for the insects grown on diets with a high amount of CF in WS diets. The mean mealworm weight for all diets in the final week was 95.6 mg f.m. The lowest mealworm weight (14.8 mg f.m.) was found for the individuals grown on WS 100. The composition of feed (Table 1) is an important factor for the growth and development of insects. The WS diet had probably inadequate nutritional composition for mealworms, having only 10.56% d.m. protein, and the highest fibre content (22.64% d.m.) compared to other diets. Harsányi et al. [40] found that the mass of yellow mealworm, giant mealworm and house cricket was significantly higher when insects were reared on chicken feed than on low quality organic wastes. As reported by Li et al. [39] mealworms fed wheat bran for the initial 10 days, followed by fermented straw and old cabbage leaves, demonstrate higher fresh weights (64.4 mg per larvae) than for mealworms fed the WS diet in the present study.

Individual larval weight gain (LWG) (Figure 2) was another parameter determined in the present experiment. The highest LWG was noted for mealworms fed diets containing WB until week 8 of the experiment, especially for larvae fed WB 100. However, the LWG of mealworms fed CF diets increased from week 7 and from week 8 they demonstrated higher LWG than the insects fed WB diets. It can be observed that mealworms fed CF diets displayed higher LWG with a surprisingly rapid increase, i.e., from 14.93 mg f.m. in week 7 to 34.47 mg f.m. in week 8, while the insects fed WB diets demonstrated a more uniform increase in weight (18.20 mg f.m. in week 6, 28.55 and 35.73 mg f.m. in week 7 and 8, respectively). In the final week of the experiment, some irregularities in LWG were observed for mealworms fed some diets. This can be confirmed in the final week when LWG decreased to 32.63 and 28.37 mg f.m. for mealworms fed diets WB 100 and WB 75/CF 25, respectively. A similar decrease and difference between week 8 and 9 were observed for mealworms fed diets RM 25/CF 75 and WS 100 (from 36.07 mg f.m. to 24.27 mg f.m. and from 4.32 to 2.85 mg f.m., respectively). The other diets did not cause significant LWG decreases during the experimental period and the growth values were uniform. It may be assumed that mealworms fed diets RM 25/CF 75, WB 100, and WB 75/CF 25 grew sufficiently and their metabolism slowed down as they prepared for pupation [41]. In the penultimate week, the mealworms' weight gain was no longer significant since their metabolism slowed down and they prepared for the pupation stage. Urrejola et al. [41] presented the bodyweight of *Tenebrio molitor* insects during different developmental stages fed various diets. It was shown that during the final larval stages there was no significant weight gain. This also could be explained based on the observation of Weaver [42] that further developmental stages are coprophagous. As previously mentioned, the experiment was stopped when the first pupae were obtained (diet WB 100 week 9). Consequently, it was assumed that further pupations would occur for the rest of the mealworms fed other diets. High LWGs for the WS 25/CF 75 diet were observed from week 7. This may be explained by an excess of chicken feed in the diet which was more willingly consumed by larvae instead of WS.

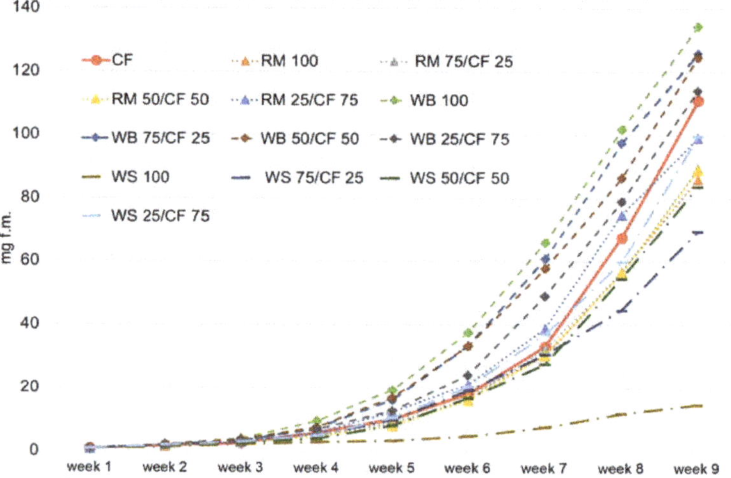

Figure 1. Individual larvae weight depending on the type of diet and week of rearing; CF (chicken feed), RM (rapeseed meal), WB (wheat bran) and WS (willowleaf sunflower).

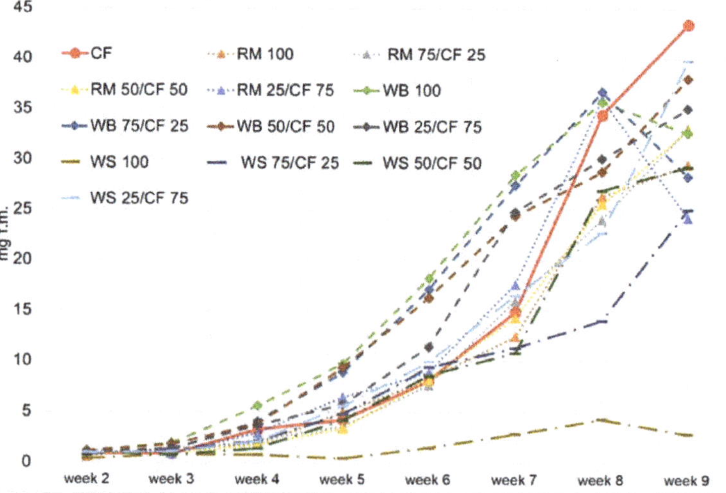

Figure 2. Individual larval weight gain depending on the type of diet and week of rearing; CF (chicken feed), RM (rapeseed meal), WB (wheat bran) and WS (willowleaf sunflower).

The mean dry individual larval weight for all diets used in the present experiment was 31.44 mg d.m. (Figure 3). The significantly highest dry weight was noted for mealworms reared on WB 100 and WB 75 diets and was 47.9 and 44.0 mg d.m., respectively. Other mixes of WB and CF produced very similar results. Mealworm reared on CF (control diet) were classified in the homogeneous group b together with diets RM 25/CF 75 and WS 25/CF 75 and the dry weights ranged from 32.3 to 34.5 mg d.m. It was noted that a high CF share in WS diets increased the dry larval weight. Mealworms fed RM diets (except RM 25/RM 75) and the WS 50/CF 50 diet demonstrated the same dry individual larval weight (25.9–28.1 mg d.m.). In the present study, mealworms grown on WB diets displayed the highest dry weight. This is contrary to the observations made by Li et al. [39], who found that the dry weight of larvae fed wheat bran was approximately 35.5 and 16.6 mg for larvae grown on plant waste diet (fermented straw and old cabbage leaves). For mealworms fed only a wheat bran diet, without any moisture supplementation, larvae dry weight was 35.9 mg [37]. In the current study, the mealworms were supplemented with carrot, which contributed to the higher individual larvae weight. In the present study, the significantly lowest dry individual larval weight was reported for diets based solely on WS, which was only 3.9 mg d.m. after week 9 of the experiment. It is possible that for other WS/CF mixes, mealworms fed more on CF and only accidentally on WS, based on self-selection of feed. The self-selection of feed was described by Morales-Ramos et al. [38], who noted that larvae selected the best components of mixed diets [41] to optimise their nutrition and growth. This habit was observed in other tenebrionids as well [43]. Willowleaf sunflower residues may be difficult to digest, possibly because of the high fibre content and low amount of non-fibrous carbohydrates (Table 1). Presumably, the residues could also be toxic for mealworms. The toxicity of some feed components for mealworm larvae has been mentioned in other research [31]. Another explanation is that when insects experience a lack of nutrients they tend to consume less and, as a result, they gain less weight [44]. The observations made during the present study revealed that the survival rate of mealworm fed WS diets was the same as for other diets (mean value for all feeds 96.5%), although the nutritional composition of the WS diet was probably insufficiently optimised for insect development and growth.

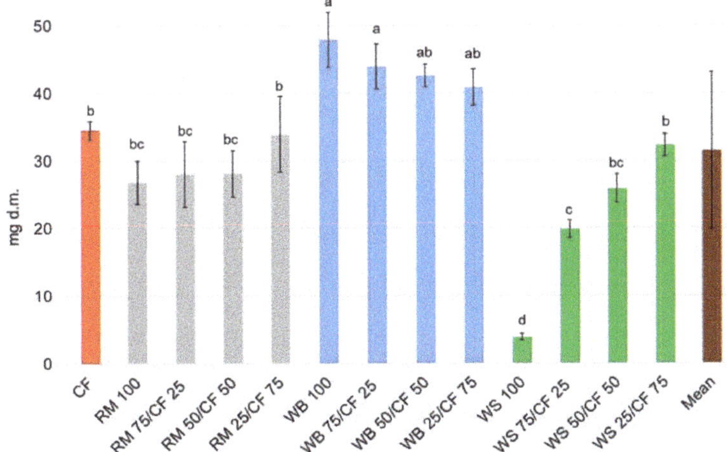

Figure 3. Final dry individual larva weight depending on the type of diet; CF (chicken feed), RM (rapeseed meal), WB (wheat bran) and WS (willowleaf sunflower); a, b, c … - letters mean that values are statistically different (Tukey's test at $p < 0.05$).

In the present study, the FCR for all diets except WS 100 (Figure 4) was statistically the same and ranged from 1.57 to 2.08 for CF and WB 100, respectively. It was observed that mealworm fed diets containing WB showed a slightly higher FCR than other diets. The significantly highest FCR

(4.42) was found for mealworm reared on the WS 100 diet. The experimental insects are cold-blooded mini-livestock and they do not need to maintain constant body temperature. In a study by Smil [45], the FCR for chicken was 2.5, for swine was 5.0 and it doubled for cattle (when FCR is expressed in kg of feed per kg of live weight). These values for livestock are much higher if expressed in kg of feed per kg of edible weight and double or even higher for cattle (4.5, 9,4 and 25, for chicken, swine, cattle, respectively) [45], while the mealworm larvae can be consumed entirely without any subsequent waste. Thus, insects have a better feed conversion ratio compared with other livestock [46]. The FCR can also differ between different insect species. Thus, the insects used for human consumption have a higher FCR than those used for fodder purposes, such as BSF (black soldier fly), which can be reared on wet substrates like fruits and vegetables and their derived products [32,47]. As mealworms are often used for food and feed purposes, their FCR can differ, depending on the quality of used feed for their growth. The FCR values for mealworms in the present experiment are similar to those obtained in a study by van Broekhoven et al. [31], in which the FCR of mealworms ranged from 2.62 to 6.05, depending on the feed mix quality. Thus, mealworms fed an HPLS (high protein low starch) diet produced the lowest FCR values. The highest FCR values were obtained for insects grown on an LPHS (low protein high starch) diet. It was concluded that an LPHS diet lacks nutrients and might contain some compounds that are toxic or hard to digest by mealworms. Oonincx et al. [32] found that mealworms had high FCRs (ranging from 3.8 to 19.1) when fed various quality feeds. The highest FCR values calculated for fresh weight (19.1 and 10.0) were reported for mealworms fed diets with low protein content (LPHF—low protein and high fat, and LPLF—low protein and low fat) supplemented with carrots. However, these results in the replication without carrot supplementation demonstrated almost half of these values, i.e., 5.3 and 6.1, respectively. The diets were prepared from food by-products, such as cookies and breadcrumbs, potato peels and beet molasses. For comparison, the same diets were fed to BSF (insects widely used for fodder purposes) and the FCR values obtained were considerably lower and varied from 2.3 to 2.6 [32].

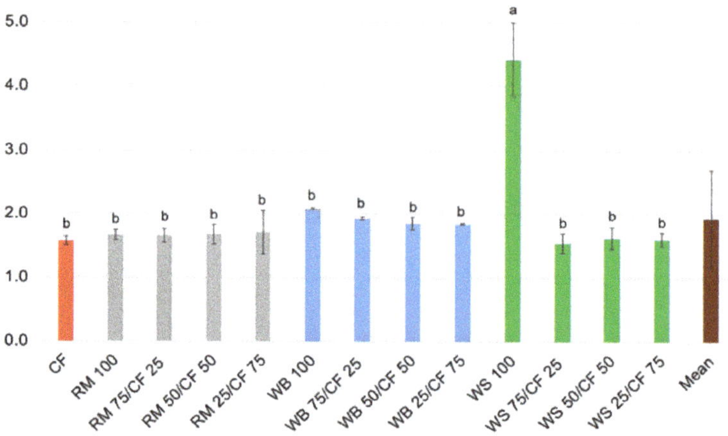

Figure 4. Feed-conversion ratio (FCR) depending on the type of feed, results on a fresh basis; CF (chicken feed), RM (rapeseed meal), WB (wheat bran) and WS (willowleaf sunflower); a, b, c … letters mean that the values are statistically different (Tukey's test at $p < 0.05$).

The results of the present study on ECI of yellow mealworm larvae are significantly different for various feed types (Figure 5). The mean value for mealworms fed in this experiment was 20.1%. The highest ECI (23.9%) was noted for mealworms fed the RM 25/CF 75 diet. A very similar value (23.3%) was reported for mealworms fed the WS 25/CF 75 diet. A slightly (but statistically significant) lower ECI value was calculated for mealworms fed the RM 25/CF 50 diet. As for FCR, the ECI value for mealworms fed the WS 100 diet was significantly lower and was only 5.9%.

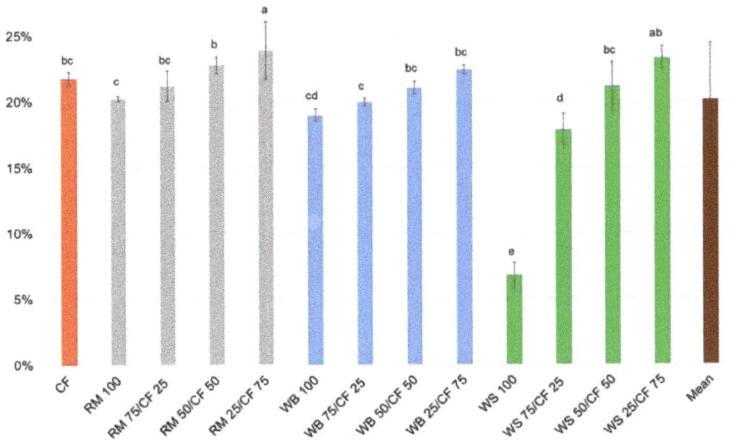

Figure 5. Efficiency of Conversion of Ingested feed (ECI) depending on the type of feed, results on a dry basis; CF (chicken feed), RM (rapeseed meal), WB (wheat bran) and WS (willowleaf sunflower); a, b, c ... - letters mean that the values are statistically different (Tukey's test at $p < 0.05$).

The ECI values reported for insects vary considerably from species to species, depending on the quality of feed. The ECI of the most commonly used insect reared on by-products (spent grains, beer yeasts, cookie crumbs, potato peels, beet molasses and bread) solely for animal feed purposes ranged from 3% to 9% (12% control) for house cricket, 16% to 30% (14% control) for Argentinean cockroach and from 17% to 24% for black soldier fly [32]. Collavo et al. [48] reported much higher ECI values for house crickets (59%) fed a human refuse diet. Van Broekhoven et al. [31] reported that ECI values for yellow mealworm fed bakery by-products ranged from 16.8% to 28.9%. In the same study, the authors investigated two other mealworm species fed the same types of diets and reared under the same conditions. Giant mealworm (*Zophobas morio*) demonstrated slightly higher ECI values (15.8–33.3%) than the yellow mealworm. The ECI of the lesser mealworm (*Alphitobius diaperinus*) was higher for high protein diets (23.0–34.4%) and much lower (6.4%) when the species was fed low protein diets [31].

4. Conclusions

Yellow mealworm is a species easily adaptable to different feed types. The present study tested the growth, development, FCR and ECI of yellow mealworm larvae fed 13 different diets. Generally, mealworms developed well during the experiment for all diets, except those containing a large share of willowleaf sunflower residues. The individual dry larvae weight was the highest for mealworms fed WB diets, followed by those fed CF and RM diets, which confirm that the best "common food" for mealworms should be cereal residues. The WS diet containing 75% of chicken feed also demonstrated high dry weight and it can probably be attributed to the self-selection of the chicken feed component rather than the WS residues. Larvae grown on a WS 100 diet displayed very low final dry weights (only 3.9 mg d.m.). The FCR for all diets, except for WS 100, was at a similar level and ranged from 1.55 to 2.08. This is a satisfactory result and can be compared with the FCR of poultry. The highest FCR was found for mealworms fed the WS 100 diet (4.42). Therefore, it can be concluded that willowleaf sunflower residues are not recommended as mealworm feed. It can be stated that the industrial residues investigated in this study can be successfully used for mealworm farming, excluding the willowleaf sunflower. These results will be used in subsequent research to estimate the best diets and time of insect harvesting using various industrial residues as feed. Moreover, to obtain more reliable results, especially for FCR and ECI, the above research should be continued.

Author Contributions: Conceptualization, M.K. and M.J.S.; Data curation, A.B.; Formal analysis, A.B., M.K. and D.P.; Funding acquisition, M.K.; Investigation, A.B. and D.P.; Methodology, M.K.; Supervision, M.K. and M.J.S.; Validation, M.J.S.; Visualization, A.B. and M.K.; Writing—original draft, A.B. and M.K.; Writing—review and editing, A.B., M.K., M.J.S. and D.P. All authors have read and agreed to the published version of the manuscript.

Funding: This paper is the result of a study carried out at the University of Warmia and Mazury in Olsztyn, Faculty of Environmental Management and Agriculture, Department of Plant Breeding and Seed Production, project number 20.610.008-300 and it was co-financed by the National (Polish) Centre for Research and Development (NCBiR), entitled "Environment, agriculture and forestry", project: BIOproducts from lignocellulosic biomass derived from MArginal land to fill the Gap In Current national bioeconomy, No. BIOSTRATEG3/344253/2/NCBR/2017. Anna Bordiean and Dumitru Peni received scholarships under the Interdisciplinary Doctoral Programme in Bioeconomy (POWR.03.02.00-00-I034/16-00) funded by the European Social Fund.

Conflicts of Interest: The authors declare no conflict of interest.

References

1. FAO. *The Future of Food and Agriculture–Trends and Challenges*; Food and Agriculture Organization: Rome, Italy, 2017.

2. Godfray, H.C.J.; Beddington, J.R.; Crute, I.R.; Haddad, L.; Lawrence, D.; Muir, J.F.; Pretty, J.; Robinson, S.; Thomas, S.M.; Toulmin, C. Food security: The challenge of feeding 9 billion people. *Science* **2010**, *327*, 812–818. [CrossRef] [PubMed]

3. FAO. *Global Food Losses and Food Waste—Extent, Causes and Prevention*; Food and Agriculture Organization: Rome, Italy, 2011.

4. Steinfeld, H.; Gerber, P.; Wassenaar, T.D.; Castel, V.; Rosales, M.; De Haan, C. *Livestock's Long Shadow: Environmental Issues and Options*; Food & Agriculture Organization: Rome, Italy, 2006.

5. World Health Organization. *The State of Food Security and Nutrition in the World 2018: Building Climate Resilience for Food Security and Nutrition*; Food & Agriculture Organization: Rome, Italy, 2018.

6. Alexander, P.; Brown, C.; Arneth, A.; Finnigan, J.; Moran, D.; Rounsevell, M.D.A. Losses, inefficiencies and waste in the global food system. *Agric. Syst.* **2017**, *153*, 190–200. [CrossRef] [PubMed]

7. Stehfest, E.; Bouwman, L.; Van Vuuren, D.P.; Den Elzen, M.G.J.; Eickhout, B.; Kabat, P. Climate benefits of changing diet. *Clim. Chang.* **2009**, *95*, 83–102. [CrossRef]

8. De Vries, M.; De Boer, I.J.M. Comparing environmental impacts for livestock products: A review of life cycle assessments. *Livest. Sci.* **2010**, *128*, 1–11. [CrossRef]

9. Wegier, A.; Alavez, V.; Pérez-López, J.; Calzada, L.; Cerritos, R. Beef or grasshopper hamburgers: The ecological implications of choosing one over the other. *Basic Appl. Ecol.* **2018**, *26*, 89–100. [CrossRef]

10. Józefiak, D.; Józefiak, A.; Kierończyk, B.; Rawski, M.; Świątkiewicz, S.; Długosz, J.; Engberg, R.M. 1. Insects–a natural nutrient source for poultry—A review. *Ann. Anim. Sci.* **2016**, *16*, 297–313. [CrossRef]

11. Wijkstrom, U. Future demand for, and supply of, fish and shellfish as food. In *Developments in Food Science*; Elsevier: Amsterdam, The Netherlands, 2004; Volume 42, pp. 9–23.

12. Chemello, G.; Renna, M.; Caimi, C.; Guerreiro, I.; Oliva-Teles, A.; Enes, P.; Biasato, I.; Schiavone, A.; Gai, F.; Gasco, L. Partially Defatted *Tenebrio molitor* Larva Meal in Diets for Grow-Out Rainbow Trout, *Oncorhynchus mykiss* (Walbaum): Effects on Growth Performance, Diet Digestibility and Metabolic Responses. *Animals* **2020**, *10*, 229. [CrossRef]

13. Beretta, C.; Stoessel, F.; Baier, U.; Hellweg, S. Quantifying food losses and the potential for reduction in Switzerland. *Waste Manag.* **2013**, *33*, 764–773. [CrossRef]

14. Secondi, L.; Principato, L.; Laureti, T. Household food waste behaviour in EU-27 countries: A multilevel analysis. *Food Policy* **2015**, *56*, 25–40. [CrossRef]

15. Parfitt, J.; Barthel, M.; Macnaughton, S. Food waste within food supply chains: Quantification and potential for change to 2050. *Philos. Trans. R. Soc. B Biol. Sci.* **2010**, *365*, 3065–3081. [CrossRef]

16. FAO. *Food Wastage Footprint: Impacts on Natural Resources: Summary Report*; Food and Agriculture Organization: Rome, Italy, 2013.

17. Monier, V.; Shailendra, M.; Escalon, V.; O'Connor, C.; Gibon, T.; Anderson, G.; Hortense, M.; Reisinger, H. *Preparatory Study on Food Waste across EU 27—European Commission (DG ENV) Directorate C-Industry*; Final Report; European Commission: Brussels, Belgium, 2010; ISBN 978-92-79-22138-5.

18. Sun-Waterhouse, D.; Waterhouse, G.I.N.; You, L.; Zhang, J.; Liu, Y.; Ma, L.; Gao, J.; Dong, Y. Transforming insect biomass into consumer wellness foods: A review. *Food Res. Int.* **2016**, *89*, 129–151. [CrossRef] [PubMed]

19. Van Huis, A. Did Early Humans Consume Insects? *J. Insects Food Feed.* **2017**, *3*, 161–163. [CrossRef]

20. Jongema, Y. *List of Edible Insects of the World, 2017*; Wageningen University: Wageningen, The Netherlands, 2018.

21. EFSA Scientific Committee. Scientific Opinion on a risk profile related to production and consumption of insects as food and feed. *EFSA J.* **2015**, *13*, 4257. [CrossRef]

22. Ramos-Elorduy, J. Anthropo-entomophagy: Cultures, evolution and sustainability. *Entomol. Res.* **2009**, *39*, 271–288. [CrossRef]

23. Makkar, H.P.S.; Tran, G.; Heuzé, V.; Ankers, P. State-of-the-art on use of insects as animal feed. *Anim. Feed Sci. Technol.* **2014**, *197*, 1–33. [CrossRef]

24. Tran, G.; Heuzé, V.; Makkar, H.P.S. Insects in fish diets. *Anim. Front.* **2015**, *5*, 37–44. [CrossRef]

25. Loponte, R.; Nizza, S.; Bovera, F.; De Riu, N.; Fliegerova, K.; Lombardi, P.; Vassalotti, G.; Mastellone, V.; Nizza, A.; Moniello, G. Growth performance, blood profiles and carcass traits of Barbary partridge (*Alectoris barbara*) fed two different insect larvae meals (*Tenebrio molitor* and *Hermetia illucens*). *Res. Vet. Sci.* **2017**, *115*, 183–188. [CrossRef]

26. Bovera, F.; Piccolo, G.; Gasco, L.; Marono, S.; Loponte, R.; Vassalotti, G.; Mastellone, V.; Lombardi, P.; Attia, Y.A.; Nizza, A. Yellow mealworm larvae (*Tenebrio molitor* L.) as a possible alternative to soybean meal in broiler diets. *Br. Poult. Sci.* **2015**, *56*, 569–575. [CrossRef]

27. Gasco, L.; Dabbou, S.; Trocino, A.; Xiccato, G.; Capucchio, M.T.; Biasato, I.; Dezzutto, D.; Birolo, M.; Meneguz, M.; Schiavone, A. Effect of dietary supplementation with insect fats on growth performance, digestive efficiency and health of rabbits. *J. Anim. Sci. Biotechnol.* **2019**, *10*, 4. [CrossRef]

28. Bordiean, A.; Krzyżaniak, M.; Stolarski, M.J.; Czachorowski, S.; Peni, D. Will Yellow Mealworm Become A Source of Safe Proteins for Europe? *Agriculture* **2020**, *10*, 233. [CrossRef]

29. Sogari, G.; Amato, M.; Biasato, I.; Chiesa, S.; Gasco, L. The potential role of insects as feed: A multi-perspective review. *Animals* **2019**, *9*, 119. [CrossRef] [PubMed]

30. Caparros Megido, R.; Sablon, L.; Geuens, M.; Brostaux, Y.; Alabi, T.; Blecker, C.; Drugmand, D.; Haubruge, É.; Francis, F. Edible Insects Acceptance by Belgian Consumers: Promising Attitude for Entomophagy Development. *J. Sens. Stud.* **2014**, *29*, 14–20. [CrossRef]

31. Van Broekhoven, S.; Oonincx, D.G.A.B.; Van Huis, A.; Van Loon, J.J.A. Growth performance and feed conversion efficiency of three edible mealworm species (Coleoptera: Tenebrionidae) on diets composed of organic by-products. *J. Insect Physiol.* **2015**, *73*, 1–10. [CrossRef] [PubMed]

32. Oonincx, D.G.A.B.; Van Broekhoven, S.; Van Huis, A.; Van Loon, J.J.A. Feed conversion, survival and development, and composition of four insect species on diets composed of food by-products. *PLoS ONE* **2015**, *10*. [CrossRef]

33. European Commission. Regulation (EU) 2015/2283 of the European Parliament and of the Council of 2015 on novel foods, amending Regulation (EU) No 1169/2011 of the European Parliament and of the Council of 25 November and repealing Regulation (EC) No 258/97 of the European Parliament and of the Council and Commission Regulation (EC) No 1852/2001. *Off. J. Eur. Union* **2015**, *L327*, 1–12.

34. European Commission. European Commission Regulation (EU) 2017/893 of 23 May 2017 amending Annexes I and IV of Regulation (EC) No 999/2001 of the European Parliament and of the Council and Annexes X, XIV and XV to Commission Regulation (EU) No 142/2011 as regards the provisions on processed animal protein. *Off. J. Eur. Union* **2017**, *L138*, 92–116.

35. Waldbauer, G.P. The consumption and utilization of food by insects. In *Advances in Insect Physiology*; Elsevier: Amsterdam, The Netherlands, 1968; Volume 5, pp. 229–288.

36. Miech, P.; Berggren, Å.; Lindberg, J.E.; Chhay, T.; Khieu, B.; Jansson, A. Growth and survival of reared Cambodian field crickets (*Teleogryllus testaceus*) fed weeds, agricultural and food industry by-products. *J. Insects Food Feed* **2016**, *2*, 285–292. [CrossRef]

37. Liu, C.; Masri, J.; Perez, V.; Maya, C.; Zhao, J. Growth Performance and Nutrient Composition of Mealworms (Tenebrio Molitor) Fed on Fresh Plant Materials-Supplemented Diets. *Foods* **2020**, *9*, 151. [CrossRef]

38. Morales-Ramos, J.A.; Rojas, M.G.; Shapiro-Ilan, D.I.; Tedders, W.L. Self-selection of two diet components by *Tenebrio* molitor (Coleoptera: Tenebrionidae) larvae and its impact on fitness. *Environ. Entomol.* **2011**, *40*, 1285–1294. [CrossRef]

39. Li, L.; Zhao, Z.; Liu, H. Feasibility of feeding yellow mealworm (*Tenebrio molitor* L.) in bioregenerative life support systems as a source of animal protein for humans. *Acta Astronaut.* **2013**, *92*, 103–109. [CrossRef]

40. Harsányi, E.; Juhász, C.; Kovács, E.; Huzsvai, L.; Pintér, R.; Fekete, G.; Varga, Z.I.; Aleksza, L.; Gyuricza, C. Evaluation of organic wastes as substrates for rearing *Zophobas morio*, *Tenebrio molitor*, and *Acheta domesticus* larvae as alternative feed supplements. *Insects* **2020**, *11*, 604. [CrossRef] [PubMed]

41. Urrejola, S.; Nespolo, R.; Lardies, M.A. Diet-induced developmental plasticity in life histories and energy metabolism in a beetle. *Rev. Chil. Hist. Nat.* **2011**, *84*, 523–533. [CrossRef]

42. Weaver, D.K.; McFarlane, J.E. The effect of larval density on growth and development of *Tenebrio molitor*. *J. Insect Physiol.* **1990**, *36*, 531–536. [CrossRef]

43. Waldbauer, G.P.; Bhattacharya, A.K. Self-selection of an optimum diet from a mixture of wheat fractions by the larvae of *Tribolium confusum*. *J. Insect Physiol.* **1973**, *19*, 407–418. [CrossRef]

44. House, H.L. Effects of low levels of the nutrient content of a food and of nutrient imbalance on the feeding and the nutrition of a phytophagous larva, *Celerio euphorbiae* (Linnaeus) (Lepidoptera: Sphingidae). *Can. Entomol.* **1965**, *97*, 62–68. [CrossRef]

45. Smil, V. Eating meat: Evolution, patterns, and consequences. *Popul. Dev. Rev.* **2002**, *28*, 599–639. [CrossRef]

46. Van Huis, A. Edible insects contributing to food security? *Agric. Food Secur.* **2015**, *4*. [CrossRef]

47. IPIFF (International Platform of Insects for Food and Feed). Guide on Good Hygiene Practices for European Union (EU) Producers of Insects as Food and Feed 2019. Available online: https://ipiff.org/good-hygiene-practices/ (accessed on 29 November 2020).

48. Collavo, A.; Glew, R.H.; Huang, Y.-S.; Chuang, L.-T.; Bosse, R.; Paoletti, M.G. House cricket small-scale farming. In *Ecological Implications of Minilivestock: Potential of Insects, Rodents, Frogs and Snails*; Science Publisher: Enfield, NH, USA, 2005; Volume 27, pp. 515–540.

Publisher's Note: MDPI stays neutral with regard to jurisdictional claims in published maps and institutional affiliations.

Communication

Adapting Syntropic Permaculture for Renaturation of a Former Quarry Area in the Temperate Zone

Moritz von Cossel [1,*], Heike Ludwig [2], Jedrzej Cichocki [1], Sofia Fesani [3], Ronja Guenther [1], Magnus Thormaehlen [1], Jule Angenendt [1], Isabell Braunstein [1], Marie-Luise Buck [1], Maria Kunle [1], Magnus Bihlmeier [1], David Cutura [1], AnnSophie Bernhard [1], Felicitas Ow-Wachendorf [1], Federico Erpenbach [1], Simone Melder [1], Meike Boob [1] and Bastian Winkler [1]

[1]   Biobased Resources in the Bioeconomy (340b), Institute of Crop Science, University of Hohenheim, Fruwirthstr. 23, 70599 Stuttgart, Germany

[2]   endstufe—Konzeption und Gestaltung, Reuchlinstraße 31, 70176 Stuttgart, Germany

[3]   Alma Mater Studiorum—Department of Agricultural and Food Sciences (DISTAL), University of Bologna, 40126 Bologna, Italy

*   Correspondence: moritz.cossel@uni-hohenheim.de; Tel.: +49-711-459-23557

Received: 9 November 2020; Accepted: 2 December 2020; Published: 4 December 2020

Abstract: In Southwest Germany, the renaturation of quarry areas close to settlements is usually based on the planting of native species of trees and shrubs, which are then neither cultivated nor used. This study investigates whether a species-rich agroforestry system based on Ernst Goetsch's syntropic agriculture approach would be suitable for both renaturation in the form of soil fertility improvement and diverse food crop production under temperate climate. The quarry syntropy project was launched in summer 2019. Two shallow stony sections of a spoil heap of the quarry in Ehningen, Southwest Germany were available for demonstration plots. An interdisciplinary project team was set up both to obtain the official permits from five governmental institutions and to begin the study. The demonstration plots were each divided into three broad strips, which differ in three vegetation types: trees, shrubs, and annual food crops. The tree and shrub areas are mainly used for biomass production for a continuous mulch supply on the entire cultivated area in order to rapidly increase soil fertility. The food crops and also partly the trees and shrubs were intended to provide organically produced food (vegetables, fruit, berries and herbs). Most of the trees (eleven species) were planted in November 2019. In March 2020, soil samples were taken (0–30 cm), and a solar-powered water storage system was installed. Currently, the shrub and annual food crop strips are under preparation (pre-renaturation phase). In this initial phase, the priority is fertility improvement of the topsoil through intensive mulching of the existing grassland stock dominated by top grasses and the legumes hybrid alfalfa (*Medicago × varia* Martyn) and common bird's-foot trefoil (*Lotus corniculatus* L.). The food crop strip should then start in 2021 after one year of mulching. Depending on the success of growth, the tree strips should then also gain in importance for mulch application in the following years. The strategy is to gradually build up food crop cultivation under organic low-input agricultural practices while simultaneously enhancing the biophysical growth conditions guided by natural succession.

Keywords: agroforestry; biodiversity; bioeconomy; biomass supply; circular economy; organic farming; perennial crops; quarry; syntropy; vegetation restoration

---

1. Introduction

Agroforestry is considered one of the most important cultivation methods, which not only serves the production of biomass but can also provide many other ecosystem services, such as carbon storage [1,2], crop yield increase [3–5], climate change adaptation [6], biodiversity conservation [7,8], habitat functions [9] and landscape improvement. While there is a lot of information on interspecific interactions within agroforestry systems in the temperate zone [10], the vegetation restoration of quarries by agroforestry approaches has not yet been investigated. In view of the increasing competition for land use [11], it might be important to deal with this in the future, as the areas that are released for renaturation by quarries every year must be enormous in view of the exorbitant production volumes, for example 10 billion tons of concrete per year [12]. So far, these areas are to be renaturalized following the local nature conservation guidelines, i.e., they are left to controlled succession through controlled planting of native tree and shrub species.

Recently, it has been shown that natural succession can be the most reasonable approach for maximum carbon sequestration within the first 20–50 years after establishment [13]. Hence, there might be more complex yet cost-effective renaturation strategies based on natural succession that allow for a much broader spectrum of ecosystem service provision and even food production [14]. This would be of great advantage for the agricultural sector (the production of biomass for food, bio-based products, and bioenergy) as well as for society and the pollinator populations, for example. With increasing land-use conflicts [11], the idea is to set up the establishment of multifunctional agricultural systems on this land. A role model for such an approach could be the syntropic agriculture of Goetsch and Colinas [14]. According to Goetsch and Colinas, "syntropic agriculture" is to be understood as an agricultural strategy according to which soil-physical cultivation parameters such as humus cover, soil fertility, water infiltration and erosion mitigation of agroforestry systems are improved through (i) the introduction of a high diversity of plant species and (ii) an intensive cutting and pruning frequency to foster both light penetration into the system and photosynthesis rate. Following Andrade et al. 2020, this triggers an interplay of natural succession and stratification, resulting in an increased number of vegetation layers consisting of various plant species and stages [15]. Due to this spatial and temporal stratification, syntropic agriculture could lead to an increased biomass productivity compared to standardized mono-cropping-based cultivation methods [14,15]. Syntropic agriculture aims at regaining agricultural productivity of marginal agricultural land in the long term, so that food crops can be cultivated on those areas in the near future [15]. Moreover, this high diversity of plant species and age structures means that syntropic agriculture is more of a transforming agroecology than a conforming agroecology. Therefore, it must be kept in mind that syntropic agriculture requires not only agricultural but also social (socio-economic) adaptation by bridging the gap between agriculture and ecosystems.

The findings by Goetsch and Colinas, on which the methodological approach of this study is rooted, are derived from field trials in the tropics. This means two major differences compared to cultivation strategies for the implementation of syntropic agroforestry systems in the temperate zone: (i) a much shorter growing period induced by a cooler and drier climate, and (ii) different (species-poorer) indigenous plant communities. Therefore, the aim of this study is to share, communicate and may subsequently discuss the concept of developing a syntropic agroforestry approach for restoring a stone quarry in temperate zones based on a preliminary field study conducted in Southwest Germany with the wider scientific community.

2. Materials and Methods

This section presents all the steps of field trial preparation, starting from location selection of the demonstration plots, the basic site characterization in terms of climatical conditions, pedological characteristics and floral biodiversity as well as the selection and installation of the basic infrastructure.

2.1. Location

The experiment is being conducted in the area of the quarry company in Ehningen, Southwest Germany (Figure 1a). Ehningen is located near the river Wuerm, marking the largest glacier extension from the previous ice age. The bedrock is upper shell limestone (Landratsamt Freiburg, 2019), which the quarry company is mining for gravel. An older part of the mine is currently renaturalized and integrated into the typical landscape, consisting of 47% agricultural fields and 34% forest, following the regulations of the federal state Baden-Württemberg. In this area, two demonstration plots (V1, 48°39′57.6″ N 8°55′53.3″ E, and V2, 48°39′51.8″ N 8°55′59.7″ E) (Figure 1b) are available for this experiment. In addition, another area located on a slope (V3, 48°39′52.3″ N 8°55′55.8″ E) (Figure 1b) was selected to account for topographical peculiarities of the given site, but this area was not large enough to establish another demonstration plot. The integration of this field test into the given renaturation plan required official authorization by five regional government departments.

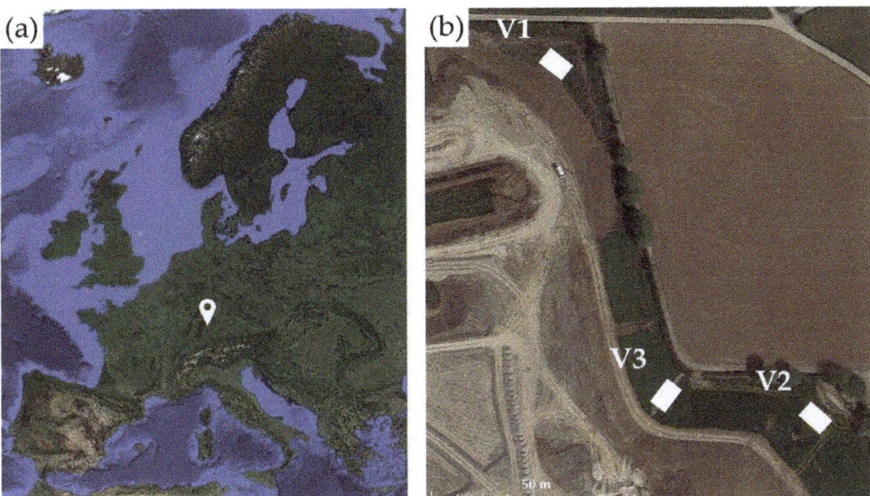

Figure 1. Location of the stone quarry (**a**), and arrangement of the demonstration plots V1–V3 at the stone quarry (**b**).

The region is characterized by temperate oceanic climate (Cfb) with 15 ice days per year, 80 frost days per year, an average annual temperature of 9.6 °C and an average annual precipitation of 672 mm. The latter tends to decline over the past five years due to climate change (Table 1). Initially, soil samples (0–30 cm) were taken in spring 2020 to describe the soil conditions at the beginning of the trial. The samples were frozen until further preparation for laboratory analyses. In addition to the pH value and texture, the contents of macronutrients (C, N, P, K, Mg and Fe) and trace elements (Mo, Co, Cu, Mn, Zn, Cl, Na, B, Ni, Si, Cl, Cd, Pb, As) of the soil samples are to be determined. The content (weight %) of stones in the topsoil was separately determined twice per area. The depth of the crumb was determined mechanically. In terms of soil type, both areas consist of several meters of gravel, sand, and overburden from the construction industry. Over time, a patchy grassland sward has developed on both areas. The botanical composition of this grassland sward was determined in July 2020 (18 July 2020). First, all vascular plant species were determined by species level (except for the genus *Taraxacum*) within a randomly selected sample area of 2 m² × 2 m². Then, the cover of each vascular plant species (%) was estimated visually. This estimation depends on the stage of phenology.

Table 1. Overview of climatical conditions in the region of the stone quarry over the past eight years (data taken from www.wetter-bw.de).

Year	Average Annual Temperature (°C)	Annual Precipitation (mm)	Climatical Water Balance (mm)	Annual Global Radiation (kWh m^{-2})	Frost Days (d)	Ice Days (d)
2012	9.3	726.7	201.4	928.7	76	20
2013	8.8	922.9	358.2	817.3	93	32
2014	10.4	763.3	173.5	861.3	39	2
2015	10.1	544.9	39.2	747.3	81	7
2016	9.1	646.5	−11.2	1078.3	88	8
2017	9.2	653.9	−51.1	1147.6	90	20
2018	10.2	525.9	−264.3	1234.8	81	19
2019	9.5	591.0	−131.4	1216.1	90	12
Average	9.6	671.9	39.3	1003.9	80	15

The color scaling within the parameters depends on the value.

2.2. Treatments, Field Trial Design and Experimental Stages

Two management methods based on the same establishment procedure were selected as treatments. This means that a mixed culture stand of tree and shrub species was first established on the trial plots, which was then divided into two sub-plots from the third trial year onwards, where two management treatments will be compared: (i) intensive organic farming, and (ii) syntropic agroforestry. The trial is thus divided into two experimental stages:

1. Establishment;
2. Main trial.

The actual start and methodology of long-term experiment evaluation within stage 2 depend on the success of the establishment process. This will be dependent on whether the tree and shrub rows are adequate to provide the vegetable strips with enough mulch material to enable cultivation in accordance with low-input agricultural practices (e.g., no irrigation, no technical weeding measures except mulching etc.), especially in syntropic agroforestry treatment.

Since the size of the three available experimental plots would not have allowed a meaningful integration of repetitions, only one treatment was planned for each plot. This field study therefore serves as a demo experiment. Furthermore, differences in the topographical conditions require a different management method at plot V3 compared with plots V1 and V2, such as those described by Fernandes and Gontijo (2020) [16]. Thus, a different management method was developed for plot V3, the first step of which was the planting of willow cuttings. If the willows grow well, terrace structures are to be designed and built in V3 within the next two years, which are intended to correspond to a syntropic agroforestry treatment. The exact procedure for V3 has not yet been determined.

2.3. Selection of Tree and Shrub Species

The tree and shrub species were compiled on the basis of literature data and the tree species available from the quarry's obligations under nature conservation law (in addition to the trial areas, there were about three hectares of land to be renaturalized) (Table 2). Since there was basically a free choice within these plant species, intensive group work was initially used to determine the objectives to be pursued by the perennial plant society. A multifunctional approach similar to syntropic agriculture [14,15] was chosen, according to which mulch material as well as food and biodiversity support (e.g. pollinators, biocontrol agents) are guaranteed. In accordance with this multifunctional approach, a closer selection of tree and shrub species was then made.

Table 2. Overview of the tree species available for planting in autumn 2019, which have been ordered by the company Baresel (Ehningen, Germany) within the framework of the renaturation. The references used for selecting suitable tree species for this project are also listed.

Trivial Name	Botanical Name	Height of the Young Plant (cm)	References
Sessile oak	*Quercus petraea* (MATTUSCHKA) LIEBL.	>80	[17,18]
European hornbeam [a]	*Capinus betulus* L.	80–120	[19,20]
Wild cherry [a]	*Prunus avium* L.	120–150	[21–23]
Sweet chestnut [a]	*Castanea sativa* Mill.	50–80	[24]
Common whitebeam [a]	*Sorbus aria* (L.) CRANTZ	50–80	[25–28]
Wild Service Tree [a]	*Sorbus torminalis* (L.) CRANTZ	30–50	[29–32]
Wild apple [a]	*Malus sylvestris* MILL. (1768)	80–120	[33–36]
European pear [a]	*Pyrus communis* L.	80–120	[37]
Field maple [a]	*Acer campestre* L.	30–50	[38]
Common dogwood [a]	*Cornus sanguinea* L.	40–70	[39]
Midland hawthorn	*Crataegus laevigata* (POIR) DC.	50–80	[40]
Wayfarer [a]	*Viburnum lantana* L.	40–70	[41]
Red honeysuckle	*Lonicera xylosteum* L.	50–80	[42]
Wild privet	*Ligustrum vulgare* L.	40–70	[43]
Dog Rose	*Rosa canina* L.	50–80	[44]
Buckthorn	*Rhamnus catharticus* L.	50–80	[45]
Blackthorn [a]	*Prunus spinosa* L.	50–80	[46]
Hazel [a]	*Coryllus avelana* L.	50–80	[47]

[a] selected for the tree rows within the demonstration plots V1 and V2.

2.4. Planting Plan and Maintenance of the Trees

The planting plan was developed based on the characteristics of native tree and shrub species. The individual tree and overall arrangement were designed in such a way that the adjacent tree species complement or benefit each other in their above-ground (e.g., crown shape) and underground (e.g., rooting depth, tolerance to temporal water-logging) growth as well as in time. For example, fast growing pioneer tress with high cutting tolerance provide mulch and shelter for emerging fruit trees. In addition, the tree stands were arranged with respect to the cardinal direction with the tree height increasing from southeast to northwest (Table 3).

Table 3. Schematic overview of the arrangement of tree species within areas V1 and V2.

Row 1	Row 2
Walnut (*Juglans regia* L.)	Quince (*Cydonia oblonga* MILL.)
Common dogwood	Willow (*Salix* L.)
European hornbeam	Hazel
Wild Service Tree	Sea-buckthorn (*Hippophae rhamnoides* L.)
Sweet chestnut	Wild apple
European hornbeam	Blackthorn
Field maple	European pear
European pear	Wayfarer
Wild Service Tree	Wild apple
Wild cherry	Common whitebeam
Wayfarer	Common dogwood

The first plantation of native plant tree species was conducted at the experimental sites V1 and V2 on 7 December 2019. For each tree, a hole with a diameter of about 40 cm and a depth of 40 cm was dug with a spade or pickaxe (Figure 2a). The planting distances between trees account for 3 m. Each planting hole was then watered with about 2 L of rainwater (Figure 2b) before the roots of the young plants were placed in the holes and covered with loose soil (Figure 2c).

Figure 2. Impressions from the planting event in December 2019. The planting holes were dug (**a**) and irrigated (**b**) before the seedlings (which were delivered as usual without soil from the nursery) were positioned in the holes and covered with loosened soil (**c**).

The trees were checked at regular intervals and kept free of grasses approximately 50 cm around the stems. The water supply was ensured by the installation of a solar-powered water storage system for irrigating the plants during establishment (Figure 3). Rainwater draining off near the surface is temporarily stored in two 1000-L intermediate bulk containers.

Figure 3. Installation of a solar-powered water pump at plot V2 at the quarry in Ehningen, Southwest Germany during lock-down in April 2020.

In addition, the tree pits were covered with grass mulch regularly (as often as possible) to lower evaporation losses. The management of the trees and shrubs only consist of the formative pruning in the first two to three years.

2.5. Preparation of the Sub-Plots for the Shrubs and Food Crops

The sub-plots for the shrubs and food crops consisted of a grassland sward at the beginning of stage 1. From May 2020, the grassland was mown at regular intervals and the mulch evenly distributed within the sub-plots in order to increase soil cover in the long term. In total, four cuts were performed (cutting height 10 cm) on 7 May, 24 June, 10 August, and 14 October 2020. The change in the species' composition of the grassland sward and its impact on the further course of stage 1 will be assessed in spring 2021.

The demo plots will remain unfertilized by external chemical fertilizers. The nutrient supply through the mulch material should be sufficient to ensure the low-input character of the agroforestry approach. The nutrient status of the soil will be checked annually, beginning with soil sampling in spring 2021. The soil sample analyses should serve to verify the potential use of the site for food crop production. If high levels of heavy metals are present, the use of the plants for human nutrition would be prohibited.

3. Preliminary Results and Discussion

At the time of vegetation analysis of the grassland community, ribwort plantain (*Plantago lanceolata* L.) and the common legume bird's-foot trefoil (*Lotus corniculatus* L.), hybrid alfalfa (*Medicago* × *varia*Martyn) and red clover (*Trifolium pratense* L.) were flowering. Ribwort plantain, black medick (*Medicago lupulina* L.) and dandelions (*Taraxacum* sect. *Ruderalia* Kirschner, H.Øllg. & Štěpánek) were at the stage of seed ripening. The cover of vascular plants accounted for 129% and 133% of the plots V1 and V2, respectively (Table 4).

The vegetation of plot V1 was dominated by the common legume bird's-foot trefoil (Figure 4a), whereas the vegetation of plot V2 was dominated by perennial competitor species such as hybrid alfalfa and couch grass (*Elymus repens* (L.) Gould) (Figure 4b; Table 4). Hybrid alfalfa was sown as part of the renaturation plan on the entire area. The species density of both plots was 17 species 4 m^{-2}. Through a targeted search, each five additional species (outside the sample area) were found in plot V1 (*Medicago lupulina* L., *Galium mollugo* L., *Lotus corniculatus* L., *Lolium multiflorum* Lam., *Arctium tomentosum* Mill.) and in plot V2 (*Potentilla anserina* (L.) Rydb., *Myosotis* spec., *Prunella vulgaris* L., *Linaria vulgaris* Mill., *Achillea millefolium* L.).

Table 4. Cover of plant species (%) within the sample areas (4 m^2) of the grassland communities of plot V1 and V2 on 18 July 2020.

	Plot V1	Plot V2
Grasses	15	15
Alopecurus myosuroides Huds.	0	0.2
Arrhenatherum elatius (L.) P.Beauv. ex J.Presl & C.Presl	8	2
Dactylis glomerata L.	2	2
Elymus repens (L.) Gould	0	5
Festuca pratensis Huds.	0	1
Lolium perenne L.	2	5
Poa annua L.	0.5	0
Poa angustifolia L.	2	0
Poa trivialis L.	0.5	0
Legumes	106	63
Lotus corniculatus L.	100	0
Medicago lupulina L.	0.2	0
Medicago × *varia* Martyn	2	60
Trifolium pratense L.	1	1
Trifolium repens L.	3	2
Vicia sepium L.	0	0.2
Forbs	8	55
Achillea millefolium L.	0	0.2
Daucus carota L.	0.2	0
Galium mollugo L.	0.2	0
Hieracium L.	0	0.2
Hypochaeris radicata L.	0.2	0
Plantago lanceolata L.	0	25
Potentilla anserina (L.) Rydb.	0	0.2
Potentilla reptans	0	25
Rosa spec.	0.2	0
Rumex obtusifolius	1	0.2
Taraxacum Sect. *Ruderalia* Kirschner, H.Øllg. & Štěpánek	6	4
Total number of species (4 m^{-2})	17	17

Figure 4. Impressions of the grassland communities on demonstration plots V1 (**a**) and V2 (**b**), 18 July 2020.

The results of the soil analyses of V1 revealed a soil bulk density of 2.1 g cm^{-3} and a high topsoil volume share of stones of 56.7% on site V1 (determined) (Table 5). The topsoil volume of stones on V2 accounted for approximately 40–50% on site V2 (estimated). According to the Joint Research Centre [48], topsoil volumes of coarse material, including rock outcrop or boulders above 15%, are defined as unsuitable for generic agricultural activity. This means that the site can be defined as marginal agricultural land. [49]

Table 5. Overview of topsoil (0–30 cm depth) analyses on demonstration plots V1 and V2. Samples were taken on 18 March 2020.

	Clay	Sand	Silt	Total Carbon	Soil Type	Stones	Earth	Organic Residues
	Weight-% of Dry Matter [a]			Weight-% of Total Dry Matter		Volume-% of Fresh Matter		
V1	20.5	32.2	47.2	9.6	Loam	56.7	42.9	0.4
V2	31.4	14.9	53.7	3.1	Silty clay loam	n.d.	n.d.	n.d.

[a] after destruction of the organic substance.

Eight of the twelve tree species (67%) planted, such as blackthorn (Figures 5 and 6), field maple and wild cherry, survived the plantation and seem to establish well within both demonstration plots (Table 4). On both demonstration plots, the average number of surviving trees (not tree species) per row was seven trees.

However, roe deer have selectively damaged the bark of some of the trees, resulting in the death of these tree species (Table 6). Due to the unexpected presence of roe deer on the heavily fenced quarry area, a countermeasure was applied to help minimize the damage caused by wild animals (roe deer). For this purpose, an electric fence was installed and put into operation around both test areas in early July 2020 (Figures 7 and 8).

Drought conditions in spring 2020 may be responsible for the death of some of the trees. The planting conditions were also sub-optimal in December 2019, given the fact that the weather was very cold and windy, so that many of the fine roots of the plantlets may have been destroyed shortly before planting. Another reason for the death of some of the trees may be the absence of soil and fungal symbionts from the pot in the nursery.

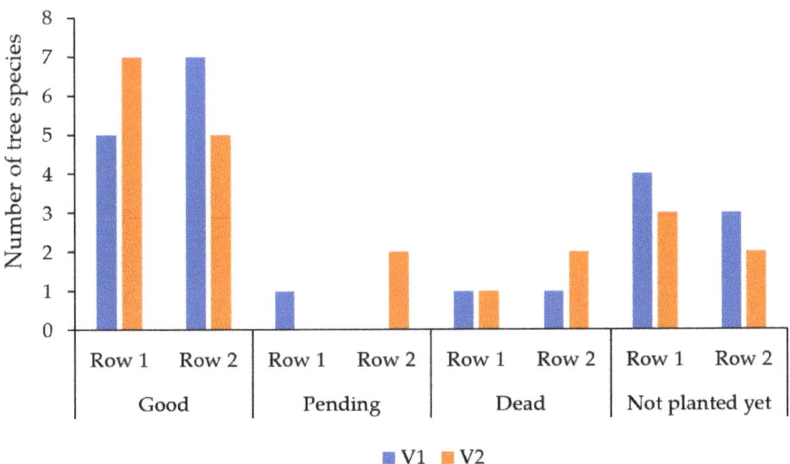

Figure 5. General status of the trees (planted in December 2019) within the two rows of the demonstration plots V1 and V2 on 19 May 2020.

Figure 6. Well-established blackthorn plantlet within demonstration plot V1 on 19 May 2020.

Table 6. Detailed overview of preliminary status of the trees per plot (V1, V2) and row (good = in healthy condition, pending = condition currently unclear, dead = no signs of vitality to be seen, not planted yet = in November 2019, there were no or not enough tree seedlings available).

Tree Species	Plot V1	Plot V2
Row 1		
Walnut	Not planted yet	Not planted yet
Common dogwood	Good	Good
European hornbeam	Dead	Good
Wild Service Tree	Good	Not planted yet
Sweet chestnut	Not planted yet	Not planted yet
European hornbeam	Not planted yet	Good
Field maple	Good	Good
European pear	Not planted yet	Good
Wild Service Tree	Pending	Dead
Wild cherry	Good	Good
Wayfarer	Good	Good
Row 2		
Quince	Not planted yet	Not planted yet
Willow	Good	Good
Hazel	Good	Good
Sea-buckthorn	Not planted yet	Not planted yet
Wild apple	Good	Good
Blackthorn	Good	Good
European pear	Not planted yet	Pending
Wayfarer	Good	Good
Wild apple	Good	Dead
Common whitebeam	Dead	Dead
Common dogwood	Good	Pending

It remains unclear whether the trees would have better established through sowing instead of planting. Here, the trees were planted instead of sown because most of the desired tree species were delivered anyway due to the renaturation of the surrounding area. Furthermore, the plantation of trees usually enables a much faster establishment of the wooden species, which under natural succession conditions could take 25 years and more, depending on the surrounding vegetation [50].

The plantation of willow on all plots, including V3 (Figure 9a), in March 2020 was successful (Figure 9b,c; Table 4) until the drought conditions became too severe in August 2020. Then, most of the willow plantlets showed senescent leaf tissue.

Figure 7. Preparations for the installation of the electric fence at site V2 on 19 May 2020.

Figure 8. View of the demonstration plot V2 with electric fence. Sheep were grazing outside of the plot a few days before this photograph was taken (18 July 2020).

Figure 9. Location of plot V3 (**a**) and impressions of willow plantlets on V3 (**b,c**). The stone gutter marks the middle of the area. Willows were planted crescent-shaped, indicated by the green vegetation left over from the last mowing process. The photographs were taken on 18 July 2020.

4. Conclusions

The present study describes the initial stage of the development of a syntropic permaculture concept in the temperate zone of Central Europe and the first steps of its establishment with native woody species. The site selected for this demonstration experiment exhibits marginal agricultural conditions in the sense of shallow stony soil. Following the principle of syntropy, these marginal growth conditions are to be defied by the diversity of the woody stand, which in the ideal case will lead to a win-win situation: the land will be renaturalized and at the same time made reusable for the production of food, fodder and renewable raw materials for material and energy use—depending on which plant species prevails the most.

Of course, the present study has limitations with regard to both the expressiveness of the data and the species diversity within the woody stand. This means that in the case of sowing, it would have been possible to test a much higher number of species on this site for individual and community adaptability. This is also the case for the tropical zone. Nevertheless, important insights can be gained regarding how the selected species will cope with the marginal growth conditions in close association. This is also related to the fact that under temperate conditions the woody species grow much slower and a head start through the establishment by planting helps to gain first insights much earlier compared with an establishment by sowing.

However, the preliminary results presented here already show that some tree species can survive despite the severe rooting and drought conditions. First empirical results on the holistic performance of the management methods investigated here may not be possible for five to ten years. Only then will the woody species begin to provide more and more raw materials (mulch material) and microclimatic effects for the adjacent shrub and vegetable strips. These mutually beneficial interactions between the shrub, bush and vegetable strips would then be one of the main features of the syntropic permaculture in the temperate zone.

Author Contributions: Conceptualization, M.V.C., H.L., J.C., and B.W.; methodology, M.V.C., H.L., J.C., S.F., R.G., M.T., J.A., I.B., M.-L.B., M.K., M.B. (Magnus Bihlmeier), D.C., A.B., F.O.-W., F.E., S.M., M.B. (Meike Boob), and B.W.; validation, M.V.C. and B.W.; formal analysis, M.V.C., J.C., M.B. (Meike Boob) and B.W.; investigation, M.V.C., H.L., J.C., S.F., R.G., M.T., J.A., I.B., M.-L.B., M.K., M.B. (Magnus Bihlmeier), D.C., A.B., F.O.-W., F.E., S.M., M.B. (Meike Boob), and B.W.; resources, M.V.C., H.L., J.C., S.F., R.G., M.T., J.A., I.B., M.-L.B., M.K., M.B. (Magnus Bihlmeier), D.C., A.B., F.O.-W., F.E., S.M., M.B. (Meike Boob), and B.W.; data curation, M.V.C.; writing—original draft preparation, M.V.C., J.C., M.B. (Meike Boob) and B.W.; writing—review and editing, M.V.C. and B.W.; visualization, M.V.C., M.B. (Meike Boob), J.C., H.L. and B.W.; supervision, M.V.C., H.L. and B.W.; project administration, M.V.C., H.L. and B.W.; funding acquisition, M.V.C., H.L. and B.W. All authors have read and agreed to the published version of the manuscript.

Funding: This research received funding from the Federal Ministry of Education and Research (01PL16003), the European Union's Horizon 2020 research and innovation program under grant agreement No 727698, and the University of Hohenheim. The responsibility for the content lies with the authors.

Acknowledgments: The authors would like to thank the company Baresel (Ehningen, Germany) for their support.

Conflicts of Interest: The authors declare no conflict of interest. The funders had no role in the design of the study; in the collection, analyses, or interpretation of data; in the writing of the manuscript, or in the decision to publish the results.

References

1. Montagnini, F.; Nair, P.K.R. Carbon sequestration: An underexploited environmental benefit of agroforestry systems. *Agrofor. Syst.* **2004**, *61–62*, 281–295. [CrossRef]
2. Gruenewald, H.; Brandt, B.K.V.; Schneider, B.U.; Bens, O.; Kendzia, G.; Hüttl, R.F. Agroforestry systems for the production of woody biomass for energy transformation purposes. *Ecol. Eng.* **2007**, *29*, 319–328. [CrossRef]
3. Muchane, M.N.; Sileshi, G.W.; Gripenberg, S.; Jonsson, M.; Pumariño, L.; Barrios, E. Agroforestry boosts soil health in the humid and sub-humid tropics: A meta-analysis. *Agric. Ecosyst. Environ.* **2020**, *295*, 106899. [CrossRef]
4. Möndel, A. Ertragsmessungen in Winterroggen-der Ertragseinfluss einer Windschutzanlage in der oberrheinischen Tiefebene. Verbundprojekt: Agroforst—neue Optionen für eine nachhaltige Landnutzung, LAP Forchheim, Germany. 2007. Available online: http://docplayer.org/38193544-Ertragsmessungen-inwinterroggen-der-ertragseinfluss-einer-windschutzanlage-in-der-oberrheinischen-tiefebene.html (accessed on 9 November 2020).

5.   Thevathasan, N.V.; Gordon, A.M. Ecology of tree intercropping systems in the North temperate region: Experiences from southern Ontario, Canada. *Agrofor. Syst.* **2004**, *61*, 257–268. [CrossRef]

6.   Sheppard, J.P.; Bohn Reckziegel, R.; Borrass, L.; Chirwa, P.W.; Cuaranhua, C.J.; Hassler, S.K.; Hoffmeister, S.; Kestel, F.; Maier, R.; Mälicke, M.; et al. Agroforestry: An Appropriate and Sustainable Response to a Changing Climate in Southern Africa? *Sustainability* **2020**, *12*, 6796. [CrossRef]

7.   Nair, P.K.R.; Gordon, A.M.; Rosa Mosquera-Losada, M. Agroforestry. In *Encyclopedia of Ecology*; Jørgensen, S.E., Fath, B.D., Eds.; Academic Press: Oxford, UK, 2008; pp. 101–110. ISBN 978-0-08-045405-4.

8.   Mosquera-Losada, M.R.; Borek, R.; Balaguer, F.; Mezzarila, G.; Ramos-Font, M.E. *EIP-AGRI Focus Group Agroforestry*; European Commission: Brussels, Belgium, 2017.

9.   Batáry, P.; Matthiesen, T.; Tscharntke, T. Landscape-moderated importance of hedges in conserving farmland bird diversity of organic vs. conventional croplands and grasslands. *Biol. Conserv.* **2010**, *143*, 2020–2027. [CrossRef]

10.   Jose, S.; Gillespie, A.R.; Pallardy, S.G. Interspecific interactions in temperate agroforestry. *Agrofor. Syst.* **2004**, *61–62*, 237–255. [CrossRef]

11.   Tilman, D.; Socolow, R.; Foley, J.A.; Hill, J.; Larson, E.; Lynd, L.; Pacala, S.; Reilly, J.; Searchinger, T.; Somerville, C. Beneficial biofuels—The food, energy, and environment trilemma. *Science* **2009**, *325*, 270–271. [CrossRef]

12.   Meyer, C. The greening of the concrete industry. *Cement Concrete Compos.* **2009**, *31*, 601–605. [CrossRef]

13.   Kalt, G.; Mayer, A.; Theurl, M.C.; Lauk, C.; Erb, K.-H.; Haberl, H. Natural climate solutions versus bioenergy: Can carbon benefits of natural succession compete with bioenergy from short rotation coppice? *GCB Bioenergy* **2019**, *11*, 1283–1297. [CrossRef] [PubMed]

14.   Goetsch, E.; Colinas, F.T. Natural Succession of Species in Agroforestry and in Soil Recovery. 1992. Available online: www.agrofloresta.net/artigos/agroforestry_1992_gotsch.pdf (accessed on 9 November 2020).

15.   Andrade, D.; Pasini, F.; Scarano, F.R. Syntropy and innovation in agriculture. *Curr. Opin. Environ. Sustain.* **2020**, *45*, 20–24. [CrossRef]

16.   Fernandes, A.C.S.A.; Gontijo, L.M. Terracing field slopes can concurrently mitigate soil erosion and promote sustainable pest management. *J. Environ. Manag.* **2020**, *269*, 110801. [CrossRef] [PubMed]

17.   Michiels, H.-G. Die Standorte der Traubeneiche. *LWF Wissen* **2014**, *75*, 25–29.

18.   Herndler, D. Traubeneiche (Quercus petraea) Trauben-Eiche Steckbrief. Available online: https://www.baumlexikon.at/eiche/traubeneiche/ (accessed on 22 August 2020).

19.   Matula, R.; Svátek, M.; Kůrová, J.; Úradníček, L.; Kadavý, J.; Kneifl, M. The sprouting ability of the main tree species in Central European coppices: Implications for coppice restoration. *Eur. J. For. Res.* **2012**, *131*, 1501–1511. [CrossRef]

20.   Gurk, C. Hainbuche (*Carpinus betulus*). Available online: https://www.baumkunde.de/Carpinus_betulus/ (accessed on 22 August 2020).

21.   Raftopoulo, J. Die Rolle der Vogelkirsche in einheimischen Waldgesellschaften. *LWF Wissen* **2010**, *65*, 57–62.

22.   Häne, K. Baum des Jahres 2010: Die Vogelkirsche. *Bündnerwald* **2010**, *63*, 63–66.

23.   Roloff, A. Baum des Jahres 2010: Die Vogelkirsche. *Jahrbuch Baumpflege* **2010**, 9–18.

24.   Häne, K. Die Edelkastanie. Baum des Jahres 2018. *Wald Holz* **2018**, *99*, 13–14.

25.   Jobling, J.; Stevens, F.R.W. *Establishment of Trees on Regraded Colliery Spoil Heaps—A Review of Problems and Practice*; Forestry Commission: Edinburgh, UK, 1980.

26.   Richardson, J.A. Whitebeam (*Sorbus Aria*) as a Nature Reserve Tree on Poor Soils. *Arboric. J.* **1992**, *16*, 99–102. [CrossRef]

27.   Grubb, P.J.; Kollmann, J.; Lee, W.G. A Garden Experiment on Susceptibility to Rabbit-Grazing, Sapling Growth Rates, and Age at First Reproduction for Eleven European Woody Species. *Plant Biol.* **1999**, *1*, 226–234. [CrossRef]

28.   Gurk, C. Echte Mehlbeere (*Sorbus aria*). Available online: https://www.baumkunde.de/Sorbus_aria/ (accessed on 22 August 2020).

29.   Häne, K. Die Elsbeere (*Sorbus torminalis*)—Die kostbare Unbekannte. Available online: https://www.waldwissen.net/wald/baeume_waldpflanzen/laub/wsl_elsbeere/index_DE (accessed on 22 August 2020).

30.   Kunz, J.; Löffler, G.; Bauhus, J. Minor European broadleaved tree species are more drought-tolerant than Fagus sylvatica but not more tolerant than *Quercus petraea*. *Forest Ecol. Manag.* **2018**, *414*, 15–27. [CrossRef]

31.   Werres, J.; Blanke, M. Die Elsbeere (Sorbus torminalis)—Königin der Wildfrüchte. *Erwerbs-Obstbau* **2019**, *61*, 165–177. [CrossRef]

32. Hemery, G.E.; Clark, J.R.; Aldinger, E.; Claessens, H.; Malvolti, M.E.; O'connor, E.; Raftoyannis, Y.; Savill, P.S.; Brus, R. Growing scattered broadleaved tree species in Europe in a changing climate: A review of risks and opportunities. *Forestry* **2010**, *83*, 65–81. [CrossRef]

33. Häne, K. Der Wildapfel—Eine hölzerne Rarität. *Schweiz. Briefmarken Zeitung* **2013**, *4*, 194–196.

34. Delate, K.; McKern, A.; Turnbull, R.; Walker, J.T.S.; Volz, R.; White, A.; Bus, V.; Rogers, D.; Cole, L.; How, N.; et al. Organic Apple Systems: Constraints and Opportunities for Producers in Local and Global Markets: Introduction to the Colloquium. *HortScience* **2008**, *43*, 6–11. [CrossRef]

35. Wagner, I.; Maurer, W.D.; Lemmen, P.; Schmitt, H.P.; Wagner, M.; Binder, M.; Patzak, P. Hybridization and Genetic Diversity in Wild Apple (*Malus sylvestris* (L.) MILL.) from Various Regions in Germany and from Luxembourg. *Silvae Genet.* **2014**, *63*, 81–93. [CrossRef]

36. Delate, K.; McKern, A.; Turnbull, R.; Walker, J.T.S.; Volz, R.; White, A.; Bus, V.; Rogers, D.; Cole, L.; How, N.; et al. Organic Apple Production in Two Humid Regions: Comparing Progress in Pest Management Strategies in Iowa and New Zealand. *HortScience* **2008**, *43*, 12–21. [CrossRef]

37. Häne, K. Baum des Jahres 1998: Die Wildbirne. *Thema Int.* **1998**, 64–66.

38. Häne, K. Der Feldahorn. Der Baum des Jahres 2015. *Schweizer Briefmarken Zeitung* **2015**, *5*, 200–201.

39. Schultes, A. Blutroter Hartriegel—*Cornus sanguinea*. Available online: https://www.pflanzen-deutschland.de/Cornus_sanguinea.html (accessed on 22 August 2020).

40. Gurk, C. Zweigriffeliger Weißdorn (*Crataegus laevigata*). Available online: https://www.baumkunde.de/Crataegus_laevigata/ (accessed on 22 August 2020).

41. Gurk, C. Wolliger Schneeball (*Viburnum lantana*). Available online: https://www.baumkunde.de/Viburnum_lantana/ (accessed on 22 August 2020).

42. Mayrhofer, R. Rote Heckenkirsche—Bestimmen/Erkennen. Available online: https://www.pflanzen-vielfalt.net/b%C3%A4ume-str%C3%A4ucher-a-z/heckenkrische-rote/ (accessed on 22 August 2020).

43. Gurk, C. Gewöhnlicher Liguster (*Ligustrum vulgare*). Available online: https://www.baumkunde.de/Ligustrum_vulgare/ (accessed on 22 August 2020).

44. Gurk, C. Hunds-Rose (*Rosa canina*). Available online: https://www.baumkunde.de/Rosa_canina/ (accessed on 22 August 2020).

45. Gurk, C. Purgier-Kreuzdorn (*Rhamnus cathartica*). Available online: https://www.baumkunde.de/Rhamnus_cathartica/ (accessed on 22 August 2020).

46. Gurk, C. Schlehe (Prunus spinosa). Available online: https://www.baumkunde.de/Prunus_spinosa/ (accessed on 22 August 2020).

47. Ruhm, W. Die Baumhasel—trockenresistent und wertvoll. *Die Landwirtschaft* **2013**, *2013*, 22–23.

48. Van Orshoven, J.; Terres, J.-M.; Tóth, T. Updated common bio-physical criteria to define natural constraints for agriculture in Europe—Definition and scientific justification for the common biophysical criteria. *JRC Sci. Policy Rep.* **2014**, *JRC89982*, 1–67. [CrossRef]

49. Von Cossel, M.; Lewandowski, I.; Elbersen, B.; Staritsky, I.; Van Eupen, M.; Iqbal, Y.; Mantel, S.; Scordia, D.; Testa, G.; Cosentino, S.L.; et al. Marginal agricultural land low-input systems for biomass production. *Energies* **2019**, *12*, 3123. [CrossRef]

50. Novák, J.; Prach, K. Vegetation succession in basalt quarries: Pattern on a landscape scale. *Appl. Veg. Sci.* **2003**, *6*, 111–116. [CrossRef]

Publisher's Note: MDPI stays neutral with regard to jurisdictional claims in published maps and institutional affiliations.

 *agriculture*

Article

Productivity and Biometric Characteristics of 11 Varieties of Willow Cultivated on Marginal Soil

Mariusz Matyka * and Paweł Radzikowski

Department of Systems and Economics of Crop Production, Institute of Soil Science and Plant Cultivation—State Research Institute, Czartoryskich 8, 24-100 Puławy, Poland; pradzikowski@iung.pulawy.pl
* Correspondence: mmatyka@iung.pulawy.pl; Tel.: +48-81-4786800

Received: 2 November 2020; Accepted: 7 December 2020; Published: 9 December 2020

Abstract: In response to the growth in the global population and climate change concerns, questions remain regarding the adaptation of production systems to meet increasing food and energy demands. The aim of the paper is to present the production potential and biometric features of 11 willow varieties bred and cultivated mainly in Europe. The experiment was set up on marginal soil. The research was conducted in 2016–2020 and concerned 11 varieties of willow harvested in a three-year cycle. The dry matter yield of the examined willow varieties ranged from 6.5 to 13.8 Mg ha^{-1} year^{-1}. Varieties Tur, Sven, Olof, Torhild, and Tordis were characterized by a relatively low level of yield (7.2–8.2 Mg ha^{-1} year^{-1}). The highest dry matter yield was obtained for the varieties Ekotur and Żubr, respectively, of 11.5 and 13.8 Mg ha^{-1} year^{-1}. The assessed varieties differed in both the level of obtained dry matter yield and biometric features. The Żubr variety produced the smallest number of shoots (three), but with the greatest height (4.8 m) and diameter (29.6 mm). Varieties with high production potential develop fewer shoots, but are taller and have a larger diameter than other varieties.

Keywords: SRC; willow; varieties; yield; biometric features; marginal soil

---

1. Introduction

In response to the growth in the global population and climate change concerns, questions remain regarding the adaptation of production systems to meet increasing food and energy demands, and the identification of the most efficient and sustainable production schemes. Interest in the biobased economy has, therefore, been increased by aligning policies relating to central and interdependent sectors, including agriculture and forestry, food, feed, bioenergy, and bio-chemistry [1,2]. Because the EU's growing priority for renewable energy sources is driven by increased energy dependency, rising fossil fuel prices, and the desire to protect the environment, the demand for energy from biomass, mainly wood, is increasing. Waste wood from forestry and municipal energy management may in future be supplemented with wood acquired from agricultural land as fuel for power or heating plants [3–6]. Although the primary role of agriculture is food production, a portion of agriculture land has typically been devoted to non-food products, mainly within the framework of emerging technologies. Such uses include the production of bioenergy and various biomaterials [7].

Fast-growing trees for biomass production are proposed as an economical and ecological solution to meet increasing demand for energy, and also for the anticipated shortage of raw materials for wood-based industries [8–11]. One of the short rotation woody coppices (SRWCs) is willow, which has been widely accepted as a renewable energy source. Willow biomass can be converted by a wide range of technologies, such as combined heat and power hydrothermal upgrading, into a variety of energy forms and carriers [12,13] Willow has several advantages as an SRWC, including a wide range of genetic diversity, easy reproduction, tolerance to a wide range of site conditions, ability to rapidly regrow after harvests, possibility of harvests in different rotations, and resistance to disadvantageous environmental conditions, e.g., morning frost, wet snow, and strong winds [14–16]. Willow may be taken into account in a concept of extensive agriculture, especially on marginal soils, which are of little use or unusable for growing food crops, soils that are periodically excluded from use, and some fallow lands [6,17].

A significant novelty of this study is the provision of a comprehensive assessment of (1) dry matter yield, (2) biomass moisture, and (3) biometric characteristics of 11 willow varieties, bred and cultivated in various European countries. An additional advantage of the presented results is that the studied varieties were cultivated in commercial plantations on marginal soil.

2. Materials and Methods

2.1. Experiment Setup

A one-factor field experiment was conducted in 2016–2020 at the Experimental Farm of the Institute of Soil Science and Plant Cultivation—State Research Institute located in Osiny, Poland (51°28′21.34″ N, 22°3′7.65″ E). The experiment was set up on marginal soil. It was a clay soil comprising light clayey sands transformed into light clay at a depth of 70–80 cm, class IV on a VI scale in the Polish classification. The average pH in KCl is 5.5, i.e., it was slightly acidic. The humus content was low at 1.06%. At the time of experiment, the average nitrogen content in soil was 378 mg NO3N kg^{-1}, phosphorus was 224 mg P_2O_5 kg^{-1}, and potassium was 124 mg K_2O kg^{-1}. The content of magnesium in soil was 49 mg MgO kg^{-1} and calcium was 70 mg CaO kg^{-1}. The content of nutrients in the soil on which the experiments were conducted was at a medium or high (for phosphorus) level. Therefore, the tested plants were optimally supplied with nutrients. The annual sum of precipitation for the locations where the experiment was conducted is 568 mm on average, and the average annual air temperature is 8 °C. Willow cuttings were planted in April 2016 at a density of 18,000 per ha. The cuttings were planted in two-row strips with rows spaced every 0.75 m apart. The strips were spaced every 1.5 m. The cuttings in each row were planted every 0.50 m. In the first year of vegetation, weeds from plantations were removed mechanically. The research covered 11 varieties of willow (Table 1).

Table 1. Varieties of willow tested in the field experiment.

Variety	Species	Country of Breeding
Gigantea	*Salix viminalis*	Denmark
Inger	*S. viminalis* × *S. triandra*	Sweden
Linnea	*S. viminalis*	Sweden
Olof	*S. viminalis*× *S. schwerinii*	Sweden
Sven	*S. viminalis*× *S. schwerinii*	Sweden
Tora	*S. viminalis*× *S. schwerinii*	Sweden
Tordis	*S. viminalis*× *S. schwerinii*	Sweden
Torhild	*S. viminalis*× *S. schwerinii*	Sweden
Ekotur	*S. viminalis*	Poland
Tur	*S. viminalis*	Poland
Żubr	*S. viminalis*	Poland

The cultivation area of each variety was 270 m^2; moreover, for the varieties located on the edge of the field, an additional strip of plants (2 rows) was planted in order to eliminate the edge effect. The total area of the experimental field was 3150 m^2. After the first year of vegetation, in January 2017, a maintenance cut was made. Willow biomass was harvested in January 2020 after three years of vegetation. In the year the plantation was established, mineral fertilization was applied in the dose of 30 kg ha^{-1} N, 60 kg ha^{-1} P_2O_5, and 90 kg ha^{-1} K_2O. After maintenance, a cut of 80 kg ha^{-1} N, 60 kg ha^{-1} P_2O_5, and 90 kg ha^{-1} K_2O was applied. No pesticides were used on the plantation.

2.2. Yield and Biometric Analysis

After three years of vegetation, in January 2020 the biomass yield and their biometric features of the tested willow varieties were determined. For this purpose, 5 whole plants of each variety were harvested in 5 replications, giving a total of 25 plants. Each of the replicates was randomized, then 5 consecutive plants in a row were harvested. No samples were taken from the edge of the experimental field to eliminate the edge effect. Sampling was performed manually using a chainsaw. Each plant was weighed immediately after harvest. Two shoots with average values of biometric features were taken from each plant and ground for determination of the share of dry matter. In the laboratory the biomass humidity was determined with the drying method at 100 °C for a period of 14 days. Dry matter yield was determined from the ratio of fresh matter yield and and dry matter content. The yield from the area of 1 ha was calculated from the product of the actual density and the unit weight of plant.

Measurements of biometric parameters of the tested plants were also performed. The following were determined: number of shoots per plant, diameter of plant shoots at a height of 15 cm, and Ø (mm) and height of shoots (m).

2.3. Statistical Analysis

The results were subjected to analysis of variance for a single factor (ANOVA) using the software Statistica v13.3 (TIBCO Software Inc., Palo Alto, CA, USA). The significance of differences between varieties for the variable, such as dry matter yield, number of shoots for plant, plant height, and shoot diameter was identified using the NIR test ($p < 0.05$). Cluster analysis was performed using the method of k means. The grouping variables were dry matter yield, number of shoots per plant, plant height, and shoot diameter. Initial cluster centers were established on the basis of a random selection of k observations using a standardized distance measure (Euclidean distance). A k-fold cross-validation was used to determine the number of clusters.

3. Results and Discussion

The dry matter yield of the tested willow varieties harvested in the three year cycle ranged from 6.5 Mg ha^{-1} year^{-1} for the Tora variety to 13.8 Mg ha^{-1} year^{-1} for the Żubr variety (Figure 1). The highest dry matter yield was obtained for the varieties Ekotur and Żubr, respectively, of 11.5 and 13.8 Mg ha^{-1} year^{-1}, which are bred at UWM in Olsztyn.

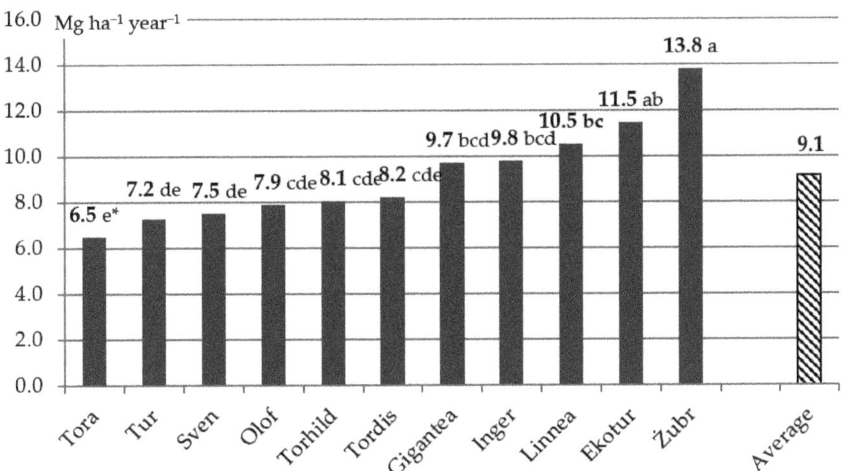

Figure 1. Dry matter yield (Mg ha^{-1}·year^{-1}) of tested willow varieties. * Explanations: data marked with the same letters do not differ significantly at $\alpha = 0.05$.

The Żubr variety obtained a significantly higher result than all other varieties. In second place in this respect was the variety Ekotur. The yield of Żubr variety was 51% higher than the average for varieties and 84–112% higher than the worst varieties. Żubr variety also yielded 20% higher than the second-best variety. Variety Ekotur yielded 26% higher than the average and 76% higher than the weakest variety (Table 2).

Table 2. Matrix of differences (%) between yields of dry mass of tested varieties of willow.

Variety	(1)	(2)	(3)	(4)	(5)	(6)	(7)	(8)	(9)	(10)	(11)	(12)
Tora (1)		−10	−13	−17	−19	−20	−33	−34	−38	−43	−53	−28
Tur (2)	11		−3	−8	−10	−11	−25	−26	−31	−37	−47	−20
Sven (3)	15	3		−5	−7	−8	−23	−23	−29	−35	−46	−18
Olof (4)	21	8	5		−2	−4	−19	−20	−25	−32	−43	−14
Torhild (5)	24	11	8	3		−2	−17	−18	−24	−30	−42	−11
Tordis (6)	26	13	9	4	2		−16	−16	−22	−29	−41	−10
Gigantea (7)	49	34	29	23	20	19		−1	−8	−15	−30	7
Inger (8)	50	35	31	25	22	20	1		−7	−15	−29	8
Linnea (9)	62	45	41	34	31	29	9	8		−8	−24	16
Ekotur (10)	76	58	53	46	42	40	18	17	9		−17	26
Żubr (11)	112	90	84	75	71	68	42	41	31	20		51
Average (12)	40	26	21	16	13	11	−6	−7	−14	−21	−34	

Average yields of dry matter (9.7–10.5 Mg ha^{-1} year^{-1}) were obtained for the varieties Gigantea, Inger, and Linnea. Varieties Tur, Sven, Olof, Torhild, and Tordis were characterized by a relatively low level of yield (7.2–8.2 Mg ha^{-1} year^{-1}). Results obtained from Olof, Torhild, Tordis, Gigantea, Inger, and Linnea varieties were close to the average value, with a deviation of up to 15%. The results of this group of varieties were from 5% to 62% better than those of the three lowest yielding varieties. Within this group Inger yielded 25% higher than Olof. Three varieties yielded much lower than average, Tora (−28%), Tur (−20%), and Sven (−18%). These varieties yielded about half as much as the Żubr variety. Despite the wide range of results, there were no statistically significant differences between the varieties Tora, Tur, Sven, Olof, Torhild, and Tordis.

Reported biomass yields of willow harvested in a three-year cycle, depending on the location and variety, ranged from 6.4 to 20.0 Mg ha^{-1} year^{-1}, and were generally similar to those obtained in our own research [15,18,19]. Stolarski et al. [16] compared 15 willow varieties and genotypes in a field experiment. As in the current research, the authors showed that the Żubr variety produces a very high dry matter yield, which, depending on the location, ranged from 18.0 to 20.3 Mg ha^{-1} year^{-1} in a three-year rotation. This yield was higher than that obtained in our own experiment (13.8 Mg ha^{-1} year^{-1}) because the willow was grown on better-quality soils. It should be noted that regardless of the soil conditions, the Żubr variety yield was clearly the highest. In the study of Stolarski et al. [16] in relation to the Żubr variety, the dry matter yield of the next three varieties was lower by 13%, and for the next five by 17%. The yields obtained in our own research were also similar to those obtained in other parts of Europe, which in the UK ranged from 4.9 to 15.4 Mg ha^{-1} year^{-1}. It should be noted that the highest yields were obtained from Ashton Stott and Ashton Parfitt varieties (15.4 and 13.9 Mg ha^{-1} year^{-1}, respectively) grown in the UK, which were not included in the own research [20–22]. In contrast, the yields obtained in Scandinavia were lower and ranged from 6.9 to 7.7 Mg ha^{-1} year^{-1} in Sweden and 3.1 to 8.6 Mg ha^{-1} year^{-1} in Denmark, depending on the variety and site [23,24]. Furthermore, studies conducted in the USA and Canada indicate that the dry matter yields of willow genotypes were significantly differentiated both in terms of genotype and site. The dry matter yield of willow in these studies ranged from 3.5 to 13.6 Mg ha^{-1} year^{-1} [25,26].

The factor that determines the quality of willow biomass and, indirectly, the level of the yield, is biomass moisture during harvest. For the studied varieties, it was very diverse and ranged from 42.3–42.9% for the varieties Olof, Torhild and Tora, to 50.1–51.4 for the varieties Żubr, Linnea, Inger, and Tur. The varieties Tordis, Ekotur, Sven, and Gigantea were characterized by the average level of humidity (46.5–49.3%). In studies conducted in north-eastern Poland, humidity of willow biomass during harvest, depending on the variety, ranged from 47.5% to 52.0% [27].

The assessed varieties differed in both the level of the obtained dry matter yields and biometric features (Table 3).

Table 3. Biometric features of tested willow varieties.

Variety	Number of Shoots for Plant	Plant Height (m)	Shoot Diameter (mm)
Gigantea	8.4 a *	3.3 d	15.6 f
Inger	7.7 ab	3.2 d	17.1 def
Linnea	6.8 b	3.4 d	18.5 cde
Olof	7.1 b	3.3 d	17.8 cdef
Sven	5.1 c	3.5 cd	18.4 cde
Tora	5.0 c	3.5 cd	19.9 bc
Tordis	4.9 c	3.9 bc	19.3 bcd
Torhild	6.7 b	3.1 d	16.3 ef
Ekotur	4.9 c	4.1 b	21.7 b
Tur	6.5 b	3.1 d	15.7 f
Żubr	2.9 d	4.8 a	29.6 a
Average	6.0	3.6	19.1

* Explanations: data marked with the same letters do not differ significantly at $\alpha = 0.05$.

The varieties Ekotur, Tordis, Tora, and Sven developed five shoots per plant, and Tur developed six. The varieties Gigantea and Inger had, on average, eight shoots per plant, and Linnea and Olof had seven. Those four varieties had the highest number of shoots per plant. The height of the shoots of the studied varieties also varied. The Żubr variety achieved the longest shoots, with an average length of 4.8 m. The varieties Ekotur, Tordis, Tora, and Sven grew shoots ranging from 3.5 to 4.1 m. Other varieties developed shoots 3.2 to 3.4 m long. The diameter of the shoots was significantly differentiated. The lowest value of this parameter (15.6–16.3 mm) was recorded for the Gigantea, Tur, and Torhild varieties. In addition to the Żubr variety, the Ekotur, Tora, and Tordis varieties had the largest shoot diameter (19.3–21.7mm). The Żubr variety, which was characterized by the highest dry matter yield, produced the smallest number of shoots (three), but had the greatest height (4.8 m), and the biggest diameter (29.6 mm). The second highest yielding variety, Ekotur, was also characterized by a small number of shoots, good height, and large diameter of shoots. However, the trend changed for the next three varieties—Gigantea, Inger, and Linnea—which also had a high yield level (9.7–10.5 Mg ha^{-1} year^{-1}). These were characterized by the highest number of shoots throughout the experiment. Linea had seven shoots on average, and Inger and Gigantea had eight. All three varieties were similar in height, but the diameter of the shoots differed. The higher yielding varieties had the thickest shoot diameter. The closest to the average values of these parameters were Tordis and Tur, however, they yielded much lower than average. Olof and Torhild varieties developed the same number of shoots as Linea, however they obtained much lower height and thickness, so the yield of these varieties was lower than the average and by almost 6 Mg ha^{-1} year^{-1} lower than the yield of the best variety. Low yielding varieties, Tora and Sven, were characterized by a small number of shoots and average values of their height and thickness, and Tur had a high average number of shoots but small height and diameter. The presented results concerning biometric features of willow varieties are similar to those obtained by other researchers. They also confirm the thesis that parameters of those features are significantly different, depending on the variety [18,28–31]. It should be emphasized that both the varieties included in the current research and those presented by other authors are characterized by more favorable values and relationships of biometric features compared to wild forms of willow growing in natural stands [32,33].

Characteristics of the features of individual varieties allow for a reliable selection of varieties most appropriate to biomass yield. Due to the growing interest in diversification of crops to compensate for other crops' flaws, mixed crops are often used. For this purpose, clusters of varieties that have similar biometric characteristics were distinguished. Different mixtures of varieties may contain varieties with the same characteristics, or varieties with extremely different parameters, depending on the purpose of cultivation. The cluster analysis confirmed that the tested willow varieties can be categorized in terms of the level of yield and biometric features into three clusters (Table 4).

Table 4. Yield of biomass and biometric features of willow crops in separated clusters.

Cluster	Variety	DM Yield (Mg ha^{-1} year^{-1})	Number of Shoots for Plant	Plant Height (m)	Shoot Diameter (mm)
1	Ekotur, Żubr	12.6	4.0	4.5	25.7
2	Gigantea, Inger, Linnea	10.0	7.7	3.3	18.5
3	Olof, Sven, Tora, Tordis, Torhild, Tur	7.6	5.8	3.4	17.9

The first cluster includes the Żubr and Ekotur varieties, which stand out primarily because of a high yield, obtained through specific biometric features. Both varieties were characterized by an average low number of shoots but had large height and diameter. It was also shown in other studies that the highest dry matter yields are obtained from plants that develop a small number of shoots, but were tall and had a larger diameter [16,34]. By comparison, a relatively high yield was obtained by varieties from the second cluster, which were characterized by completely different biometric characteristics. Gigantea, Inger, and Linnea varieties had an average of 26.7% shorter shoots and 28.0% smaller diameter of shoots than the varieties from the first cluster, which was compensated for by a large number of shoots. Furthermore, plants from the second cluster were a group combining the characteristics of varieties from the first and third cluster. The third group of varieties was characterized by parametric features that were highly similar to the second cluster, but with a much smaller number of shoots, resulting in the lowest yield of the three mentioned clusters. The results obtained suggest that for energy purposes, varieties from the first cluster are definitely preferred. Large biomass obtained from less numerous shoots can provide proportionally more wood in relation to bark. For purposes other than energy production, for example, as a substrate for the extraction of active substances, varieties with a higher proportion of bark in the yield could be preferred [35]. Ideal for this purpose would be varieties from the second cluster, which have a large number of shoots and provide a relatively high yield. In contrast, the lowest yielding varieties could have a higher content of active substances, which would reduce the importance of total biomass yield. The productivity of individual varieties in terms of substrates used for different purposes should be further investigated.

The biometric features of different varieties of the willow *Salix viminalis* L. differ due to its crossbreeding with other species of willows. In the case of the willow, several dozen hybrid varieties are available on the European market. The most frequently cultivated domestic varieties are Start, Sprint, Turbo, Tur, Bison, and Ekotur. In Poland, Swedish varieties of Torhild, Sven, Olof, Gudrun, Tordis, Tora, and Inger are also often cultivated. Due to the high variability of yields of individual varieties in following years, it is recommended to use mixtures of several different clones in cultivation [35–37]. Different varieties, in addition to the clusters of similar varieties distinguished in the current work, may have different yield levels depending on the geographical location of cultivation, soil type, and cultivation technology used. It cannot be precluded that climatic differences, in addition to the progression of climate warming, will result in changes in the productive potential of varieties. The key in this aspect will be the sum and distribution of precipitation and the average summer temperatures. Willow is quite predictable in this respect: the biomass yield is linearly related to the annual total precipitation. In many regions, recent years have been dominated by periods of dry years with annual perception below 600 mm, interrupted by years of very large amounts of precipitation. Therefore, regardless of location, drought-resistant varieties will be preferred. In this case, features such as strong development of the root system will be crucial, probably at the expense of the aboveground biomass in the first years of cultivation. As average temperatures increase, the pressure of diseases and pests will become more severe, so high-yielding varieties could be unreliable in some years. On the basis of previous studies, no impact of climate change on willow cultivation in Poland has been observed to date. In the conditions of north-eastern Poland, the best yielding varieties were Ekotur (17.8 Mg ha^{-1} year^{-1}), Żubr (15.9 Mg ha^{-1} year^{-1}), and Start (16.7 Mg ha^{-1} year^{-1}), which coincides with the results obtained in the current experiment. It may prove that the climatic gradient of the Polish area is of little importance for the selection of varieties [35–37].

In the current paper we focus on the cultivation of willows on marginal land. Such areas will have some production limitations, which will directly translate into the yield and its quality. In Poland, land is marginalized for various reasons. In more than half of the cases, the reason for abandonment of land is low soil productivity potential, caused by low humus content, low fertility, and poor water holding capacity [38]. On such soils, the highest yielding varieties with favorable biometric features may be found to be weaker than on soils which are optimal for the production of willow. Furthermore, good quality soils are also often located in marginal areas. Most commonly, good soils in Poland are abandoned for organizational reasons. In this category, most of the land is set aside in the vicinity of large cities and main roads. This land was abandoned, among other reasons, as a result of industrial transformations. This often leads to extension of the distance from the farm to the field due to the construction of a highway, which eliminates local roads. It is possible to use such land for willow cultivation, however, management of these fields will be limited due to the long distance from the farm. There is also potential in degraded and polluted land, on which willow cultivation can be carried out successfully, however, use of its yield will be limited to energetic and technical purposes. In the current study, it was pointed out that the most productive varieties of willow have also obtained a decent yield on marginal soils. However, it should be remembered that the plantation was professionally cared for and carried out with the best scientific knowledge. Many of the key measures that are provided in experimental plantations are often lacking in commercial crops [36]. On marginal land, often some kind of agrotechnical negligence occurs, so it is advisable to select varieties that require as little input as possible. For example, harvesting may be delayed by organizational problems and therefore the production cycle may be extended by an additional year. It is often observed that hybrid varieties, designed for high yields, can cope less well with environmental pressure and the consequences of agrotechnical negligence. It is also unclear how the tested varieties would survive without any fertilization and chemical protection on poor soils. Rich soils can ensure the survival of plantations without fertilization and even allow a decent yield. On marginal soils, such cultivation can carry a risk of failure [35,39].

The yields of individual varieties will also be quite different depending on the cultivation technology used. Disease and insect resistance are also important. To obtain the highest yield of biomass, a three year cycle is optimal, in which the highest yields are obtained by Ekotur and Żubr varieties. In Poland the cultivation was tested in harvest cycles ranging from annual to over five years. It was indicated that some willow varieties, considered as low yielding, may yield quite well in longer cultivation cycles. The experiment was also conducted with plant density modifications. The smallest stocking density is used in the Eco-Salix system, where willow is planted from large stakes in a small stocking density of less than 10,000 per ha. In the Eco-salix system, the best yielding variety was UWM 043. A very large planting density of about a hundred thousand individuals per ha was also tested in case of seedlings from annual offshoots, planted mechanically. In the case of the cultivation of willow grown in high density (48,000–96,000 per ha), there was a significant decrease in yield in the long-term perspective for all varieties tested (Tur, Turbo, UWM 046, UWM 200, and UWM 095), however, the UWM 095 variety provided the most stable yield. Depending on the height of the harvest cycle, the highest yield was obtained from Tora and 1023 clone in a one year cycle, Tur and 1054 clone in a two year cycle, Żubr in a three year cycle, and Ekotur and 1058 clone in a four year cycle [35,36,40].

The biomass yield depends not only on the biometric features, but also on the physiological characteristics of the plant. In this case, the characteristics of resistance to diseases, e.g., rust caused by *Melampsora sp.*, which can reduce the biomass yield by up to 40%, may be crucial. Resistance to willow rust was obtained in most varieties with high yield potential, however, other biological limitations remain. An important and often underestimated feature is the resistance of willows to pests, whose pressure may increase with climate warming. Unfortunately, pest-resistant willow species are characterized by very slow growth in comparison to the willow *Salix viminalis* L., whereas hybrids between species do not provide satisfactory resistance and yield [41]. Obtaining resistance features, which reduces the pressure of pests, can be supported by molecular techniques. Gene identification and gene transfer application in industrial plants is less controversial than in the case of food plants. Modification of willow varieties can have a positive impact on mitigating the effects of climate crisis, allowing substantial amounts of carbon dioxide to be sequestered in the crop biomass. By comparison, the genetic purity of wild populations of willows, which can easily cross-pollinate with modified crops, cannot be assured. It is unlikely that elaborate and expensive genetic engineering methods will be applied to the willow breeding process because willow is not considered to be an important crop for the human economy [41].

A high level of biomass yield is crucial if willow is converted to energy production. However, the use of lower yielding varieties for industrial cropping is also possible. Willow is grown for various purposes, including traditional wicker products. In this case the yield of dry matter is not as crucial as the quality of the material, so short and shrubby varieties could be preferred. However, varieties often damaged by pests will be avoided. The content of active substances is another feature of the varieties, not necessarily related to the yield. In this case, varieties with a higher proportion of bark in the biomass, i.e., varieties that are vigorously shrubby, will be preferred for medicinal and cosmetic products. An important branch of the economy in the future may be natural construction, due to its growing interest. It also uses products from willow plantations in addition to other industrial crops. In this case, high yield does not have to be a key value, which should relate to the strength and durability of the obtained material [35,41].

4. Conclusions

This study provides an assessment of the productivity and biometric features of some of the most common willow varieties in Europe. The presented results clearly show that the key element that determines the production effects is the appropriate selection of varieties. Varieties with high production potential develop fewer shoots, but are taller and larger in diameter than other varieties. The obtained results also indicate that it is possible to effectively use marginal land for production purposes other than food production, including, for example, the cultivation of willow intended for industrial purposes.

Author Contributions: Conceptualization, M.M.; methodology, M.M.; validation, M.M. and P.R.; investigation, M.M. and P.R.; resources, M.M.; data curation, M.M. and P.R.; writing—original draft preparation, M.M. and P.R.; writing—review and editing, P.R.; visualization, P.R.; supervision, M.M.; project administration, M.M.; funding acquisition, M.M. All authors have read and agreed to the published version of the manuscript.

Funding: This paper is the result of a long–term study carried out at the Institute of Soil Science and Plant Cultivation—State Research Institute, Department of Systems and Economics of Crop Production, and it was co-financed by the National (Polish) Centre for Research and Development (NCBiR), entitled "Environment, agriculture and forestry", project: BIOproducts from lignocellulosic biomass derived from MArginal land to fill the Gap In Current national bioeconomy, No. BIOSTRATEG3/344253/2/NCBR/2017.

Acknowledgments: We would also like to thank the staff of the Department of Systems and Economics of Crop Production and Experimental Farm "Kępa" for their technical support during the experiments.

References

1. El-Chichakli, B.; von Braun, J.; Lang, C.; Barben, D.; Philp, J. Policy: Five cornerstones of a global bioeconomy. *Nature* **2016**, *535*, 221–223. [CrossRef] [PubMed]
2. Ben Fradj, N.; Rozakis, S.; Borzęcka, M.; Matyka, M. Miscanthus in the European bio-economy: A network analysis. *Ind. Crops Prod.* **2020**, *148*, 112281. [CrossRef]
3. Volk, T.A.; Abrahamson, L.P.; Nowak, C.A.; Smart, L.B.; Tharakan, P.J.; White, E.H. The development of short-rotation willow in the northeastern United States for bioenergy and bioproducts, agroforestry and phytoremediation. *Biomass Bioenergy* **2016**, *30*, 715–727. [CrossRef]
4. Walle, I.V.; Van Camp, N.; Van De Casteele, L.; Verheyen, K.; Lemeur, R. Short-rotation forestry of birch, maple, poplar and willow in Flanders (Belgium) II. Energy production and CO_2 emission reduction potential. *Biomass Bioenergy* **2007**, *31*, 276–283. [CrossRef]
5. Borkowska, H.; Molas, R. Yield comparison of four lignocellulosic perennial energy crop species. *Biomass Bioenergy* **2013**, *51*, 145–153. [CrossRef]
6. Stolarski, J.M.; Szczukowski, S.; Tworkowski, J.; Krzyżaniak, M. Extensive Willow Biomass Production on Marginal Land. *Pol. J. Environ. Stud.* **2019**, *6*, 4359–4367. [CrossRef]
7. Diamantidis, N.D.; Koukis, E.G. Agricultural crops and residues as feedstock for non-food products in Western Europe. *Ind. Crops Prod.* **2000**, *11*, 97–106. [CrossRef]
8. Tharakan, P.J.; Volk, T.A.; Abrahamson, L.P.; White, E.H. Energy feedstock characteristics of willow and hybrid poplar clones at harvest age. *Biomass Bioenergy* **2003**, *25*, 571–580. [CrossRef]
9. Licht, L.A.; Isebrands, J.G. Linking phytoremediated pollutant removal to biomass economic opportunities. *Biomass Bioenergy* **2005**, *28*, 203–218. [CrossRef]
10. Fillion, M.; Brisson, J.; Teodorescu, T.I.; Sauve, S.; Labrecque, M. Performance of *Salix viminalis* and *Populus nigra* x *Populus maximowiczii* in short rotation intensive culture under high irrigation. *Biomass Bioenergy* **2009**, *33*, 1271–1277. [CrossRef]
11. Cunniff, J.; Purdy, S.J.; Barraclough, T.J.; Castle, M.; Maddison, A.L.; Jones, L.E.; Karp, A. High yielding biomass genotypes of willow (*Salix* spp.) show differences in below ground biomass allocation. *Biomass Bioenergy* **2015**, *80*, 114–127. [CrossRef] [PubMed]
12. Kuzovkina, Y.A.; Volk, T.A. The characterization of willow (*Salix* L.) varieties for use in ecological engineering applications: Coordination of structure, function and autecology. *Ecol. Eng.* **2009**, *35*, 1178–1189. [CrossRef]
13. Cao, Y.; Lehto, T.; Piirainen, S.; Kukkonen, J.V.K.; Pelkonen, P. Effects of planting orientation and density on the soil solution chemistry and growth of willow cuttings. *Biomass Bioenergy* **2012**, *46*, 165–173. [CrossRef]
14. Djomo, S.N.; Kasmioui, O.E.; Ceulemans, R. Energy and greenhouse gas balance of bioenergy production from poplar and willow: A review. *GCB Bioenergy* **2011**, *3*, 181–197. [CrossRef]
15. Zamora, S.D.; Apostol, G.K. Biomass production and potential ethanol yields of shrub willow hybrids and native willow accessions after a single 3-year harvest cycle on marginal lands in central Minnesota, USA. *Agrofor. Syst.* **2014**, *88*, 593–606. [CrossRef]
16. Stolarski, J.M.; Krzyżaniak, M.; Załuski, D.; Tworkowski, J.; Szczukowski, S. Effects of site, genotype and subsequent harvest rotation of willow productivity. *Agriculture* **2020**, *10*, 412. [CrossRef]
17. Stolarski, J.M.; Szczukowski, S.; Krzyżaniak, M.; Tworkowski, J. Energy Value of yield and biomass quality in a 7-year rotation of willow cultivated on marginal soil. *Energies* **2020**, *13*, 2144. [CrossRef]
18. Hinton-Jones, M.; Valentine, J. Variety and altitude effects on yields and other characters of SRC willow in Wales. *Biomass Energy Crops III Asp. Appl. Biol.* **2009**, *90*, 67–73.
19. El Bassam, N. *Handbook of Bioenergy Crops*; Earthscan Ltd.: London, UK, 2010; pp. 392–398.
20. Aylott, M.J.; Casella, E.; Tubby, I.; Street, N.R.; Smith, P.; Taylor, G. Yield and spatial supply of bioenergy poplar and willow short-rotation coppice in the UK. *New Phytol.* **2008**, *178*, 358–370. [CrossRef]
21. Lindegaard, K.N.; Carter, M.M.; McCracken, A.; Shield, I.; MacAlpine, W.; Jones, M.H.; Larsson, S. Comparative trials of elite Swedish and UK biomass willow varieties 2001–2010. *Asp. Appl. Biol.* **2011**, *112*, e66.
22. Pei, M.H.; Lindegaard, K.; Ruiz, C.; Bayon, C. Rust resistance of some varieties and recently bred genotypes of biomass willows. *Biomass Bioenergy* **2008**, *32*, 453–459. [CrossRef]
23. Dimitriou, I.; Mola-Yudego, B. Poplar and willow plantations on agricultural land in Sweden: Area, yield, groundwater quality and soil organic carbon. *Biomass Bioenergy* **2017**, *383*, 99–107. [CrossRef]

24. Larsen, S.U.; JØrgensen, U.; Laerke, P.E. Willow yield is Highly dependent on clone and site. *Bioenerg. Res.* **2014**, *7*, 1280–1292. [CrossRef]
25. Serapiglia, M.J.; Cameron, K.D.; Stipanovic, A.J.; Abrahamson, L.P.; Volk, T.A.; Smart, L.B. Yield and woody biomass traits of novel shrub willow hybrids at two contrasting sites. *Bioenergy Res.* **2013**, *6*, 533–546. [CrossRef]
26. Fabio, E.S.; Volk, T.A.; Miller, R.O.; Serapiglia, M.J.; Gauch, H.G.; Van Rees, K.C.J.; Hangs, R.D.; Amichev, B.Y.; Kuzovkina, Y.A.; Labrecque, M.; et al. Genotype x environment interaction analysis of North American shrub willow yield trials confirms superior performance of triploid hybrids. *GCB Bioenergy* **2017**, *9*, 445–459. [CrossRef]
27. Stolarski, M.J.; Szczukowski, S.; Tworkowski, J.; Klasa, A. Willow biomass production under conditions of low-input agriculture on marginal soils. *For. Ecol. Managm.* **2011**, *262*, 1558–1566. [CrossRef]
28. Stolarski, M.J.; Krzyzaniak, M.; Szczukowski, S.; Tworkowski, J.; Śnieg, M. Willow productivity on a commercial plantation in triennial harvest cycle. *Bulg. J. Agric. Sci.* **2016**, *22*, 65–72.
29. Nord-Larsen, T.; Sevel, L.; Raulund-Rasmussen, K. Commercially grown short rotation coppice willow in Denmark: Biomass production and factors affecting production. *Bioenergy Res.* **2015**, *8*, 325–339. [CrossRef]
30. Castaño-Díaz, M.; Barrio-Anta, M.; Afif-Khouri, E.; Cámara-Obregón, A. Willow short rotation coppice trial in a former mining area in northern Spain: Effects of clone, fertilization and planting density on yield after five years. *Forests* **2018**, *9*, 154. [CrossRef]
31. Hernea, C.; Trava, I.D.; Borlea, G.F. Biomass production of some Swedish willow hybrids on the West of Romania. A case study. *J. Hortic. For. Biotechnol.* **2015**, *19*, 103–106.
32. Maděra, P.; Kovářová, P. Primary succession of white willow communities in the supraregional biocorridor in the Middle water reservoir of Nové Mlýny. *Ekológia* **2004**, *23*, 191–204.
33. Maděra, P.; Packová, P.; Manjarrés, D.R.L.; Štykar, J.; Simanov, V. The model of potential biomass production in Odra R. basin. *Ekológia* **2009**, *28*, 170–190. [CrossRef]
34. Amichev, B.Y.; Volk, T.A.; Hangs, R.D.; Bélanger, N.; Vujanovic, V.; Van Rees, K.C.J. Growth, survival, and yields of 30 short-rotation willow cultivars on the Canadian Prairies: 2nd rotation implications. *New For.* **2018**, *49*, 649–665. [CrossRef]
35. Stolarski, M.J.; Niksa, D.; Krzyżaniak, M.; Tworkowski, J.; Szczukowski, S. Willow productivity from small-and large-scale experimental plantations in Poland from 2000 to 2017. *Renew. Sust. Energy Rev.* **2019**, *101*, 461–475. [CrossRef]
36. Stolarski, M.J.; Szczukowski, S.; Tworkowski, J.; Krzyżaniak, M.; Załuski, D. Willow production during 12 consecutive years—The effects of harvest rotation, planting density and cultivar on biomass yield. *GCB Bioenergy* **2019**, *11*, 635–656. [CrossRef]
37. Stolarski, M.J.; Śnieg, M.; Krzyżaniak, M.; Tworkowski, J.; Szczukowski, S. Short rotation coppices, grasses and other herbaceous crops: Productivity and yield energy value versus 26 genotypes. *Biomass Bioenergy* **2018**, *119*, 109–120. [CrossRef]
38. Pudełko, R.; Kozak, M.; Jędrejek, A.; Gałczyńska, M.; Pomianek, B. Regionalisation of unutilised agricultural area in Poland. *Pol. J. Soil Sci.* **2018**, *51*, 119. [CrossRef]
39. Scholz, V.; Ellerbrock, R. The growth productivity, and environmental impact of the cultivation of energy crops on sandy soil in Germany. *Biomass Bioenergy* **2002**, *23*, 81–92. [CrossRef]
40. Peacock, L.; Hunter, T.; Turner, H.; Brain, P. Does host genotype diversity affect the distribution of insect and disease damage in willow cropping systems? *J. Appl. Ecol.* **2001**, *38*, 1070–1081. [CrossRef]
41. Karp, A.; Hanley, S.J.; Trybush, S.O.; Macalpine, W.; Pei, M.; Shield, I. Genetic improvement of willow for bioenergy and biofuels free access. *J. Integr. Plant. Biol.* **2011**, *53*, 151–165. [CrossRef]

Publisher's Note: MDPI stays neutral with regard to jurisdictional claims in published maps and institutional affiliations.

 *agriculture*

Article

Biodiversity of Weeds and Arthropods in Five Different Perennial Industrial Crops in Eastern Poland

Paweł Radzikowski *, Mariusz Matyka and Adam Kleofas Berbeć

Department of Systems and Economics of Crop Production, Institute of Soil Science and Plant Cultivation—State Research Institute, Czartoryskich 8, 24-100 Puławy, Poland; mmatyka@iung.pulawy.pl (M.M.); aberbec@iung.pulawy.pl (A.K.B.)
* Correspondence: pradzikowski@iung.pulawy.pl; Tel.: +48-81-4786820

Received: 28 October 2020; Accepted: 9 December 2020; Published: 14 December 2020

Abstract: A growing interest in the cultivation of non-food crops on marginal lands has been observed in recent years in Poland. Marginal lands are a refuge of agroecosystems biodiversity. The impact of the cultivation of perennial industrial plants on the biodiversity of weeds and arthropods have been assessed in this study. The biodiversity monitoring study, carried out for three years, included five perennial crops: miscanthus *Miscanthus* × *giganteus*, cup plant *Silphium perfoliatum*, black locust *Robinia pseudoacacia*, poplar *Populus* × *maximowiczii*, and willow *Salix viminalis*. As a control area, uncultivated fallow land was chosen. The experiment was set up in eastern Poland. A decrease in plant diversity was found for miscanthus and black locust. The diversity of arthropods was the lowest for the cup plant. No decrease in the number of melliferous plants and pollinators was observed, except for the miscanthus. The biodiversity of plants and arthropods was affected by the intensity of mechanical treatments, the fertilization dose, and the use of herbicides. The biodiversity also decreased with the age of plantation.

Keywords: biological diversity; marginal land

1. Introduction

There are many commonly accepted definitions of marginal (fallow) lands. This category includes a wide range of lands that historically has been used for agricultural purposes, but for some reason, their cultivation has been abandoned. The reasons for land marginalization in Poland are of different nature, including local agricultural decline, infrastructural changes, and soil pollution. However, the most important factor of land abandonment is the low productivity potential of the soils. The area of agricultural land excluded from agricultural use reached over two million hectares in Poland in 2018, more than half of which are located on low-quality soils [1]. Perennial industrial crops might be a promising alternative to marginal land utilization. The cultivation of the most productive species, such as willow and poplar, on marginal lands, can be economically viable and bring some environmental benefits [2,3]. On the other hand, some studies indicate perennial industrial crops lead to environmental changes and a decline in biodiversity [4–6]. The impact on biodiversity may vary on the species grown and the used technology. The biodiversity of agroecosystems is largely related to the landscape and crop diversity. On a homogeneous agricultural landscape, perennial industrial crops can introduce an element of landscape enrichment and create new habitats [7–9]. On the other hand, cultivation of such crops might be carried out on lands that are valuable from the perspective of biodiversity conservation. Marginal lands are known to be the biodiversity hotspots in agroecosystems, therefore, the way of their management should take it into consideration [10–13].

Biodiversity plays an important role for agriculture by providing ecosystem services. Marginal lands are the habitat of melliferous plants that provide pollen and nectar for the honeybee *Apis mellifera* L., as well as for wild pollinators. The feed collected from fallow lands is not contaminated with pesticides, in contrast to that collected from most crops and orchards [14]. The fallow lands are also a living habitat for wild pollinators. Wild bees are crucial for certain crops and there are endangered by habitat loss. Many beneficial organisms such as ladybird beetles *Coccinellidae* and parasitic wasps *Ichneumonidae, Aphidiinae* live and reproduce in such habitats [15]. Simultaneously, the management of these lands can be quite challenging. The establishment of the new perennial industrial crop plantation requires intensive weed management both by mechanical and chemical means in the year before plantation establishment. This results in an immediate loss of biodiversity. In general, regardless of the choice of the crop species, it is recommended to increase the density of plantings, due to the expected losses of plants (survival rate). There is no doubt that the cultivation of industrial crops on marginal lands (mostly of poor agricultural quality) will require more inputs than on the land of better agricultural quality [16–18].

The introduction of industrial plants to fallow land does not necessarily lead to a loss of biodiversity in the long term. In most cases, perennial crops require less input than annual crops. In particular, species cultivated in long production cycles reduce the use of plant protection products, fertilizers, and soil disturbances such as tillage. Native crops, such as willow, are the basis of the trophic chain, feeding numerous arthropod taxa. Most fast-growing plants provide benefits for bees and wild pollinators. The litter, formed under perennial crops, can be a convenient habitat for animals that are not able to survive on arable lands [19–21]. The impact of the cultivation of these plants on biodiversity depends on many factors, including the adopted indicators and assessment methods used [22].

The study aims at closing existing knowledge gaps about the impact of industrial perennial crops on biodiversity. The biodiversity of such crops is usually compared to other industrial crops or other annual crops. Moreover, the evaluation of different crops is often made at different sites. Additionally, studies rarely use marginal lands as a control object. Consequently, the added value of the work is the comparison of biodiversity impact of industrial species cultivation under the marginal ground conditions, including the variant with no agricultural management.

As the first hypothesis (H_1), it was presumed that perennial industrial crops can significantly affect the diversity of flora and fauna of marginal lands. This hypothesis is strongly supported by previous studies, especially in the case of floral diversity [6,7]. However, the outputs may be different on marginal lands, as such sites are of insufficient agricultural quality. Cultivation of common agricultural crops on soils of such quality carries the risk of crop cultivation failure. This might open the living space for wild flora, and other wild organisms that benefit biodiversity. On the other hand, the cultivation of crops on marginal lands with increased cultivation intensity may lead to biodiversity loss. The commonly known and accepted dependencies may prove to be missed in those specific conditions.

The study aims at closing knowledge gaps on biodiversity levels on marginal lands which most likely differs under different industrial crop cultivation conditions [16–18]. Marginal lands are usually managed in a different way than agricultural lands of good quality. The impact of the cultivation of crops commonly considered to be beneficial for wild biodiversity, such as willow, can differ when grown under optimal conditions, or under marginal land conditions. Therefore, the second hypothesis was constructed (H_2): The diversity of plants and arthropods will be different in each of the tested crops under the conditions of marginal lands. Therefore, choosing the right crop species should mitigate the loss of biodiversity.

The agricultural management used in specific crops is entirely dependent on farmer management strategy. In order to obtain an optimal yield, basic agrotechnical recommendations must be followed, as well as the adaptation of agrotechnology to marginal soil conditions. Literature study shows that cultivation technology and the intensification of agrotechnical treatments significantly impact biodiversity [4–8]. However, it is not certain whether the same mechanical and chemical treatments will have a similar effect on marginal lands as on other agricultural lands. These uncertainties were the foundations of the third hypothesis (H$_3$): The level of biodiversity of fauna and flora will depend on the intensity of agricultural management practices of perennial industrial crops cultivated on marginal lands.

2. Materials and Methods

2.1. Experiment Setup

The research was conducted in the years of 2018–2020 in the Experimental Station of Institute of Soil Science and Plant Cultivation in Osiny near Puławy in Lublin Province in the eastern part of Poland (51°27′59.2″ N 22°02′48.7″ E). Soil-water conditions were typical for marginal land in Poland. The soil was classified to the 5th class (1st—the best soil class, 6th—the poorest soil class), weak rye soil complex. The groundwater surface was located at an average depth of 2–2.5 m. The acidity (pH in KCl) of cultivated soils near the direct vicinity of the experiment was on average 5.5; however, the soil under industrial crops had lower values of soil acidity (pH = 4.5). The soil was with the prevalence of medium and weakly clay sands as well as small deposits of gravel. In the soil profile, there was a poorly permeable layer of clay at the depth from 50 to 100 cm. The average humus content was 1.06%. At the time of the experiment, the average nitrogen content in soil was 37.8 mg N-NO$_3$ 100 g^{-1}, phosphorus 22.4 mg P$_2$O$_5$100 g^{-1}, and potassium 12.4 mg K$_2$O 100 g^{-1}. The content of magnesium in soil was 4.9 mg MgO 100g^{-1} and calcium 70 mg CaO 100 g^{-1}. The maximum water capacity at the depth of 15 cm was 36%, but during the drought, soil humidity was below 5%. The sum of rainfall in the growing season, which in Eastern Poland is from April to late October was 492.2 mm in 2018 and only 380.9 mm in 2019.

2.2. Structure of Field Experiment

The experiment was carried out on 5 plots with industrial species cultivation and on one control plot. *Miscanthus × giganteus* Greef and Deuter was cultivated on a 280 m^2 plot, at a distance of 135 m from the control area. Prior to the beginning of the experiment, different miscanthus cultivars were grown on that plot for the last 10 years (and were still there when the experiment started). The density of crops at the time of planting was 15 thousand plants ha^{-1}, in a 0.75 × 1 m spacing. Over time, the surface of the experiment was almost completely covered with miscanthus. There were few spots covered with perennial weeds as a result of frost damage to miscanthus. In each year of cultivation, the sum of NPK fertilization was 78-60-90. The fertilization consisted of two treatments. In early April, a multicomponent fertilizer Polifoska 6 (NPK: 6-20-30) was applied at a dose of 300 kg ha^{-1}. Another fertilization was applied in mid-May at a dose of 188 kg ha^{-1} ammonium nitrate (N = 32). At the end of April, chemical weed control using the herbicide MCPA (Chwastox Extra 300SL) was applied at a dose of 1 kg ha^{-1}. The was no mechanical weed control between rows of miscanthus during the time of the experiment. The plant biomass was harvested in the first week of March. An average dry biomass yield was 17.7 Mg ha^{-1} in 2018 and 10.3 Mg ha^{-1} in 2019.

The cup plant *Silphium perfoliatum* L. plot was of 300 m^2 and was not differentiated into varieties. The cultivation of cup plant was directly adjacent to the control object. The year 2018 was the first year of growing this plant. The plantings were obtained from seeds. They were grown under greenhouse conditions from January 2017. In mid-April of 2018, the plants were planted in the field. A spacing of 0.5 × 1m was applied, in order to reach 20 thousand plants per ha. Losses of seedlings on the plantation were replenished from the reserves. The seedlings were irrigated, a total of 1000 dm^3 of water per 300 m^2 of cultivation was used. In the first and second year of cultivation, inter-row cultivation was performed with a soil tiller. Weeds were also removed by hand. The fertilization scheme was exactly the same as in the miscanthus cultivation. The biomass was harvested in January. In the first year, 2.4 Mg ha^{-1} of dry mass was obtained, and in the second year 7.5 Mg ha^{-1} was obtained.

Black locust *Robinia pseudoacacia* L. was cultivated on an area of 3300 m^2. Cultivation of black locust was separated from control by a distance of 850 m, but the soil conditions were very similar.

The cultivation was established 9 years before the beginning of the monitoring. The trees were spaced at 1.5 × 0.5 m, achieving a density of 13 thousand plants per hectare. No mineral fertilization or chemical plant protection treatments were used in black locust cultivation. In May, weeding was carried out between the rows with a mulching mower and by a power scythe. The field was cultivated in a 9-year cycle, the yield was harvested in three phases. Over a period of 3 years, one-third of the plantation was harvested annually. An average annual dry biomass yield was 12.0 Mg ha^{-1}.

Poplar *Populus* × *maximowiczii* Henry plantation was grown on 4500 m^2 area and consisted of 14 varieties. Cultivation of poplar was directly adjacent to the control object and to the cup plant plot. The plantation was established in April 2017 at a density of 18 thousand plants per ha. The cuttings were planted in two-row strips with rows spaced every 0.75 m. The strips were spaced every 1.5 m. The cuttings in each row were planted every 0.50 m. At the beginning of May of the first year of the cultivation, 30 kg N ha^{-1} of ammonium nitrate was applied. In 2018, NPK fertilization in the dose of 78-60-90 ha^{-1}, using Polifoska 6 fertilizer and ammonium nitrate was applied, along with miscanthus and cup plant fields fertilizations. In 2019 and 2020, no fertilizers were used. In the beginning of May 2018, the herbicide Quizalofop-p-ethyl (Supero 05EC) at dose of 50 g ha^{-1} was applied. In the late May, the treatment was repeated using Quizalofop-p-ethyl at the same dose, combined with neonicotinoid insecticide Acetamiprid (Mospilan 20SP) at a dose of 20 g ha^{-1}. Between 2019 and 2020, chemical weed control was no longer necessary. The biomass harvest was performed in February, dry biomass yield was 9.6 Mg ha^{-1} in 2018 and 11.1 Mg ha^{-1} in 2019.

Willow *Salix viminalis* L. was cultivated on an area of 3400 m^2. Distance between willow crop and control plot was 150 m. The plantation was established in April 2016 and consisted of 11 varieties. The same planting technology was used as for poplar. In May 2018, a dose of 300 kg ha^{-1} of Polifoska 6 multi-compound fertilizer was applied, introducing a total of 18 kg N, 60 kg P$_2$O$_5$ and 90 kg K$_2$O per ha. In 2019 no fertilization was used. In 2020, fertilization in the same dose as in 2018 was applied again. In 2018and 2019, no chemical plant protection products or agrotechnical treatments were used in willow cultivation. In January 2020, biomass was collected from the entire field, so there was a risk of higher weed infestation during the regrowing of the willow. Soil tillage was carried out in the paths using a soil tiller in the mid-April and mowing between rows using power scythe in the first decade of June. In mid-June, the insecticide deltamethrin (Decis Mega 50 EW) was applied at a dose of 12.5 g ha^{-1}, due to strong pressure from the *Amphimallon solstitiale* L. beetle on the regrowing willow. In the first week of May, a herbicide quizalofop-p-ethyl (Labrador Extra) at a dose of 50 g ha^{-1} was applied.

An uncultivated area in the direct vicinity of industrial crops has been selected as a control plot. The area of the fallow land was 450 m^2. In the recent years, it was an arable land, the use of which was suspended in 2017, for the purpose of the experiment. There were no fertilization or chemical weed control treatments on this area during the experiment. The spontaneous development of wild vegetation was allowed. Only in the mid-June 2019 cultivation was a tractor flail mulcher carried out. The control plot was designed to simulate marginal land conditions.

2.3. Fauna and Flora Monitoring

The species richness of the flora was examined in mid-June from 2018 to 2020. Five measurements were taken annually in each crop and in the control plot, using a 0.5 × 1 m botanical frame. Measurements were taken every 5 m. Samples were taken in one, 30-meter long transect. In smaller plots, e.g., in a miscanthus and a cup plant, the transect was taken across the plot. All plant specimens in the frame were identified to the species at the place of the experiment. In total, 15 repetitions were obtained for each variant.

The study of arthropod fauna took place over a period of two weeks in mid-June in three consecutive years. Monitoring of the fauna consisted of a system of two types of traps. The fauna living on the soil surface was collected by pitfall traps, 8 cm in diameter and 300 cm^3 capacity. Each trap was filled up with 100 cm^3 of glycol with a detergent, which was grey liquid soap. The container was covered with a 12 cm diameter canopy to protect against precipitation. In each crop, three traps were set up, with 10 m spacing between traps. The traps were set for a period of two weeks. After each invertebrate collection (harvests) the traps were removed and washed. Three harvest cycles were made, one per year. In each year the traps were set over again, in the same locations. In three years, nine repetitions were obtained from each plot. The traps were set in the place where vegetation measurements were taken (Table 1).

Table 1. Structure of monitoring of flora and fauna in the plot.

Distance	Weeds Frame [0.5 × 1 m]	Arthropods Pitfall Traps	Arthropods Yellow Bowls
5 m	Edge of field	–	–
5 m	Frame 1	Trap 1	–
5 m	Frame 2	–	Trap 1
5 m	Frame 3	Trap 2	–
5 m	Frame 4	–	Trap 2
5 m	Frame 5	Trap 3	–
5 m	Edge of field	–	–

The second type of traps were yellow bowls (30 cm wide). The bowls were filled with 1:1 solution of glycol and water in the amount of 2 dm^3. As a detergent, gray liquid soap was added. The solution was supplemented if the hot weather caused evaporation. Yellow traps were placed on a platform. The height of placing of the traps was adjusted to the height of the vegetation, (it varied from 30 to 100 cm). In each crop, two yellow bowls were placed at a distance of 10 m from each other (Table 1). The collection was carried out for two weeks, just as in the case of pitfall traps. After monitoring, bowls and platforms were removed from the fields. During three years of the experiment, six samples were obtained from each plot. After a period of two weeks, the collected material was filtered through a sieve (1 × 1 mm) and preserved with 75% ethanol solution. The collected invertebrates were sorted and determined into orders, families, and species. The material was determined using standard identification keys.

2.4. Statistical Analysis

Species were assigned to families and orders. Ecological functions of individual species were determined. Identified plant species were divided according to their suitability for pollinators, as melliferous and non-melliferous. The arthropods were divided due to their impact on crops as pests and natural enemies of them (serving as biological pest control). For bowl traps also a category of pollinators was added, for pitfall traps detritivores category was added. The data were accompanied by crop parameters such as: the age of the crop, the plot area, the fertilizer doses, the number of mechanical and chemical treatments, and the dry mass yield (DM) of the crop. The results of flora monitoring were also compared with the results of fauna monitoring, in order to identify the impact of wild plant diversity on the arthropod population. The biodiversity indicators, including Shannon's Index and the Dominance Index, were counted using the Past3 software. For most data sets, a distribution different from normal was observed, therefore non-parametric tests were chosen for statistical analysis. The differences between the data distributions were analyzed using the Kruskal–Wallis test, followed by the post-hoc Tukey's test in STATISTICA 10 software. The values of Shannon's Index and Simpson's Dominance Index in subsequent years were compared using the Hutcheson's *t*-test. The influence of factors on the obtained results was described by means of a multiple regression model. Only statistically significant results were presented. The control object was excluded from factor analyses.

3. Results

3.1. Diversity of Weeds

A total of 90 weed flora measurements were made during the three consecutive years of the study. A total of 66 species of weeds were found, representing 22 families within 16 orders. Almost all specimens were identified to species level (Table 2).

Table 2. Frequency of the weed species in five different perennial industrial crops: FL—Fallow land (control), MG—*Miscanthus × giganteus* Greef and Deuter, SP—*Silphium perfoliatum* L., RP—*Robinia pseudoacacia* L., PM—*Populus × maximowiczii* Henry, SW—*Salix viminalis* L.

Taxa			Average Number in the Sample					
Order	Family	Species	FL	MG	SP	RP	PM	SW
Asterales	Asteraceae	*Achillea millefolium*	–	–	–	–	–	0.2
		Anthemis arvensis	14	–	–	–	–	–
		Artemisia vulgaris	–	–	–	–	–	10.4
		Centaurea cyanus	0.6	–	–	–	–	–
		Erigeron annuus	1.2	–	8.2	–	–	3.4
		Erigeron canadensis	12.6	–	5	–	7.8	2.6
		Galinsoga parviflora	–	–	0.6	–	0.4	–
		Galinsoga quadriradiata	–	–	0.8	–	–	–
		Gnaphalium sylvaticum	–	–	–	0.4	–	–
		Lapsana communis	–	2.4	–	–	–	–
		Senecio vulgaris	–	–	4.2	–	–	–
		Solidago canadensis	8	–	–	26	–	9.4
		Sonchus arvensis	4.2	–	4.6	–	0.2	0.8
		Taraxacum officinale	2	–	8.2	–	0.2	6.4
		Tragopogon pratensis	1.4	–	–	–	–	–
		Tripleurospermum maritimum	20	–	1.8	–	20.6	–
Boraginales	Boraginaceae	*Anchusa arvensis*	1.4	–	–	–	–	–
		Anchusa officinalis	–	–	0.2	–	–	–
Brassicales	Brassicaceae	*Camelina sativa*	–	–	0.2	–	–	–
		Capsella bursa-pastoris	12.2	2.4	2.6	–	1.8	0.6
Caryophyllales	Amaranthaceae	*Amaranthus retroflexus*	18.2	–	–	–	0.2	0.6
		Chenopodium album	90.8	0.2	7.2	0.8	43.2	0.4
	Caryophyllaceae	*Cerastium arvense*	–	–	50.2	–	–	12
		Silene latifolia	0.8	–	–	–	–	0.4
		Stellaria media	6.6	–	16.2	0.4	–	14.4
	Polygonaceae	*Fallopia convolvulus*	–	–	0.4	–	–	0.4
		Polygonum aviculare	–	–	2.4	–	0.4	0.2
		Polygonum lapathifolium	–	–	–	–	1.2	–
		Polygonum persicaria	2.6	–	–	–	–	–
Fabales	Fabaceae	*Trifolium repens*	–	–	0.6	–	–	0.8
		Vicia cracca	–	–	0.6	–	0.2	–
		Vicia villosa	0.2	–	–	–	–	–
Gentianales	Rubiaceae	*Galium aparine*	–	–	–	45.2	–	29.8
Geraniales	Geraniaceae	*Erodium cicutarium*	7.2	–	1.4	–	6.8	2.4
		Geranium pratense	0.2	–	3.8	–	–	–
		Geranium pusillum	–	–	0.6	–	–	–
		Geranium robertianum	–	–	–	3.2	–	–
Lamiales	Lamiaceae	*Galeopsis tetrahit*	–	–	1	–	0.6	–
	Plantaginaceae	*Plantago major*	–	–	–	–	0.2	–
		Veronica arvensis	–	0.4	–	–	–	0.2
		Veronica chamaedrys	–	1.4	0.6	–	0.4	0.2
		Veronica persica	1.4	–	–	–	–	–
Malpighiales	Violaceae	*Viola arvensis*	3	2.6	1.8	1.4	5	7
Malvales	Malvaceae	*Malva sylvestris*	0.2	–	0.2	–	–	–
		Sida hermaphrodita	–	–	–	–	–	1.4
Oxalidales	Oxalidaceae	*Oxalis acetosella*	–	–	–	–	0.2	–

Table 2. *Cont.*

| Taxa | | | Average Number in the Sample | | | | | |
Order	Family	Species	FL	MG	SP	RP	PM	SW
Ranunculales	*Papaveraceae*	*Papaver rhoeas*	2.6	–	–	–	–	–
Rosales	*Rosaceae*	*Geum rivale*	–	–	–	–	–	0.4
	Urticaceae	*Urtica dioica*	–	–	–	2.6	–	–
Solanales	*Convolvulaceae*	*Convolvulus arvensis*	3.2	–	1	0.2	–	–
	Cyperaceae	*Carex* sp.	–	–	–	2	–	–
	Juncaceae	*Luzula campestris*	–	–	–	–	0.2	–
		Agrostis capillaris	2	–	–	–	0.8	–
		Apera spica-venti	2.6	23.2	–	–	2.8	–
		Bromus inermis	89.2	22.8	–	–	–	–
		Calamagrostis epigejos	–	–	–	4.8	–	–
		Echinochloa crus-galli	19.8	–	0.4	–	–	–
Poales		*Elymus repens*	69.2	58.8	1	16	2	4
	Poaceae	*Elymus elongatus*	–	–	4	–	–	–
		Festuca rubra	–	–	–	–	–	24
		Holcus mollis	–	–	–	–	–	0.2
		Lolium perenne	–	–	0.2	–	–	–
		Poa annua	29.6	13.6	66.6	–	2.4	11.2
		Secale cereale	–	–	–	0.2	–	–
		xTriticosecale	12	–	–	–	1.6	–
Equisetales	*Equisetaceae*	*Equisetum arvense*	1.2	–	1.6	0.2	1.4	0.4

The number of species in different crops was found to vary significantly (chi^2 = 62.06, $p < 0.01$). In the control site, which was fallow land, 32 species were found in total, while the median of the species number was 9. The average (median) number of species was the highest on the control site. The total number of species found in cup plant plantation was the same as on the control site (32), but on average (median) there were 2 weed species less in the cup plant than on fallow land. Ten species were found in total in miscanthus (with a median value of 3 species). Fourteen weed species were found in black locust, with a median value of 3. In poplar, 24 species were found in total, with an average (median) of five. Twenty-eight species were found in total in willow, with 7 species average (median). Cup plant and willow did not differ significantly in weed species number between with the control object (fallow land). The average number of species found in miscanthus, black locust and poplar were significantly lower than those obtained in the control. In addition, some significant differences in weed number were observed for different crops (chi^2 = 23.68, $p < 0,01$). The average number of species was significantly greater for the cup plant than for the miscanthus, black locust, and poplar. The number of weed species of willow was significantly greater than for miscanthus and black locust. There was significantly more weed species found in Poplar than in miscanthus. There were no significant differences in species number between willow and poplar, black locust and poplar, and black locust and miscanthus (Table 2). Significant differences were found between miscanthus and other tested objects, with the exception of the black locust. The results indicate that weed diversity of willow, poplar, and cup plant is lower than the control object, but the differences are low (Table 2). In the case of all crops, the number of wild plants (weeds) per m^2 was significantly lower than in the control object (fallow land). This is undoubtedly the effect of the domination of crops over weeds. The number of melliferous plants in the tested crops was also analyzed (chi^2 = 35.99, $p < 0.01$). Only the miscanthus had a significantly lower number of melliferous plants than the control. Additionally, significantly more melliferous plants were found in the cup plant than in the miscanthus. The remaining crops did not differ significantly. Significantly more plants which were not beneficial for pollinators were found on the unmanaged fallow land than in the cultivation of perennial industrial plants (chi^2 = 41.13, $p < 0.01$). On the basis of these observations, it can be concluded that there are fewer wild plants in perennial crops than on fallow land, but weeds of fallow land represent mostly species that are not suitable for pollinating insects. This means that tested perennial crops, with the exception of miscanthus, should provide at least the same level of food for pollinators as fallow land.

The results of the comparison of plant biodiversity indicators using the Kruskal-Wallis ANOVA test are shown in Table 3.

Table 3. Median and median error values of selected weed diversity indicators in perennial industrial crops in three-year monitoring. Significantly different results ($p < 0.05$) were marked with different letters.

Flora by the Frame Method	Crop					
	FL	MG	SP	RP	PM	SW
Number of weed species	9 ± 1 a	3 ± 0.3 d	7± 0.7 a	3 ± 0.4 cd	5 ± 0.5 bc	7 ± 0.7 ab
Number of weeds [m²]	288 ± 67 a	74 ± 20 b	34 ± 45 b	66 ± 9 b	46 ± 16 b	68 ± 19 b
Number of melliferous [m²]	62 ± 21 a	2 ± 3 b	14 ± 35 a	52 ± 6 ab	18 ± 9 ab	54 ± 11 ab
Number of non-melliferous [m²]	190 ± 63 a	60 ± 20 b	18 ± 27 b	2 ± 7 b	18 ± 13 b	12 ± 18 b

FL—Fallow land (control), MG—*Miscanthus × giganteus.* Greef and Deuter, SP—*Silphium perfoliatum* L., RP—*Robinia pseudoacacia* L., PM—*Populus × maximowiczii* Henry, SW—*Salix viminalis* L.

The results from each year were tested with Hutcheson's t-test for Shannon's biodiversity index. Significant differences ($p < 0.01$) were found between tested crops and control in individual years. In the first year of the study, lower plant biodiversity was found in miscanthus, black locust, and poplar than in cup plant, willow, and control plot. The Shannon diversity index in the cup plant was lower than in the control and willow cultivation. In the second year of the study, a decrease in plant diversity on the control plot was observed which resulted in significantly higher results of Shannon index in the cup plant, poplar, and willow (than on fallow land). The Shannon diversity index of weeds of miscanthus and black locust crops was lower than in the control in 2019. Shortly after biodiversity monitoring in 2019, inter-row cultivation was carried out in control and in black locust, which increased the diversity of weeds in the next year. In 2020, fallow land and plantation of willow had the highest level of Shannon index diversity of weeds (Figure 1). The diversity of weeds also increased in black locust, after the application of inter-row cultivation. Miscanthus had the lowest Shannon diversity index each year. The increase in plant diversity in poplar can be explained by the drought observed in 2019, which caused litter damage and exposing the soil, which allowed for weed seeds germination. The decrease in weed diversity in willow cultivation could be explained by the biomass harvest at the beginning of 2020. In spring 2020, it was possible to undertake attempts to mechanically limit weed infestation in the re-growing willow. In 2018 and 2019, in the cup plant field, the inter-row cultivation was applied, which was not carried out in 2020 due to the rapid growth of the plant. Most likely, the observed decrease in diversity was associated with the abandonment of weed management treatments or with the soil shading by the crop (or both). The impacts of agrotechnical factors on the diversity of the weed flora is analyzed in more detail in Figures 3–8 in this study.

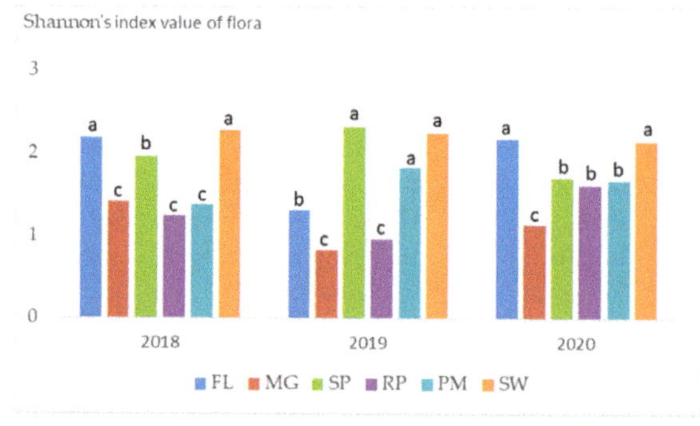

Figure 1. Shannon's Index values of plant diversity compered using Hutcheson's t-test in three consecutive years. Statistically significant differences ($p < 0.01$) were marked with different letters. FL—Fallow land (control), MG—*Miscanthus × giganteus* Greef and Deuter, SP—*Silphium perfoliatum* L., RP—*Robinia pseudoacacia* L., PM—*Populus × maximowiczii* Henry, SW—*Salix viminalis* L.

The Simpson's Dominance Index indicates low plant diversity in the habitat or a strong dominance of one of the species. In 2018 the lowest value (thus the best biodiversity "performance") of this indicator was observed in the control plot and plantation of willow. The remaining objects had significantly higher domination of single species of weeds. The highest domination index values were found in plantations of poplar and black locust. In 2019 and 2020 the highest values of domination index were found for weeds of miscanthus. The value of this indicator varied strongly over the years (Figure 2).

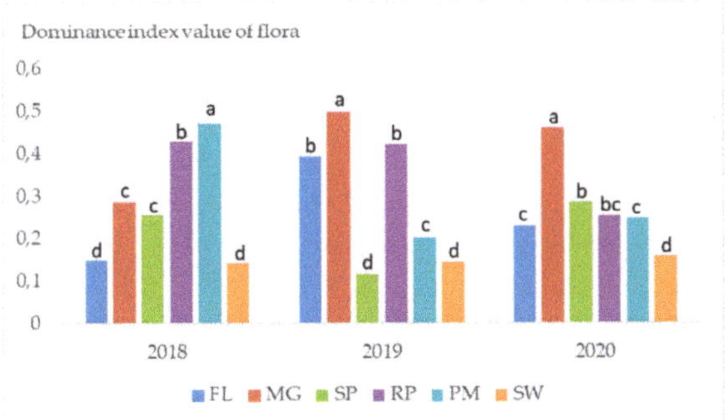

Figure 2. Simpson's Dominance Index values of plants compared using Hutcheson's t-test in three consecutive years. Statistically significant differences ($p < 0.01$) were marked with different letters. FL—Fallow land (control), MG—*Miscanthus × giganteus* Greef and Deuter, SP—*Silphium perfoliatum* L., RP—*Robinia pseudoacacia* L., PM—*Populus × maximowiczii* Henry, SW—*Salix viminalis* L.

Different weed species dominated in different plots. In control object, *Poa annua* L., *Elymus repens* L., dominated in 2018. In 2019, *Amaranthus retroflexus* L. and *Chenopodium album* L. were observed in large numbers. In the third year, grasses such as *Bromus inermis* Leyss and *Elymus repens* L were the most dominant species. The same three species of grasses were dominant in miscanthus in all 3 years of the study. Changes in the dominance of species were much more dynamic in cup plant due to intensive cultivation of this plantation. *Stellaria media* L. and *Chenopodium album* L. were the most dominant species in 2018. Species distribution of weeds was very even. The most common species were *Equisetum arvense* L., *Taraxacum officinale* Wigg., *Poa annua* L., and *Erigeron canadensis* L. *Galium aparine* L., *Elymus repens* L., *Solidago canadensis* L., and *Calamagrostis epigejos* L. were present in high numbers in every year on the plantation of black locust. A similar situation was observed for poplar. Four species were the most dominant in different years: *Apera spica-venti*, *Chenopodium album*, *Erigeron canadensis* and *Tripleurospermum maritimum*. Soil under poplar and under black locust was mostly covered by litter, leaving little space for weeds. Willow showed no domination of any weed species as the weed community was very diverse. The most common species were: *Festuca rubra* L., *Galium aparine* L., *Cerastium arvense* L., *Solidago canadensis* L., *Elymus repens* L., *Poa annua* L., and *Stellaria media* L.

In order to describe the strength of the impact of the agrotechnical factors on plant biodiversity, multiple regression was performed. It was found that the number of species is negatively affected by two factors: the age of plantation and the number of agrotechnical treatments (Figure 3). Despite the lack of correlation between plant density and agrotechnical factors, the analysis of variance revealed five factors, which significantly affected the number of weeds per m^2. The use of herbicides, the age of the plantation, and the use of phosphorus fertilizer had a positive effect on weed densities (Figure 4). Nitrogen fertilization and crop yield had a negative effect on weed population density, especially in the case of melliferous plants (Figure 5). Factors such as the level of potassium fertilization and, surprisingly, pesticide use also had a positive effect on the number of melliferous plants. Herbicides in perennial industrial crops probably had a negative impact mainly on grasses, with little or no effect on flowering plants. However, the use of herbicides and the age of plantation had a positive effect on the density of non-melliferous plants (Figure 6). The age of the plantation and herbicide use largely explain the decrease of Shannon's diversity index (Figure 7). It was also found the value of the Dominance Index increases with the age of plantation and the number of herbicide treatments applied (Figure 8).

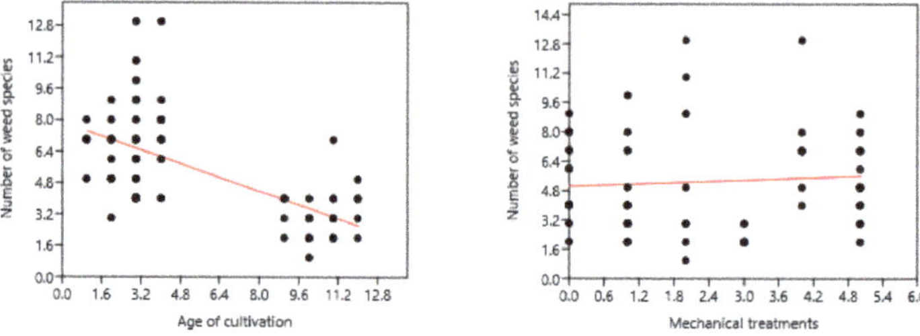

Figure 3. Two independent variables affecting number of weed species in perennial industrial crops: x_1—Year of cultivation, x_2—Mechanical treatments. $Y = 9.073 - 0.557x_1 - 0.303x_2$. $R^2 = 0.630$, $p < 0.001$.

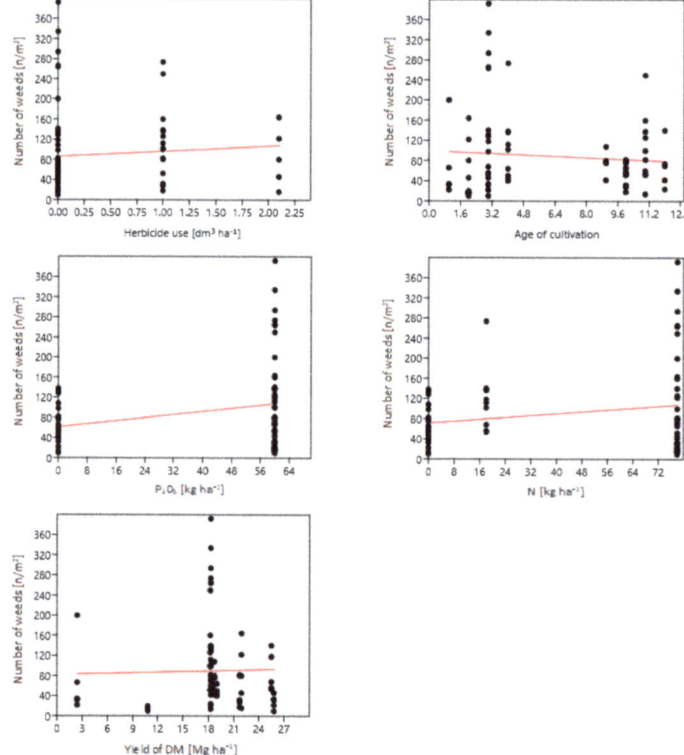

Figure 4. Five independent variables affecting number of weeds in perennial industrial crops: x_1—Herbicide use [dm^3 ha^{-1}], x_2—Age of cultivation, x_3—P$_2$O$_5$ [kg ha^{-1}], x_4—N [kg ha^{-1}], x_5—Yield of DM [Mg ha^{-1}]. $Y = 113.94 + 53.86x_1 + 5.28x_2 + 1.85x_3 - 2.00x_4 - 4.61x_5$. $R^2 = 0.289$, $p = 0.08$.

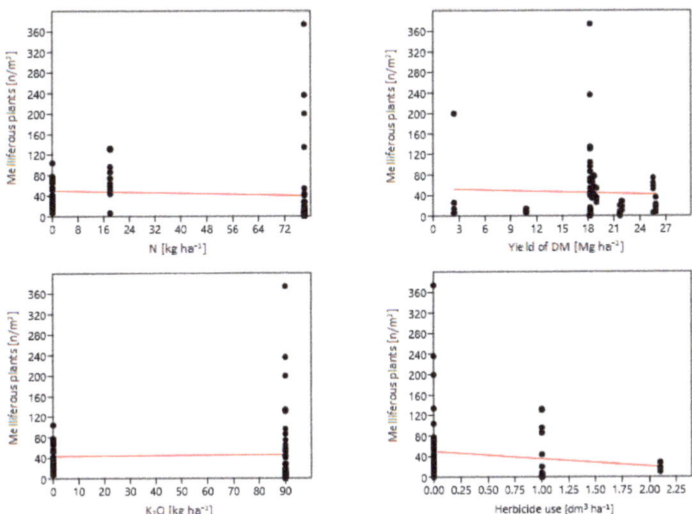

Figure 5. Four independent variables affecting melliferous plants in perennial industrial crops: x_1—N [kg ha^{-1}], x_2—Yield of DM [Mg ha^{-1}], x_3—K$_2$O [kg ha^{-1}], x_4—Herbicide use [dm^3 ha^{-1}]. $Y = 125.93 - 1.61x_1 - 4.05x_2 + 0.63x_3 + 25.15x_4$. $R^2 = 0.350$, $p = 0.001$.

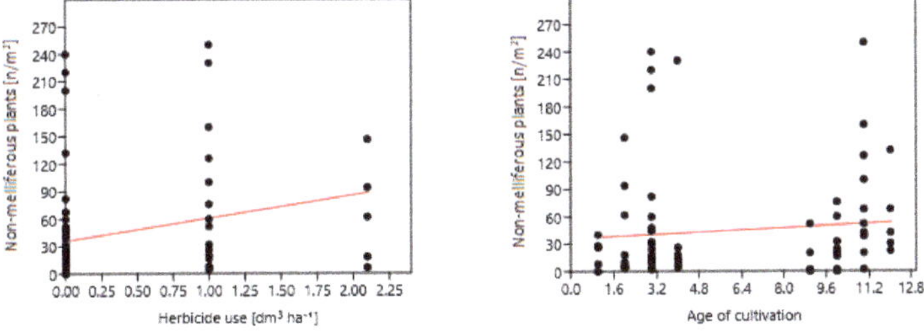

Figure 6. Two independent variables affecting non-melliferous plants in perennial industrial crops: x_1—Herbicide use [dm^3 ha^{-1}], x_2—Age of cultivation, x_3—P$_2$O$_5$ [kg ha^{-1}]. $Y = -16.75 + 23.11x_1 + 4.24x_2 + 0.52x_3$. $R^2 = 0.386$, $p < 0.001$.

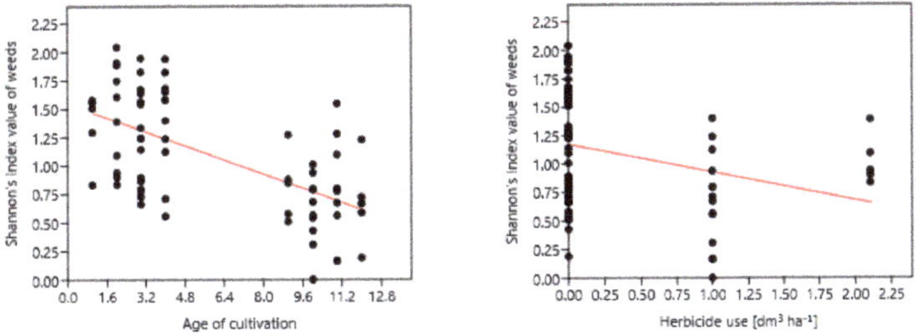

Figure 7. Two independent variables affecting Shannon's Index value of weeds in perennial industrial crops: x_1—Age of cultivation, x_2—Herbicide use [dm^3 ha^{-1}]. $Y = 1.763 - 0.098x_1 - 0.248x_2$. $R^2 = 0.634$, $p < 0.001$.

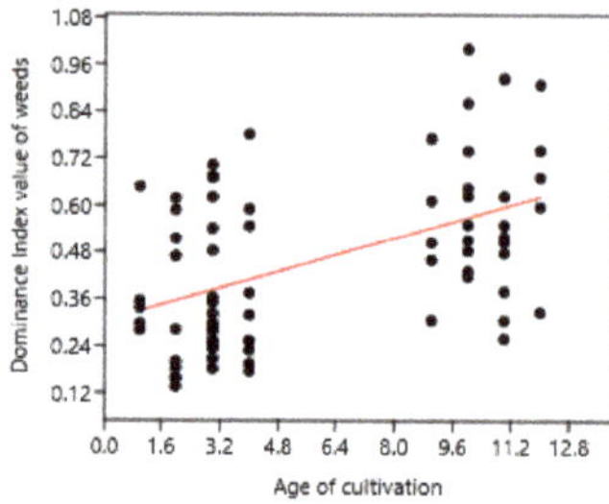
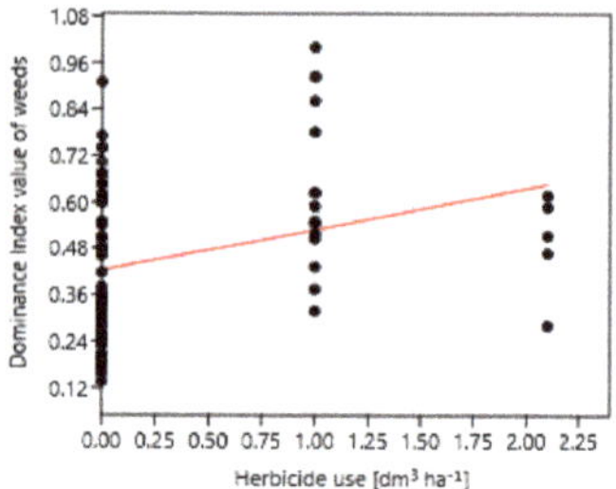

Figure 8. Two independent variables affecting Dominance Index value of weeds in perennial industrial crops: x_1—Age of cultivation, x_2—Herbicide use [dm^3 ha^{-1}]. Y = 0.157 + 0.038x_1 + 0.084x_2. R^2 = 0.529, $p < 0.00$.

3.2. Diversity of Arthropods in Pitfall Traps

During the monitoring, a total of 54 pitfall trap samples were collected, giving 9 repetitions in each crop. Over 13 000 arthropod specimens belonging to 19 orders were identified: *Acari, Araneae, Blattodea, Coleoptera, Dermaptera, Diptera, Entomobryomorpha, Hemiptera, Hymenoptera, Isopoda, Julida, Lepidoptera, Lithobiomorpha, Neuroptera, Opiliones, Orthoptera, Siphonaptera, Symphypleona,* and *Thysanoptera*. Sixty-seven arthropod families and 85 taxa have been identified. Few invertebrates were identified to species level (Table 4.)

Table 4. Frequency of the arthropod taxa in pitfall traps in five different perennial industrial crops: FL—Fallow land (control), MG—*Miscanthus × giganteus* Greef and Deuter, SP—*Silphium perfoliatum* L., RP—*Robinia pseudoacacia* L., PM—*Populus × maximowiczii* Henry, SW—*Salix viminalis* L.

Taxa			Average Number in the Sample					
Order	Family	Species	FL	MG	SP	RP	PM	SW
Acari	–	–	2.9	0.2	0.7	3.1	1.4	1.8
	Trombidiidae	*Trombidium holosericeum*	–	–	–	–	0.1	–
Araneae	*Agelenidae*	–	0.8	0.1	0.1	0.1	0.2	0.2
	Araneidae	–	0.1	0.1	0.1	0.3	0.2	1.2
	Clubionidae	–	0.6	0.1	0.4	1.3	0.9	0.2
	Eutichuridae	*Cheiracanthium punctorium*	–	–	–	–	–	0.1
	Linyphiidae	–	2.6	2.4	4.4	3.2	47.0	38.4
	Lycosidae	–	25.2	18.2	1.9	19.4	5.4	5.7
	Salticidae	–	–	0.1	0.1	–	–	–
	Tetragnathidae	–	–	0.1	–	–	–	–
	Thomisidae	–	0.4	1.2	–	2.3	0.4	1.3
Blattodea	*Ectobiidae*	*Ectobius lapponicus*	–	–	–	0.2	–	–
Coleoptera	*Cantharidae*	-	0.1	0.2	–	–	–	–
	Carabidae	-	52.0	38.7	32.6	30.9	66.8	52.2
		Cicindela hybrida	0.3	–	–	–	–	–
	Chrysomelidae	*Chrysomela populi*	–	–	–	–	0.1	–
		–	6.4	0.2	1.6	2.2	1.3	1.3
	Coccinellidae	–	4.6	1.7	6.1	2.2	1.4	1.2
	–	–	1.8	4.4	3.9	5.0	2.0	0.4
	Curculionidae	–	0.3	0.1	2.2	–	–	0.8
		Otiorhynchus sp.	0.2	5.8	–	0.1	0.1	1.3
	Dytiscidae	–	–	–	–	–	0.1	–
	Elateridae	–	–	1.0	0.2	0.1	1.4	0.6
	Geotrupidae	*Geotrupes stercorarius*	–	–	–	0.3	–	–
	Histeridae	–	0.7	–	–	0.1	–	–

Table 4. *Cont.*

Taxa			Average Number in the Sample					
Order	Family	Species	FL	MG	SP	RP	PM	SW
	Scarabaeidae	*Amphimallon solstitiale*	0.1	–	–	0.1	–	2.2
		Anomala dubia	1.4	–	0.1	–	0.1	–
		Cetoniinae sp.	–	–	–	0.1	–	–
		Melolonthinae sp.	–	0.2	–	–	0.1	–
		Phyllopertha horticola	0.1	–	–	–	–	–
	Silphidae	–	–	–	–	–	0.6	–
		Sphecidae sp.	0.3	–	–	–	–	–
	Staphylinidae	*Staphylinidae* sp.	7.6	6.6	36.6	10.9	7.6	15.3
	Tenebrionidae	–	0.2	6.1	1.1	–	0.7	–
Dermaptera	Forficulidae	*Forficula auricularia*	–	3.6	0.7	0.1	1.3	2.7
	Asilidae	–	0.2	–	–	–	0.1	–
	Bibionidae	–	0.3	2.2	6.8	3.6	0.6	5.9
	Culicidae	–	–	–	–	–	0.1	–
Diptera	Muscidae	–	34.1	23.4	30.4	26.3	25.4	41.9
	Empididae	–	–	0.1	–	0.1	–	–
	Syrphidae	–	15.0	0.6	0.9	2.3	0.4	0.4
	Tipulidae	–	–	–	–	0.3	1.4	0.4
Entomobryomorpha	Entomobryidae	–	120.1	5.4	14.2	31.8	15.0	10.2
	Aphididae	–	6.6	3.3	20.3	1.9	1.1	5.9
	Cicadellidae	–	2.8	2.9	2.8	7.7	1.3	7.6
	–	–	–	–	2.9	0.8	0.9	1.1
Hemiptera	Miridae	–	27.3	0.6	3.0	1.8	1.2	0.6
	Nabidae	–	0.2	–	–	–	–	–
	Pentatomidae	–	0.4	–	–	–	–	–
	Scutelleridae	–	0.2	–	–	–	–	–
	Apidae	–	0.6	0.1	0.1	–	–	0.1
		Apis mellifera	0.1	–	–	–	–	–
	Cephidae	–	–	–	–	–	–	0.1
		Formica fusca	–	0.2	–	0.1	–	0.1
		Formica rufa	0.1	–	–	–	–	–
	Formicidae	–	0.4	–	–	0.2	–	–
Hymenoptera		*Lasius niger*	44.8	55.6	50.8	8.8	4.8	16.0
		Myrmica rubra	0.3	11.1	0.4	0.6	–	2.7
	Sphecidae	–	0.1	–	–	–	–	–
	Tenthredinidae	–	–	0.9	1.8	1.0	3.1	3.4
		Aphidiinae sp.	2.2	1.0	3.0	4.1	2.8	3.9
	Vespidae	*Ichneumonidae* sp.	0.7	0.1	–	0.6	0.3	0.6
		–	3.2	0.1	-	–	–	–
		Vespinae sp.	5.4	0.1	0.2	–	0.1	–
Isopoda	Porcellionidae	*Porcellio scaber*	–	–	0.1	–	–	–
Julida	Blaniulidae	–	–	–	–	0.1	0.2	–
	Julidae	–	–	–	–	0.1	–	–
	–	–	2.6	0.9	0.8	3.7	1.7	0.6
	Nymphalidae	–	–	–	0.3	0.2	–	–
Lepidoptera	Pieridae	–	–	–	0.1	–	–	–
	Silphidae	–	–	–	–	–	–	0.1
	Tineidae	–	–	-	0.1	1.0	–	–
Lithobiomorpha	Lithobiidae	*Lithobius forficatus*	–	1.7	0.7	0.3	3.3	3.7
Neuroptera	Chrysopidae	*Chrysopa* sp.	–	–	–	–	0.2	0.1
	Hemerobiidae	2212	–	–	–	–	0.1	0.1
Opiliones	2212	2212	0.7	3.0	3.8	0.2	9.9	2.9

Table 4. *Cont.*

Taxa			Average Number in the Sample					
Order	Family	Species	FL	MG	SP	RP	PM	SW
Orthoptera	Acrididae	*Acridinae* sp.	–	0.1	–	–	–	–
		Chorthippus albomarginatus	–	–	0.1	–	–	–
		Chrysochraon dispar	–	0.1	–	–	–	–
	Gryllidae	*Gryllus campestris*	0.4	–	–	–	–	–
	Tetrigidae	*Tetrix* sp.	–	–	–	–	–	0.2
		Tetrix subulata	0.2	–	–	–	–	–
	Tettigoniidae	*Pholidoptera griseoaptera*	–	0.1	–	–	–	–
		–	–	0.1	–	0.2	–	–
Siphonaptera	Pulicidae	–	0.2	–	–	0.1	–	–
Symphypleona	Sminthuridae	–	0.8	1.4	0.1	0.8	–	1.9
Thysanoptera	–	–	1.4	0.3	0.2	–	0.1	1.4

No significant differences in the number of invertebrates were found between tested plots (Table 5). Statistically significant differences occurred in the number of identified species (chi^2 = 12.29, p = 0.03). The highest number of taxa was found in the control plot, where the number of taxa (species) was 50 and the average (median) value was 21. Significantly lower values (than for fallow land) were found in the cultivation of cup plant and poplar. The remaining crops did not differ from each other. There were no differences in the number of natural enemies of pests between the studied objects nor for the detritivores. The number of taxa that were classified as potential crop pests was significantly lower in poplar than in the control plot, miscanthus, and cup plant (chi^2 = 30.03, p < 0.01). It should be noted that the same taxonomic groups may be identified as pests or detritivores, depending on the crop they are found in. On average, the highest abundance of pests, natural enemies, and saprophytes were found on the control plot.

Table 5. Median and median error values of selected indexes of ground arthropod diversity in perennial industrial crops in three-year monitoring. Significantly different results (p < 0.05) were marked with different letters.

Arthropods in Pitfall Traps	Crop					
	FL	MG	SP	RP	PM	SW
Number of arthropods	357 ± 98 a	214 ± 41 a	192 ± 43 a	186 ± 24 a	125 ± 75 a	193 ± 57 a
Number of species	21 ± 1.5 a	19 ± 1.3 ab	18 ± 0.6 b	19 ± 1.1 ab	19 ± 0.6 b	22 ± 1.5 ab
Number of natural enemies of pests	109 ± 27 a	93 ± 18 a	67 ± 25 a	73 ± 13 a	61 ± 65 a	59 ± 51 a
Number of potential pests	86 ± 25 a	64 ± 29 a	92 ± 21 a	34 ± 7 ab	20 ± 4 b	59 ± 7 ab
Number of detritivores	161 ± 94 a	65 ± 29 a	105 ± 21 a	79 ± 19 a	61 ± 32 a	99 ± 32 a

FL—Fallow land (control), MG—*Miscanthus* × *giganteus.* Greef & Deuter, SP—*Silphium perfoliatum* L., RP—*Robinia pseudoacacia* L., PM—*Populus* × *maximowiczii* Henry, SW—*Salix viminalis* L.

A large variation in the value of Shannon's diversity index in different years was observed. At the beginning of the experiment, the greatest diversity of terrestrial arthropods was found in the control plot. Some differences were also found between the crops. In poplar and willow in 2018, the diversity of arthropods was found to be lower than in the miscanthus, cup plant, and black locust. The lowest diversity of arthropods was found in the control field in the second year of the study. The greatest diversity was found in the black locust, poplar, and willow. In the third year of the study, the diversity of arthropods decreased in the control plot. Higher values were observed in the cup plant crop, and even greater in black locust and poplar. In 2020, the greatest value of Shannon's diversity index was found for the willow and miscanthus. Large fluctuations in the diversity of invertebrates were observed over the years. A gradual decrease of the diversity of the control was observed (Figure 9).

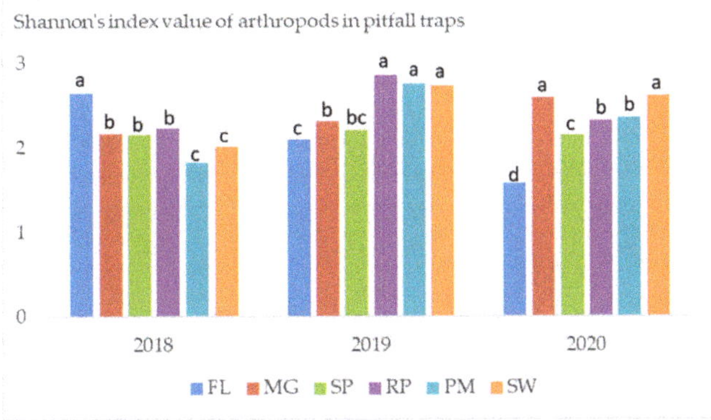

Figure 9. Shannon's Index values of arthropod diversity in pitfall traps compered using Hutcheson's t-test in three following years. Statistically significant differences ($p < 0.01$) were marked with different letters. FL—Fallow land (control), MG—*Miscanthus × giganteus* Greef and Deuter, SP—*Silphium perfoliatum* L., RP—*Robinia pseudoacacia* L., PM—*Populus × maximowiczii* Henry, SW—*Salix viminalis* L.

Simpson's Dominance Index value for ground arthropods was low in most of the samples. Exceptionally high Dominance Index value was found in the control plot in 2020 and it was caused by very high number of *Entomobryidae* springtails. Gradual growth of Dominance Index value was observed in control, while in perennial industrial crops it varied between years (Figure 10). There were significant differences between values of domination according to Hutcheson's t-test, but none of the crops had the lowest nor the highest index value across all three years of the experiment. Ground beetles *Carabidae* and black garden ants *Lasius niger* L. were also very numerous in control object, as well as in miscanthus and cup plant. The rove beetles *Staphylinidae* were the most abundant taxa of arthropods in willow, polar and black locust, which could be relevant from the perspective of natural protection of crops. Another taxas which were present in large numbers in every crop and in the control were wolf spiders *Lycosidae*.

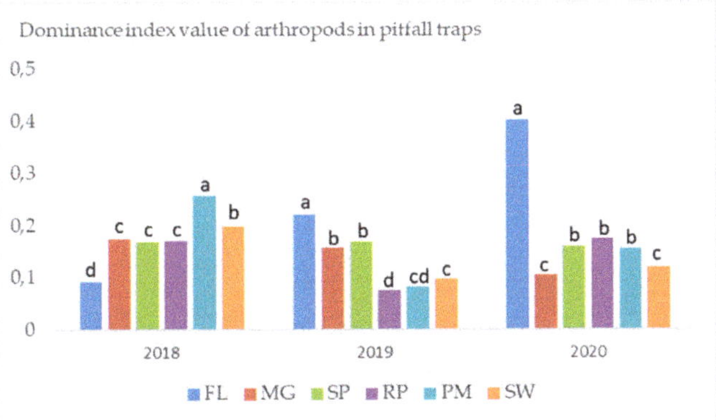

Figure 10. Simpson's Dominance Index values of arthropods in pitfall traps compared using Hutcheson's t-test in three following years. Statistically significant differences ($p < 0.01$) were marked with different letters. FL—Fallow land (control), MG—*Miscanthus × giganteus* Greef and Deuter, SP—*Silphium perfoliatum* L., RP—*Robinia pseudoacacia* L., PM—*Populus × maximowiczii* Henry, SW—*Salix viminalis* L.

None of the agrotechnical factors explained differences in the number of taxa. A higher level of phosphorus fertilization was a factor increasing the number of arthropod individuals in the samples

(Figure 11), including natural enemies of pests (Figure 12). The number of pests decreased with the size of plantations (Figure 13). Phosphorus fertilization had a positive effect on the number of detritivores (Figure 14). A higher level of phosphorus fertilization also influenced lower values of Shannon's index of arthropod population (Figure 15). The level of potassium fertilization increased the value of the Dominance Index (Figure 16). The obtained results indicate that the diversity of ground insects may decrease with increasing intensity of agrotechnical treatments, however, some beneficial arthropod groups such as predators and detritivores may become more numerous under these conditions.

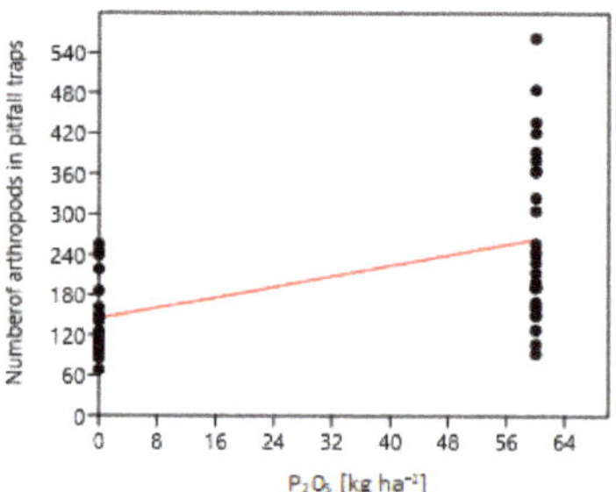

Figure 11. Independent variable x_1—P_2O_5 [kg ha^{-1}] affecting number of arthropods in pitfall traps in perennial industrial crops. $Y = 132.50 + 2.97x_1$. $R^2 = 0.430, p < 0.001$.

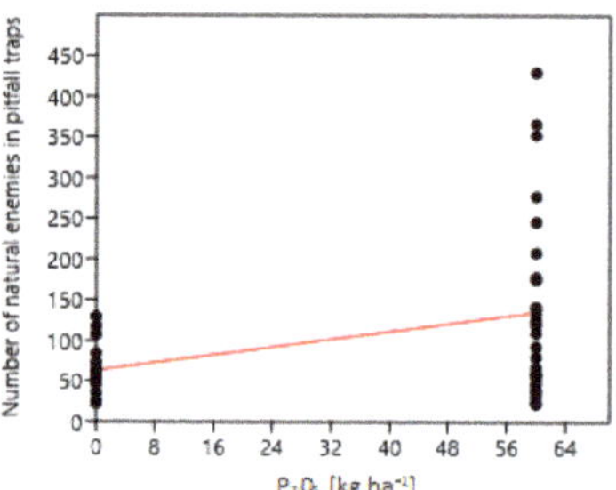

Figure 12. Independent variable x_1—P_2O_5 [kg ha^{-1}] affecting number of natural enemies of pests in pitfall traps in perennial industrial crops. $Y = 64.250 + 1.985x_1$, $R^2 = 0.320, p = 0.001$.

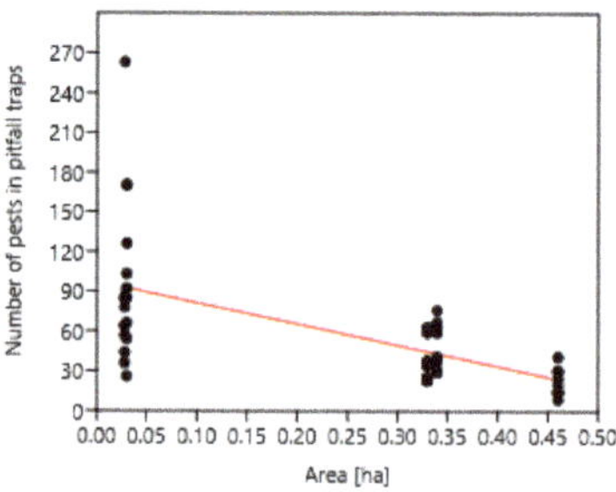

Figure 13. Independent variable x_1—Area [ha] affecting number of pests in pitfall traps in perennial industrial crops. $Y = 112.10 - 195.14x_1$. $R^2 = 0.408, p < 0.001$.

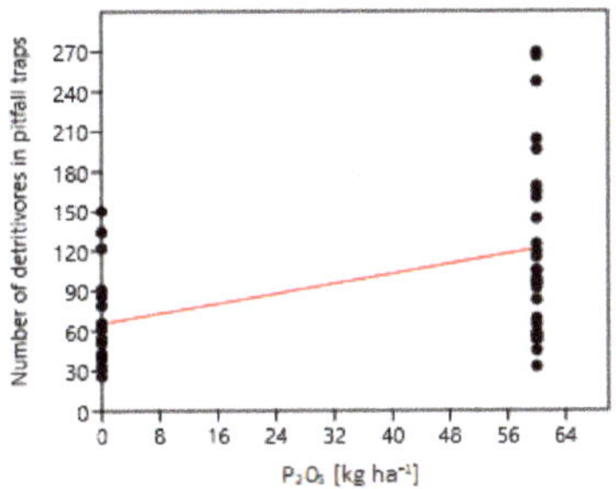

Figure 14. Independent variable x_1—P_2O_5 [kg ha^{-1}] affecting number of detritivores in pitfall traps in perennial industrial crops. $Y = 49.583 + 1.434x_1$. $R^2 = 0.341$, $p < 0.001$.

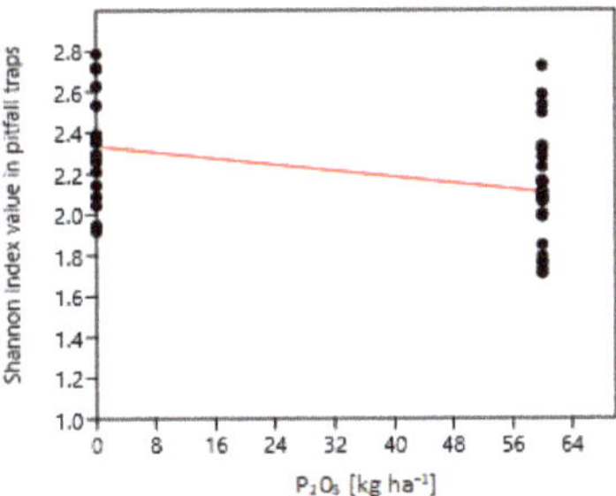

Figure 15. Independent variable x_1—P_2O_5 [kg ha^{-1}] affecting Shannon's Index value of arthropods in pitfall traps in perennial industrial crops. $Y = 2.385 - 0.007x_1$. $R^2 = 0.397$, $p < 0.001$.

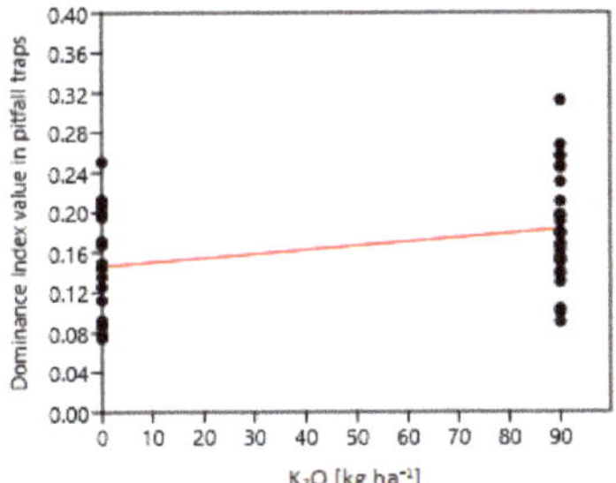

Figure 16. Independent variable x_1—K_2O [kg ha^{-1}] affecting Dominance Index value in pitfall traps in perennial industrial crops. $Y = 0.133 + 0.001x_1$. $R^2 = 0.315$, $p < 0.001$.

3.3. Diversity of Arthropods in Yellow Bowl Traps

The total number of specimens collected from yellow bowl traps was similar to the total number of specimens collected from pitfall traps, despite the smaller number of samples of yellow bowl traps. Representatives of 17 orders were identified, including: *Acari, Araneae, Coleoptera, Dermaptera, Diptera, Entomobryomorpha, Hemiptera, Hymenoptera, Lepidoptera, Mecoptera, Neuroptera, Opiliones, Orthoptera, Raphidioptera, Symphypleona Thysanoptera, Zygoptera,* as well as 63 families and 83 taxa at species level (Table 6).

Table 6. Frequency of the arthropod taxa in Yellow bowl traps in five different perennial industrial crops: FL—Fallow land (control), MG—*Miscanthus* × *giganteus* Greef and Deuter, SP—*Silphium perfoliatum* L., RP—*Robinia pseudoacacia* L., PM—*Populus* × *maximowiczii* Henry, SW—*Salix viminalis* L.

| Taxa | | | Average Number in the Sample | | | | | |
Order	Family	Species	FL	MG	SP	RP	PM	SW
Acari	–	–	–	1.0	–	6.8	0.5	0.2
	Trombidiidae	*Trombidium holosericeum*	–	0.2	–	–	–	–
Araneae	*Araneidae*	–	1.3	2.2	0.8	0.5	0.7	0.5
	Clubionidae	–	0.2	0.2	–	0.3	0.3	0.2
	Linyphiidae	–	1.0	0.2	0.3	–	1.0	1.5
	Lycosidae	–	0.3	–	0.2	–	–	–
	Pisauridae	*Pisaura mirabilis*	–	0,2	0.2	–	–	–
	Salticidae	–	–	–	–	–	–	0.2
	Tetragnathidae	–	0.7	1.3	0.2	0.2	0.2	–
	Thomisidae	–	0.2	0.3	0.2	–	0.2	–
Coleoptera	*Anthicidae*	*Notoxus monoceros*	0.2	–	0.2	–	–	0.3
	Cantharidae	–	1.0	–	–	0.3	1.2	1.2
	Carabidae	–	2.0	1.5	0.2	2.0	1.2	0.8
		Cicindela hybrida	0.2	–	–	–	–	–
	Cerambycidae	–	–	–	0.2	0.2	–	–
	Chrysomelidae	*Alticinae* sp.	–	–	–	–	–	0.7
		Chrysomela populi	–	–	–	–	0.5	–
		–	26.8	16.3	3.2	5.5	4.0	16.5
		Oulema sp.	–	–	–	–	0.7	0.8
	Coccinellidae	–	1.2	1.2	2.7	1.7	2.7	0.3
	Curculionidae	–	24.5	5.2	52.5	18.5	24.5	7.2
		–	1.7	0.3	–	0.7	0.7	12.3
		Otiorhynchus sp.	–	0.3	–	–	0.2	0.8
	Elateridae	–	0.7	–	–	0.3	–	0.3
	Histeridae	–	0.3	–	–	–	–	–
	Scarabaeidae	*Amphimallon solstitiale*	0.5	0.7	–	–	–	16.2
		Anomala dubia	0.8	0.2	0.2	–	–	0.2
		Cetoniinae sp.	0.5	0.8	–	0.2	-	0.2
		Melolonthinae sp.	–	–	0.2	0.2	0.8	–
		Oxythyrea funesta	3.2	7.0	0.2	–	–	1.0
		Phyllopertha horticola	0.2	–	–	0.2	–	0.7
		Rutelinae sp.	–	–	–	0.2	–	–
	Silphidae	*Nicrophorus vespillo*	–	–	0.5	–	–	–
	Staphylinidae	–	4.5	3.0	6.0	2.8	3.5	2.0
Dermaptera	*Forficuliidae*	*Forficula auricularia*	–	0.5	0.7	–	–	0.7
Diptera	*Asilidae*	–	1.8	0.7	0.2	–	0.3	0.7
	Bibionidae	–	5.8	6.2	10.8	4.5	17.2	33.8
	Bombyliidae	*Bombylius major*	–	–	–	0.2	–	–
	Cecidomyiidae	–	0.3	1.2	0.7	3.5	0.8	2.5
	Culicidae	–	0.5	–	–	0.3	0.2	1.0
	Muscidiale	–	116.2	77.2	91.0	105.0	290.0	191.0
	Empididae	–	0.2	0.8	–	0.7	–	0.8
	Syrphidae	*Eristalis tenax*	–	–	–	–	–	0.3
		–	70.2	11.3	7.5	41.2	10.0	20.5
	Tabanidae	–	0.3	0.2	–	–	–	0.2
	Tipulidae	–	1.3	0.2	–	–	5.7	1.3
Entomobryomorpha	*Entomobryidae*	–	5.7	0.5	2.3	1.0	1.8	0.5

Table 6. *Cont.*

Taxa			Average Number in the Sample					
Order	**Family**	**Species**	**FL**	**MG**	**SP**	**RP**	**PM**	**SW**
Hemiptera	*Aphididae*	–	42.0	30.0	44.3	27.2	31.8	38.3
	Cicadellidae	–	5.3	2.0	2.0	5.7	63.3	15.8
	–	–	0.2	0.5	1.5	–	1.5	0.8
	Miridae	–	12.5	0.7	0.8	0.8	0.5	7.3
	Nabidae	–	0.5	0.2	–	0.2	0.2	0.3
Hymenoptera	*Apidae*	–	11.8	2.7	6.5	3.0	1.7	3.5
		Apis mellifera	1.0	-	1.2	0.8	0.5	2.7
		Bombus sp.	0.7	0.2	2.2	0.2	0.7	0.5
	Cephidae	–	1.0	1.3	1.3	1.3	3.5	6.3
	Formicidae	–	3.5	–	–	–	2.2	–
		Lasius niger	2.8	4.8	1.0	2.0	3.7	0.8
		Myrmica rubra	0.2	0.2	–	0.2	0.5	–
	Pamphiliidae	–	0.2	–	–	1.3	0.2	0.2
	Tenthredinidae	–	14.8	6.5	4.5	16.3	13.7	7.2
	Vespidae	*Aphidiinae* sp.	10.7	30.0	11.7	30.3	30.2	30.5
		Ichneumonidae sp.	5.7	3.2	1.0	3.8	9.0	10.7
		–	1.2	0.3	–	–	0.3	0.7
		Vespinae sp.	4.5	2.2	5.0	5.0	15.5	5.3
Lepidoptera	*Lycaenidae*	–	–	–	-	1.5	–	–
	–	–	0.7	0.7	0.7	9.0	2.5	2.8
	Noctuidae	–	-	–	0.7	–	–	–
	Nymphalidae	–	-	0.2	–	6.0	0.8	0.8
	Tineidae	–	0.2	–	–	5.7	–	–
Mecoptera	–	–	0.2	–	–	–	–	–
Neuroptera	*Chrysopidae*	*Chrysopa* sp.	0.2	–	0.3	1.3	1.8	0.7
	Hemerobiidae	–	–	0.2	0.3	–	0.5	–
Opiliones	–	–	–	–	0.8	–	–	–
Orthoptera	*Acridinae*	–	0.2	–	–	–	–	–
	Tettigoniidae	*Tettigonia viridissima*	–	–	0.2	–	–	–
		–	–	0.2	–	–	–	–
Raphidioptera	–	–	–	–	–	–	–	0.2
Symphypleona	*Sminthuridae*	–	–	0.2	–	0.7	0.2	–
Thysanoptera	–	–	14.3	14.2	6.5	3.8	4.5	8.8
Zygoptera	*Calopterygidae*	–	–	–	–	–	0.2	–

No significant differences between the crops were found, which was probably related to the low number of repetitions. Most insects were captured in the control plot (Table 7). Pollinators also included taxa, which may be as well crop pests or natural enemies of pests in part of their life cycle.

Table 7. Median and median error values of selected indicators of the diversity of arthropods collected in yellow bowl traps in perennial industrial crops in three-year monitoring.

Arthropods in Yellow Bowls	Crop					
	FL	**MG**	**SP**	**RP**	**PM**	**SW**
Number of arthropods	404 ± 101 a	213 ± 91 a	151 ± 175 a	318 ± 37 a	299±342 a	374 ± 156 a
Number of species	27 ± 3.4 a	19 ± 3.6 a	16 ± 1 a	21 ± 4.2 a	24 ± 1.9 a	26 ± 3.7 a
Number of natural enemies	121 ± 64 a	35 ± 29 a	29 ± 12 a	58 ± 36 a	86 ± 11 a	51 ± 29 a
Number of potential pests	118 ± 59 a	85 ± 39 a	87 ± 82 a	111 ± 16 a	152 ± 91 a	95 ± 92 a
Number of pollinators	74 ± 68 a	26 ± 29 a	28 ± 17 a	116 ± 42 a	77 ± 15 a	78 ± 24 a

FL—Fallow land (control), MG—*Miscanthus* × *giganteus* Greef & Deuter, SP—*Silphium perfoliatum* L., RP—*Robinia pseudoacacia* L., PM—*Populus* × *maximowiczii* Henry, SW—*Salix viminalis* L.

The diversity of arthropods obtained by bowl trap method varied from each year to the next. In 2018, the diversity in poplar cultivation was significantly lower than in the cultivation of black locust (Figure 17). In the following year, all crops did not differ in Shannon's diversity index, which was high for all crops. In 2020, the greatest diversity of arthropods was observed in poplar than for the control, cup plant, and willow.

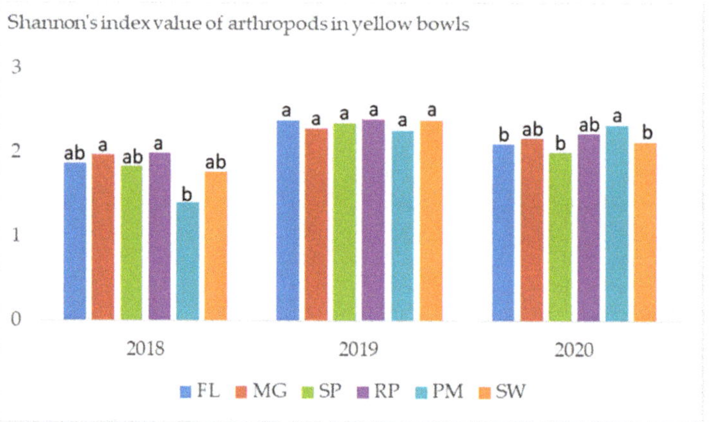

Figure 17. Shannon's Index values of arthropod diversity in yellow bowl traps compared using Hutcheson's t-test in three following years. Statistically significant differences ($p < 0.01$) were marked with different letters. FL—Fallow land (control), MG—*Miscanthus × giganteus* Greef and Deuter, SP—*Silphium perfoliatum* L., RP—*Robinia pseudoacacia* L., PM—*Populus × maximowiczii* Henry, SW—*Salix viminalis* L.

In yellow bowl traps high values of Dominance Index were observed only in the first year of the study. The highest level of domination was found in the samples from poplar and willow in 2018 and 2019 (Figure 18). *Diptera insects* were the most abundant in all crops. Hover flies *Syrphidae* were dominant pollinator taxa in control and in black locust. Relatively few pollinators were caught, which could be explained by poorly diversified landscape, dominated mostly by cereal crops. Few representatives of bees *Apidae* were found, mostly in the control plot. On the other hand, there was good representation of parasitic wasps both *Ichneumonidae* and *Aphidiinae*. Aphid parasitoids accounted for 12.5% of the insect population in miscanthus, 9.4% in black locust, and 6.6% in willow. Parasitoids were dominant taxa in all perennial crops, while in control they accounted only for 2.6% of all arthropods from bowl traps. Aphid population was the most numerous in the control plot (10.3%), being second dominant only after taxa after flies *Diptera*.

Several significant relationships between results obtained from bowl traps and agricultural practices were found. The number of arthropods in the samples was significantly influenced by the herbicide use, the age of the plantation, and the dose of potassium fertilization (Figure 19). The number of predators decreased with the increasing dose of phosphorus (Figure 20). The number of potential pests was positively influenced by the plantation area, contrary to pest species from pitfall traps. The use of herbicides was also a positive factor for the abundance of pest species. The occurrence of pests was negatively related to the yields (Figure 21). A high level of phosphorus fertilization negatively affected the number of pollinators possibly by favoring plants that do not benefit them. (Figure 22). The value of Shannon's diversity index was negatively affected by the use of herbicides in crops (Figure 23). The domination of species in the samples increased on larger plantations and with higher dose of potassium fertilization (Figure 24).

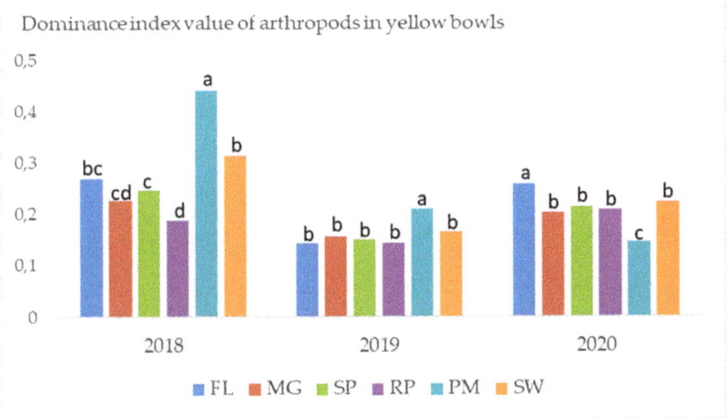

Figure 18. Simpson's Dominance Index values for arthropods in yellow bowl traps compared using Hutcheson's t-test in three following years. Statistically significant differences ($p < 0.01$) were marked with different letters. FL—Fallow land (control), MG—*Miscanthus × giganteus* Greef and Deuter, SP—Silphium perfoliatum L., RP—Robinia pseudoacacia L., PM—*Populus × maximowiczii* Henry, SW—*Salix viminalis* L.

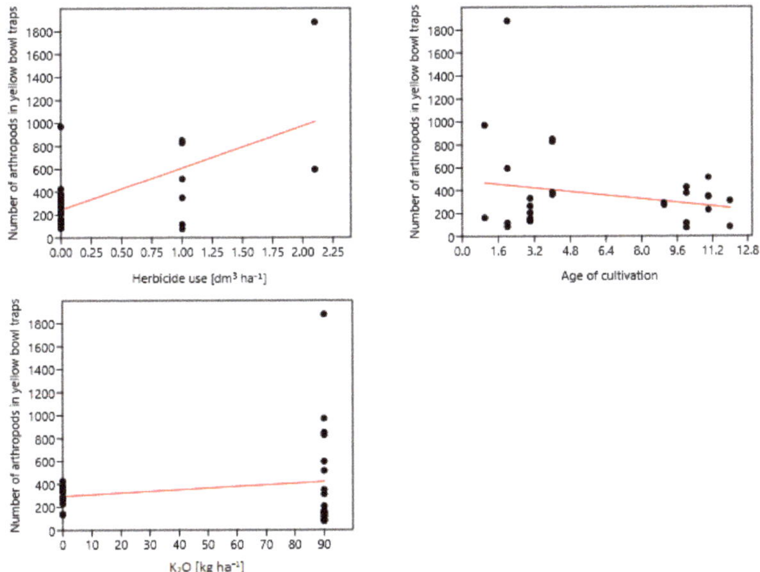

Figure 19. Three independent variables affecting number of arthropods in yellow bowl traps in perennial industrial crops: x_1—Herbicide use [dm^3 ha^{-1}], x_2—Age of cultivation, x_3—K$_2$O [kg ha^{-1}]. Y= 525.7 + 424.8x_1 − 34.9x_2 − 2.4x_3. $R^2 = 0.431$, $p = 0.025$.

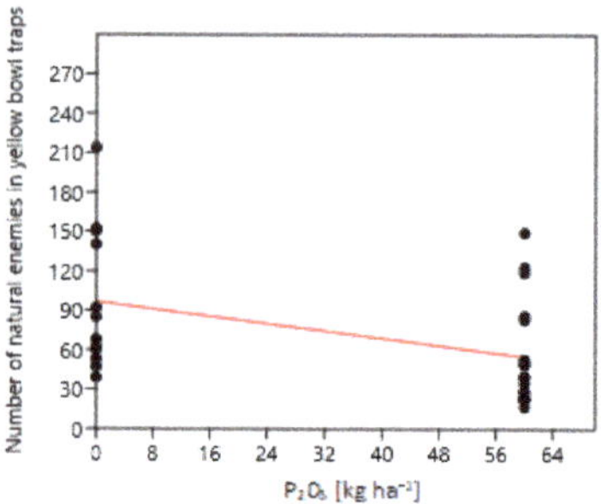

Figure 20. Independent variable x_1—P_2O_5 [kg ha^{-1}] affecting number of natural enemies of pests in yellow bowl traps in yellow bowl traps in perennial industrial crops. $Y = 115.5 - 0.832x_1$. $R^2 = 0.209$, $p = 0.042$.

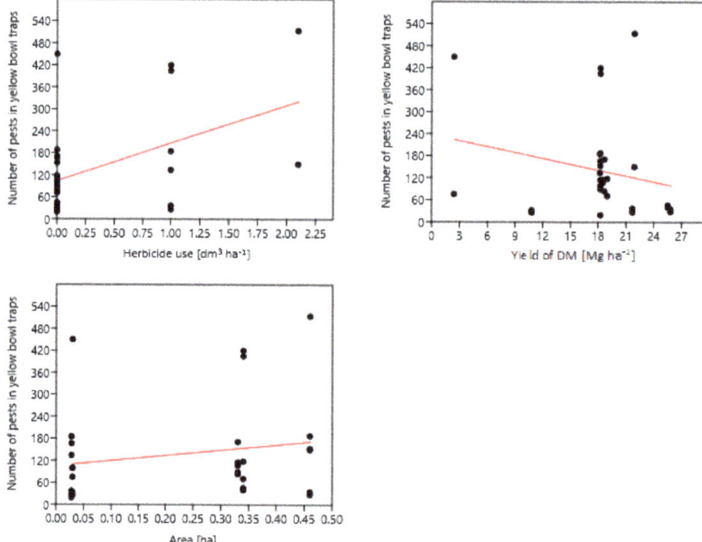

Figure 21. Three independent variables: x_1—Herbicide use [dm^3 ha^{-1}], x_2—Yield of DM [Mg ha^{-1}], x_3—Area [ha] affecting number of pests in yellow bowl traps in perennial industrial crops. $Y = 252.8 + 112.6x_1 - 14.8x_2 + 391.5x_3$. $R^2 = 0.503$, $p = 0.009$.

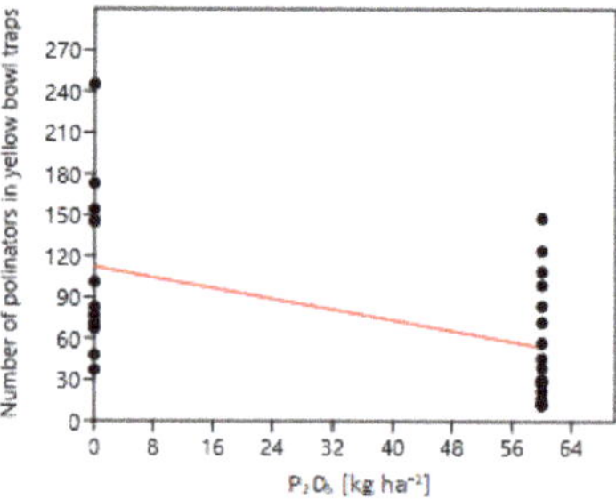

Figure 22. Independent variable x_1—P_2O_5 [kg ha^{-1}] affecting number of pollinators in yellow bowl traps in perennial industrial crops. $Y = 136.25 - 0.81x_1$. $R^2 = 0.355$, $p = 0.006$.

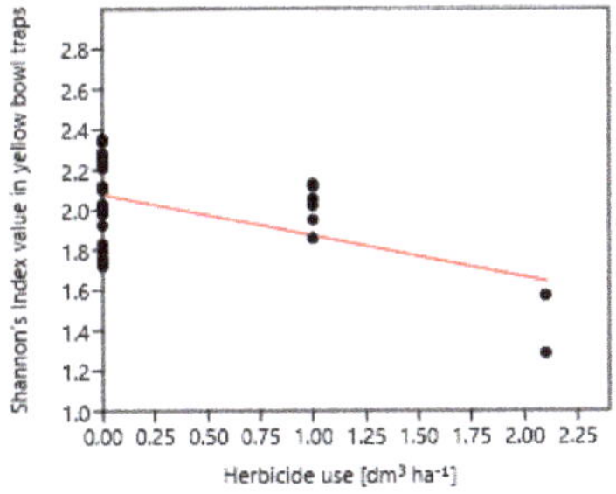

Figure 23. Independent variable x_1—Herbicide use [dm^3 ha^{-1}] affecting Shannon's index value of arthropods in yellow bowl traps in perennial industrial crops. $Y = 1.932 - 0.229x_1$. $R^2 = 0.462, p = 0.005$.

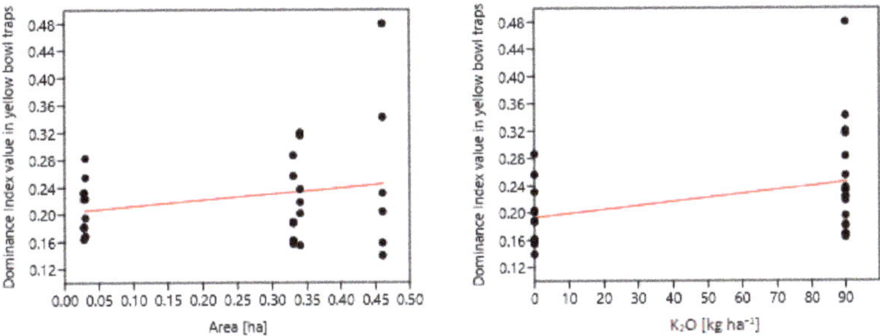

Figure 24. Two independent variables affecting Dominance Index value of arthropods in yellow bowl traps in perennial industrial crops: x_1—Area [ha], x_2—K$_2$O [kg ha^{-1}]. $Y = 0.121 + 0.57x_1 + 0.002x_2$, $R^2 = 0.807, p < 0.001$.

4. Discussion

Three research hypotheses have been tested in this work, all of which have been partly confirmed. The first hypothesis (H$_1$: perennial industrial crops can significantly affect the diversity of flora and fauna of marginal lands) was confirmed only in the case of floral diversity. The biodiversity of weeds in miscanthus and black locust was significantly lower than on marginal land. For the arthropod diversity, results differ across the years of the study, so there was no clear evidence that the cultivation of perennial industrial crops affects this kind of fauna. In other studies, information about a change in the structure of the flora population under miscanthus crops and black locust stands can be found, however, a decrease in the population of arthropods is rarely reported [6,17,21]. Willow, poplar, and cup plant had similar results of floral species richness as the control site.

The key aspect of the present research was the selection of the proper control object. The tested perennial industrial crops were cultivated on a soil of poor agricultural quality, which could be qualified as marginal land [23]. The best control object would be a fallow land in the same, or near the same location as tested industrial crops. Such land was not located anywhere in the vicinity of the experiment, and thus a piece of field of 450 m^2 was excluded from agricultural cultivation in order to simulate the conditions of fallow land. As a consequence, a very species-rich flora habitat was created, which was maintained for 3 consecutive years. In the first year of the study, results were misleading, as one-year fallow was the most attractive object for weeds and arthropods, presenting a new space to colonize. In the second year of the study, the diversity of plants and arthropods started to decrease there, so the results of biodiversity monitoring were higher in the crops of willow, poplar, and cup plant. That also proves the assumptions of the second hypothesis (H$_2$), that results of different crops will differ significantly and the right choice of the species of the perennial industrial crop can mitigate the loss of flora and fauna diversity on marginal lands. In this case, willow provided the most suitable habitat both for flora and arthropods.

A persistence of high density of flowering, melliferous plants was a positive phenomenon observed in the study. The abundance of melliferous plants was low only in miscanthus cultivation, which was most likely due to the use of selective herbicides, directed against dioecious plants. In all other tested crops, mainly herbicides against grasses were used (once every few years—at the time of planting or after harvesting). Limited herbicide treatments make industrial plants an excellent source of feed for bees and wild pollinators. It should be noted that industrial plants chosen for the present experiment are beneficial for honeybees. Although miscanthus does not produce nectar, bees can collect pollen from its blossoms. Plantations of miscanthus were linked with flowering weeds: *Viola arvensis* Murray, *Capsella bursa-pastoris* (L.) Medik, *Lapsana communis* L. However, those species were found at very low densities of about 2 plants per m^2 on average. In the case of miscanthus cultivation, unfavorable changes in the diversity of flora and its usefulness for pollinators were observed. It cannot be excluded that extensive or organic cultivation of miscanthus could provide a greater diversity of melliferous plants. The cultivation carried out on marginal lands usually requires greater inputs than cultivation on soils of better quality [17].

The biodiversity of arthropods was at a similar level in all tested crops. Fewer taxa were found only in cup plant. The cup plant is the most productive melliferous plant among tested, producing up to 550 kg of nectar per ha. It blooms relatively late, from July to October. A numerous melliferous weeds were found in cup plant crop: *Cerastium arvense* L., *Stellaria media* (L.) Vill., *Taraxacum officinale* Kirschner, H.Øllg. and Štěpánek and *Erigeron annuus* (L.) Desf. A surprise was the low abundance of arthropod species found in this crop. The low abundance of arthropods found in cup plant might have been due to the monitoring date, which was always in June. The cup plant blooms relatively late, so at the time of insect collection, it didn't attract many pollinators. Cultivation of cup plant was related to a large diversity of wild plants, also including melliferous plants. The low number of taxa and the generally low abundance of arthropods in the cup plant is difficult to explain because it contradicts the results of other authors [7]. It may be due to the fact that the cup tree in present study was the youngest of the tested crops. As a result, the number and intensification of agrotechnical treatments in this plant were relatively larger than in other tested plants.

An important physical feature of the crop that affects arthropod diversity is the plant conformation, its density, and height [24]. It is possible that a dense vegetation structure of the cup plant blocked the access of light to the surface of the soil, thus reducing the occurrence of some heliophile weed species. The decrease in the diversity of arthropods was not recorded in the black locust. This tree also provides an efficient bee feed (up to 65 kg of nectar per hectare). In Poland, it blooms in late May and early June, which was partly the time of biodiversity monitoring. Numerous melliferous plants, of which the most common were *Galium aparine* L. and *Solidago canadensis* L., were found on black locust plantation. Weeds that occurred in black locust were of low diversity but were able to provide feed for bees from June to October, long after black locust blooming has ended. Poplar does not provide nectar benefits to bees, but it is an important source of plant exudates, used by bees to produce propolis. The bees can also collect poplar's pollen. Numerous mellifluous weeds species were found in poplar cultivation: *Tripleurospermum* maritimum (L.) W.D.J.Koch, *Erigeron canadensis* (L.) Cronquist and *Erodium cicutarium* (L.) L'Hér. ex Aiton. The diversity of arthropods in poplar cultivation varied over the years and depended on the collection method, representing both the highest and the lowest results. This only confirms the conclusions of other authors that different assessment methods should be used, at different times and with as many repetitions as possible [22,25,26]. The willow provides early pollen and nectar (from March to May) which can be used by bees. Willow plantations are also a great source of honeydew of aphids *Aphididae* which is collected by honey bees. In addition, the weeds found in the willow plantation bloomed at different times. The most numerous weed species which can be beneficial for bees were: *Galium aparine* L., *Stellaria media* (L.) Vill., *Cerastium arvense* L., *Solidago canadensis* L., and *Taraxacum officinale* Kirschner, H.Øllg. and Štěpánek. Willow can also be cultivated in more extensive production systems than other crops, which increases its environmental value for pollinators and other beneficial arthropods [27].

Despite the low biodiversity of the plants, the cultivation of perennial industrial crops can bring significant benefits to the diversity of pollinators and honey production. The third hypothesis (H$_3$), which assumed the agrotechnical treatments have a key impact on biodiversity, also has been proved. The assumption was confirmed for both weeds and arthropods. Factors such as fertilization and herbicide use negatively affected the biodiversity of flora and fauna, while increasing the number of some pest species. The impact of intensive farming on arthropod biodiversity is a well-explored phenomenon [28–30]. This was confirmed by the increased diversity of flora in control plot compared to miscanthus and poplar cultivations. The lowest number of pollinators was observed in the miscanthus and in the cup plant. In the case of miscanthus, it could result from the low diversity of flora and low number of mellifluous plants. In the case of the cup plant, it is more difficult to explain the small number of pollinators. Although the monitoring of arthropods wasn't done at the time of its flowering, there were a lot of melliferous weeds there. The cultivation of the cup plant was directly adjacent to the control object. In 2019 a large field of oil rapeseed was located in the direct vicinity of the plantation. Plants of oil rapeseed could attract pollinators more than cup plant in June.

The results could also be influenced by the choice of monitoring methods for arthropods. The number of detected species increases with the number of traps used. Studies by other authors indicate that the number of samples from yellow bowls used in their experiment may have also been too low [22,25,31]. In the present study, the size of some fields did not allow us to use a larger number of traps without avoiding the side effect, when the neighboring habitat affects the results. In the case of pollinators, which can see the traps from the above, it would require to have a field area of at least one hectare. The traps and yellow bowls were set for a period of 2 weeks, which resulted in a very large amount of material to be collected. Handling more yellow bowls traps could also prove to be too much of a challenge, as the identification of taxa from yellow bowls takes more man-hours than from pitfall traps. The risk of depleting local pollinator populations, especially wild bees, was also considered. The use of more traps is not justified if subsequent handling of the collected material cannot be ensured. Each collection method is to some extent imperfect and selective, so it is worthwhile to use a combination of several methods instead of a large number of repetitions of the same method [22,25].

5. Conclusions

Perennial industrial crops did not cause a decline in wild biodiversity in comparison with unmanaged marginal land. Nevertheless, the cultivation of some crop species can cause a decrease in diversity of flora and fauna in long term. Miscanthus and black locust cultivation were linked with a decrease in the number of plant species.

The greatest biodiversity of plants and animals among crops was linked with the cultivation of willow, however, other crops also provided a good habitat for arthropods. No significant decrease of abundance of pollinators or natural enemies of pests were found in any perennial industrial crop.

The intensity of cultivation negatively affected the diversity of weeds and arthropods. The factors negatively affecting the biodiversity were: mechanical treatments, the level of nitrogen, phosphorus and potassium fertilization, and the use of herbicides. There is a possibility that other, more extensive management strategies or the choice of species requiring low inputs will allow for the maintenance of a high biodiversity of perennial industrial crops.

Author Contributions: Conceptualization, P.R.; methodology, P.R.; validation, M.M. and P.R.; investigation, P.R. and M.M.; resources, M.M.; data curation, P.R.; writing—original draft preparation, P.R.; writing—review and editing, P.R. and A.K.B.; visualization, P.R.; supervision, M.M.; project administration, M.M.; funding acquisition, M.M. All authors have read and agreed to the published version of the manuscript.

Funding: This paper is the result of a long–term study carried out at the Institute of Soil Science and Plant Cultivation—State Research Institute, Department of Systems and Economics of Crop Production, and it was co-financed by the National (Polish) Centre for Research and Development (NCBiR), entitled "Environment, agriculture and forestry", project: BIOproducts from lignocellulosic biomass derived from MArginal land to fill the Gap In Current national bioeconomy, No. BIOSTRATEG3/344253/2/NCBR/2017.

Acknowledgments: We would also like to thank the staff of the Department of Systems and Economics of Crop Production and Experimental Farm "Kępa" for their technical support during the experiments.

Conflicts of Interest: The authors declare no conflict of interest. The funders had no role in the design of the study; in the collection, analyses, or interpretation of data; in the writing of the manuscript, or in the decision to publish the results.

References

1. Pudełko, R.; Kozak, M.; Jędrejek, A.; Gałczyńska, M.; Pomianek, B. Regionalisation of unutilised agricultural area in Poland. *Pol. J. Soil Sci.* **2018**, *51*, 119. [CrossRef]
2. Stolarski, M.J.; Krzyianiak, M.; Szczukowski, S.; Tworkowski, J.; Bieniek, A. Short rotation woody crops grown on marginal soil for biomass energy. *Pol. J. Environ. Stud.* **2014**, *23*, 1727–1739.
3. Stolarski, M.J.; Krzyżaniak, M.; Załuski, D.; Tworkowski, J.; Szczukowski, S. Effects of Site, Genotype and Subsequent Harvest Rotation on Willow Productivity. *Agriculture* **2020**, *10*, 412. [CrossRef]
4. Dauber, J.; Miyake, S. To integrate or to segregate food crop and energy crop cultivation at the landscape scale? Perspectives on biodiversity conservation in agriculture in Europe. *Energy Sustain. Soc.* **2016**, *6*, 25. [CrossRef]
5. Fletcher, R.J., Jr.; Robertson, B.A.; Evans, J.; Doran, P.J.; Alavalapati, J.R.; Schemske, D.W. Biodiversity conservation in the era of biofuels: Risks and opportunities. *Front. Ecol. Environ.* **2011**, *9*, 161–168. [CrossRef]
6. Feledyn-Szewczyk, B.; Matyka, M.; Staniak, M. Comparison of the Effect of Perennial Energy Crops and Agricultural Crops on Weed Flora Diversity. *Agronomy* **2019**, *9*, 695. [CrossRef]
7. Chmelíková, L.; Wolfrum, S. Mitigating the biodiversity footprint of energy crops—A case study on arthropod diversity. *Biomass Bioenergy* **2019**, *125*, 180–187.
8. Gansberger, M.; Montgomery, L.F.; Liebhard, P. Botanical characteristics, crop management and potential of Silphium perfoliatum L. as a renewable resource for biogas production: A review. *Ind. Crop. Prod.* **2015**, *63*, 362–372. [CrossRef]
9. Purtauf, T.; Roschewitz, I.; Dauber, J.; Thies, C.; Tscharntke, T.; Wolters, V. Landscape context of organic and conventional farms: Influences on carabid beetle diversity. *Agr. Ecosyst. Environ.* **2005**, *108*, 165–174. [CrossRef]
10. Dauber, J.; Purtauf, T.; Allspach, A.; Frisch, J.; Voigtländer, K.; Wolters, V. Local vs. landscape controls on diversity: A test using surface-dwelling soil macroinvertebrates of differing mobility. *Global Ecol. Biogeogr.* **2005**, *14*, 213–221. [CrossRef]
11. Pokluda, P.; Hauck, D.; Cizek, L. Importance of marginal habitats for grassland diversity: Fallows and overgrown tall-grass steppe as key habitats of endangered ground-beetle Carabus hungaricus. *Insect Conserv. Diver.* **2012**, *5*, 27–36. [CrossRef]
12. Poschlod, P.; Bakker, J.P.; Kahmen, S. Changing land use and its impact on biodiversity. *Basic Appl. Ecol.* **2005**, *6*, 93–98. [CrossRef]
13. Werling, B.P.; Dickson, T.L.; Isaacs, R.; Gaines, H.; Gratton, C.; Gross, K.L.; Robertson, B.A. Perennial grasslands enhance biodiversity and multiple ecosystem services in bioenergy landscapes. *Proc. Natl. Acad. Sci. USA* **2014**, *111*, 1652–1657. [CrossRef] [PubMed]
14. Decourtye, A.; Mader, E.; Desneux, N. Landscape enhancement of floral resources for honey bees in agro-ecosystems. *Apidologie* **2010**, *41*, 264–277. [CrossRef]
15. Holzschuh, A.; Steffan-Dewenter, I.; Tscharntke, T. How do landscape composition and configuration, organic farming and fallow strips affect the diversity of bees, wasps and their parasitoids? *J. Anim. Ecol.* **2010**, *79*, 491–500. [CrossRef]
16. Blanco-Canqui, H. Growing dedicated energy crops on marginal lands and ecosystem services. *Soil Sci. Soc. Am. J.* **2016**, *80*, 845–858. [CrossRef]
17. Scholz, V.; Ellerbrock, R. The growth productivity, and environmental impact of the cultivation of energy crops on sandy soil in Germany. *Biomass Bioenergy* **2002**, *23*, 81–92. [CrossRef]
18. Soldatos, P. Economic aspects of bioenergy production from perennial grasses in marginal lands of South Europe. *Bioenergy Res.* **2015**, *8*, 1562–1573. [CrossRef]
19. Paoletti, M.G.; Hassall, M. Woodlice (Isopoda: Onisci-dae): Their potential for assessing sustainability and use asbioindicators. *Agric. Ecosyst. Environ.* **1999**, *74*, 157–165. [CrossRef]

20. Semere, T.; Slater, F.M. Invertebrate populations in miscanthus (Miscanthus × giganteus) and reed canary-grass (Phalaris arundinacea) fields. *Biomass Bioenergy* **2007**, *31*, 30–39. [CrossRef]

21. Vítková, M.; Müllerová, J.; Sádlo, J.; Pergl, J.; Pyšek, P. Black locust (Robinia pseudoacacia) beloved and despised: A story of an invasive tree in Central Europe. *Forest. Ecol. Manag.* **2017**, *384*, 287–302. [CrossRef] [PubMed]

22. McCravy, K.W. A review of sampling and monitoring methods for beneficial arthropods in agroecosystems. *Insects* **2018**, *9*, 170. [CrossRef]

23. Shortall, O.K. "Marginal land" for energy crops: Exploring definitions and embedded assumptions. *Energy Policy* **2013**, *62*, 19–27. [CrossRef]

24. Li, X.; Liu, Y.; Duan, M.; Yu, Z.; Axmacher, J.C. Different response patterns of epigaeic spiders and carabid beetles to varying environmental conditions in fields and semi-natural habitats of an intensively cultivated agricultural landscape. *Agric. Ecosyst. Environ.* **2018**, *264*, 54–62. [CrossRef]

25. Shapiro, L.H.; Tepedino, V.J.; Minckley, R.L. Bowling for bees: Optimal sample number for "bee bowl"sampling transects. *J. Insect Conserv.* **2014**, *18*, 1105–1113. [CrossRef]

26. Spence, J.R.; Niemelä, J.K. Sampling carabid assemblages with pitfall traps: The madness and the method. *Can. Entomol.* **1994**, *126*, 881–894. [CrossRef]

27. Stolarski, J.M.; Szczukowski, S.; Tworkowski, J.; Krzyżaniak, M. Extensive Willow Biomass Production Marginal Land. *Pol. J. Environ. Stud.* **2019**, *6*, 4359–4367. [CrossRef]

28. Biesmeijer, J.C.; Roberts, S.P.M.; Reemer, M.; Ohlemüller, R.; Edwards, M.; Peeters, T.; Schaffers, A.P.; Potts, S.G.; Kleukers, R.; Thomas, C.D.; et al. Parallel declines in pollinators and insect-pollinated plants in Britain and the Netherlands. *Science* **2006**, *313*, 351–354. [CrossRef]

29. Landis, D.A.; Wratten, S.D.; Gurr, G.M. Habitat management to conserve natural enemies of arthropod pests in agriculture. *Annu. Rev. Entomol.* **2000**, *45*, 175–201. [CrossRef]

30. Losey, J.E.; Vaughan, M. The economic value of ecological services provided by insects. *BioScience* **2006**, *56*, 311–323. [CrossRef]

31. Bąkowski, M.; Piekarska-Boniecka, H.; Dolańska-Niedbała, E.; Michaud, J.P. Monitoring of the red-belted clearwing moth, Synanthedon myopaeformis, and its parasitoid Liotryphon crassiseta in apple orchards in yellow Moericke traps. *J. Insect Sci.* **2013**, *13*, 1–11. [CrossRef] [PubMed]

Publisher's Note: MDPI stays neutral with regard to jurisdictional claims in published maps and institutional affiliations.

Review

Silphium perfoliatum—A Herbaceous Crop with Increased Interest in Recent Years for Multi-Purpose Use

Dumitru Peni [1,*], Mariusz Jerzy Stolarski [1], Anna Bordiean [1], Michał Krzyżaniak [1] and Marcin Dębowski [2]

[1] Department of Plant Breeding and Seed Production, Faculty of Environmental Management and Agriculture, University of Warmia and Mazury in Olsztyn, Plac Łódzki 3, 10-724 Olsztyn, Poland; mariusz.stolarski@uwm.edu.pl (M.J.S.); anna.bordiean@uwm.edu.pl (A.B.); michal.krzyzaniak@uwm.edu.pl (M.K.)

[2] Department of Environmental Engineering, Faculty of Geoengineering, University of Warmia and Mazury in Olsztyn, Warszawska 117, 10-719 Olsztyn, Poland; marcin.debowski@uwm.edu.pl

* Correspondence: dumitru.peni@uwm.edu.pl

Received: 17 November 2020; Accepted: 14 December 2020; Published: 16 December 2020

Abstract: *Silphium perfoliatum* is a perennial crop native to North America that has been the subject of increased scientific interest in recent years, especially in Europe. It is drought- and frost-resistant, which makes it suitable for cultivation in Europe on marginal lands that are not used for growing other crops. This review analyzed the distribution and purposes of the cultivation of *Silphium perfoliatum* worldwide, as well as its biomass yields and characteristics as a feedstock for biogas production and other purposes. A total of 121 scientific publications on *Silphium perfoliatum* were identified, with the highest number (20 papers) published in 2019. It was found that higher biomass yields can be obtained at higher precipitation levels, with the use of fertilizers and an adequate type of plantation. The mean dry matter yield of *Silphium perfoliatum* was 13.3 Mg ha^{-1} DM (dry matter), and it ranged from 2 to over 32 Mg ha^{-1} DM. In some countries, *Silphium* is used as a forage crop mainly due to its high crude protein content (from 4.9% to 15% DM), depending on the vegetation phase. *Silphium perfoliatum* is a promising perennial crop in terms of energy and other benefits for biodiversity, soil quality and applications in medicine and pharmacology.

Keywords: cup plant; perennial energy crop; energy expenses; biogas; biomass yield

---

1. Introduction

Since climate change is caused mainly due to the wide use of fossil fuels, it is necessary to mitigate their impact and prevent their depletion. Under these conditions, it is very important to replace fossil fuels with renewable energy sources (RES), to stop the increase in greenhouse gases in the atmosphere [1] and reduce pollution caused by burning fossil fuels. Researchers are trying to identify different biomass sources that are acceptable for production, cheap, easy to grow and offer high yields that will be suitable for the commonly used conversion technologies. Moreover, there is a need to adapt some of the possible biomass sources to specific climatic conditions of some regions or try to use local sources. Anaerobic digestion can be seen as the most attractive renewable energy pathway to convert organic material into green fuel [2]. These complex processes provide several additional environmental benefits, since digestate from biogas production can be used as a biofertilizer and its utilization can reduce the amount of fertilizers needed [1,3].

Biomass is considered to be an organic, non-fossil material of biological origin, such as manure, wastewater sludge, food and organic waste, municipal organic waste, animal waste, agricultural residues, and forest and industrial wood waste. The biological wastes are classified as a renewable energy resource due to the possibility of the incorporation of solar energy [4,5].

Perennial crops are seen as a promising feedstock for renewable energy technologies [5,6], including biogas production, due to the large amount of biomass that can be obtained from a relatively small area compared to annual crops. Moreover, perennial crops offer other benefits for biodiversity and the environment [7–10].

The *Silphium perfoliatum* (cup plant) is considered a promising alternative substrate for biogas production [11,12] that could replace the current use of maize silage [13–15]. It is characterized by low production costs and can be grown on less productive or polluted soil [16]. The crop is resistant to winter frost and summer drought and is less dependent on atmospheric precipitation [17]. Moreover, in Central Europe, it is not very susceptible to pests and diseases that affect biomass productivity or yield [11,18]. Existing pests are unlikely to limit the production of seeds or biomass of *Silphium perfoliatum* [19]. In the USA, there have been cases reported of plantation damage caused by larvae, birds and mammals (moths, turkey and deer) [20]. In this case, damage was caused by natural biodiversity factors, which are beyond human control and can occur in every country. Silphium was studied in different countries, for different purposes: as a potential forage crop in New Zealand [21], Ukraine [22], USA [23,24] and Romania [25]; and as a forage crop in Belarus [26], Chile [27], China [28] and Russia [29–33]. In Austria [12], Czech Republic [15] and Poland [34,35], biomass has been investigated for biogas production. It is also being studied for future possible use in Kazakhstan as a forage crop [36]. Another use of *Silphium perfoliatum* was proposed as a raw material for particleboards that can successfully replace wood and reduce wood shortages [37], which is important for industrial applications. Seeds of *Silphium* L. species, including *S. perfoliatum*, *S. integrifolium* and *S. trifoliatum*, were studied by using different parameters for different purposes [35]. *Silphium perfoliatum* was also cultivated as an ornamental plant [25] with high nectar productivity [38–41].

The current review analyzes general interest in the use of *Silphium perfoliatum* around the world for different purposes over the last 20 years. The existing information on its description, productivity, utilization and chemical composition is summarized, along with energy efficiency and biogas yield, to analyze its use as a promising future RES and a future field of research.

2. Materials and Methods/Data Collection and Selection

The current study screened papers for *Silphium perfoliatum* utilization, energy, biomass production, yielding, biogas yields, etc. The search was conducted by using ScienceDirect, Scopus, Google Scholar and other sources, including Russian websites. The search focused on original papers, research articles, reports and short communiqués issued between 2000 and 2020, mostly in Europe, Asia and North America. The search involved specific keywords: cup plant, *Silphium perfoliatum*, sylphia, silphium and сильфиа (Russian). A total of 121 papers were identified that matched the search described above. Most of these papers were published in English (101), but also in Russian (12), Polish (3), German (4) and French (1). The papers were scanned for a description of *Silphium perfoliatum* and different species/subspecies of Silphium, their use around the world and the different requirements for planting. Information on biomass productivity in different years from different locations, soil types and environmental conditions was collected. Moreover, information about the biomass quality, chemical composition and biogas properties of *Silphium perfoliatum* studied by researchers was gathered. The paper also presents economic and energy information found in various research papers on *Silphium perfoliatum* biomass production.

3. Results and Discussion

3.1. Scientific Papers Related to Silphium perfoliatum

An increased interest in the use of *Silphium perfoliatum* was observed from 2000 to June 2020 (Figure 1). In 2000, only four papers were found that discuss *Silphium perfoliatum*, and there were no papers in two years (2001 and 2006). The number of publications on this topic increased to 12 published papers in 2015 and 2017. The maximum number of such papers (20) was released in 2019, and they continue to be published in 2020. Over the studied period, a total of 118 scientific publications were identified related to *Silphium perfoliatum*.

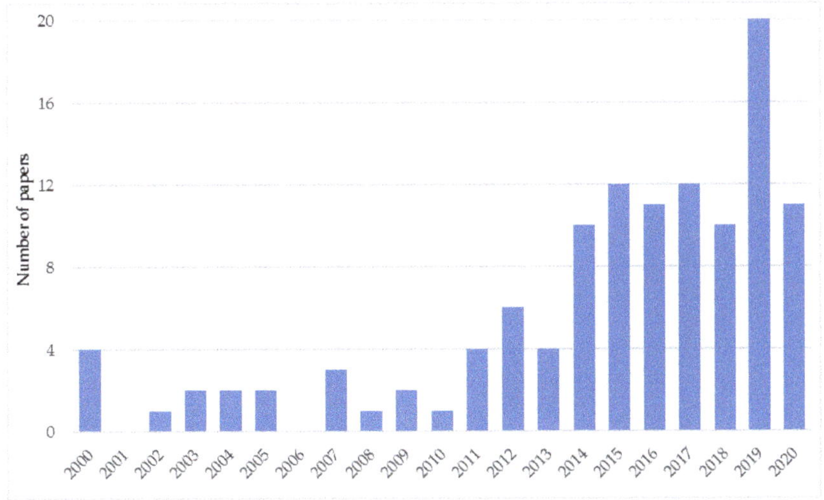

Figure 1. The number of peer-reviewed *Silphium perfoliatum*–related scientific papers in 2000–2020.

The general interest of these papers was mostly based on the use of *Silphium perfoliatum* as feed for livestock (mostly in post–Soviet Union countries) and as a possible future RES due its high biomass yield. There is different interest in using *Silphium perfoliatum* around the world. There were 96 papers published in Europe, including Russia, in the selected period (Figure 2). The highest number of research papers was found in Germany (31 papers), Poland (30 papers) and Lithuania (11). Some studies were also conducted in Russia (7), Republic of Moldova (4), Belarus (4), Austria (3) and Czech Republic (3), as well as in Romania, Ukraine, Kazakhstan and The Netherlands (1). A high interest/number of papers was also found in the USA (12 papers), which was due to the North American origin of *Silphium perfoliatum* [42,43]. There were seven papers found from China, while Chile and Egypt published one paper each.

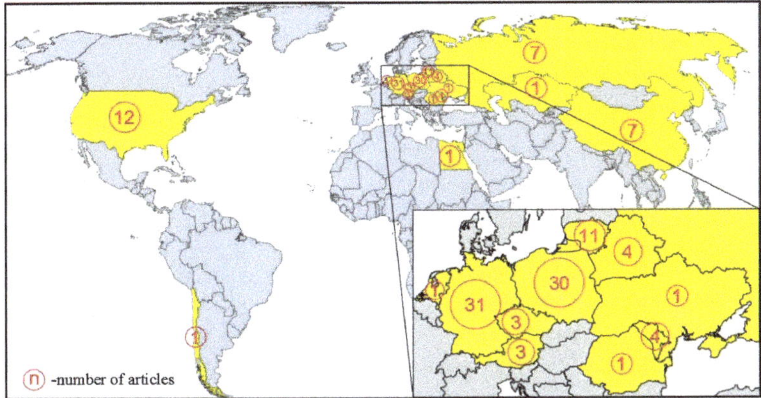

Figure 2. Map of published *Silphium perfoliatum*–related scientific articles, by country, in 2000–2020.

3.2. Description and Origin

Silphium L. is a genus of the sunflower family (*Asteraceae*) found in the woodlands and prairies of North America [42,43]. The name of the Silphium genus originates from the Greek *silphion* and means a plant secreting resin, used earlier by Hippocrates [44]. In the cited work, it is mentioned that Silphium was imported to Europe around the 18th century as an ornamental plant. The best-known species are *Silphium perfoliatum* L., *Silphium trifoliatum* L., *Silphium integrifolium* Michx and *Silphium laciniatum* L., although there are more varieties and subspecies [45–47].

The *Silphium perfoliatum* species is a part of *Heliantheae* and is a perennial herb, also called a cup plant, Indian cup [48], Rosinweed [49], sylph or cap-plant [46]. It is native to the eastern half of the USA and Canada and grows on moist soils [50]. It is tolerant of low winter temperatures and slightly drained soil [27,28]. Some species of Silphium can grow from 1.8 m in May to 2.5 m in July [46,47] in university/botanical gardens [46,47,51] with short winters and warm springs, moderate precipitation levels and on good-quality black soils (chernozems).

In a study conducted in the USA, the genetic differentiation between five populations of *Silphium perfoliatum* was verified. The aim was to identify the genetic variation between populations and identify the traits related to higher biomass yield and methane productivity. It was found that four plant populations (Russia, USA, East Germany and Northern Europe) have almost similar genetic characteristics and their yields are almost the same. A fifth plant population from Ukraine had a lower biomass yield, methane hectare yield (MHY) and represented a clear example of population stratification [52].

The most recommended method of cultivating *Silphium perfoliatum* is by seedlings, but this method is still limited by high costs. The best method is to plant seedlings by using the pattern of 70×30 cm [53]. Some experimental studies used a plant density of 20,000 plants ha^{-1} with a 100 cm distance between each row and 50 cm distance between the seedlings in each row [17,54,55]. A similar plant density of 28,000 plants ha^{-1} can be obtained using the 100×40 and 90×40 cm method [45,56]. Another pattern of planting was used by Stolarski et al. [18,57], who manually planted seedlings, leaving 100 cm between rows and strips [18] and obtained 10,000 plants ha^{-1}. Other studied plant densities, including 35,000 and 70,000 plant ha^{-1}, were described by Stankevich et al. [58].

To use *Silphium perfoliatum* as a substrate for biogas, it is necessary to increase the cultivated area. As planting by seedlings is the most expensive agricultural operation, plant cultivation by sowing was also studied. It was proposed to improve the seeder, as well as to place seeds at least 15 mm deep in soil to prevent them from drying out. It is necessary to cover seeds by using different types of rollers for a more precise sowing, but in a way which avoids compromising germination [59]. It is also necessary to improve seed quality to increase productivity [60]. The size of the seed was found to be important for uniform sowing to extend the period of a *Silphium perfoliatum* plantation use (over 10 years). This is important for the improvement/optimization of sowing precision and for choosing the right seeder [60,61].

Some studies involved experiments using seed sterilization and micropropagation processes in improving sowing material [62–64]. Micropropagation also eliminates the possibility for *Silphium perfoliatum* to become invasive, due to achenes (fruits) that do not contain seeds and do not present a danger to native biodiversity [65]. For some purposes (e.g., pharmaceutical), it is recommended that all raw material should be standardized and maintained under controlled in vitro conditions. Different component concentrations were tested by using drip irrigation, which influenced the height of plants, the length and number of internodes, the number of flowers, leaf quantity and area, fresh weight of the shoots and stalk thickness [64,66], which are all important when extracting active components for the purposes listed above. There are reviews [67] detailing information on the metabolites and phytochemicals contained in *Silphium perfoliatum* parts. Many of the studies on these phytochemicals were conducted in Poland [34,44,68].

A study conducted in Poland found that seed production per *Silphium perfoliatum* plant is higher compared to other Silphium species [35]. It is the most valuable species for seed productivity.

It produced 19.02 g of seeds per shoot, 185.06 g of seeds per plant and the weight of 1000 seeds was 21.45 g [35]. The application of nitrogen fertilizers in soil (N_{90} and N_{120}) in an experiment involving sowing seeds produced a higher yield of seeds, ranging from 0.36 to 0.39 Mg ha^{-1} [53].

The germination capacity of *Silphium perfoliatum* seeds is high and is influenced by environmental conditions, such as light, temperature, pretreatment of seeds and pre-chilling. The germination capacity can be considerably increased when seeds are pretreated with a chemical solution (GA_3 and KNO_3) rather than only supplied with water for germination. The same study found that the best germination was achieved when a dark–light cycle (12:12) was applied with a temperature alternating between 20 and 30 °C, preceded by wet stratification, for seven days, at 0 °C [12]. Even though it is recommended to plant Silphium by using seedlings, to reduce costs, the planting by sowing can also be used. Another study found that the best seed germination results can be obtained by using pelleted seeds. In the planting of Silphium by sowing, pelleted seeds were pretreated with gibberellic acid. The most recommended time of year for sowing Silphium seeds in Central Europe is the end of April [69].

3.3. Utilization and Benefits/Advantages of Silphium perfoliatum

Silphium genus was used by native North American tribes for different medicinal purposes [48]. Due to the presence of active compounds in *Silphium perfoliatum*, its application in the production of drugs [70,71] and cosmetics is currently being investigated [72]. Some active components from Silphium leaves in combination with other chemical solutions can even contribute to improving the germination of common wheat seeds (*Tritium aestivum* L.) [73]. Around 16 compounds have now been isolated from *Silphium perfoliatum* for different pharmacological uses [48,74]. Some of them have shown immunosuppressive activity [48], with potential use in organ transplants. In addition, some extracts from leaves, inflorescences and rhizomes have been separated and found to display antibacterial activity on Gram-positive and Gram-negative bacteria [44]. The antifungal activity of 5% and 10% *Silphium perfoliatum* leaf extracts, that was applied on different types of fungi (*Alternaria alternata, Botrytis cinerea, Colletotrichum coccodes, Fusarium oxysporum, Penicillium expansum* and *Trichoderma harzianum*) found usually on some plants [75] was tested as an antifungal solution for pepper (*Capsicum annuum* L.). It was found that different plant parts, such as leaves, inflorescences and rhizomes contain phenolic acids (caffeic, p-coumaric, ferulic, vanillic, etc.). Due to their antioxidant activity, *Silphium perfoliatum* may offer medicinal benefits [76–78] as well as provide essential oils [68,79,80]. A very detailed analysis was recently performed on the active compounds of different *Silphium perfoliatum* parts [80,81]. These compounds (L-ascorbic acid; chlorophyll a and b; tannins; microelements—K, Ca, Mg, Zn, Fe, Mn and Se) are of interest to the pharmaceutical and food industries [80].

Other studies have analyzed the concentration of various compounds in *Silphium perfoliatum* that are important for fodder purposes [34]. There are also studies that have investigated *Silphium perfoliatum* silage [24] and its digestibility [23] as forage for livestock. In a study conducted in Moldova, it was found that a variety of *Silphium perfoliatum* (Vital variety) could be recommended for use as fodder for livestock, due to its characteristics (high protein content) and high biomass yield [46,47]. The same species was tested for renewable energy purposes, such as feedstock for biogas and briquette production [51,82]. It was also found that the presence of pollen and nectar on the *Silphium perfoliatum* disc florets has a positive effect on biodiversity [39,40,83]. It is also recommended for remediation of intensively managed soils [84–88] due to an increasing number of earthworms and long periods without any agro-technical work (plowing, rest and cultivation) [85]. Likewise, these fields became a home for arthropods [89] and can be an option for honeybees due to the long flowering period of the studied plant [38].

It should be emphasized, however, that prior to the establishment of a *Silphium perfoliatum* plantation (as in the case of other perennial plants) it is very important to adequately prepare the cultivation site, which includes removal of perennial weeds and soil cultivation. In the first year of the plantation, particular attention should be paid to weed removal by two or three mechanical weedings. Moreover, due to its high resistance in severe conditions (winter frosts, drought and acidic soils), a *Silphium perfoliatum* field does not require recultivation for 20 years. Other advantages of using a field for longer periods include reduced expenses on pesticides or herbicides due to shadow created by leaves. The annual litter residues (leaves) and the extensive root system of *Silphium perfoliatum* enrich valuable humus in the soil and prevent soil erosion. However, soil moisture significantly influences the density/biomass of the roots. Thus, in the case of soil excess moisture, the largest branching of roots (biomass roots) were found in the lower layers of soil, with 32% of roots compared to 12.3% of roots found in soil without excess moisture [90]. Therefore, soil moisture plays an important role in increasing the *Silphium perfoliatum* root biomass [91]. It was found that the biodiversity of soil is improved and regenerated on Silphium plantations [7,84,86]. Due to the presence of earthworms, for example, water infiltration in the soil can be improved [92]. One study found that the degradation of litter from Silphium leaves by insects, worms and other larvae help to increase nitrogen levels in soils (depending on the soil type) [93]. The effect of fertilizers on *Silphium perfoliatum* fields was also investigated. The use of lime and N fertilizers for *Silphium perfoliatum* plantations has a positive impact on soil. It was found that the soil under *Silphium perfoliatum* accumulated a high concentration of C_{tot}, N_{tot} and S_{tot} in soil layers, due to improving the Silphium root system (especially by liming) [94]. The abundant flowering from July to September promotes insect species and pollinators [39,40,84,89], which are vital insects for the world economy. It was established that although the *Silphium perfoliatum* is resistant to drought, the amount of nectar sugar obtained per hectare depends on the rainfall amount. Thus, insects more often visit plantations with higher soil moisture, e.g., that are irrigated with a higher water level (58 kg ha^{-1}), than plantations with lower soil moisture levels (rainfed plots) (20 kg ha^{-1}) [39]. It was also found that Silphium is a suitable permanent catch crop (melliferous plant) [95] and that it may become infested by fungi such as *Ascochyta silphia* (which can be found on *Silphium perfoliatum* leaves) [96]. This can be prevented by careful observation over time, without using any protective measures, but only by removing the infected crop [95]. However, in a recent study [20] conducted in North Dakota, the authors found the activity of the giant eucosma moth, *Eucosma giganteana* (Riley), which can inflict serious damage on a plantation. At the investigated sites, 25% of the plants were affected by the moth larvae and the biomass yield was affected in a similar proportion. These results depend mostly on the location of the sites/plantations. Some sites were situated on moist soils and some on soil with lower precipitation. Thus, it was found that larvae size depends on Silphium shoot size which, in turn, depends on soil moisture [20]. A study in Lithuania, ranked *Silphium perfoliatum* in third place as one of the most promising energy crops from a list of 22 annual and perennial crops (photosynthesis type C4 and C3) with low environmental pressure, suitable for the Lithuanian climate. The studied crops were evaluated by using a multi-criteria framework, such as photosynthesis type, soil carbon sequestration, erosion control, water adaptation, N input requirement and dry matter [97].

Silphium perfoliatum can be used for phytoremediation of soil polluted with heavy metals, such as Cd, which do not influence the total biomass and could be seen as a good candidate for phytostabilization [16]. The concentration of heavy metals in *Silphium perfoliatum* is important when it is used as a forage crop and must meet food standards concerning heavy metal contents. The influence of heavy metals on the yield of overground biomass of *Silphium perfoliatum* was investigated in Polish studies [98]. Different heavy metal concentrations were applied on Silphium plots. The value of the overground biomass yield ranged from 12.3% to 23.1%. It was also found that some Silphium overground parts met the food standards for heavy metal pollution (chromium), but some did not (manganese), depending on applied doses [98]. Nevertheless, the use of biochar from *Silphium perfoliatum* phytoremediation is safe when high-temperature pyrolysis (750 °C) is used for production processes to reduce the leaching risk of potentially toxic metals such as Zn, Cd and Pb [99]. The use of *Silphium perfoliatum* as a renewable energy source to produce pellets and briquettes was also studied, particularly their quality and composition [82,100–102].

Several studies have analyzed the best conditions for growing *Silphium perfoliatum* as feedstock for biogas production [11,12,103]. It is also seen as a promising alternative substrate for silage maize, due to its reduced agricultural requirements, which enable it to be grown on marginal or less productive soil [16]. Growing *Silphium perfoliatum* reduces the soil compaction and improves regional biodiversity [85–87].

3.4. *Silphium perfoliatum* Yields

3.4.1. Biomass Yields

The possibility of increasing the volume of biomass to meet energy needs has been widely studied [104]. It was estimated that when planting *Silphium perfoliatum* with 70 cm between rows and 50 cm distance between plants, around 28,500 plants ha^{-1} would be harvested, from which more than 40 Mg ha^{-1} fresh matter (FM) of biomass would be obtained 15 Mg ha^{-1} dry matter (DM) [105]. In other studies, the yield of green biomass of *Silphium perfoliatum* at the end of the flowering period was 62.8 Mg ha^{-1} FM, 24.6% DM, resulting in 15.4 Mg ha^{-1} DM [15]. A six-year experiment in Poland (2012–2017) on different types of biomass, including *Silphium perfoliatum*, found that the results of fresh mass yield were higher in the second year of growing and increased in 2016 and 2017, when the highest yields were obtained of all six years [106].

Silphium perfoliatum is a tolerant plant in environmentally unfavorable conditions [27]. However, good soil quality is necessary to achieve high biomass yield [17,107]. In Germany, yields were tested depending on the different origin (regions/continents) of the *Silphium perfoliatum* seeds. The most representative features/traits that influence the growth and yield were tested. It was found that there are some differences between the dry matter yield (DM) (from 13.6 to 17.2 Mg ha^{-1} DM) and other biomass traits. The lowest yield was achieved from Silphium seeds from Ukraine. It was assumed that the yield would be lower because the different development of the plant due to climate/environmental factors, as well as the different field of use (since it is used more as a forage crop it has features of a forage crop and less for biogas production and energy purposes) [52]. In another study, the biomass yield of progenies of 33 half-sib families of *Silphium perfoliatum* were tested in the USA. The yields showed different values in three consecutive years of growth. In the second year, the mean biomass yield was considerably lower than the first and third years (first = 4.7–7.5 Mg ha^{-1} DM; second = 1.5–3.4 Mg ha^{-1} DM; third = 5.2–10.5 Mg ha^{-1} DM). It is well-known that yields are usually higher in second and subsequent years, as compared to the first year of cultivation [45]. Another study observed that biomass yield was higher during earlier harvesting (at the end of August (17.2 Mg ha^{-1} DM) and two weeks later (17.1 Mg ha^{-1} DM)) and can decrease significantly when harvesting was conducted two weeks later (end of September and mid-October) (15.6 and 13.0 Mg ha^{-1} DM, respectively) [14]. In other field research conducted in Dagestan (Russia), the highest biomass yield was harvested in the third year of growth (27.9 Mg ha^{-1} DM) and the lowest was in the fourth year of growth (21.7 Mg ha^{-1} DM) (sum of two harvests in June and the end of September to the beginning of October) [29].

Figure 3 presents the biomass yield of *Silphium perfoliatum* in the ten countries analyzed in the current review [13–15,17,20,27,29,33,36,41,54,86,99,102,107–114]. The lowest dry matter yield was found in a study from the USA (1.6 Mg ha^{-1} DM) and Lithuania (4.4 Mg ha^{-1} DM). The highest yield was obtained in a study conducted in Russia (32.1 Mg ha^{-1} DM). High yields were also found in Poland (26.6 Mg ha^{-1} DM), Chile (22.3 Mg ha^{-1} DM) and Lithuania (21.9 Mg ha^{-1} DM). The mean dry matter yield of *Silphium perfoliatum* from the presented studies was 13.3 Mg ha^{-1} DM.

Although biomass yield can be influenced by planting density, the relationships are unclear because of other important factors. A study conducted in the USA found that higher planting density (68,000 plants ha^{-1}) produced higher biomass yields (14.2 Mg ha^{-1} DM) than lower planting density (17,000 plants ha^{-1} yielded 10.8 Mg ha^{-1} DM). Other results were obtained in another study conducted by the same researchers, in which the second year produced a very low biomass yield. The reported biomass yield value from all three density sites (17,000, 34,000 and 68,000 plants ha^{-1}) for the first year (2011) was 7.5 Mg ha^{-1} DM and 1.6 Mg ha^{-1} DM, respectively (2012) (the mean for all densities) [20]. However, the highest yields obtained in Germany (17.2 Mg ha^{-1} DM) were from a density of 40,000 plants ha^{-1} [52] and in Lithuania (21.9 Mg ha^{-1} DM) from a density of 20,000 plants ha^{-1} [54].

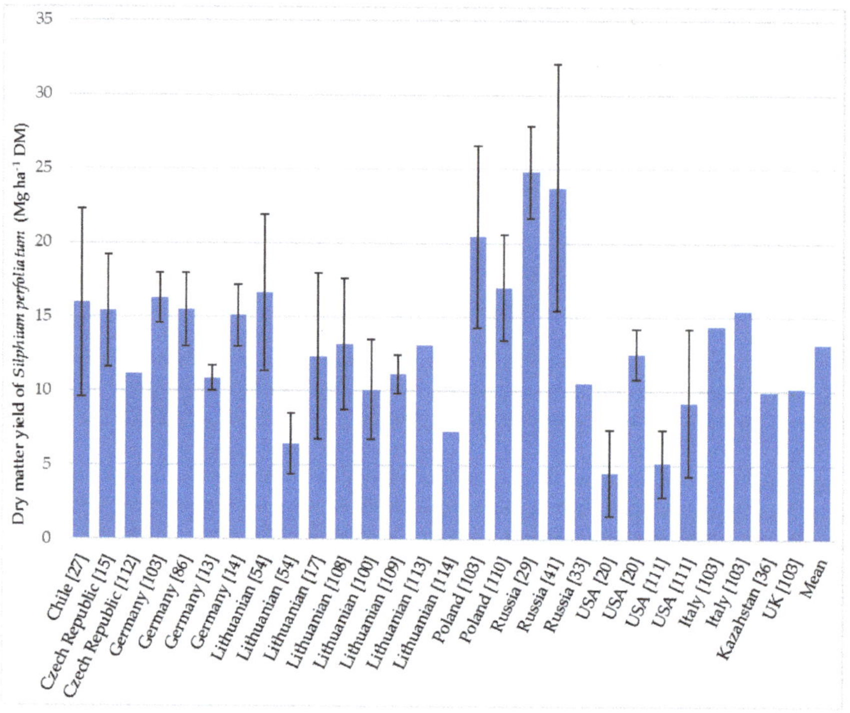

Figure 3. Dry matter yield of *Silphium perfoliatum* in different countries influenced by different factors (fertilizers, growing areas and different years of growth) and the mean for all countries; error bars represent the minimum and maximum values.

Table 1 presents the different biomass yields depending on the year of growth of *Silphium perfoliatum*. It can be seen that harvested yields depend on the soil, climatic and weather conditions, as well as other environmental factors in each country/region. The harvesting strategy implemented (once or twice per growing season) in some studies provides information on the choice of the optimal method for harvesting and the harvest period (month).

In a study conducted in Chile [27], the dry matter yield obtained in the first year ranged from 9.6 to 22.3 Mg ha^{-1} DM, and for the second year, it ranged from 14.6 to 19.4 Mg ha^{-1} DM, under similar climatic conditions (Table 1). It was observed that, with a longer growing period, the yields were higher and the plants were taller [27]. The cultivation of Silphium in combination with other crops was also tested, to increase the biomass yields [32,111].

Testing different fertilizers is not uncommon and is very often used to increase the *Silphium perfoliatum* DM yield. The use of lime fertilizers (CaCO$_3$) has shown good results for dry matter yield [17,54,100,109,113]. Jasinskas et al. [17] showed that lower soil acidity increased the amount of dry biomass yield. The use of fertilizers (6 Mg ha^{-1} CaCO$_3$) to increase soil pH when planting on acid soil pH (4.2–4.4) obtained 11.8 Mg ha^{-1} DM and increased to 15.0 Mg ha^{-1} DM when the soil pH was 5.6–5.7 [17]. Lime fertilizers also increased the DM productivity of *Silphium perfoliatum* (by 15.5% in the first year, 44.2% in the second year and 22.9% in the third year) [54,115]. Other studies averaged 22.7–27.3% [100,109,113]. Liming considerably increases dry matter productivity from the first year, at 4.4–8.5 Mg ha^{-1} DM, to 11.37–21.9 Mg ha^{-1} DM in the second year [54]. Similar results were obtained in other Lithuanian studies [100,109,113]. Another study assessed the influence of the application of sewage sludge on the biomass yield on Silphium plantations. Thus, the application of 45 Mg ha^{-1} of granulated sewage sludge showed the best results (10.4 Mg ha^{-1} DM average yield), compared to an application of a dose of 90 Mg ha^{-1}, which produced 9.3 Mg ha^{-1} DM yield. However, this yield was better than the yield obtained with the use of mineral fertilizers ($N_{60}P_{60}K_{60}$) 7.3 Mg ha^{-1} and from control plots 5.9 Mg ha^{-1} [108]. The application of 120 kg ha^{-1} of N fertilization increased the biomass yield from 11.0 to 15.0 Mg ha^{-1} DM (by 26.7%) [17].

A study using Albeluvisols and Fluvisols also applied fertilization by liming and N fertilizers, and it was found that liming did not result in a significant influence on stem numbers per plant, which was completely different for N fertilizers, which produced 5.7 stems per plant in the first year of growth, and up to 12.2 stems per plant in subsequent years. A German study by Mueller and Dauber [83] reported similar results of 6.3 stems per plant on average (1–15 stems), while Polish researchers reported 10–25 stems, which increased in subsequent years [101]. It was also found that *Silphium perfoliatum* biomass productivity (FM and DM yields) was lower in the first harvest year (19.2 Mg ha^{-1} FM and 6.7 Mg ha^{-1} DM, respectively), but increased significantly in the next two years (45.2 Mg ha^{-1} FM or 13.5 Mg ha^{-1} DM, on average) for the control *Silphium perfoliatum* plantation (no liming and no nitrogen) [55]. The biomass yield of *Silphium perfoliatum* in some cases showed better results than maize. On different types of soils that are predisposed to compaction (e.g., stagnosoils, planosols and related soils), *Silphium perfoliatum* biomass yield was considerably higher in the second year of harvest, when maize showed a considerably lower biomass yield [87].

Table 1. Dry matter yield of *Silphium perfoliatum* in different years of growth.

Year of Harvesting	Yield (Mg ha^{-1} DM)	Country	Observations and Harvesting Details	Reference
1st year	22.3	Chile	Chahuilco site (Trumao soils): very fertile acidic soils, pH 5.6, high Al and Mn concentration, 18% organic matter. Average rainfall 1303 mm. Harvested at flower initiation.	[27]
	9.6	Chile	La Unión site (Red clay soils): pH 5.2, organic matter 8.0%, suffered extreme water deficit during the summer season. Average precipitation was 1277 mm. Harvested at flower initiation.	[27]
	10.0 *–15.5 **	Germany	Basic fertilization and nitrogen fertilizers. The values depend on the watering regime. Harvest on different dates, depending on watering: August.	[13]
	13.0 ***–17.2 ⁑	Germany	Soil of experimental fields was Cambisol with a heavy loam texture overlaid with loess loam. Average precipitation 591 mm (2011) and 727 mm (2012).	[14]
	4.4–8.5	Lithuania	The values of DM yields differ due to the amount and types of fertilization: by liming and nitrogen treatments. Average precipitation during the vegetation period 437 mm. Harvest at maturity stage on September 16.	[54]
	7.5	USA	Mean biomass for three planting densities (17,000 34,000 and 680,000 plants ha^{-1}) at Brookings on soil considered marginal for conventional crop production due to poor drainage. No fertilizer, previous planted crop was soybean (*Glycine max* L. Merr.) with long-term rotation with wheat (*Triticum aestivum* L.). Planting was executed one year earlier in June by seedlings (2010). Harvesting time: in October.	[20]

Table 1. *Cont.*

Year of Harvesting	Yield (Mg ha^{-1} DM)	Country	Observations and Harvesting Details	Reference
2nd year	19.4	Chile	Chahuilco site (see above).	[27]
	14.6	Chile	La Unión site (see above).	[27]
	17.8	Chile	Nochaco site (Ñadi soils): iron and aluminum hardpan layer between the soil, pH 5.4. Average rainfall 1420 mm, dry period 1–2 months. Harvested at flower initiation.	[27]
	11.7 *–16.8 **	Germany	The values depend on the watering regime (see from the first year).	[13]
	13.4–15.7 ****; 19.4–21.7 ***	Poland	The experiment was conducted on light rust-brown sandy soil of poor rye complex. Average rainfall during the growing season was 359 mm.	[110]
	11.4–21.9	Lithuania	Different types and rates of fertilization. Average precipitation during the growing season was 620 mm. Plants planted in 2008, the first harvest was in 2009, the second harvest was on 30 September 2010.	[54]
	1.6	USA	Brookings site (see above).	[20]
	10.8–14.3	USA	Soil at Arlington location: Huntsville silt loam (fine–silty, mixed, mesic Cumulic Hapludoll in a low-lying area with a capability class of II because of potential flood damage from water retention. Nitrogen fertilizer (180 kg N ha^{-1}). The previous crop was alfalfa (*Medicago sativa* L.) Cultivation was in June 2010. Harvesting time: in September 2011.	[20]
	1.8 7.4–10.8	USA	The experiment was conducted at two different sites on prime and marginal cropland (Brookings and Arlington) with a well-detailed description of soil. Lower value was obtained on marginal cropland at Brookings (2013–second year) due to severe drought during the 2012 growing season. Hand-harvesting in November.	[111]
3rd year	19.1–20.6	Poland	Harvesting was conducted in the second (2017) and third years (2018) of growth.	[110]
	8.8–17.6	Lithuania	The experiment was started in 2013. Soil of the experimental site was naturally acid moraine loam *Bathygleyic Dystric Glossic Retisol*. Different treatments were tested: not fertilized, fertilized N60P60K60, and fertilized with different amounts of granulated sewage sludge (45 and 90 Mg ha^{-1}). The precipitation was 526 mm and temperatures were close to perennial values. Harvest by a rotary mower at the end of September.	[108]
	15.4	Czech Republic	This average value of cup-plant is achieved in long-term experiments under the same agro-ecological conditions and conventional agro-techniques. Harvested at the end of the flowering period.	[15]
	11.2	Czech Republic	Different plants were tested, only under experimental conditions, and only on a very small scale; only several species have been tested under field conditions.	[112]

Table 1. *Cont.*

Year of Harvesting	Yield (Mg ha^{-1} DM)	Country	Observations and Harvesting Details	Reference
Mean	15.5	Germany	The experimental fields were situated in two different locations with N and P fertilizers applied. Mean for six biomass traits over three years (2014–2016). Plants were harvested each year in August.	[52]
	13.2	Lithuania	Three different concentrations of fertilization were used. Mean for the period between 2009 and 2014. Harvesting time: end of September.	[113]

* rainfed; ** and irrigation; harvest on different dates: *** August and $^{\#}$ October; Method of planting by **** seeds (seed) and $^{*}_{**}$ seedlings (plantation).

3.4.2. Biogas and Biomethane Yields

Research conducted in the Czech Republic confirmed that biomass yields were the highest for *Silphium perfoliatum* and lower for maize, but the quality needed for biogas production is still higher for maize than for *Silphium perfoliatum* [15]. There are several studies that compared the biogas yields of maize and *Silphium perfoliatum* [116–118]. Maize is the most popular agricultural crop and shows the best results when used for biogas production, and it has frequently been used as a reference plant to compare different agricultural residues/crops [117]. Specific biogas yield (SBY) was studied by Siwek et al. [110], who investigated the effect of the planting type on some properties of *Silphium perfoliatum*. Thus, although the SBY was analyzed for the planting and sowing techniques, no significant results were obtained. They also tested the effect of the number of harvests per year on the biogas yield. It was found that following two harvests per year (June and October) in the second year of growing (the first harvest), the mean of SBY ranged from 484 to 490 NL kg^{-1} VS (VS-volatile solids), and this value was 504 NL kg^{-1} VS when the plants were harvested once a year (October). In the third year of growing (the second harvest), the values ranged from 485 to 502 NL kg^{-1} VS and 502 NL kg^{-1} VS for two harvests per year and one harvest per year, respectively [110]. Almost similar values were obtained by Țîței et al. [51], who found that the gas-forming potential of organic dry matter (ODM) in *Silphium perfoliatum* silage was 471 NL kg^{-1} VS (52.4%) of methane. However, in a two-harvest and a single-harvest system, the biogas yield per hectare was 11,076 and 7262 m^3 ha^{-1}, respectively [110]. In some experiments, *Silphium perfoliatum* was used as a reference crop for biogas production as well as with maize. However, in this case, it was assumed that Silphium is less suitable for anaerobic digestion, which is indicated by the lower biogas yields compared to maize [117].

According to research conducted in Lithuania, the biogas volume obtained from *Silphium perfoliatum* can exceed 5000 m^3 ha^{-1} and contains 58.6% methane. It was also estimated that such an amount of biogas is equal to 113.0 GJ of energy ha^{-1} [113]. One study in Germany focused on the impact of different harvest dates on biogas potential and found that MHY for *Silphium perfoliatum* may be higher if harvesting is performed earlier and the MHY was almost similar to the yields of maize and, in some cases, even higher than some perennial crops. Thus, the specific methane yield (SMY) in the total biogas yield was higher for *Silphium perfoliatum* (53.9%) than for other perennial crops (energy dock *Rumex schavnat*—52.8%, Szarvasi *Elymus elongatus* and Igniscum *Falopia sachalinensis*—53.2%) and maize (52.6% of methane) [14]. The methane content in biogas from the *Silphium perfoliatum* silage was higher (56.3%) than in maize silage (55%) [119]. The share of biomethane in biogas ranged from 51.1% to 52.9%. The highest values were obtained in an earlier harvest in the summer (June) and were slightly lower for a harvest made later in the autumn (October) [110].

The biochemical methane potential (BMP) of *Silphium perfoliatum* was analyzed in another study in Germany. On average, the BMP was 260 NL kg^{-1} VS [87], and it was almost 7.7% higher than in other previously published data, namely 251 NL kg^{-1} VS [120] and 252 NL kg^{-1} VS [14].

The medium gross energy yield for *Silphium perfoliatum* silage was 25.7 kWh ha^{-1} (92.5 GJ ha^{-1}) and was 46.5 kWh ha^{-1} lower, compared with the maize silage (167.4 GJ ha^{-1}). Therefore, the conclusion was that the new energy crops (Silphium) cannot compete with the specific methane yield of maize [120], due to the high yields and low market prices of the latter.

It was found that lignin is a very important component of biomass intended for biogas production since it influences specific methane yields. With lower contents of a lignin and fiber fractions, higher methane yields can be obtained [117,119].

Table 2 presents the values for SMY and MHY. The values differ when the crop is exposed to different factors, such as soil moisture, growing period and the number of harvests, or the type/origin of some varieties of Silphium. Even if *Silphium perfoliatum* is tolerant of drought, a reduction in water/rainfall reduces the methane yield per hectare [13,118]. The highest values for specific methane yield were estimated to be 321 NL kg^{-1} VS and for methane hectare yield to be 5399 Nm^3 ha^{-1} [13].

Table 2. The specific methane yield (SMY) and methane hectare yield (MHY) of *Silphium perfoliatum* in different studies.

Country	Specific Methane Yield SMY (NL kg^{-1} VS)	Methane Hectare Yield MHY (Nm^3 ha^{-1})	Observations	Reference
Southern Germany	232 * 275 ***	4301 [a] 3318 [b]	The highest and the lowest yields obtained from four different harvested batches, two doses of N fertilizer: 80 and 100 kg ha^{-1}.	[14]
Germany	236–2450 * 273–282 **	n.d.	Different excess and non-excess soil moisture were tested. Higher values for excess moisture.	[87]
Germany	290 *–303 ** 310 *–321 **	2889 *–3543 ** 4789 *–5399 **	Rainfed and irrigated, first and second-year harvest, respectively.	[13]
Southwest Germany	260	4856	Average for four years of harvesting (2015–2018). The first year (2014, n.d.) was excluded from the calculation of average.	[117]
Republic of Moldova	275	4235	The local variety Vital was tested for biomethane productivity.	[51]
Czech Republic	276	3921	Study based on long-term experiments and mixed samples with different fertilization rates.	[15]
Germany	258 [c]–273 [d]	3697 [d]–4634 [c]	Study of five countries of origin: USA, East Germany, Russia, Northern Europe, Ukraine over three years without fertilizer and the application of 100–150 kg ha^{-1} N.	[52]

* (1st year); ** (2nd year); *** (3rd year); harvested, [a] (19 September) and [b] (4 October); country of origin, [c] (Russia) and [d] (Ukraine); n.d. (no data).

3.5. Composition of Biomass and Chemical Characteristics

The composition of Silphium biomass depends on many factors, such as species, variety, harvest period (months, vegetative period or after the end of the vegetative period), soil types and climatic factors. The crude protein content in the early vegetative stages is higher (13% to 15% DM) than in the bud stage (10% DM) and continues to decline at seed setting (5% to 8% DM), due to very low N level in the steams [27]. It was observed that higher protein content can be obtained in wet years than in those with low precipitation [28]. In another study, the crude protein content of *Silphium perfoliatum* in vegetation phases ranged from 4.9% to 8.5% DM, the content of crude fat ranged from 2.1% to 2.5% DM and crude fiber from 23.1% to 29.7% DM. The hemicellulose content ranged from 5.4% DM at the vegetative phase to 10.1% DM at the beginning of seed setting [121]. Another study reported similar results of the crude protein content of *Silphium perfoliatum* (end of flowering) of 7.5% DM, crude fat content of 2.6% DM, crude fiber content of 26.4% DM and crude ash content of 8.0% DM [15]. A detailed analysis of the biomass properties of *Silphium perfoliatum* and other crops harvested after the vegetation period (in February) was conducted by Stolarski et al. [122]. The raw *Silphium perfoliatum* biomass had the cellulose content of 51.8% DM, the hemicellulose content of 20.3% DM and the lignin content of 11.5% DM. In the same study, the elemental composition of Silphium biomass was tested. It was found that the carbon content was 51.9% DM, the hydrogen content was 5.75% DM and the nitrogen content was below 0.49% DM. The sulfur content was 0.043% DM and for chlorine it was below 0.026% DM [122].

In a study conducted in Germany, the *Silphium perfoliatum* silage parameters were as follows: 27.1% DM, ODM 88.4% DM, pH of silage at 5.1, lactic acid 4.2% DM, acetic acid 1.9% DM, butyric acid 1.2% DM and alcohols 0.5% DM [119]. The chemical characteristics of *Silphium perfoliatum* silage were 42.5% DM for NFE (nitrogen-free extracts), 52.3% DM for NDF (neutral detergent fiber), 36.5% DM for ADF (acid detergent fiber), 7.9% DM for ADL (acid detergent lignin) and the C:N ratio was 39 [119].

The lower heating value (LHV), moisture and ash content depend on the harvesting time of *Silphium perfoliatum*, while a higher heating value (HHV), carbon, hydrogen and sulfur contents are less influenced by this factor [57,123]. Jasinskas et al. [17] found that the highest HHV of *Silphium perfoliatum* was recorded in the third year of growing (17.5 MJ kg^{-1} DM) and it was higher than the two previous years. In another study, a slightly higher HHV (18.3 MJ kg^{-1} DM) was reported [122]. However, in the studied literature, Stolarski et al. [5] obtained the highest HHV (18.8 MJ kg^{-1} DM). This value was calculated as the mean for six consecutive months, starting from November. In the same study, the calculated mean of LHV for six months was 11.1 MJ kg^{-1} FM [5]. In a recent study by Stolarski et al. [122], the LHV of *Silphium perfoliatum* raw biomass was higher (15.7 MJ kg^{-1} FM) because of a very low moisture content reaching 7.19%. As an energy plant, Silphium was also investigated for the production of pellets and briquettes. Thus, it has been established that the pressure of compaction and moisture plays an important role in the process of pressing and forming the pellets. A moisture level of 8% and pressure of 262 MPa provide improved density and durability of Silphium pellets [102].

A recent study of the chemical composition of the Silphium genus, including *Silphium perfoliatum*, found that it is a valuable nutritional source, and it can compete with other forage crops. Moreover, *Silphium perfoliatum* offers the highest content of ash and phosphorus (7.82% and 0.35%, respectively) [22]. A detailed amino acid profile has been evaluated in several studies. Figure 4 presents the mean amino acid profile [46,81] obtained for the whole plant and the results obtained from leaves of *Silphium perfoliatum* [81]. It contains all types of amino acids which are important for use as a forage crop or extraction of some active compounds for a specific purpose. For some cases, leaves can contain a higher percentage of some components than whole plants, as shown in Figure 4. The content of cysteine was found to reach 0.8% in herbs (whole plant) [81], while an earlier publication [46] did not provide any data on this amino acid.

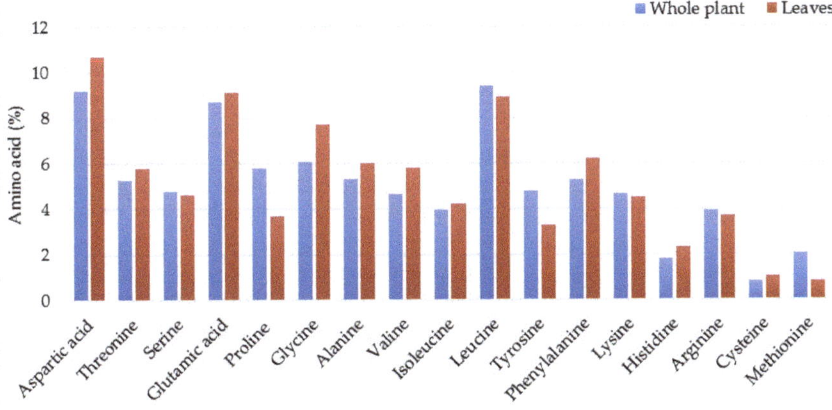

Figure 4. Percentage of amino acids from *Silphium perfoliatum*, compared to its leaves. Source: References [46,81].

4. Economic Aspects and Energy Efficiency of Biomass Production

Economic factors are very important in the use of *Silphium perfoliatum* as a feedstock for energy and other purposes. Since this plant is tolerant of unfavorable environmental conditions, it offers high biomass yields and the possibility of being planted on poor quality soils where other crops would be unprofitable. However, due to the high price of seedlings and the very high establishment costs for the farmers, *Silphium perfoliatum* cultivation is low and limited to small scale and trial plantations [124]. It was estimated that to be able to obtain reasonable yields, seed processing is necessary, which costs 1700 Euro per hectare. This operation is considered to be time- and cost-consuming due to the different and unequal seed size and maturity. The seed amount required per hectare is 2.5 kg, whose market price is approximatively 680 Euro kg^{-1}. Thus, the cultivation by seedlings can be replaced with sowing. To obtain a high yield level, sowing 1 ha with *Silphium perfoliatum* seeds is estimated to cost about 3150 Euro. This cost is estimated to be around 60% of the total costs of maintaining a plantation. If the plantation is used for ten years, these costs are divided to this period and total around 400 Euro per hectare per year. Of course, this cost may decrease if the plantation is operated for longer periods, such as 15 years (290 Euro per hectare per year) [125]. The cost of using *Silphium perfoliatum* as a feedstock for biogas production was discussed by Gerstberger et al. [84]. It was estimated that, together with additional compensation, the costs are almost the same as for maize crops [84]. Moreover, compared with maize fields, *Silphium perfoliatum* fields facilitate water infiltration and nutrient circulation [92] and are well-suited for the diversification of bioenergy farming landscapes [84,85]. *Silphium perfoliatum* may contribute to reducing the need for nitrogen fertilizers since its well-developed root system stops the leaching of nutrients from the soil. Thus, in subsequent years of growing, this plant may reduce the amount of fertilizers and prevent the contamination of waters with nitrogen due to reduced erosion [92]. To further reduce the costs, in some cases it was proposed to sow different crops, including forage crop [26,32] between plant rows for the first year of growing, when the *Silphium perfoliatum* biomass yields are significantly lower than in the following years [92].

Another important issue is the energy efficiency of *Silphium perfoliatum* biomass production. The goal is to obtain much higher energy output than energy input level in the biomass production of this species. The energy input depends on the production technology used. The amount of applied mineral fertilizers, mainly nitrogen fertilizers, is of particular importance. In studies conducted in Lithuania [54], the energy input for the cultivation of this species without mineral fertilization was 7.4 GJ ha^{-1} (Table 3). The use of liming and nitrogen fertilization increased these inputs up to 28 GJ ha^{-1}, depending on the variant and the amount of fertilizers used. It must be stressed that the applied fertilization also increased the energy accumulated in biomass from 187.6 to 361.9 GJ ha^{-1} both with and without the highest fertilization doses, respectively. Therefore, the energy gain from biomass harvested at the end of September was high and ranged from 180.2 to 333.9 GJ ha^{-1}, respectively. However, the energy ratio of *Silphium perfoliatum* biomass production was the highest (25.3) without fertilization, although this index was only 12.9 in the variant with the highest fertilization dose [54].

In other studies, in which *Silphium perfoliatum* was produced as a feedstock for solid biofuels for six years and the biomass was harvested at the end of March, the mean energy input was 8.1 GJ ha^{-1} (Table 4). The energy accumulated in the biomass increased from 82.3 to 184.4 (GJ ha^{-1}) in the second and seventh years of cultivation, respectively, and the energy gain ranged from 74.7 to 175.3 GJ ha^{-1}, respectively. The energy ratio of *Silphium perfoliatum* biomass production was the highest in the seventh year of cultivation (20.3). However, this index was only 10.8 in the second year of cultivation. In turn, the average energy ratio of cup plant cultivation as solid biofuel for six successive years was 13.5. In connection with the above, it should be stated that both energy gain and energy ratio increased significantly with the age of the plantation. Moreover, the accumulated energy in biomass and energy gain were higher when *Silphium perfoliatum* was harvested in September rather than in March [106].

Table 3. Energy-efficiency indicators of *Silphium perfoliatum* for different types and rates of fertilization, harvested at the end of September; based on Šiaudinis et al. [54].

Fertilization Rate	Energy Input (GJ ha^{-1})	Energy Accumulated in Biomass (GJ ha^{-1})	Energy Gain (GJ ha^{-1})	Energy Ratio
N0 (not limed)	7.4	187.6	180.2	25.3
N120 (not limed)	17.2	299.4	282.2	17.4
N0 + 0.5 liming rate	12.8	267.1	254.3	20.9
N120 + 0.5 liming rate	22.6	290.5	267.9	12.9
N0 + 1.0 liming rate	18.2	332.6	314.4	18.3
N120 + 1.0 liming rate	28.0	361.9	333.9	12.9

N—nitrogen; 0.5 liming rate (3.0 Mg ha^{-1} CaCO$_3$); 1.0 liming rate (6.0 Mg ha^{-1} CaCO$_3$).

Table 4. Energy-efficiency indicators of *Silphium perfoliatum* production as feedstock for solid biofuels in six successive harvest rotations, harvested at the end of March; based on Stolarski et al. [106].

Year/ Plantation Age	Energy Input (GJ ha^{-1})	Energy Accumulated in Biomass (GJ ha^{-1})	Energy Gain (GJ ha^{-1})	Energy Ratio
2012/2	7.6	82.3	74.7	10.8
2013/3	8.2	93.2	85.0	11.4
2014/4	7.6	84.0	76.4	11.0
2015/5	7.9	99.3	91.5	12.7
2016/6	8.2	119.8	111.7	14.6
2017/7	9.1	184.4	175.3	20.3
Mean	8.1	110.5	102.4	13.5

5. Conclusions

The reviewed information about *Silphium perfoliatum* indicates that it can be used for different purposes, such as a fodder crop in Eastern Europe (Russia, Belarus and Ukraine) and China, as a renewable energy source in Central and Northeastern Europe (Germany, Czech Republic, Austria, Poland and Lithuania) and even as a promising competitor to maize for biogas production (Germany and Austria). It was interesting to find that different types/varieties and different environmental factors/features (different soil types, level of precipitations, drought and irrigation level) have been tested around the world.

It is necessary to investigate the possibility of cultivating Silphium on marginal and degraded soils, after extensive agriculture, that cannot be used for other crop plantations. In addition, long-term cultivation of *Silphium perfoliatum* will contribute to improving the soil quality, controlling erosion, improving water infiltration, enriching soil mineral compounds and enhancing biodiversity, especially regarding honeybees and other pollinators.

However, more studies are needed on the costs of maintaining a Silphium plantation for future possible large-scale implementation. It is well-known that cultivation by sowing rather than by seedlings can considerably reduce the cost of a *Silphium perfoliatum* plantation. It is very important to improve growing characteristics to obtain sufficient biomass yields to enable the replacement of a maize crop by *Silphium perfoliatum* for biogas production. Although the reviewed studies show that the use of fertilizers can improve biomass yields, these operations can generate other costs which reduce energy efficiency. Therefore, further long-term research is needed to evaluate the use of Silphium biomass in comparison with other substrates, taking into account the full cultivation cycle and economic, environmental and energy efficiency. The results of such studies would help to determine whether *Silphium perfoliatum* is competitive to other crops and which ecosystem services are the most important and the most reliable.

Author Contributions: Conceptualization, M.J.S. and D.P.; methodology, M.J.S. and D.P.; formal analysis, M.K. and M.J.S.; validation, M.D., M.J.S. and M.K.; investigation, D.P., M.J.S. and A.B.; resources, M.J.S.; data curation, D.P. and M.J.S.; writing—original draft preparation, D.P., A.B., M.J.S. and M.K.; writing—review and editing, D.P., M.K. and M.J.S., A.B. and M.D.; visualization, D.P., A.B. and M.J.S.; supervision, M.J.S., M.D. and M.K.; funding acquisition, M.J.S. All authors have read and agreed to the published version of the manuscript.

Funding: This paper was co-financed by the National (Polish) Centre for Research and Development (NCBiR), entitled "Environment, agriculture and forestry", project BIOproducts from lignocellulosic biomass derived from MArginal land to fill the Gap in Current national bioeconomy, No. BIOSTRATEG3/344253/2/NCBR/2017. Dumitru Peni and Anna Bordiean are recipients of a scholarship from the Program of Interdisciplinary Doctoral Studies in Bioeconomy (POWR. 03.02.00-00-I034/16-00), which is funded by the European Social Fund.

Conflicts of Interest: The authors declare no conflict of interest.

References

1. Paolini, V.; Petracchini, F.; Segreto, M.; Tomassetti, L.; Naja, N.; Cecinato, A. Environmental impact of biogas: A short review of current knowledge. *J. Environ. Sci. Health Part A* **2018**, *53*, 899–906. [CrossRef]

2. Appels, L.; Baeyens, J.; Degreve, J.; Dewil, R. Principles and potential of the anaerobic digestion of waste-activated sludge. *Prog. Energy Combust. Sci.* **2008**, *34*, 755–781. [CrossRef]

3. Herrmann, A. Biogas production from maize: Current state, challenges and prospects. 2. Agronomic and environmental aspects. *Bioenergy Res.* **2013**, *6*, 372–387. [CrossRef]

4. Wu, X.; McLaren, J.; Madl, R.; Wang, D. Biofuels from Lignocellulosic Biomass. In *Sustainable Biotechnology: Sources of Renewable Energy*; Singh, O.V., Harvey, S.P., Eds.; Springer: Dordrecht, The Netherlands, 2010; pp. 19–41.

5. Stolarski, M.J.; Krzyzaniak, M.; Snieg, M.; Slominska, E.; Piórkowski, M.; Filipkowski, R. Thermophysical and chemical properties of perennial energy crops depending on harvest period. *Int. Agrophysics* **2014**, *28*, 201–211. [CrossRef]

6. Romanowska-Duda, Z.; Grzesik, M.; Pszczółkowski, W.; Piotrowski, K.; Pszczółkowska, A. The didactic and environmental functions of the collection of energy crops in the transfer technology center in Konstantynów Łódzki. *Acta Innov.* **2014**, *13*, 37–48.

7. Van Tassel, D.L.; Albrecht, K.A.; Bever, J.D.; Boe, A.A.; Brandvain, Y.; Crews, T.E.; Gansberger, M.; Gerstberger, P.; González-Paleo, L.; Hulke, B.S. Accelerating *Silphium* domestication: An opportunity to develop new crop ideotypes and breeding strategies informed by multiple disciplines. *Crop Sci.* **2017**, *57*, 1274–1284. [CrossRef]

8. Landis, D.A.; Gratton, C.; Jackson, R.D.; Gross, K.L.; Duncan, D.S.; Liang, C.; Meehan, T.D.; Robertson, B.A.; Schmidt, T.M.; Stahlheber, K.A. Biomass and biofuel crop effects on biodiversity and ecosystem services in the North Central US. *Biomass Bioenergy* **2018**, *114*, 18–29. [CrossRef]

9. Langhammer, M.; Grimm, V. Mitigating bioenergy-driven biodiversity decline: A modelling approach with the European brown hare. *Ecol. Model.* **2020**, *416*, 108914. [CrossRef]

10. Fiedler, A.K.; Landis, D.A. Attractiveness of Michigan native plants to arthropod natural enemies and herbivores. *Environ. Entomol.* **2007**, *36*, 751–765. [CrossRef]

11. Gansberger, M.; Montgomery, L.F.R.; Liebhard, P. Botanical characteristics, crop management and potential of *Silphium perfoliatum* L. as a renewable resource for biogas production: A review. *Ind. Crop. Prod.* **2015**, *63*, 362–372. [CrossRef]

12. Gansberger, M.; Stüger, H.-P.; Weinhappel, M.; Moder, K.; Liebhard, P.; von Gehren, P.; Mayr, J.; Ratzenböck, A. Germination characteristic of *Silphium perfoliatum* L. seeds. *Bodenkult. J. Land Manag. Food Environ.* **2017**, *68*, 73–79. [CrossRef]

13. Schoo, B.; Kage, H.; Schittenhelm, S. Radiation use efficiency, chemical composition, and methane yield of biogas crops under rainfed and irrigated conditions. *Eur. J. Agron.* **2017**, *87*, 8–18. [CrossRef]

14. Mast, B.; Lemmer, A.; Oechsner, H.; Reinhardt-Hanisch, A.; Claupein, W.; Graeff-Hönninger, S. Methane yield potential of novel perennial biogas crops influenced by harvest date. *Ind. Crop. Prod.* **2014**, *58*, 194–203. [CrossRef]

15. Ustak, S.; Munoz, J. Cup-plant potential for biogas production compared to reference maize in relation to the balance needs of nutrients and some microelements for their cultivation. *J. Environ. Manag.* **2018**, *228*, 260–266. [CrossRef] [PubMed]

16. Zhang, X.; Xia, H.; Li, Z.; Zhuang, P.; Gao, B. Potential of four forage grasses in remediation of Cd and Zn contaminated soils. *Bioresour. Technol.* **2010**, *101*, 2063–2066. [CrossRef]

17. Jasinskas, A.; Simonavičiūtė, R.; Šiaudinis, G.; Liaudanskienė, I.; Antanaitis, Š.; Arak, M.; Olt, J. The assessment of common mugwort (*Artemisia vulgaris* L.) and cup plant (*Silphium perfoliatum* L.) productivity and technological preparation for solid biofuel. *Zemdirb. Agric.* **2014**, *101*, 19–26. [CrossRef]

18. Stolarski, M.J.; Śnieg, M.; Krzyżaniak, M.; Tworkowski, J.; Szczukowski, S. Short rotation coppices, grasses and other herbaceous crops: Productivity and yield energy value versus 26 genotypes. *Biomass Bioenergy* **2018**, *119*, 109–120. [CrossRef]

19. Reinert, S.; Hulke, B.S.; Prasifka, J.R. Pest potential of *Neotephritis finalis* (Loew) on *Silphium integrifolium* Michx., *Silphium perfoliatum* L., and interspecific hybrids. *Agron. J.* **2020**, *112*, 1462–1465. [CrossRef]

20. Boe, A.; Albrecht, K.A.; Johnson, P.J.; Wu, J. Biomass Production of Cup Plant (*Silphium perfoliatum* L.) in Response to Variation in Plant Population Density in the North Central USA. *Am. J. Plant Sci.* **2019**, *10*, 904–910. [CrossRef]

21. Douglas, J.A.; Follett, J.M.; Halliday, I.R.; Hughes, J.W.; Parr, C.R. Silphium: Preliminary research on a possible new forage crop for New Zealand. *Proc. Agron. Soc. N. Z.* **1987**, *17*, 51–53.

22. Rakhmetov, D.B.; Vergun, O.M.; Stadnichuk, N.O.; Shymanska, O.V.; Rakhmetova, S.O.; Fishchenko, V.V. Biochemical study of plant raw material of *Silphium* L. SPP. in M.M. Gryshko National Botanical Garden of the NAS of Ukraine. *Plant Introd.* **2019**, *83*, 80–86. [CrossRef]

23. Han, K.J.; Albrecht, K.A.; Mertens, D.R.; Kim, D.A. Comparison of in vitro digestion kinetics of cup-plant and alfalfa. *Asian-Australas. J. Anim. Sci.* **2000**, *13*, 641–644. [CrossRef]

24. Han, K.J.; Albrecht, K.A.; Muck, R.E.; Kim, D.A. Moisture effect on fermentation characteristics of cup-plant silage. *Asian-Australas. J. Anim. Sci.* **2000**, *13*, 636–640. [CrossRef]

25. Puia, I.; Szabo, A.T. Experimental cultivation of a new forage species-*Silphium perfoliatum* L.-in the Agrobotanical Garden from Cluj-Napoca. *Not. Bot. Horti Agrobot. Cluj-Napoca* **1985**, *15*, 15–20.

26. Pastukhova, M.A.; Sheliuto, B.V. Возделываниесильфии пронзеннолистной под покровом сельскохозяйственных культур [Cultivation of *Silphium perfoaliatum* under agricultural crops cover]. *Bull. Belarus. State Agric. Acad.* **2019**, *3*, 83–87.

27. Pichard, G. Management, production, and nutritional characteristics of cup-plant (*Silphium perfoliatum*) in temperate climates of southern Chile. *Cienc. Investig. Agrar.* **2012**, *39*, 61–77. [CrossRef]

28. Guoyan, P.; Ouyang, Z.; Qunying, L.; Qiang, Y.; Jishun, W. Water consumption of seven forage cultivars under different climatic conditions in the North China plain. *J. Resour. Ecol.* **2011**, *2*, 74–82. [CrossRef]

29. Gimbatov, A.S.; Alimirzaeva, G.A. Optimization of technology for the cultivation of new feed crops in the irrigated conditions of the lowland zone of Dagestan. *Probl. Dev. Agric. Ind. Apk Reg.* **2011**, *5*, 8–11.

30. Zaynullina, K.S.; Ruban, G.A.; Mikhovich, J.E.; Portnyagina, N.V.; Potapov, A.A.; Punegov, V.V.; Fomina, M.G. Prospects of introduction in culture in the North some resource plants (Komi Republic). *Bull. Samara Sci. Cent. Russ. Acad. Sci.* **2015**, *17*, 121–126.

31. Novichikhin, A.M.; Piscareva, L.A. The study of elements of the cultivation technology of *Silphium Perfoliatum*. *Symb. Sci.* **2016**, *10*, 38–41.

32. Semerenko, M.V.; Chupina, M.P.; Stepanov, A.F. The influence of compacting crops on the productivity of perfoliate sylphs. *Bull. Omsk State Agric. Univ.* **2016**, *2*, 61–65.

33. Chupina, M.P.; Stepanov, A.F. The application of *Silphium perfoliatum* in the system of green and raw material conveyors. *Agric. Sci. J. Omsk* **2017**, *4*, 92–97.

34. Kowalski, R. Analysis of lipophilic fraction from leaves, inflorescences and rhizomes of *Silphium perfoliatum* L. *Acta Soc. Bot. Pol.* **2005**, *74*, 5–10. [CrossRef]

35. Kowalski, R.; Wierciński, J. Evaluation of chemical composition of some *Silphium* L. species seeds as alternative foodstuff raw materials. *Pol. J. Food Nutr. Sci.* **2004**, *13*, 349–354.

36. Danilov, K.P. The effect of the method and sowing rate on the yield of *Silphium perfoliatum*. *News Orenbg. State Agrar. Univ.* **2013**, *4*, 37–39.

37. Klímek, P.; Meinlschmidt, P.; Wimmer, R.; Plinke, B.; Schirp, A. Using sunflower (*Helianthus annuus* L.), topinambour (*Helianthus tuberosus* L.) and cup-plant (*Silphium perfoliatum* L.) stalks as alternative raw materials for particleboards. *Ind. Crop. Prod.* **2016**, *92*, 157–164. [CrossRef]

38. Pastukhova, M.A. The place of *Silphium perfoliatum* in the nectariferous conveyor in the conditions of the south-west of Belarus. *Bull. Belarus. State Agric. Acad.* **2019**, *3*, 88–92.

39. Mueller, A.L.; Berger, C.A.; Schittenhelm, S.; Stever-Schoo, B.; Dauber, J. Water availability affects nectar sugar production and insect visitation of the cup plant *Silphium perfoliatum* L. (Asteraceae). *J. Agron. Crop Sci.* **2020**, *206*, 529–537. [CrossRef]

40. Mueller, A.L.; Biertümpfel, A.; Friedritz, L.; Power, E.F.; Wright, G.A.; Dauber, J. Floral resources provided by the new energy crop, *Silphium perfoliatum* L. (Asteraceae). *J. Apic. Res.* **2020**, *59*, 232–245. [CrossRef]

41. Savin, A.P.; Gudimova, N.A. The influence of mineral fertilizers on the nectar, forage and seed productivity sylphs standardized. Современные проблемы пчеловодства иапитерапии: монография/под, Рыбное: фГБНУ«фНЦ пчеловодства. In *Modern Problem of Beekeeping and Apitherapy*; Brandorf, A.Z., Lebedev, V.I., Kharitonova, M.N., Savina, A.P., Savushkina, L.N., Lizunova, A.S., Eds.; Publisher: Rybnoe, Russia, 2019; pp. 186–191.

42. Clevinger, J.A. New combinations in Silphium (Asteraceae: Heliantheae). *Novon* **2004**, *14*, 275–277.

43. Clevinger, J.A.; Panero, J.L. Phylogenetic analysis of Silphium and subtribe Engelmanniinae (Asteraceae: Heliantheae) based on ITS and ETS sequence data. *Am. J. Bot.* **2000**, *87*, 565–572. [CrossRef] [PubMed]

44. Kowalski, R.; Kędzia, B. Antibacterial activity of *Silphium perfoliatum*. Extracts. *Pharm. Biol.* **2007**, *45*, 494–500. [CrossRef]

45. Assefa, T.; Wu, J.; Albrecht, K.A.; Johnson, P.J.; Boe, A. Genetic variation for biomass and related morphological traits in cup plant (*Silphium perfoliatum* L.). *Am. J. Plant Sci.* **2015**, *6*, 1098–1108. [CrossRef]

46. Țîței, V.; Teleuță, A.; Muntea, A. The perspective of cultivation and utilization of the species *Silphium perfoliatum* L. and *Helianthus tuberosus* L. *Mold. Bull. UASMV Agric.* **2013**, *70*, 160–166.

47. Țîței, V. Biological peculiarities of cup plant (*Silphium perfoliatum* L.) and utilization possibilities in the Republic of Moldova. *Agron. Ser. Sci. Res. Lucr. Stiintifice Ser. Agron.* **2014**, *57*, 289–293.

48. Feng, W.-S.; Pei, Y.-Y.; Zheng, X.-K.; Li, C.-G.; Ke, Y.-Y.; Lv, Y.-Y.; Zhang, Y.-L. A new kaempferol trioside from *Silphium perfoliatum*. *J. Asian Nat. Prod. Res.* **2014**, *16*, 393–399. [CrossRef]

49. Franzaring, J.; Holz, I.; Kauf, Z.; Fangmeier, A. Responses of the novel bioenergy plant species *Sida hermaphrodita* (L.) Rusby and *Silphium perfoliatum* L. to CO_2 fertilization at different temperatures and water supply. *Biomass Bioenergy* **2015**, *81*, 574–583. [CrossRef]

50. Stanford, G. *Silphium perfoliatum* (cup-plant) as a new forage. In Proceedings of the Twelfth North American Prairie Conference: Recapturing a Vanishing Heritage, Cedar Falls, Iowa, 5–9 August 1990; pp. 33–37.

51. Țîței, V. The evaluation of biomass of the *Sida hermaphrodita* and *Silphium perfoliatum* for renewable energy in Moldova. *Sci. Pap. Ser. A Agron.* **2017**, *60*, 534–540.

52. Wever, C.; Höller, M.; Becker, L.; Biertümpfel, A.; Köhler, J.; van Inghelandt, D.; Westhoff, P.; Pude, R.; Pestsova, E. Towards high-biomass yielding bioenergy crop *Silphium perfoliatum* L.: Phenotypic and genotypic evaluation of five cultivated populations. *Biomass Bioenergy* **2019**, *124*, 102–113. [CrossRef]

53. Shalyuta, B.V.; Kostitskaya, E.V. The yield of *Silphium perfoliatum* L. depending on the conditions of cultivation. *Agric. Eng.* **2018**, *22*, 91–97. [CrossRef]

54. Šiaudinis, G.; Jasinskas, A.; Šlepetienė, A.; Karčauskienė, D. The evaluation of biomass and energy productivity of common mugwort (*Artemisia vulgaris* L.) and cup plant (*Silphium perfoliatum* L.) in *Albeluvisol*. *Agriculture* **2012**, *99*, 357–362.

55. Šiaudinis, G.; Skuodienė, R.; Repšienė, R. The investigation of three potential energy crops: Common mugwort, cup plant and Virginia mallow on Western Lithuania's Albeluvisol. *Appl. Ecol. Environ. Res.* **2017**, *15*, 611–620. [CrossRef]

56. Assefa, T.; Wu, J.; Boe, A. Genetic variation for achene traits in cup plant (*Silphium perfoliatum* L.). *Open J. Genet.* **2015**, *5*, 71–82. [CrossRef]

57. Stolarski, M.J.; Śnieg, M.; Krzyżaniak, M.; Tworkowski, J.; Szczukowski, S.; Graban, Ł.; Lajszner, W. Short rotation coppices, grasses and other herbaceous crops: Biomass properties versus 26 genotypes and harvest time. *Ind. Crop. Prod.* **2018**, *119*, 22–32. [CrossRef]

58. Stankevich, S.I.; Kiselev, A.A.; Nesterenko, T.K. The influence of the method of spreading on productivity of cup plant. *Bull. Belarusian State Agric. Acad.* **2017**, *3*, 77–80.

59. Schäfer, A.; Meinhold, T.; Damerow, L.; Lammers, P.S. Crop establishment of *Silphium perfoliatum* by precision seeding. *Agric. Eng.* **2015**, *70*, 254–261. [CrossRef]

60. Schäfer, A.; Leder, A.; Graff, M.; Damerow, L.; Lammers, P.S. Determination and sorting of cup plant seeds to optimize crop establishment. *Agric. Eng.* **2018**, *73*, 97–105.

61. Schäfer, A.; Damerow, L.; Lammers, P.S. Determination of the seed geometry of cup plant as requirement for precision seeding. *Agric. Eng.* **2017**, *72*, 122–129.
62. Tomaszewska-Sowa, M.; Figas, A. Optimization of the processes of sterilization and micropropagation of cup plant (*Silphium perfoliatum* L.) from apical explants of seedlings in in vitro cultures. *Acta Agrobot.* **2011**, *64*, 3–10. [CrossRef]
63. Figas, A.; Rolbiecki, R.; Tomaszewska-Sowa, M. Influence of drip irrigation on the height of cup plant (*Silphium perfoliatum* L.) cultivated on the very light soil from the micropagation seedlings. *Infrastruct. Ecol. Rural Areas* **2011**, *10*, 245–253.
64. Figas, A.; Rolbiecki, R.; Tomaszewska-Sowa, M. Influence of drip irrigation on the height of the biennial cup plant (*Silphium perfoliatum* L.) from the micropropagation seedlings. *Infrastruct. Ecol. Rural Areas* **2015**, *3*, 779–786. [CrossRef]
65. Figas, A.; Sawilska, A.K.; Rolbiecki, R.; Tomaszewska-Sowa, M. Morphological characteristics of achenes and fertility plants of cup plant (*Silphium perfoliatum* L.) obtained from micropropagation growing under irrigation. *Infrastruct. Ecol. Rural Areas* **2016**, *4*, 1363–1372. [CrossRef]
66. Figas, A.; Siwik-Ziomek, A.; Rolbiecki, R. Effect of irrigation on some growth parameters of cup plant and dehydrogenase activity in soil. *Ann. Wars. Univ. Life Sci. Land Reclam.* **2015**, *47*, 279–288. [CrossRef]
67. Oleszek, M.; Kowalska, I.; Oleszek, W. Phytochemicals in bioenergy crops. *Phytochem. Rev.* **2019**, *18*, 893–927. [CrossRef]
68. Kowalski, R.; Wolski, T. The chemical composition of essential oils of (*Silphium perfoliatum* L.). *Flavour Fragr. J.* **2005**, *20*, 306–310. [CrossRef]
69. Von Gehren, P.; Gansberger, M.; Mayr, J.; Liebhard, P. The effect of sowing date and seed pretreatments on establishment of the energy plant *Silphium perfoliatum* by sowing. *Seed Sci. Technol.* **2016**, *44*, 310–319. [CrossRef]
70. Guo, Y.; Shang, H.; Zhao, J.; Zhang, H.; Chen, S. Enzyme-assisted extraction of a cup plant (*Silphium perfoliatum* L.) Polysaccharide and its antioxidant and hypoglycemic activities. *Process Biochem.* **2020**, *92*, 17–28. [CrossRef]
71. Piszczek, P.; Kuszewska, K.; Błaszkowski, J.; Sochacka-Obruśnik, A.; Stojakowska, A.; Zubek, S. Associations between root-inhabiting fungi and 40 species of medicinal plants with potential applications in the pharmaceutical and biotechnological industries. *Appl. Soil Ecol.* **2019**, *137*, 69–77. [CrossRef]
72. Tomaszewska-Sowa, M.; Figas, A. Environmental and ecological aspects of cultivation of selected energy and herbal plants propagated by in vitro culture. *Infrastruct. Ecol. Rural Areas* **2014**, *4*, 1457–1465. [CrossRef]
73. Davidyants, E.S. The effect of the purified amount of triterpene glycosides and their enriched extract from leaves of *Silphium perfoliatum* L. on the growth and activity of nitrate reductase of winter wheat plants. *Chem. Plant Raw Mater.* **2019**, *4*, 441–448. [CrossRef]
74. El-Sayed, N.H.; Wojcińska, M.; Drost-Karbowska, K.; Matławska, I.; Williams, J.; Mabry, T.J. Kaempferol triosides from *Silphium perfoliatum*. *Phytochemistry* **2002**, *60*, 835–838. [CrossRef]
75. Jemiołkowska, A.; Kowalski, R. In vitro estimate of influence of *Silphium perfoliatum* L. leaves extract on some fungi colonizing the pepper plants. *Acta Sci. Pol. Hort. Cult* **2012**, *11*, 43–55.
76. Wu, H.; Shang, H.; Guo, Y.; Zhang, H.; Wu, H. Comparison of different extraction methods of polysaccharides from cup plant (*Silphium perfoliatum* L.). *Process Biochem.* **2020**, *90*, 241–248. [CrossRef]
77. Kowalski, R.; Wolski, T. Evaluation of phenolic acid content in *Silphium perfoliatum* L. leaves, inflorescences and rhizomes. *Electron. J. Pol. Agric. Univ.* **2003**, *6*, 1–10.
78. Shang, H.-M.; Zhou, H.-Z.; Li, R.; Duan, M.-Y.; Wu, H.-X.; Lou, Y.-J. Extraction optimization and influences of drying methods on antioxidant activities of polysaccharide from cup plant (*Silphium perfoliatum* L.). *PLoS ONE* **2017**, *12*, e0183001. [CrossRef] [PubMed]
79. Wolski, T.; Kowalski, R.; Mardarowicz, M. Chromatographic analysis of essential oil occurring in inflorescences, leaves and rhizomes of *Silphium perfoliatum* L. *Herba Pol.* **2000**, *46*, 235–242.
80. Kowalski, R. *Silphium* L. extracts-composition and protective effect on fatty acids content in sunflower oil subjected to heating and storage. *Food Chem.* **2009**, *112*, 820–830. [CrossRef]
81. Kowalska, G.; Pankiewicz, U.; Kowalski, R. Evaluation of Chemical Composition of Some *Silphium* L. Species as Alternative Raw Materials. *Agriculture* **2020**, *10*, 132. [CrossRef]
82. Țîței, V. The potential growth and the biomass quality of some herbaceous species for the production of renewable energy in Moldova. *Rev. Bot. Bot. J.* **2019**, *18*, 83–91.

83. Mueller, A.L.; Dauber, J. Hoverflies (Diptera: Syrphidae) benefit from a cultivation of the bioenergy crop *Silphium perfoliatum* L. (Asteraceae) depending on larval feeding type, landscape composition and crop management. *Agric. For. Entomol.* **2016**, *18*, 419–431. [CrossRef]

84. Gerstberger, P.; Asen, F.; Hartmann, C. Economy and ecology of cup plant (*Silphium perfoliatum* L.) compared with silage maize. *J. Kult.* **2016**, *68*, 372–377. [CrossRef]

85. Schorpp, Q.; Schrader, S. Earthworm functional groups respond to the perennial energy cropping system of the cup plant (*Silphium perfoliatum* L.). *Biomass Bioenergy* **2016**, *87*, 61–68. [CrossRef]

86. Schorpp, Q.; Schrader, S. Dynamic of nematode communities in energy plant cropping systems. *Eur. J. Soil Biol.* **2017**, *78*, 92–101. [CrossRef]

87. Ruf, T.; Emmerling, C. Site-adapted production of bioenergy feedstocks on poorly drained cropland through the cultivation of perennial crops. A feasibility study on biomass yield and biochemical methane potential. *Biomass Bioenergy* **2018**, *119*, 429–435. [CrossRef]

88. Ruf, T.; Makselon, J.; Udelhoven, T.; Emmerling, C. Soil quality indicator response to land-use change from annual to perennial bioenergy cropping systems in Germany. *GCB Bioenergy* **2018**, *10*, 444–459. [CrossRef]

89. Chmelíková, L.; Wolfrum, S. Mitigating the biodiversity footprint of energy crops—A case study on arthropod diversity. *Biomass Bioenergy* **2019**, *125*, 180–187. [CrossRef]

90. Ruf, T.; Audu, V.; Holzhauser, K.; Emmerling, C. Bioenergy from periodically waterlogged cropland in Europe: A first assessment of the potential of five perennial energy crops to provide biomass and their interactions with soil. *Agronomy* **2019**, *9*, 374. [CrossRef]

91. Schoo, B.; Schroetter, S.; Kage, H.; Schittenhelm, S. Root traits of cup plant, maize and lucerne grass grown under different soil and soil moisture conditions. *J. Agron. Crop Sci.* **2017**, *203*, 345–359. [CrossRef]

92. Grunwald, D.; Panten, K.; Schwarz, A.; Bischoff, W.A.; Schittenhelm, S. Comparison of maize, permanent cup plant and a perennial grass mixture with regard to soil and water protection. *GCB Bioenergy* **2020**, *12*, 694–705. [CrossRef]

93. Schorpp, Q.; Riggers, C.; Lewicka-Szczebak, D.; Giesemann, A.; Well, R.; Schrader, S. Influence of *Lumbricus terrestris* and *Folsomia candida* on N_2O formation pathways in two different soils—With particular focus on N_2 emissions. *Rapid Commun. Mass Spectrom.* **2016**, *30*, 2301–2314. [CrossRef]

94. Šiaudinis, G.; Liaudanskiene, I.; Slepetiene, A. Changes in soil carbon, nitrogen and sulphur content as influenced by liming and nitrogen fertilization of three energy crops. *Icel. Agric. Sci.* **2017**, *30*, 43–50. [CrossRef]

95. Bufe, C.; Korevaar, H. *Evaluation of Additional Crops for Dutch List of Ecological Focus Area: Evaluation of Miscanthus, Silphium perfoliatum, Fallow Sown in with Melliferous Plants and Sunflowers in Seed Mixtures for Catch Crops*; Report/WPR, no. 793; Wageningen Research Foundation (WR) Business Unit Agrosystems Research: Lelystad, The Netherlands, 2018; p. 34. [CrossRef]

96. Bedlan, G. *Ascochyta silphii* sp. nov.—A new Ascochyta species on *Silphium perfoliatum*. *J. Kult.* **2014**, *66*, 281–283. [CrossRef]

97. Balezentiene, L.; Streimikiene, D.; Balezentis, T. Fuzzy decision support methodology for sustainable energy crop selection. *Renew. Sustain. Energy Rev.* **2013**, *17*, 83–93. [CrossRef]

98. Antonkiewicz, J.; Jasiewicz, C. Effect of Soil Contamination with Heavy Metals on Element Contents in Silphium perfoliatum (*Silphium perfoliatum* L.). *Chem. Ecol. Eng.* **2003**, *10*, 205–210.

99. Du, J.; Zhang, L.; Ali, A.; Li, R.; Xiao, R.; Guo, D.; Liu, X.; Zhang, Z.; Ren, C.; Zhang, Z. Research on thermal disposal of phytoremediation plant waste: Stability of potentially toxic metals (PTMs) and oxidation resistance of biochars. *Process Saf. Environ. Prot.* **2019**, *125*, 260–268. [CrossRef]

100. Šiaudinis, G.; Jasinskas, A.; Šarauskis, E.; Steponavičius, D.; Karčauskienė, D.; Liaudanskienė, I. The assessment of Virginia mallow (*Sida hermaphrodita* Rusby) and cup plant (*Silphium perfoliatum* L.) productivity, physico-mechanical properties and energy expenses. *Energy* **2015**, *93*, 606–612. [CrossRef]

101. Wrobel, M.; Frączek, J.; Francik, S.; Slipek, Z.; Mudryk, K. Influence of degree of fragmentation on chosen quality parameters of briquette made from biomass of cup plant *Silphium perfoliatum* L. In Proceedings of the 12th International Scientific Conference Engineering for Rural Development, Jelgava, Latvia, 23–24 May 2013; pp. 653–657.

102. Styks, J.; Wróbel, M.; Frączek, J.; Knapczyk, A. Effect of Compaction Pressure and Moisture Content on Quality Parameters of Perennial Biomass Pellets. *Energies* **2020**, *13*, 1859. [CrossRef]

103. Bury, M.; Facciotto, G.; Chiocchini, F.; Cumplido-Marín, L.; Graves, A.; Kitczak, T.; Martens, R.; Morhart, C.; Możdżer, E.; Nahm, M. Preliminary results regarding yields of Virginia mallow (*Sida hermaphrodita* L. Rusby) and cup plant (*Silphium perfoliatum* L.) in different condition of Europe. In Proceedings of the 27th European Biomass Conference and Exhibition, Lisbon, Portugal, 27–30 May 2019; pp. 101–104. [CrossRef]

104. Von Cossel, M.; Wagner, M.; Lask, J.; Magenau, E.; Bauerle, A.; Von Cossel, V.; Warrach-Sagi, K.; Elbersen, B.; Staritsky, I.; Van Eupen, M. Prospects of bioenergy cropping systems for a more social-ecologically sound bioeconomy. *Agronomy* **2019**, *9*, 605. [CrossRef]

105. Jucsor, N.; Sumalan, R. Researches concerning the potential of biomass accumulation in cup plant (*Silphium perfoliatum* L.). *J. Hortic. For. Biotechnol.* **2018**, *22*, 34–39.

106. Stolarski, M.J.; Krzyżaniak, M.; Warmiński, K.; Olba-Zięty, E.; Penni, D.; Bordiean, A. Energy efficiency indices for lignocellulosic biomass production: Short rotation coppices versus grasses and other herbaceous crops. *Ind. Crop. Prod.* **2019**, *135*, 10–20. [CrossRef]

107. Franzaring, J.; Schmid, I.; Bäuerle, L.; Gensheimer, G.; Fangmeier, A. Investigations on plant functional traits, epidermal structures and the ecophysiology of the novel bioenergy species *Sida hermaphrodita* Rusby and *Silphium perfoliatum* L. *J. Appl. Bot. Food Qual.* **2014**, *87*, 36–45. [CrossRef]

108. Šiaudinis, G.; Karčauskienė, D.; Aleinikovienė, J. Assessment of a single application of sewage sludge on the biomass yield of *Silphium perfoliatum* and changes in naturally acid soil properties. *Agriculture* **2019**, *106*, 213–218. [CrossRef]

109. Šiaudinis, G.; Skuodienė, R.; Katutis, K. Biomass and energy productivity of different plant species under western Lithuania conditions. In Proceedings of the 25th Nordic View to Sustainable Rural Development Congress, Riga, Latvia, 16–18 June 2015; pp. 232–237.

110. Siwek, H.; Włodarczyk, M.; Możdżer, E.; Bury, M.; Kitczak, T. Chemical Composition and Biogas Formation potential of *Sida hermaphrodita* and *Silphium perfoliatum*. *Appl. Sci.* **2019**, *9*, 4016. [CrossRef]

111. Boe, A.; Albrecht, K.A.; Johnson, P.J.; Wu, J. Biomass production of monocultures and mixtures of cup plant and native grasses on prime and marginal cropland. *Am. J. Plant Sci.* **2019**, *10*, 911. [CrossRef]

112. Heneman, P.; Červinka, J. Energy crops and bioenergetics in the Czech Republic. *Agric. (Agric. Eng.)* **2007**, *51*, 73–78.

113. Skorupskaitė, V.; Makarevičienė, V.; Šiaudinis, G.; Zajančauskaitė, V. Green energy from different feedstock processed under anaerobic conditions. *Agron. Res.* **2015**, *13*, 420–429.

114. Slepetys, J.; Kadziuliene, Z.; Sarunaite, L.; Tilvikiene, V.; Kryzeviciene, A. Biomass potential of plants grown for bioenergy production. In Proceedings of the Renewable Energy and Energy Efficiency, Growing and Processing Technologies of Energy Crops, Jelgava, Latvia, 2012; pp. 66–72. Available online: https://www.semanticscholar.org/paper/Biomass-potential-of-plants-grown-for-bioenergy-%C5% A0lepetys-Kadz%CC%8Ciuliene%CC%87/3d0f87eb3d601ee7472c9eca3abfe8f895c0c009?p2df (accessed on 4 December 2020).

115. Šiaudinis, G.; Šlepetienė, A.; Karčauskienė, D. The evaluation of dry mass yield of new energy crops and their energetic parameters. In Proceedings of the Renewable Energy and Energy Efficiency, Growing and Processing Technologies of Energy Crops, Jelgava, Latvia, 2012; pp. 24–28. Available online: https: //www.db-thueringen.de/receive/dbt_mods_00024401 (accessed on 4 December 2020).

116. Von Cossel, M.; Amarysti, C.; Wilhelm, H.; Priya, N.; Winkler, B.; Hoerner, L. The replacement of maize (*Zea mays* L.) by cup plant (*Silphium perfoliatum* L.) as biogas substrate and its implications for the energy and material flows of a large biogas plant. *Biofuels Bioprod. Biorefining* **2020**, *14*, 152–179. [CrossRef]

117. Von Cossel, M.; Steberl, K.; Hartung, J.; Pereira, L.A.; Kiesel, A.; Lewandowski, I. Methane yield and species diversity dynamics of perennial wild plant mixtures established alone, under cover crop maize (*Zea mays* L.), and after spring barley (*Hordeum vulgare* L.). *GCB Bioenergy* **2019**, *11*, 1376–1391. [CrossRef]

118. Schittenhelm, S.; Schoo, B.; Schroetter, S. Yield physiology of biogas plants: Comparison of streaky Silphie, maize and Luzernegra. *J. Kult.* **2016**, *68*, 378–384. [CrossRef]

119. Herrmann, C.; Idler, C.; Heiermann, M. Biogas crops grown in energy crop rotations: Linking chemical composition and methane production characteristics. *Bioresour. Technol.* **2016**, *206*, 23–35. [CrossRef]

120. Haag, N.L.; Nägele, H.-J.; Reiss, K.; Biertümpfel, A.; Oechsner, H. Methane formation potential of cup plant (*Silphium perfoliatum*). *Biomass Bioenergy* **2015**, *75*, 126–133. [CrossRef]

121. Majtkowski, W.; Piłat, J.; Szulc, P.M. Prospects of cultivation and utilization of *Silphium perfoliatum* L. in Poland. *Biul. Inst. Hod. I Aklim. Roślin* **2009**, *251*, 283–291.

122. Stolarski, M.J.; Warmiński, K.; Krzyżaniak, M.; Tyśkiewicz, K.; Olba-Zięty, E.; Graban, Ł.; Lajszner, W.; Załuski, D.; Wiejak, R.; Kamiński, P. How does extraction of biologically active substances with supercritical carbon dioxide affect lignocellulosic biomass properties? *Wood Sci. Technol.* **2020**, *54*, 519–546. [CrossRef]

123. Stolarski, M.; Szczukowski, S.; Tworkowski, J. Biofuels from biomass of perennial energy crops. *Energetyka Ekol.* **2008**, *1*, 77–80.

124. Trölenberg, S.D.; Kruse, M.; Jonitz, A. Improvement of the seed quality in the Streaky Silphie (*Silphium perfoliatum* L.). In *Nachhaltigkeitsindikatoren für die Landwirtschaft: Bestimmung und Eignung, VDLUFA-Schriftenreihe. Presented at the 124th VDLUFA-Kongress*; VDLUFA-Verlag: Darmstadt, Germany, 2012; pp. 926–933.

125. Biertümpfel, A.; Conrad, M. (Eds.) *Joint Project: Increase of the Performance Potential and the Competitiveness of the Streaky Silphie as an Energy Plant through Breeding and Optimization of the Cultivation Method: Subproject 2: Optimization of the Cultivation Method and Provision of Selection Material*; Abschlussbericht; Project no. 99.05; FKZ; no.: 22012809; Friedrich-Schiller-Universität Jena: Jena, Germany, 2014.

Publisher's Note: MDPI stays neutral with regard to jurisdictional claims in published maps and institutional affiliations.

Review

Ground Beetles (*Carabidae*) in the Short-Rotation Coppice Willow and Poplar Plants—Synergistic Benefits System

Natalia Stefania Piotrowska [1,*], Stanisław Zbigniew Czachorowski [1] and Mariusz Jerzy Stolarski [2]

[1] Department of Ecology and Environmental Protection, Faculty of Biology and Biotechnology, University of Warmia and Mazury in Olsztyn, Plac Łódzki 3, 10-727 Olsztyn, Poland; stanislaw.czachorowski@uwm.edu.pl

[2] Department of Plant Breeding and Seed Production, Faculty of Environmental Management and Agriculture, University of Warmia and Mazury in Olsztyn, Plac Łódzki 3, 10-724 Olsztyn, Poland; mariusz.stolarski@uwm.edu.pl

* Correspondence: natalia.piotrowska@uwm.edu.pl

Received: 24 October 2020; Accepted: 14 December 2020; Published: 18 December 2020

Abstract: In a short period, we have observed the rapid expansion of bioenergy, resulting in growth in the area of energy crops. In Europe, willow and poplar growing in short-rotation coppices (SRC) are popular bioenergy crops. Their potential impact on biodiversity has not yet been fully investigated. Therefore, there are many uncertainties regarding whether commercial production can cause environmental degradation and biodiversity impoverishment. One of the aspects examined is the impact of these crops on entomofauna and ecosystem services. The best-studied insect group is ground beetles from the *Carabidae* family. This work gathers data on biodiversity and the functions of carabids in willow and poplar energy plants. The results of these investigations show that energy SRC plants and *Carabidae* communities can create a synergistic system of mutual benefits. Willow and poplar plants can be a valuable habitat due to the increased biodiversity of entomofauna. Additionally, SRC creates a transitional environment that allows insect migration between isolated populations. On the other hand, ground beetles are suppliers of ecosystem services and make a significant contribution to the building of sustainable agriculture by pest control, thereby ameliorating damage to field crops.

Keywords: willow SRC; energy plants; ground beetles; *Carabidae*; ecosystem services; invertebrate biodiversity

---

1. Introduction

Currently, in the EU28 (28 European Union countries), the acreage of lignocellulosic plants is estimated to be 50,000 hectares of short-rotation coppices (SRC), mainly willow (*Salix* spp.) and poplar (*Populus* spp.) plants [1]. However, their impact on biodiversity remains only partially known [2]. It is considered that treatment for the natural environment differs for different types of energy crops [3–5].

Non-edible lignocellulosic plants, including willow and poplar, are used for heat and power generation and second-generation liquid biofuel production. It is considered that use of lignocellulosic plants can reduce competition with traditional crops for land and water resources [6], especially when grown on marginal land unsuitable for food production [7]. Despite this, many uncertainties exist about the potential impacts of biomass crops on the environment and biodiversity. There is major concern that commercial production can cause environmental degradation, significantly raising the risk of habitat fragmentation, native extinction, and bio-invasion [6]. Some studies report that the large-scale homogeneous landscape of biofuel plantation has resulted in a simplified bio-community and food web, severely damaging ecosystem services and contributing to the decline in the biodiversity, in particular in areas of high nature-conservation value [8]. Therefore, although biomass can help as an energy source to reduce the world's reliance on fossil energy and mitigate global warming [9], there is growing concern about the hypothetical disturbances biofuel can have on ecosystems and biodiversity. Accordingly, in this article, we will focus on willow and poplar SRC. The impact of these plantations on abiotic factors is well understood [10–36], but they can also affect the biotope, changing the species composition of plant and animal communities that directly inhabit the cultivation area, as well as adjacent habitats. So far, however, no comprehensive research has been made that fully shows the impact of willow plantations on the natural environment. It is known that in large agrocenoses, the introduction of small environmental patches of energy crops into cultivation can contribute to an increase in landscape mosaicism, significantly reduced by monocultures [7,37,38].

Studies carried out on large-scale monoculture crops show that the decrease in landscape mosaicism causes many adverse changes at all levels of the trophic chain. These changes occur both at the local level, limited to populations living in a given area, and global, affecting the structure of whole biomes. Over the past few decades, a significant decline in total biomass and the diversity of insect clusters has been observed, especially in North America and Europe [39–41]. Recent research has indicated that flying insect biomass decline may be up to 75% in some areas [42]. The main reasons hypothesized are anthropogenic drivers including land-use change [43,44], transport routes [45–47], environmental pollution, and pesticides [44,48–50] as well as climate changes [43,44,51,52]. Since the end of World War II, we have observed great intensification of agriculture and the evolution of the entire agrocenoses. Crop intensification and excessive pesticide use and accompanying processes, such as melioration or cessation of grazing, have led to the degradation of the habitat [53]. Those changes have contributed to the dramatic loss of biodiversity—many organisms have lost their ecological niches because of lack of shelter or other environmental resources, such as nutrition base. Thus, progressive degradation and the breakdown of ecological networks has been observed.

Due to the possibility of negative environmental effects, research has also been carried out on the influence of energy crops on the fauna that inhabits them. These studies have mainly focused on birds [34,54–67], and a smaller number of them concern mammals [60,61,68–72] or invertebrates [73–93]. Experiments have also been carried out to investigate the differences between biodiversity in woody crops and herbaceous perennial crops and grasses, such as, e.g., Virginia mallow or miscanthus [94]. The most investigated insect group in energy crops are carabids, the largest family of adephagan beetles (Coleoptera: *Carabidae*) [95,96]. So far, more than 40,000 species of ground beetles have been described, including more than 2700 in Europe and over 2000 in North America [97]. Due to their high plasticity, ground beetles have acquired a wide variety of habitats. This group includes eurytopic (ubiquitous), forest (occurring in wooded environments), open area (found in fields and meadows), coastal (associated with wetlands and banks of waters), and peat bog species. They differ in preference regarding humidity of environment, development cycle, size, and eating habits. Due to the above, *Carabidae* can be divided into five main groups: large predatory species (body length over 12 mm), medium predatory species (5–12 mm), small predatory species (<5 mm), hemizoophages (half-herbivorous), and phytophages (herbivorous) [85,86,92,98,99]. Most beetles in this group are characterized by a high level of predation. Although a diversified forest ecosystem is abundant in factors reducing the presence of phytophages, highly specialized agrocenoses are exposed to an excessive increase in the number of pests. As a result, in addition to anthropogenic factors, beetles from the *Carabidae* family are one of the main groups that contribute to the control of the pest population. Moreover, their services are not limited to SRC but are also provided to adjacent crops, which plays an important role in sustainable agriculture [100]. Accordingly, the role they are playing in ecological services cannot be underestimated.

The dominance structure in the *Carabidae* population may be a reflection of the habitat conditions [92]. Sharpening the structure of dominance can be considered a result of destructive factors existing in the environment [14,101]. Meanwhile, in stable habitats with the correct structure, smooth transitions are observed between the gradually decreasing percentages of species from individual groups [86]. Similarly, the trophic structure of the carabid population changes depending on the state of their living environment. Previously, the presence of large zoophages was considered the most desirable [102], but now the important role of herbivorous species has also been emphasized [103,104].

According to research by [105] in highly intensified agrocenoses, large zoophages are replaced by smaller predatory beetles, and as the pressure increases, the proportion of granivorous carabids increases in the grouping. However, these studies were carried out for meadows and arable fields, and similar works for woodland habitats such as poplar and willow crops are lacking. Therefore, more investigations should be done to estimate this factor for SRC [92,105]. *Carabidae* were chosen as the subject of this review for several reasons:

- As a well-known group of epigeic insects, they can be treated as a monitoring group;
- They play an important role, providing valuable ecosystem services for willow plantations and adjacent crops;
- Compared to other insect groups for which the amount of data is negligible, there are more studies for *Carabidae* clusters, allowing for analysis.

The novelty of this work is that it show a new view on SRC plantations as environmental islands—areas that can be refugia and environmental corridors for endangered populations. The purpose of this work is not only to show the diversity of entomofauna but also to draw attention to the relationship between the habitat of *Carabidae* and the shape of their population and their ecologic function, investigations of which have not been extensively developed so far. This will allow a better understanding of the role not only of the ground beetles themselves, as a group providing ecosystem services, but also of the entire complex environment of the energy willow plantation.

2. Materials and Methods/Data Collection and Selection

This review presents the most important studies on carabids in energy willow and poplar plantations, mainly in Central Europe. The areas covered are presented in Figure 1.

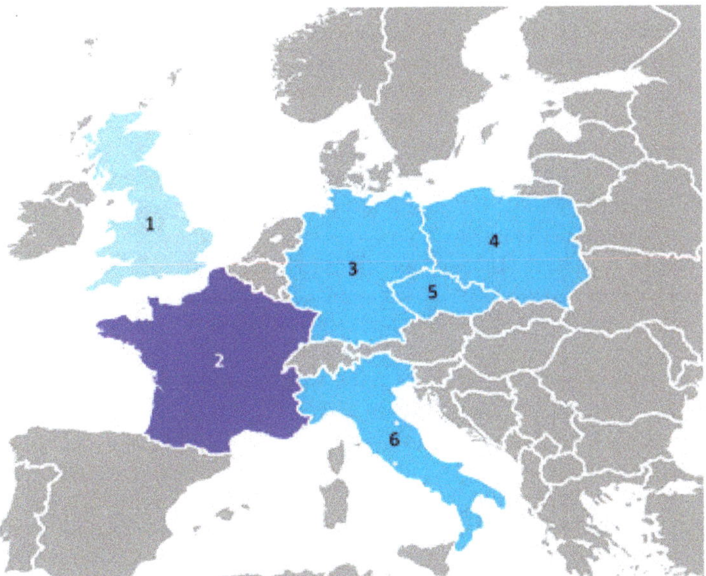

Figure 1. Countries for which original research works on *Carabidae* beetle fauna on poplar and willow plants were available in searched databases (1—Great Britain, 2—France, 3—Germany, 4—Poland, 5—Czech Republic, 6—Italy; light blue—works used in the study; blue—works used as well not used in the study; dark blue—works not used in the study).

Materials were acquired from articles published in English, German, and Polish. As the study materials, we used proceedings papers, original research studies, and review articles published between 1950 and 2020, mostly concerning plantations located in Europe. The research was conducted in three scientific databases: International Web of Science, Scopus, and CEON Biblioteka Nauki. In the final database, which is an information source of science papers published only in Polish periodicals, the searched keywords were: "wierzba *Carabidae*", "wierzba owady epigeiczne", "wierzba biegaczowate", "wierzba bioróżnorodność owadów", "wierzba bioróżnorodność entomofauny", "wierzba entomofauny", "wierzba owadów", "wierzba biegaczowatych", "wierzby biegaczowatych", "wierzby owadów epigeicznych", "wierzby *Carabidae*", "wierzby bioróżnorodność owadów", "wierzby bioróznorodność entomofauny", "wierzby entomofauna", "wierzby biegaczowate", and "wierzby owady epigeiczne". Meanwhile, from the Web of Science database, the material was acquired by searching for the following words: "energetic willow *Carabidae*", "energy crops *Carabidae*", "*Salix viminalis Carabidae*", "short rotation coppice *Carabidae*", "energetic poplar *Carabidae*", "energetic willow ground beetles", "energy crops ground beetles", "*Salix viminalis* ground beetles", "short rotation coppice ground beetles", "energetic poplar ground beetles", "energetic willow epigeic insects", "energy crops epigeic insects", "*Salix viminalis* epigeic insects", "short rotation coppice epigeic insects", "energetic poplar epigeic insects", "energetic willow insects biodiversity", "energy crops insects biodiversity", "*Salix viminalis* insects biodiversity", "short rotation coppice insects biodiversity", and "energetic poplar insects biodiversity". In Scopus, to restrict the search, we used words such as: "energetic willow *Carabidae*", "*Salix viminalis Carabidae*", "short rotation coppice *Carabidae*", "energetic poplar *Carabidae*", "energetic willow ground beetles", "*Salix viminalis* ground beetles", "short rotation coppice ground beetles", "energetic poplar ground beetles", "energetic willow epigeic insects", "*Salix viminalis* epigeic insects", "short rotation coppice epigeic insects", "energetic poplar epigeic insects", "energetic willow insects biodiversity", "*Salix viminalis* insects biodiversity", "short rotation coppice insects biodiversity", and "energetic poplar insects biodiversity".

In both databases, records were searched in all fields. In the Web of Science, our investigations included all collections. To broaden the results of the research, the "References" sections of the research articles used to prepare this review were studied, and relevant articles, if were not previously included, were added to our investigations. Furthermore, similar proceedings were applied to articles found in the Scopus database, where the "cited by" section was used. Most of the articles found were excluded from the analysis due to low relevance and duplications. The exclusion criteria for the selected articles were: no connection with the subject, the articles concerned other species of energy plants, the articles related to animals other than *Carabidae*, and the articles described plantations from outside of Europe.

The CEON Biblioteka Nauki database showed 58 records, among which only 3 records were selected as related to the topic. The International Web of Science database indicated 77 records, among which 22 were chosen as relevant. The Scopus database yielded the highest number of articles, as many as 5517 records. Due to this, it was decided to narrow down the search area and to reject key phrases containing "energy crops". This reduced the number of records to 458, out of which 24 relevant results were selected.

The articles collected were sorted according to the year of publication and the country of origin (the country was assigned based on the corresponding author). The results are presented in Figures 2 and 3. In the years 1998–2020, 32 articles on *Carabidae* biodiversity in energy willow and poplar plantations were published. Most of them, as many as 15 items, were published by German authors. Their number is three times higher than that of Polish and British publications. For Czech and Swedish authors, a database search showed two publications each. One publication was found for the Netherlands, Slovakia, and Belgium, respectively. In investigations for years from the 1950s until 1998, the databases did not show any publications described by the keywords used. In 1998, one publication was issued, and another was issued after 9 years, in 2008. An upward trend was observed for the following years, with 6 publications issued in 2012. For 2013 and 2014, there were 4 publications each. After this year, there is a clear decrease in the number of publications—in 2015, no publications were issued, in 2016, two were issued, and in the following years, only one publication per year was issued.

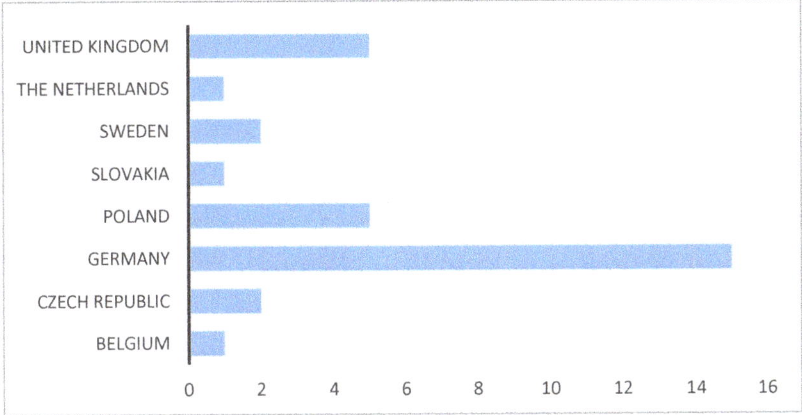

Figure 2. Publications on *Carabidae* in energy willow and poplar plantations—number per country. Articles found in three scientific databases: International Web of Science, Scopus, and CEON Biblioteka Nauki after discarding unrelated works and duplicated records.

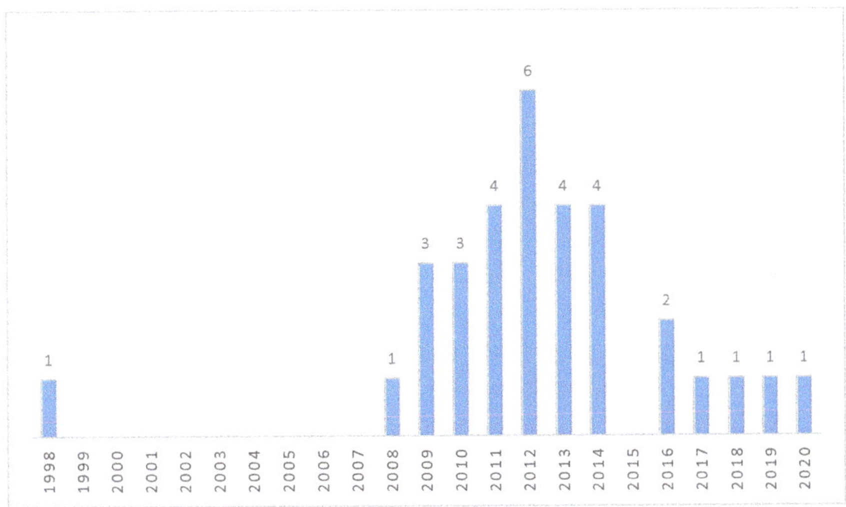

Figure 3. Publications on *Carabidae* in energy willow and poplar plantations—number per year. Articles found in three scientific databases: International Web of Science, Scopus and CEON Biblioteka Nauki after discarding unrelated works and duplicated records.

From the publications, four articles were selected that contained sufficient data to create an ecological characterization of *Carabidae* inhabiting the plantations described in them [85–87,92]. This enabled the designation of 12 research areas and the determination of the dominance structure, as well as phenology, hygro- and habitat preference, trophy group and dispersion powers. Energy plants characteristics are described in Table 1. To describe the phenology, hygropreference, trophy group and dispersion powers, individual species were assigned appropriate characteristics, and then their percentage of the total number of species was calculated. To describe the dominance structure based on abundance, species were assigned to different domination classes: eudominants, dominants, subdominants, recedents and subrecedents. Both methods were used to describe habitat preference, assessing both the percentage share of individual species in the total pool and the percentage of individuals with given environmental preferences on each of the plantations. Additionally, the number of forest species occurring on each study plot was shown concerning the age of the plantation and the size of the stand. The obtained results are presented in the form of tables and graphs in the Results and Discussion sections. Microsoft Office Excel was used for its design.

Table 1. Research articles that were the basis of this review–plantations characteristic (Local.—localization, Plant.—plantation, Years—years of investigation).

References	Investigated Issues	Local.	Plant. Type	Plant. Age	Canopy Age	Adjacent Habitat	Years Date
[86]	1. Species richness and diversity, dominance structure. 2. Community structure: trophic, habitat preference, humidity preference, development type.	Northeast Poland	willow	8–9 years	Not given	none	2004–2005
[85]	1. Species richness and diversity, dominance structure. 2. Community structure: trophic, habitat preference, humidity preference, development type.	Northeast Poland	willow	1 to 3 years	1 to 3 years	none	2005–2006
[87]	1. Species composition and abundance of ground beetles inhabiting unexploited willows plantation. 2. Population ecological characteristic and dominance structure. 3. Margalef's index, Shannon' diversity, Evenness H/log(N) Pielou.	Southeast Poland	willow	not given	8 and 9 years	none	2011–2012
[92]	1. Species richness and diversity, dominance structure. 2. The similarity to natural woodlands.	Po Valley, Italy	poplar	1 to 10 years	not given	natural woods, crops: maize, tobacco	1989–1999
[102]	1. Influence of the plantation vicinity and anthropogenic factors on *Carabidae* assemblages. 2. The structure of beetle communities.	South Bohemia (Czech Republic)	willow	2, 4, 6, 8 years	not given	Field with a stream; pond, field, alder trees and meadow; pasture, field, cultural forest;	2007

Table 1. *Cont.*

References	Investigated Issues	Local.	Plant. Type	Plant. Age	Canopy Age	Adjacent Habitat	Years Date
[81]	1. Comparison of how predation processes by ground arthropods varied between short rotation coppice (SRC) willow bioenergy plantations and alternative land-uses: arable and set-aside. 2. Predation pressure investigations: prey removal assay coupled with pitfall traps and direct searches	North Nottinghamshire, England	willow	1 to 10 years	1 to 9 years	set-aside, arable	2008
[88]	1. Simpson biodiversity index, evenness, level of anthropogenic influence 2. The influence of the length of rotation on biodiversity parameters;	Peklov, Czech Republic	poplar	9 years	1, 3, 6 years rotations	none	2003–2008
[89]	1. Taxonomy and identification. 2. Species traits and categorization. 3. Habitat preferences. 4. Endangered species. 5. Dispersal of forest species; corridor function. 6. Species traits concerning age of the SRC versus age of the SRC standing crop. 7. Factors influencing SRC biodiversity functions.	Germany (different sites) and the Czech Republic	willow poplar	1 to 23 years	1 to 9 years	different	Meta-study

3. Results

Based on the data analysis, presented in Figure 4, the predominance of *Carabidae* species preferring the environment of open areas was found. The comparison of Figure 5 and Table 2 showed that the percentage share of species preferring open areas is independent of the age of the plantation, while its dependence on the age of the stand seems impossible to assess due to insufficient data. However, their share appears to decline as the age of the canopy increases. Species preferring open areas also constituted the most numerous group in terms of the number of individuals in a given population, which is presented in Figure 5. This tendency occurred on most plantations, both willow [85–87] and poplar [92]. Species with unknown habitat preference constituted less than 15% of all recorded species. On three plantations, the niche of open ground species was occupied by eurytypical species. The share of forest species was greater, at over 18 percent (Figure 4). The number of forest species depending on the age of the plantation and the age of the canopy is presented in Table 2. In some plantations, there was a visible predominance of ground beetles representing forest species, which, however, was not correlated with the age of the plantation. Unfortunately, the amount of data allowing us to assess the influence of the surrounding environment on the number of *Carabidae* from different ecological groups was insufficient. A similar correlation concerning the age of the stand is impossible to analyze due to the lack of sufficient data, as the analyzed publications lack information enabling its determination. The number of peatland species was less than 10% that of all species collected, and the number of individuals with this preference was even lower.

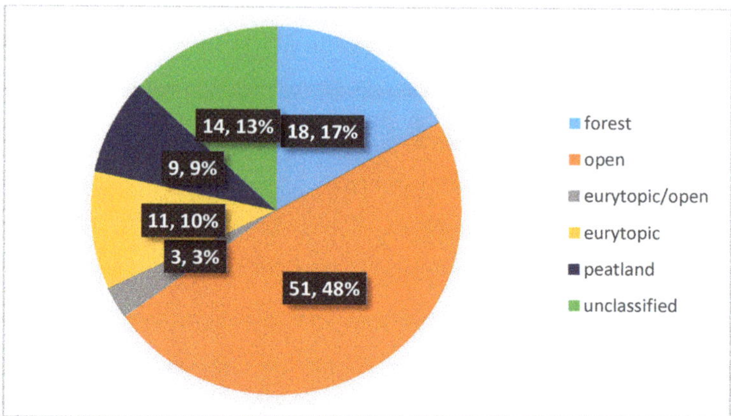

Figure 4. Percentage distribution of *Carabidae* species found in selected willow and poplar plantations depending on environmental preferences. The diagram was made based on selected publications, including data on the ecological characteristics of *Carabidae* communities inhabiting investigated plants.

Table 2. The number of *Carabidae* species preferring the forest environment, taking into account the number of species in the classes of eudominants (ED) and dominants (D). The table was made based on selected publications, including data about the ecological characteristic of *Carabidae* communities inhabiting investigated plants (W1, W2 and unmarked–willow plots; P1–P4–poplar plots).

Investigated Plots and Year of Investigation	Plantation Age (Years)	Canopy Age (Years)	Number of Forest Species (including ED and D)
[86] 2004	8	Not given	5 (1 ED)
[86] 2005	9	Not given	4 (2 ED)
[85] 2005 W1	2	2	2 (2 ED)
[85] 2006 W1	3	3	2
[85] 2005 W2	1	1	2 (2 ED)
[85] 2006 W2	2	2	2 (1 ED)
[87] 2011	Not given	8	7 (1 ED)
[87] 2012	Not given	9	5 (1 ED)
[92] 1989 P1	2	Not given	5
[92] 1999 P2	6	Not given	3 (1 ED, 1 D)
[92] 1999 P3	6	Not given	2 (1 D)
[92] 1991 P4	10	Not given	4 (2 ED)

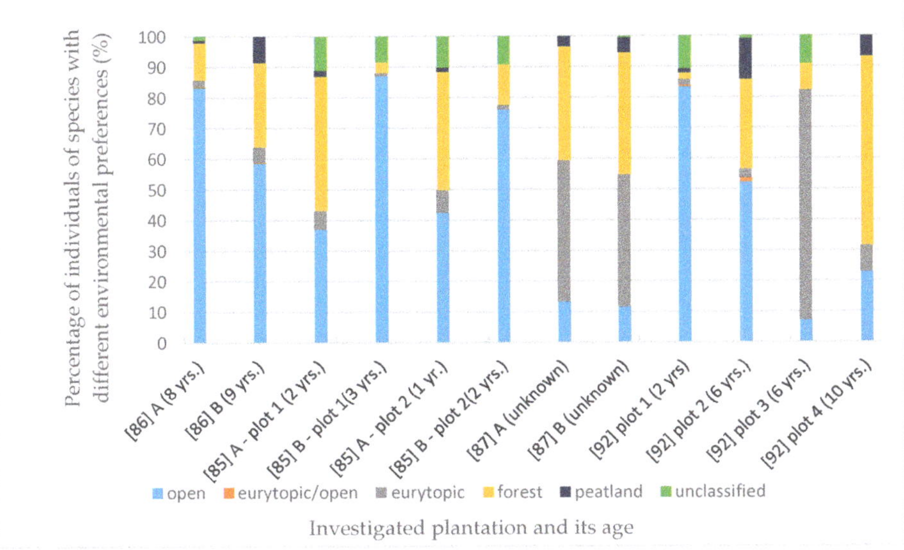

Figure 5. Habitat preferences vs. plantation age. Percentage of individuals of species with different environmental preferences depending on the age of the plantation (in brackets). The last four plots, studied by [92], are poplar crops. The remaining 8 are willow plantations [85–87], A—first year of investigations, B—second year of investigations.

A wider assessment of the carabid population in the SCR plantation, taking into account the structure of dominance, trophic structure, and species properties, indicated the predominance of poorly specialized species with wide environmental tolerance. Data analysis showed that the number of brachypterous species was smaller than that of macropterous ones (Figure 6), comprising only one fifth of all examined species. Macropterous species, characteristic for disturbed environments, constituted 64% of the investigated population. Similarly, species with spring biology were significantly more numerous than autumn biology ones, which are characteristic of well-balanced environments (Figure 7). The compilation of the collected data showed that the dominance structure of the studied populations was poorly balanced. In 11 out of 12 assessed plots, the number of eudominants predominated, amounting to over 60% in eight cases, and over 70% of all specimens in five cases. (Figure 8). Granivorous species predominated, followed by small and large zoophages (Figure 9). In terms of hygropreference (Figure 10), mesophiles were the most numerous. A large group was also species with undefined hygropreference. Poor *Carabidae* specialization may indicate an imbalance in the environment in which they live. On the other hand, there were no sharp transitions between individual trophic groups, while the percentage of large zoophages was average. This proves that even though the plantation environment is subject to intensive changes, it is possible to maintain a sustainable biocenosis there.

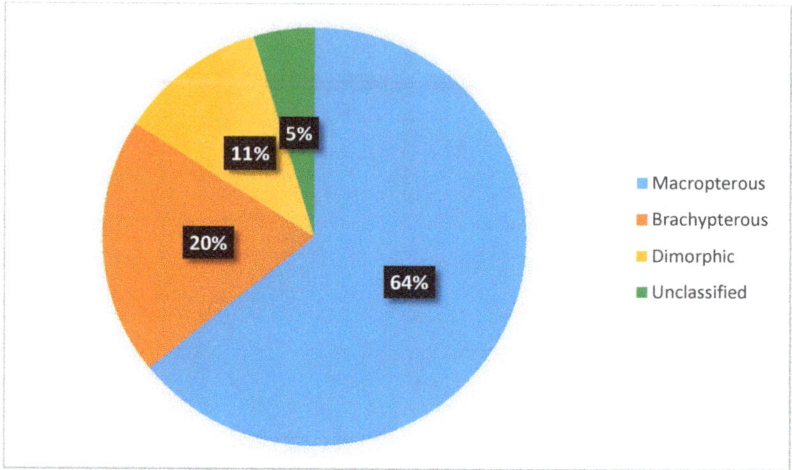

Figure 6. Percentage distribution of *Carabidae* species with different dispersibility. This diagram was made based on selected publications, including data on the ecological characteristics of *Carabidae* communities inhabiting the investigated plants.

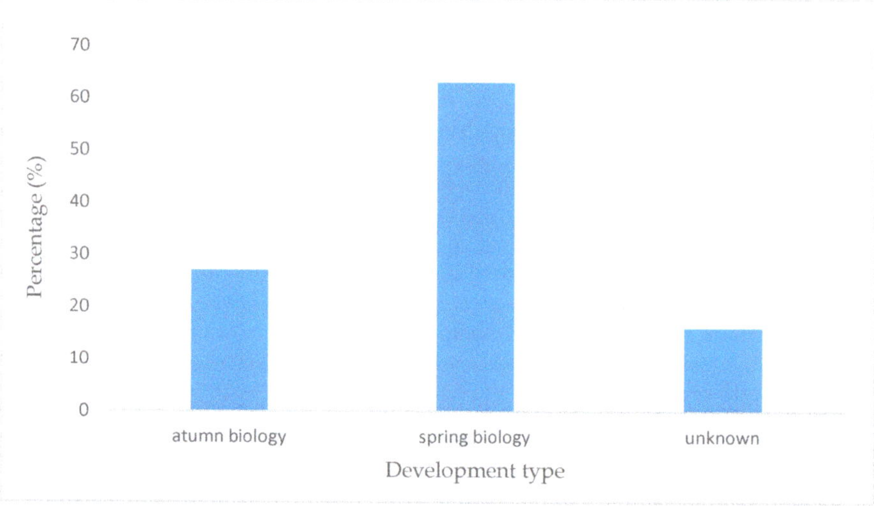

Figure 7. Per cent of *Carabidae* species with a different development type. This phenology diagram was made based on selected publications, including data on the ecological characteristic of *Carabidae* communities inhabiting the investigated plants.

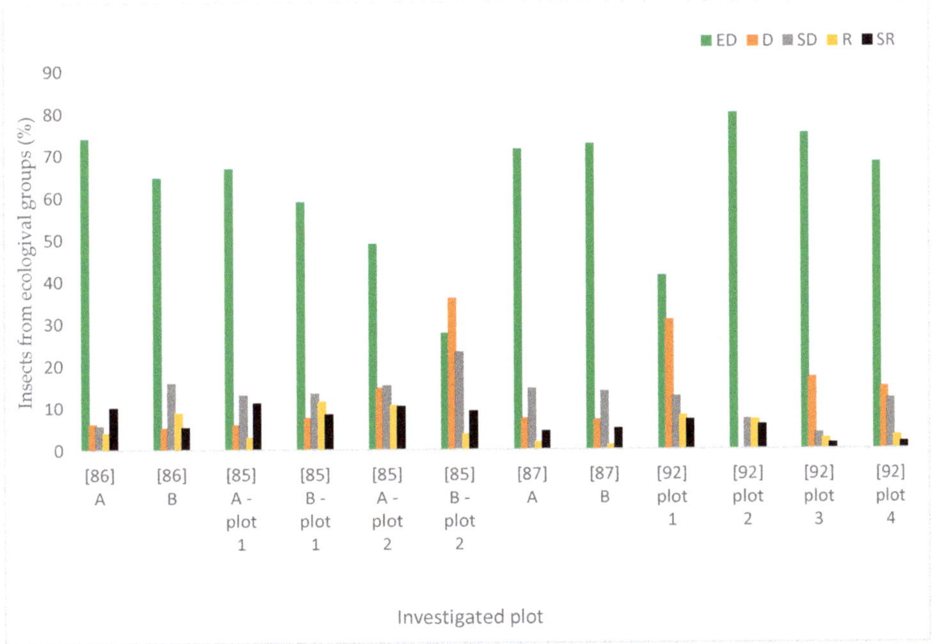

Figure 8. Structure of dominance in the *Carabidae* population based on the percentage of insects from particular groups (ED—eudominants; D—dominants; SD—subdominants; R—recedents; SR—subrecedents). This diagram was made based on selected publications, including data on the ecological characteristics of *Carabidae* communities inhabiting the investigated plants. (A—first year of investigations, B—second year of investigations).

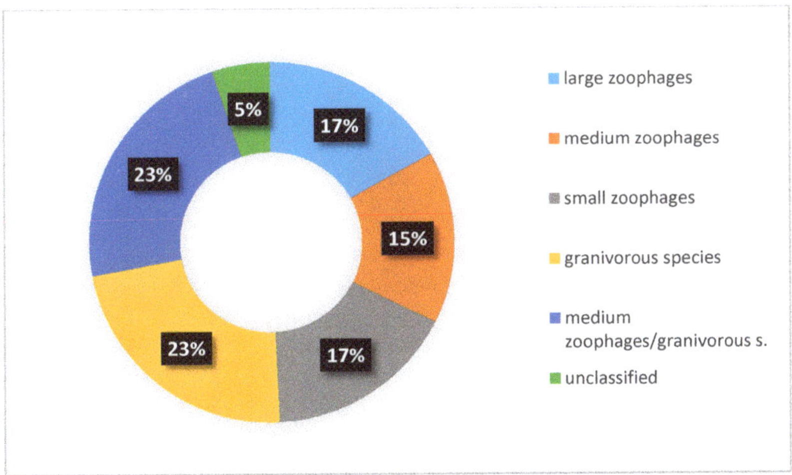

Figure 9. Percentage of *Carabidae* species from different trophy groups. This diagram was made based on selected publications, including data on the ecological characteristics of *Carabidae* communities inhabiting the investigated plants.

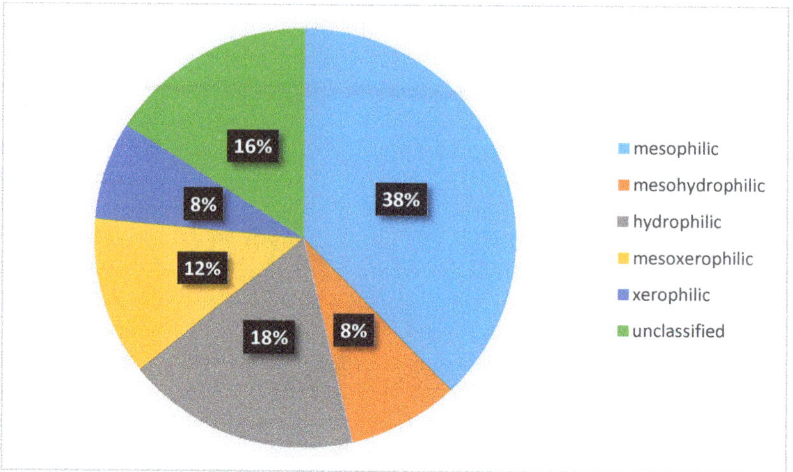

Figure 10. Per cent of *Carabidae* species with a different hygropreference. Diagram was made based on selected publications, including data about the ecological characteristic of *Carabidae* communities inhabiting the investigated plants.

4. Discussion

While the impact of willow and energy poplar plantations on abiotic factors is well-studied, there is a lack of comprehensive, multi-faceted research on the impact of plantations on biotic factors. Another problem is the lack of a homogeneous testing methodology. This makes research results difficult to interpret and compare with results obtained by other scientists. Of the many studies conducted, there were varying results, depending on the study region, the establishment phase of the SRC, its characteristics and its surroundings. Factors shaping SRC nature are adjacent land, terrain and water relations in the area where the plantation is located, the size and spatial configuration of the plantation itself, plantation age, the length of the production cycle and its phase (canopy age), and the willow strains planted. Due to a large number of variables, these results can be difficult to interpret. So far, no comprehensive research has been carried out to describe the roles of all specific factors. The impact of particular plantation features on the *Carabidae* populations inhabiting them is described below.

4.1. Factors Affecting Biodiversity

4.1.1. Rotation Length and Canopy Age vs. Plantation Age

It was found that, in general, agricultural fields and forests were both characterized by greater species richness (estimated by the species number) than willow and poplar SRC [89]. However, it was noticed that the strictly quantitative indicator "species richness" (species numbers) is only weakly correlated with qualitative biodiversity targets such as rare and endangered or specialized species [89,96].

A factor strongly influencing the biodiversity values of SRC regarding stenotopic species is rotation length. Wegner et al. [88] hypothesized that longer rotations may favor "ubiquitous" versus "stenotopic" species in terms of anthropogenic influence being less flexible. Their research indicated that poplar plantations are the habitat preferred mainly by non-specialized species, whereas increased inflow of adaptable species is observed after the harvests and is connected with differences in the re-growth dynamics (steam sprouting is slower in older plantations, which have the characteristics of an open vegetation site during this period) [88,89]. This seems to corroborate the data from articles collected by Müller-Kroehling et al. [89], indicating that in the majority of cases, only very young SRC in the establishment phase can provide an ephemeral pioneer habitat with a particular value for species protection [106]. Harvesting can at least partially renew a habitat function for open-habitat species by creating favorable conditions for the development of the herbaceous plant. Creating fallow-like conditions, they can act as a refugium for short-lived, open habitat species, especially granivorous or polyphagous species for which older SRC have a limited conservation value [89,92]. The analyzed data showed that not fully established plantations are a habitat for a high abundance of seed-eating species, the number of which diminishes with the age of the plantation. Similarly, in freshly cut stands, the number of granivorous species was higher, which means that harvesting promotes population increase for some of them, e.g., *Agonum sexpunctatum*, *Amara plebeja*, *A. similata*, *A. aenea*, *Anisodactylus binotatus*, *Harpalus affinis*, *Pseudoophonus. rufipes*, *Poecilus cupreus* and *P. versicolor*, which are eurytopic, less specialized species. Furthermore, it was proven that there is a group of species which prefers openings and seams of SRC than strictly open habitats such as fields and grassland, particularly in highly impoverished landscapes with few ecotone habitats, as was shown for *Carabus auratus* [89]. Therefore, many authors found later rotations of SRC as having a lower biodiversity value, hence they were inhabited mostly by species assemblages dominated by common, eurytopic species [89,96].

Similarly, Müller-Kroehling et al. [89] found that the abundance of the red-listed species is negatively correlated with the age of the plantation, while for the age of stand, no significant trend was observed. Meanwhile, in-depth analysis showed that while federally red-listed species occurred mostly in the newly established SRC, for the regional lists, the trend is reversed, and the red-listed species were more abundant among older plantations. Additionally, in this case, the longitude of rotation was negatively correlated with the occurrence of the endangered species. This proves that even though willow and poplar plants are uniform regarding their age, they still undergo a directed development and the age of the plantation plays a substantially larger role than the age of the canopy attained after harvesting [87,89]. It could also be anticipated that the age of the stand was a significant occurrence factor; surprisingly, the age of the plantation was even strongly positively correlated with this species' presence [89,96]. Changes taking place during plantation growth affected microclimatic conditions, depending on other factors from the accompanying plant cover [89,107–109]. Long-established SRC plantations probably provide better soil conditions, including soil moisture and soil particle size distribution, which is important especially for the development of larval carbide forms [89,92,97,110,111]. Furthermore, as a refugium for some species endangered at the local scale, SRC may help to increase regional population stability and genetic diversity across the country. Additionally, it was found that the presence of endangered species, for example *Abax ovalis*, *A. carinatus* and *A. parallelus*, *Molops elatus* and *Carabus auratus*, was also strongly positively correlated with the proximity of forest. During this research, it was found that a plant with extended time without coppicing can serve as a refugium for forest species, contributing to their protection in the agricultural landscape. Among the collected beetles, in both qualitative and quantitative terms, dendrophile species predominated. Most of them were typically forest species, while ecotone species partly associated with tree stands were also numerous. Articles juxtaposition indicated that more than half of all identified species were characteristic for open areas, which proves the role of the energy willow plantation as a reservoir for these taxa. Such a result was also obtained by Konieczna et al. [87], who examined abandoned plantations. In this work, a large number of beetles characteristic of open areas was also observed. It is worth mentioning that the time for which the investigated plants were not harvested was similar to the period without canopy cutting on EcoSalix plantations [7,112]. Unfortunately, in some publications, significant data describing the surroundings of the plantation and its spatial arrangement as well as determining the method of sampling were lacking. This reduces the possibilities of interpretation of the data obtained. The greatest disadvantage of some works was the missing information about plantation age, which makes it impossible to determine whether the presence of forest species is the result of the canopy or the age of the plantation.

4.1.2. Surrounding Area—Environmental Corridors and Islands

The plantation area is the factor for which the influence has not been fully estimated so far. Larger plantations provide habitats for a greater amount of forest species; however, most of the research was conducted in not very large SRC [89]. Therefore, some authors claimed that large plantations can contribute to a decrease in biodiversity [96,110,113,114]. Another factor examined was the surrounding crops. Many of the endangered species are forest species with low dispersal power. Among this group, there are many stenotopic brachypterous species with reduced vestigial wings. Because of the lacking development of the hind wings, they do not have the ability for dispersal flights. Therefore, the distance to the nearest forest is, in some cases, a limiting factor [89,96,110]. Additionally, it is worth mentioning that flight ability is a species adaptation often connected with inhabiting a disturbed environment, for example, cultural land [97,110].

In high-land-pressure agricultural landscapes with high fragmentation and a lack of connectivity between individual habitats, strictly forest species can be isolated in forest "islands". According to the island theory, only large habitat patches providing enough resources can serve as a stable environment. On the other hand, vast areas of monocultures and low landscape diversity can lead to species declining, especially stenotopic species or those with high environmental requirements [115,116]. One of the main reasons for this is the genetic erosion of the population [89,96]. This issue has great importance in the SLOSS (single large or several small) debate [117–119].

It is known that green corridors prevent the isolation of the population and thus the depletion of their genetic pool [107,108]. A decrease in the effective population size leads to the disclosure of the negative effects of genetic drift and coalescence, which is more visible in small, isolated populations [109,111,113,114]. It leads to an increase in homozygosity and the loss of genetic variability. In extreme cases homogeneity can depress population fitness, general resistance to environmental factors, and flexibility in coping with environmental challenges [109,115]. Therefore, one of the assumptions of sustainable agriculture is to increase the landscape mosaicism. Based on the concept of "Islands biogeography" introduced by McArthur and Wilson [120–125], landscape corridors connecting isolated habitat patches can be applied to the agrocenoses, for example by fallowing or introducing extensive agricultural crops [37,126–128]. Establishing energy willow and poplar crops on lower-class soils or on the marginal land enables one to diversify the use of arable land, maintaining an income for local farmers [7,38]. In this type of area, the additional function of SRC, especially old ones, for forest species is being the corridor habitat [89,90,129]. The scope of this role might vary largely regionally according to inhabiting species and environmental factors. It is also worth noting that the effects of climate change will exacerbate the problem of the decline of both epigeic beetles [52,130] as well as insects in general [43,44,51,131–135].

4.2. Carabidae as Bioindicators and Ecosystem Services Providers

Determining the characteristics of epigeic beetle assemblies on SRC plants and making ecological descriptions of them, taking into account trophic structure, habitat preferences, hydro preferences and phenology enables their use as bioindicators [136]. Jowett et al. [137] emphasized the role of the species profile of a given population as an important factor determining habitat maturity. One of the most important features of *Carabidae* allowing us to assess the ecological status of a given environment is food preferences. Although there is no doubt that carabids have a potential for pest control [95,97,138–144], as well as weed seeds [145–148], there is remarkably little interest in the role they play in underpinning ecosystem services in bioenergy plantations. Rowe et al. [81] compared the processes of predation and litter decomposition in willow SRC and alternative land-uses: arable and set-aside. In the described bioassay, even though willow plants had the highest abundance and diversity of ground-dwelling arthropod predators, these factors had no detectable influence on predation rates. The reason for this was that the carabids were more active in cereal crops vs. SRC. As a result, the predation rates in the investigated plots did not differ between habitats. These observations were connected to the environmental factors, characteristic for each of the habitats. As it is known, these processes are inextricably linked to crop productivity and ecosystem stability; however, so far, our understanding of them in SRC crops is limited, and further investigations are required [81]. Therefore, it would be very valuable to carry out research work that combines the assessment of carabid communities and the ecosystem services provided by them and takes into account environmental factors. Unfortunately, there are currently no such publications.

4.3. Carabidae Assemblages Structure

A detailed description of *Carabidae* groupings allows for characterizing the habitat of a young willow plantation, rich mainly in eurytopic, poorly specialized species, as a temporary environment undergoing a dynamic development [85,86]. There was a large divergence gap between the dominant and the other dominance classes, which has also been proven by other researchers (Table 2, Figure 8) and is a factor indicating the instability of the population. It was shown that the most common group were open-field carabids, characteristic for adjacent areas (Table 2, Figures 4 and 5) Walerys et al. [86] claimed that this could be the result of the small size of the plantations studied. However, such a tendency was not visible for the older plantations, most of which had a higher share of forest species (Table 2, Figure 5). In most plantations mesophilic carabids dominated [85,86], Figure 10, as the most tolerant in terms of humidity requirements. Spring breeding carabids characteristic for newly colonized areas, dominated [86], Figure 7, while autumn and wintering beetle larvae, characteristic for older developmental stages [87] of stands, were absent on newly established plants. On the other hand, Kosewska et al. [85] indicated that the domination of one of the developmental *Carabidae* types in the studied area did not depend on the canopy, but was correlated with the study year [85]. Additionally, Kosewska et al. [85] showed that in older plants, beetles were less numerous than in the newly situated ones. This confirms the thesis that plantations in the initial period of the production cycle, as a transitional environment, constitute an attractive habitat for this type of species. The conducted study also showed that the number of large zoophages was higher in the older plants. This proves that the population is stable. Boháč et al. [110] investigated the impact of adjacent habitats on *Carabidae* and *Staphylinidae* population structure. As expected, the highest number of species was found on plots least affected by humans. For communities occurring in these plots, stenotopic species (e.g., *Platynus assmilis*) and large *Carabidae* species (e.g., *Carabus hortensis hortensis*, *C. violaceus violaceus*) were typical. Additionally, psychrophilic species characterized by winter activity were the best-represented group in these plots. Their research proved that the influence of anthropogenic factors was associated with an increased prevalence of eurytopic species, as well as an increased prevalence of species with summer imago activity.

On the other hand, forest species (e.g., *Pterostichus oblongopunctatus*), as well the protected *Carabus scheidleri*, were noted only in the fields adjacent to the forest, even though these were also fields with the highest anthropopressure. In addition, in studies evaluating carabidofauna in out of use plantations with old canopies [87], the presence of endangered and rare species was noticed. In these types of plants, the most numerous carabids were mesophilic species: *Pterostichus melanarius*, *P. niger* and hydrophilic *Limodromus assimilis*. *P. niger* and *L. assimilis* were classified as a forest species. Similarly, [88] reported that species preferring light forest or forest edges, such aslike *Carabus convexus*, as with species with higher humidity requirements such as the hygrophilous *Panagaeus cruxmajor*, are listed only for the longer rotations. Similar results obtained by other investigators [91] can suggest that these species are a regular component of carabidofauna in Central and Eastern Europe [87]. These results indicate the need for further research on the structure of the *Carabidae* population and prove that not the number of beetles in a given area is less important compared to their species profile.

5. Conclusions with Remarks

Energy willow and poplar plantations may increase the biodiversity of *Carabidae* beetles. Due to the lower use of pesticides, as well as the smaller number of agrotechnical treatments carried out, SRC can be a habitat which is more stable than annual crops, such as rape or wheat. However, energy plants also undergo dynamic changes, associated not only with the production cycle and canopy age, but also with the age of the plantation itself. This is especially important for forest species, which in older SRC may find conditions similar to those in their natural habitat. Nevertheless, the data describing forest species requirements and biodiversity in the SRC are still lacking. It is known, however, that even if poplar and willow plants cannot provide a permanent habitat for forest species, they can serve as ecological corridors, connecting isolated land patches, inhabited by populations of epigeic insects with low dispersal powers. Energy willow and poplar plantations, if they meet several conditions, can be used for agriculture, as well as serving as a refugium for animals inhabiting areas with a high risk of anthropopression

SRC plantations can provide carabids with breeding and shelter places, important especially if the adjacent areas of arable land are characterized by a strongly intensified agricultural production. Maintaining a stable population of *Carabidae* is recommended in the integrated agriculture model, the assumptions of which include reducing the negative impact on the natural environment and human life. Ground beetles play a crucial role in providing valuable ecosystem services. Moreover, their services concern not only the energy willow plantations themselves, but also contribute to the protection of adjacent annual crops. *Carabidae* communities are still not studied enough on SRC plants. This is the result of, on the one hand, the small number of studies that have been carried out on these crops, compared to, for example, rape cultivation, and on the other hand, the fact that many works were not translated into English and appear only in local periodicals. Therefore, assessment of ground beetle population on willow and poplar plants based on community structure is currently impossible. For energy willow, there is still a lack of complete data at the regional scale, which would enable us to create the ground beetle community model describing the level of naturalness of the habitat. According to this, further investigations should be done to estimate community structure to obtain additional data on the state of the habitat and diversity of species inhabiting poplar and willow plants.

Author Contributions: Conceptualization, N.S.P., S.Z.C. and M.J.S.; methodology, N.S.P., S.Z.C. and M.J.S.; validation, N.S.P., S.Z.C. and M.J.S.; formal analysis, N.S.P., S.Z.C. and M.J.S.; investigation, N.S.P.; resources, N.S.P., S.Z.C. and M.J.S.; data curation, N.S.P.; writing—original draft preparation, N.S.P.; writing—review and editing, N.S.P., S.Z.C. and M.J.S.; visualization, N.S.P.; supervision, S.Z.C. and M.J.S.; project administration, M.J.S.; funding acquisition, M.J.S.; All authors have read and agreed to the published version of the manuscript.

Funding: This paper was co-financed by the National (Polish) Centre for Research and Development (NCBiR), entitled "Environment, agriculture and forestry", project: BIOproducts from lignocellulosic biomass derived from MArginal land to fill the Gap in Current national bioeconomy, No. BIOSTRATEG3/344253/2/NCBR/2017.

Acknowledgments: Natalia Piotrowska is a recipient of a scholarship from the Programme Interdisciplinary Doctoral Studies in Bioeconomy (POWR.03.02.00-00-I034/16-00), which is funded by the European Social Funds.

Conflicts of Interest: The authors declare no conflict of interest.

References

1. Lindegaard, K.N.; Adams, P.W.R.; Holley, M.; Lamley, A.; Henriksson, A.; Larsson, S.; von Engelbrechten, H.-G.; Esteban Lopez, G.; Pisarek, M. Short rotation plantations policy history in Europe: Lessons from the past and recommendations for the future. *Food Energy Secur.* **2016**, *5*, 125–152. [CrossRef] [PubMed]

2. Volk, T.A.; Verwijst, T.; Tharakan, P.J.; Abrahamson, L.P.; White, E.H. Growing fuel: A sustainability assessment of willow biomass crops. *Front. Ecol. Environ.* **2004**, *2*, 411–418. [CrossRef]

3. Groom, M.J.; Gray, E.M.; Townsend, P.A. Biofuels and Biodiversity: Principles for Creating Better Policies for Biofuel Production. *Conserv. Biol.* **2008**, *22*, 602–609. [CrossRef]

4. Mohr, A.; Raman, S. Lessons from first-generation biofuels and implications for the sustainability appraisal of second-generation biofuels. *Energy Policy* **2013**, *63*, 114–122. [CrossRef] [PubMed]

5. Saladini, F.; Patrizi, N.; Pulselli, F.M.; Marchettini, N.; Bastianoni, S. Guidelines for emergy evaluation of first, second and third-generation biofuels. *Renew. Sustain. Energy Rev.* **2016**, *66*, 221–227. [CrossRef]

6. Liu, Y.; Xu, Y.; Zhang, F.; Yun, J.; Shen, Z. The impact of biofuel plantation on biodiversity: A review. *Chin. Sci. Bull.* **2014**, *59*, 4639–4651. [CrossRef]

7. Stolarski, M.; Szczukowski, S.; Tworkowski, J.; Krzyżaniak, M. Extensive Willow Biomass Production on Marginal Land. *Pol. J. Environ. Stud.* **2019**, *28*, 4359–4367. [CrossRef]

8. Dauber, J.; Jones, M.B.; Stout, J.C. The impact of biomass crop cultivation on temperate biodiversity: Biomass crops and biodiversity. *GCB Bioenergy* **2010**, *2*, 289–309. [CrossRef]

9. Demirbas, M.F.; Balat, M.; Balat, H. Potential contribution of biomass to the sustainable energy development. *Energy Convers. Manag.* **2009**, *50*, 1746–1760. [CrossRef]

10. Jensen, J.K.; Holm, P.E.; Nejrup, J.; Larsen, M.B.; Borggaard, O.K. The potential of willow for remediation of heavy metal polluted calcareous urban soils. *Environ. Pollut.* **2009**, *157*, 931–937. [CrossRef]

11. Kacálková, L.; Tlustoš, P.; Száková, J. Phytoextraction of risk elements by willow and poplar trees. *Int. J. Phytoremediation* **2015**, *17*, 414–421. [CrossRef] [PubMed]

12. Pulford, I. Phytoremediation of heavy metal-contaminated land by trees—A review. *Environ. Int.* **2003**, *29*, 529–540. [CrossRef]

13. Ruttens, A.; Boulet, J.; Weyens, N.; Smeets, K.; Adriaensen, K.; Meers, E.; van Slycken, S.; Tack, F.; Meiresonne, L.; Thewys, T.; et al. Short rotation coppice culture of willows and poplars as energy crops on metal contaminated agricultural soils. *Int. J. Phytoremediation* **2011**, *13* (Suppl. 1), 194–207. [CrossRef]

14. Riddell-Black, D. Heavy metal uptake by fast growing willow species. In *A Biological Purification System: Conference and Workshop: Papers*; Worldcat; Aronsson, P., Perttu, K., Eds.; Swedish University of Agricultural Sciences, Uppsala (Sweden) Department of Ecology and Environmental Research: Uppsala, Sweden, 1994; p. 231. ISSN 0282-6267.

15. Punshon, T.; Lepp, N.W.; Dickinson, N.M. Resistance to copper toxicity in some British willows. *J. Geochem. Explor.* **1995**, *52*, 259–266. [CrossRef]

16. Dickinson, N.M.; Punshon, T.; Hodkinson, R.B.; Lepp, N.W. Metal tolerance and accumulation in willows. In *A Biological Purification System: Conference and Workshop: Papers*; Worldcat; Aronsson, P., Perttu, K., Eds.; Swedish University of Agricultural Sciences, Uppsala (Sweden) Department of Ecology and Environmental Research: Uppsala, Sweden, 1994; p. 231. ISSN 0282-6267.

17. Greger, M.; Landberg, T. Use of Willow in Phytoextraction. *Int. J. Phytoremediation* **1999**, *1*, 115–123. [CrossRef]

18. Pilipović, A.; Zalesny, R.S.; Rončević, S.; Nikolić, N.; Orlović, S.; Beljin, J.; Katanić, M. Growth, physiology, and phytoextraction potential of poplar and willow established in soils amended with heavy-metal contaminated, dredged river sediments. *J. Environ. Manag.* **2019**, *239*, 352–365. [CrossRef]

19. Vysloužilová, M.; Tlustoš, P.; Száková, J.; Pavlíková, D. As, Cd, Pb and Zn uptake by *Salix* spp. clones grown in soils enriched by high loads of these elements. *Plant Soil Environ.* **2011**, *49*, 191–196. [CrossRef]

20. Yang, W.; Zhao, F.; Ding, Z.; Wang, Y.; Zhang, X.; Zhu, Z.; Yang, X. Variation of tolerance and accumulation to excess iron in 24 willow clones: Implications for phytoextraction. *Int. J. Phytoremediation* **2018**, *20*, 1284–1291. [CrossRef]

21. Dimitriou, I.; Mola-Yudego, B.; Aronsson, P.; Eriksson, J. Changes in Organic Carbon and Trace Elements in the Soil of Willow Short-Rotation Coppice Plantations. *Bioenergy Res.* **2012**, *5*, 563–572. [CrossRef]

22. Dimitriou, I.; Baum, C.; Baum, S.; Busch, G.; Schulz, U.; Kohn, J.; Lamersdorf, N.; Leinweber, P.; Aronsson, P.; Weih, M.; et al. The impact of Short Rotation Coppice (SRC) cultivation on the environment. *Landbauforsch. Volkenrode* **2009**, *59*, 159–162.

23. Klang-Westin, E.; Eriksson, J. Potential of Salix as phytoextractor for Cd on moderately contaminated soils. *Plant Soil* **2003**, *249*, 127–137. [CrossRef]

24. Aronsson, P.; Perttu, K. Willow vegetation filters for wastewater treatment and soil remediation combined with biomass production. *For. Chron.* **2001**, *77*, 293–299. [CrossRef]

25. Börjesson, P.; Berndes, G. The prospects for willow plantations for wastewater treatment in Sweden. *Biomass Bioenergy* **2006**, *30*, 428–438. [CrossRef]

26. Dimitriou, I.; Mola-Yudego, B.; Aronsson, P. Impact of Willow Short Rotation Coppice on Water Quality. *Bioenergy Res.* **2012**, *5*, 537–545. [CrossRef]

27. Urbaniak, M.; Wyrwicka, A.; Tołoczko, W.; Serwecińska, L.; Zieliński, M. The effect of sewage sludge application on soil properties and willow (*Salix* sp.) cultivation. *Sci. Total Environ.* **2017**, *586*, 66–75. [CrossRef]

28. Dimitriou, I.; Eriksson, J.; Adler, A.; Aronsson, P.; Verwijst, T. Fate of heavy metals after application of sewage sludge and wood–ash mixtures to short-rotation willow coppice. *Environ. Pollut.* **2006**, *142*, 160–169. [CrossRef]

29. Mao, R.; Zeng, D.-H.; Hu, Y.-L.; Li, L.-J.; Yang, D. Soil organic carbon and nitrogen stocks in an age-sequence of poplar stands planted on marginal agricultural land in Northeast China. *Plant Soil* **2010**, *332*, 277–287. [CrossRef]

30. McIvor, I.R.; Douglas, G.B. Poplars and willows in hill country—Stabilising soils and storing carbon. *Adv. Nutr. Manag. Gains Past Goals Future* **2012**, *25*, 1–11.

31. Coleman, M.D.; Isebrands, J.G.; Tolsted, D.N.; Tolbert, V.R. Comparing Soil Carbon of Short Rotation Poplar Plantations with Agricultural Crops and Woodlots in North Central United States. *Environ. Manag.* **2004**, *33*. [CrossRef]

32. Mukherjee, A.; Coleman, M.D. Converting Conventional Agriculture to Poplar Bioenergy Crops: Soil Chemistry. *Commun. Soil Sci. Plant Anal.* **2020**, *51*, 364–379. [CrossRef]

33. Lockwell, J.; Guidi, W.; Labrecque, M. Soil carbon sequestration potential of willows in short-rotation coppice established on abandoned farm lands. *Plant. Soil* **2012**, *360*, 299–318. [CrossRef]

34. Langeveld, H.; Quist-Wessel, F.; Dimitriou, I.; Aronsson, P.; Baum, C.; Schulz, U.; Bolte, A.; Baum, S.; Köhn, J.; Weih, M.; et al. Assessing Environmental Impacts of Short Rotation Coppice (SRC) Expansion: Model Definition and Preliminary Results. *Bioenergy Res.* **2012**, *5*, 621–635. [CrossRef]

35. Acharya, B.S.; Blanco-Canqui, H. Lignocellulosic-based bioenergy and water quality parameters: A review. *GCB Bioenergy* **2018**, *10*, 504–533. [CrossRef]

36. Baum, C.; Leinweber, P.; Weih, M.; Lamersdorf, N.; Dimitriou, I. Effects of short rotation coppice with willows and poplar on soil ecology. *Landbauforsch. vTI Agric. Res.* **2009**, *59*, 183–196.

37. Weissteiner, C.J.; García-Feced, C.; Paracchini, M.L. A new view on EU agricultural landscapes: Quantifying patchiness to assess farmland heterogeneity. *Ecol. Indic.* **2016**, *61*, 317–327. [CrossRef]

38. Shortall, O.K. "Marginal land" for energy crops: Exploring definitions and embedded assumptions. *Energy Policy* **2013**, *62*, 19–27. [CrossRef]

39. Leather, S.R. "Ecological Armageddon"—More evidence for the drastic decline in insect numbers: Insect declines. *Ann. Appl. Biol.* **2018**, *172*, 1–3. [CrossRef]

40. Simmons, B.I.; Balmford, A.; Bladon, A.J.; Christie, A.P.; De Palma, A.; Dicks, L.V.; Gallego-Zamorano, J.; Johnston, A.; Martin, P.A.; Purvis, A.; et al. Worldwide insect declines: An important message, but interpret with caution. *Ecol. Evol.* **2019**, *9*, 3678–3680. [CrossRef]

41. Shortall, C.R.; Moore, A.; Smith, E.; Hall, M.J.; Woiwod, I.P.; Harrington, R. Long-term changes in the abundance of flying insects. *Insect Conserv. Divers.* **2009**, *2*, 251–260. [CrossRef]

42. Hallmann, C.A.; Sorg, M.; Jongejans, E.; Siepel, H.; Hofland, N.; Schwan, H.; Stenmans, W.; Müller, A.; Sumser, H.; Hörren, T.; et al. More than 75 percent decline over 27 years in total flying insect biomass in protected areas. *PLoS ONE* **2017**, *12*, e0185809. [CrossRef]

43. van Klink, R.; Bowler, D.E.; Gongalsky, K.B.; Swengel, A.B.; Gentile, A.; Chase, J.M. Meta-analysis reveals declines in terrestrial but increases in freshwater insect abundances. *Science* **2020**, *368*, 417–420. [CrossRef] [PubMed]

44. Sánchez-Bayo, F.; Wyckhuys, K.A.G. Worldwide decline of the entomofauna: A review of its drivers. *Biol. Conserv.* **2019**, *232*, 8–27. [CrossRef]

45. Bhattacharya, M.; Primack, R.B.; Gerwein, J. Are roads and railroads barriers to bumblebee movement in a temperate suburban conservation area? *Biol. Conserv.* **2003**, *109*, 37–45. [CrossRef]

46. Keller, I.; Nentwig, W.; Largiadèr, C.R. Recent habitat fragmentation due to roads can lead to significant genetic differentiation in an abundant flightless ground beetle: Differentiation due to habitat fragmentation. *Mol. Ecol.* **2004**, *13*, 2983–2994. [CrossRef]

47. Keller, I.; Excoffier, L.; Largiader, C.R. Estimation of effective population size and detection of a recent population decline coinciding with habitat fragmentation in a ground beetle. *J. Evol. Biol.* **2005**, *18*, 90–100. [CrossRef]

48. Goulson, D.; Nicholls, E.; Botias, C.; Rotheray, E.L. Bee declines driven by combined stress from parasites, pesticides, and lack of flowers. *Science* **2015**, *347*, 1255957. [CrossRef]

49. van Lexmond, M.B.; Bonmatin, J.-M.; Goulson, D.; Noome, D.A. Worldwide integrated assessment on systemic pesticides: Global collapse of the entomofauna: Exploring the role of systemic insecticides. *Environ. Sci. Pollut. Res.* **2015**, *22*, 1–4. [CrossRef]

50. Stokstad, E. Field Research on Bees Raises Concern about Low-Dose Pesticides. *Science* **2012**, *335*, 1555. [CrossRef]

51. Wallisdevries, M.F.; Van Swaay, C.A.M. Global warming and excess nitrogen may induce butterfly decline by microclimatic cooling. *Glob. Chang. Biol.* **2006**, *12*, 1620–1626. [CrossRef]

52. Müller-Kroehling, S.; Jantsch, M.; Fischer, H.; Fischer, A. Modelling the effects of global warming on the ground beetle (Coleoptera: Carabidae) fauna of beech forests in Bavaria, Germany. *Eur. J. Entomol.* **2014**, *111*, 35–49. [CrossRef]

53. Blanco, H.; Lal, R. *Principles of Soil Conservation and Management*; Springer: Dordrecht, The Netherlands, 2010; ISBN 978-1-4020-8709-7.

54. Fry, D.A.; Slater, F.M. Early rotation short rotation willow coppice as a winter food resource for birds. *Biomass Bioenergy* **2011**, *35*, 2545–2553. [CrossRef]

55. Sage, R.B.; Robertson, P.A. Factors affecting songbird communities using new short rotation coppice habitats in spring. *Bird Study* **1996**, *43*, 201–213. [CrossRef]

56. Londo, M.; Dekker, J.; Terkeurs, W. Willow short-rotation coppice for energy and breeding birds: An exploration of potentials in relation to management. *Biomass Bioenergy* **2005**, *28*, 281–293. [CrossRef]

57. Sage, R.; Cunningham, M.; Boatman, N. Birds in willow short-rotation coppice compared to other arable crops in central England and a review of bird census data from energy crops in the UK: Birds in short-rotation coppice. *IBIS* **2006**, *148*, 184–197. [CrossRef]

58. Riffell, S.; Verschuyl, J.; Miller, D.; Wigley, T.B. A meta-analysis of bird and mammal response to short-rotation woody crops: Meta-analysis of bird and mammal response. *GCB Bioenergy* **2011**, *3*, 313–321. [CrossRef]

59. Berg, Å. Breeding birds in short-rotation coppices on farmland in central Sweden—The importance of Salix height and adjacent habitats. *Agric. Ecosyst. Environ.* **2002**, *90*, 265–276. [CrossRef]

60. Christian, D.P.; Hoffman, W.; Hanowski, J.M.; Niemi, G.J.; Beyea, J. Bird and mammal diversity on woody biomass plantations in North America. In *Biomass and Bioenergy*; Elsevier: Amsterdam, The Netherlands, 1998.

61. Campbell, S.P.; Frair, J.L.; Gibbs, J.P.; Volk, T.A. Use of short-rotation coppice willow crops by birds and small mammals in central New York. *Biomass Bioenergy* **2012**, *47*, 342–353. [CrossRef]

62. Baxter, D.A.; Sage, R.B.; Hall, D.O. A methodology for assessing gamebird use of short rotation coppice. *Biomass Bioenergy* **1996**, *10*, 301–306. [CrossRef]

63. Dhondt, A.A.; Wrege, P.H.; Sydenstricker, K.V.; Cerretani, J. Clone preference by nesting birds in short-rotation coppice plantations in central and western New York. *Biomass Bioenergy* **2004**, *27*, 429–435. [CrossRef]

64. Dhondt, A.A.; Wrege, P.H.; Cerretani, J.; Sydenstricker, K.V. Avian species richness and reproduction in short-rotation coppice habitats in central and western New York. *Bird Study* **2007**, *54*, 12–22. [CrossRef]

65. Göransson, G. Bird fauna of cultivated energy shrub forests at different heights. *Biomass Bioenergy* **1994**, *6*, 49–52. [CrossRef]

66. Hanowski, J.M.; Niemi, G.J.; Christian, D.C. Influence of Within-Plantation Heterogeneity and Surrounding Landscape Composition on Avian Communities in Hybrid Poplar Plantations. Influencia de la Heterogeneidad Intra-plantacion y de la Composicion del Paisaje Circundante Sobre Comunidades de Aves en Plantaciones de Alamo Hibrido. *Conserv. Biol.* **1997**, *11*, 936–944. [CrossRef]

67. Wilson, J. The breeding bird community of willow scrub at Leighton Moss, Lancashire. *Bird Study* **1978**, *25*, 239–244. [CrossRef]

68. Bergström, R.; Guillet, C. Summer browsing by large herbivores in short-rotation willow plantations. *Biomass Bioenergy* **2002**, *23*, 27–32. [CrossRef]

69. Christian, D.P. Wintertime use of hybrid poplar plantations by deer and medium-sized mammals in the midwestern U.S. *Biomass Bioenergy* **1997**, *12*, 35–40. [CrossRef]

70. Christian, D.P.; Niemi, G.J.; Hanowski, J.M.; Collins, P. Perspectives on biomass energy tree plantations and changes in habitat for biological organisms. *Biomass Bioenergy* **1994**, *6*, 31–39. [CrossRef]

71. Giordano, M.; Alberto, M. Use by small mammals of short-rotation plantations in relation to their structure and isolation. *Hystrix Ital. J. Mammal.* **2010**, *20*. [CrossRef]

72. Moser, B.W.; Pipas, M.J.; Witmer, G.W.; Engeman, R.M. Small mammal use of hybrid poplar plantations relative to stand age. *Northwest. Sci.* **2002**, *76*, 158–165.

73. Björkman, C.; Bommarco, R.; Eklund, K.; Höglund, S. Harvesting disrupts biological control of herbivores in a short-rotation coppice system. *Ecol. Appl.* **2004**, *14*, 1624–1633. [CrossRef]

74. Langer, V. The potential of leys and short rotation coppice hedges as reservoirs for parasitoids of cereal aphids in organic agriculture. *Agric. Ecosyst. Environ.* **2001**, *87*, 81–92. [CrossRef]

75. Minor, M.; Norton, R. Erratum: Effects of soil amendments on assemblages of soil mites (Acari: Oribatida, Mesostigmata) in short-rotation willow plantings in central New York. *Can. J. For. Res.* **2004**, *34*, 1417–1425. [CrossRef]

76. Minor, M.A.; Volk, T.A.; Norton, R.A. Effects of site preparation techniques on communities of soil mites (Acari: Oribatida, Acari: Gamasida) under short-rotation forestry plantings in New York, USA. *Appl. Soil Ecol.* **2004**, *25*, 181–192. [CrossRef]

77. Mueller, M.; Klein, A.-M.; Scherer-Lorenzen, M.; Nock, C.A.; Staab, M. Tree genetic diversity increases arthropod diversity in willow short rotation coppice. *Biomass Bioenergy* **2018**, *108*, 338–344. [CrossRef]

78. Mueller, A.L.; Dauber, J. Hoverflies (Diptera: Syrphidae) benefit from a cultivation of the bioenergy crop *Silphium perfoliatum* L. (Asteraceae) depending on larval feeding type, landscape composition and crop management. *Agric. For. Entomol.* **2016**, *18*, 419–431. [CrossRef]

79. Peacock, L.; Batley, J.; Dungait, J.; Barker, J.H.A.; Powers, S.; Karp, A. A comparative study of interspecies mating of *Phratora vulgatissima* and *P. vitellinae* using behavioural tests and molecular markers. *Entomol. Exp. Appl.* **2004**, *110*, 231–241. [CrossRef]

80. Peacock, L.; Herrick, S.; Brain, P. Spatio-temporal dynamics of willow beetle (*Phratora vulgatissima*) in short-rotation coppice willows grown as monocultures or a genetically diverse mixture. *Agric. For. Entomol.* **1999**, *1*, 287–296. [CrossRef]

81. Rowe, R.L.; Goulson, D.; Doncaster, C.P.; Clarke, D.J.; Taylor, G.; Hanley, M.E. Evaluating ecosystem processes in willow short rotation coppice bioenergy plantations. *GCB Bioenergy* **2013**, *5*, 257–266. [CrossRef]

82. Sage, R.B.; Tucker, K. The distribution of *Phratora vulgatissima* (Coleoptera: Chrysomelidae) on cultivated willows in Britain and Ireland. *For. Pathol* **1998**, *28*, 289–296. [CrossRef]

83. Sage, R.B.; Fell, D.; Tucker, K.; Sotherton, N.W. Post hibernation dispersal of three leaf-eating beetles (Coleoptera: Chrysomelidae) colonising cultivated willows and poplars. *Agric. For. Entomol.* **1999**, *1*, 61–70. [CrossRef]

84. Reddersen, J. SRC-willow (Salixviminalis) as a resource for flower-visiting insects. *Biomass Bioenergy* **2001**, *20*, 171–179. [CrossRef]

85. Kosewska, A.; Nietupski, M.; Agnieszka, L.-D.; Dolores, C. Biegaczowate (Col. Carabidae) zasiedlające uprawy wierzby krzewiastej w okolicach Olsztyna. *Prog. Plant. Prot.* **2010**, *50*, 1504–1510.

86. Walerys, G.; Kosewska, A.; Wojciech, S. Uprawa wierzby krzewiastej *Salix* spp. miejscem bytowania drapieżnych biegaczowatych *Carabidae*. *Fragm. Agron.* **2008**, *XXV*, 158–169.

87. Konieczna, K.; Melke, A.; Olbrycht, T. Unexploited willow's plantation (*Salix viminalis*) as a reservoir of epigeic ground beetles (Col., Carabidae) and rove beetles (Col., Staphylinidae). *Prog. Plant. Prot.* **2013**, *53*, 319–326.

88. Weger, J.; Vávrová, K.; Kašparová, L.; Bubeník, J.; Komárek, A. The influence of rotation length on the biomass production and diversity of ground beetles (*Carabidae*) in poplar short rotation coppice. *Biomass Bioenergy* **2013**, *54*, 284–292. [CrossRef]

89. Müller-Kroehling, S.; Hohmann, G.; Helbig, C.; Liesebach, M.; Lübke-Al Hussein, M.; Al Hussein, I.A.; Burmeister, J.; Jantsch, M.C.; Zehlius-Eckert, W.; Müller, M. Biodiversity functions of short rotation coppice stands—Results of a meta study on ground beetles (Coleoptera: Carabidae). *Biomass Bioenergy* **2020**, *132*, 105416. [CrossRef]

90. Burel, F. Landscape structure effects on carabid beetles spatial patterns in western France. *Landsc. Ecol* **1989**, *2*, 215–226. [CrossRef]

91. Czerniakowski, Z.W.; Olbrycht, T. Ground beetles (Coleoptera, Carabidae) in the short-rotation biomass plantations. *Zesz. Nauk. Południowo Wschod. Oddziału Pol. Tow. Inżynierii Ekol. Z Siedzibą W Rzesz. I Pol. Tow. Glebozn. Oddział W Rzesz.* **2009**, *11*, 39–42.

92. Allegro, G.; Sciaky, R. Assessing the potential role of ground beetles (Coleoptera, Carabidae) as bioindicators in poplar stands, with a newly proposed ecological index (FAI). *For. Ecol. Manag.* **2003**, *175*, 275–284. [CrossRef]

93. Brauner, O.; Schulz, U. Ground beetle communities (Carabidae) on short rotation coppices and adjacent crop areas—Investigations in Saxony and Brandenburg. *Entomol. Blaetter Fuer Biol. Und Syst. Der Kaefer* **2011**, *107*, 31–64.

94. Sage, R.; Cunningham, M.; Haughton, A.J.; Mallott, M.D.; Bohan, D.A.; Riche, A.; Karp, A. The environmental impacts of biomass crops: Use by birds of miscanthus in summer and winter in southwestern England: Birds in miscanthus. *IBIS* **2010**, *152*, 487–499. [CrossRef]

95. Holland, J. Carabid Beetles: Their Ecology, Survival and Use in Agroecosystems. In *The Agroecology of Carabid Beetles*; Intercept: New York, NY, USA, 2002; pp. 1–40. ISBN 1-898298-76-9.

96. Schulz, U.; Brauner, O.; Gruss, H. Animal diversity on short-rotation coppices—A review. *Vti Agric. Res.* **2009**, *3*, 171–182.

97. Kromp, B. Carabid beetles in sustainable agriculture: A review on pest control efficacy, cultivation impacts and enhancement. *Agric. Ecosyst. Environ.* **1999**, *74*, 187–228. [CrossRef]

98. Regulska, E. *Carabidae* in landscape research on the basis on literature, 2005–2008. *Pol. J. Environ. Stud.* **2011**, *20*, 733–741.

99. Kotze, D.J.; Brandmayr, P.; Casale, A.; Dauffy-Richard, E.; Dekoninck, W.; Koivula, M.J.; Lövei, G.L.; Mossakowski, D.; Noordijk, J.; Paarmann, W.; et al. Forty years of carabid beetle research in Europe—From taxonomy, biology, ecology and population studies to bioindication, habitat assessment and conservation. *ZooKeys* **2011**, *100*, 55–148. [CrossRef]

100. Bianchi, F.J.J.A.; Booij, C.J.H.; Tscharntke, T. Sustainable pest regulation in agricultural landscapes: A review on landscape composition, biodiversity and natural pest control. *Proc. R. Soc. B* **2006**, *273*, 1715–1727. [CrossRef]

101. Thiele, H.-U. *Carabid Beetles in Their Environments: A Study on Habitat Selection by Adaptations in Physiology and Behaviour*; Springer: Berlin/Heidelberg, Germany, 1977; ISBN 978-3-642-81154-8.

102. Szyszko, J. *Próba Waloryzacji Środowisk Leśnych Przy Pomocy Biegaczowatych (Carabidea, Col.) Waloryzacja Ekosystemów Leśnych Metodami Bioidnykacyjnymi.*; SGGW: Warszawa, Poland, 1997.

103. Kulkarmi, S.; Dosdall, L.; Willenborg, C. The Role of Ground Beetles (Coleoptera: Carabidae) in Weed Seed Consumption: A Review. *Weed Sci.* **2015**, *63*, 150304133636004. [CrossRef]

104. Frei, B.; Guenay, Y.; Bohan, D.A.; Traugott, M.; Wallinger, C. Molecular analysis indicates high levels of carabid weed seed consumption in cereal fields across Central Europe. *J. Pest. Sci* **2019**, *92*, 935–942. [CrossRef]

105. Huruk, S. Comparison of structure of carabid (Coleoptera: Carabidae) communities of hay meadows and adjacent cultivated fields. *Wiadomości Entomol.* **2006**, *25*, 9–32.

106. Nakamura, M.; Kagata, H.; Ohgushi, T. Trunk cutting initiates bottom-up cascades in a tri-trophic system: Sprouting increases biodiversity of herbivorous and predaceous arthropods on willows. *Oikos* **2006**, *113*, 259–268. [CrossRef]

107. Baum, S.; Weih, M.; Busch, G.; Kroiher, F.; Bolte, A. The impact of Short Rotation Coppice plantations on phytodiversity. *Landbauforsch Volkenrode* **2009**, *59*, 163–170.

108. Briones, M.J.I.; Elias, D.M.O.; Grant, H.K.; McNamara, N.P. Plant identity control on soil food web structure and C transfers under perennial bioenergy plantations. *Soil Biol. Biochem.* **2019**, *138*, 107603. [CrossRef]

109. Kahle, P.; Janssen, M. Impact of short-rotation coppice with poplar and willow on soil physical properties. *J. Plant Nutr. Soil Sci.* **2020**, *183*, 119–128. [CrossRef]

110. Boháč, J.; Celjak, I.; Moudrý, J.; Kohout, P.; Wotavová, K. Communities of beetles in plantations of fast growing plant species for energetic purposes. *Entomol.Rom.* **2007**, *12*, 213–221.

111. Sotherton, N.W. The distribution and abundance of predatory Coleoptera overwintering in field boundaries. *Ann. Appl. Biol.* **1985**, *106*, 17–21. [CrossRef]

112. Stolarski, M.J.; Szczukowski, S.; Tworkowski, J.; Klasa, A. Yield, energy parameters and chemical composition of short-rotation willow biomass. *Ind. Crop. Prod.* **2013**, *46*, 60–65. [CrossRef]

113. Graham, J.B.; Nassauer, J.I. Wild bee abundance in temperate agroforestry landscapes: Assessing effects of alley crop composition, landscape configuration, and agroforestry area. *Agroforest Syst* **2019**, *93*, 837–850. [CrossRef]

114. Vanbeveren, S.; Ceulemans, R. Biodiversity in short-rotation coppice. *Renew. Sustain. Energy Rev.* **2019**, *111*, 34–43. [CrossRef]

115. Damschen, E.I. Landscape Corridors. In *Encyclopedia of Biodiversity*; Elsevier: Amsterdam, The Netherlands, 2013; pp. 467–475. ISBN 978-0-12-384720-1.

116. Rogan, J.E.; Lacher, T.E. Impacts of Habitat Loss and Fragmentation on Terrestrial Biodiversity. In *Reference Module in Earth Systems and Environmental Sciences*; Elsevier: Amsterdam, The Netherlands, 2018; p. B9780124095489110000. ISBN 978-0-12-409548-9.

117. Fahrig, L. Why do several small patches hold more species than few large patches? *Glob. Ecol. Biogeogr.* **2020**, *29*, 615–628. [CrossRef]

118. Rybicki, J.; Abrego, N.; Ovaskainen, O. Habitat fragmentation and species diversity in competitive communities. *Ecol. Lett.* **2020**, *23*, 506–517. [CrossRef]

119. Tjørve, E. How to resolve the SLOSS debate: Lessons from species-diversity models. *J. Theor. Biol.* **2010**, *264*, 604–612. [CrossRef]

120. Cartwright, J. Ecological islands: Conserving biodiversity hotspots in a changing climate. *Front. Ecol. Env.* **2019**, fee.2058. [CrossRef]

121. Losos, J.B.; Ricklefs, R.E.; MacArthur, R.H. *The Theory of Island Biogeography Revisited*; Princeton University Press: Princeton, NJ, USA, 2010; ISBN 978-1-4008-3192-0.

122. MacArthur, R.H.; Wilson, E.O. *The Theory of Island Biogeography*; Princeton University Press: Princeton, NJ, USA, 2001; ISBN 978-0-691-08836-5.

123. Banaszak, J. *Wyspy Środowiskowe: Bioróżnorodność i Próby Typologii*; Akademia Bydgoska im. Kazimierza Wielkiego: Bydgoszcz, Poland, 2002; ISBN 978-83-7096-459-7.

124. Warren, B.H.; Simberloff, D.; Ricklefs, R.E.; Aguilée, R.; Condamine, F.L.; Gravel, D.; Morlon, H.; Mouquet, N.; Rosindell, J.; Casquet, J.; et al. Islands as model systems in ecology and evolution: Prospects fifty years after MacArthur-Wilson. *Ecol. Lett.* **2015**, *18*, 200–217. [CrossRef] [PubMed]

125. Helmus, M.R.; Behm, J.E. Island Biogeography Revisited. In *Reference Module in Earth Systems and Environmental Sciences*; Elsevier: Amsterdam, The Netherlands, 2019; ISBN 978-0-12-409548-9.

126. Corbet, S.A. Insects, plants and succession: Advantages of long-term set-aside. *Agric. Ecosyst. Environ.* **1995**, *53*, 201–217. [CrossRef]

127. Staley, J.T.; Sparks, T.H.; Croxton, P.J.; Baldock, K.C.R.; Heard, M.S.; Hulmes, S.; Hulmes, L.; Peyton, J.; Amy, S.R.; Pywell, R.F. Long-term effects of hedgerow management policies on resource provision for wildlife. *Biol. Conserv.* **2012**, *145*, 24–29. [CrossRef]

128. Wehling, S.; Diekmann, M. Importance of hedgerows as habitat corridors for forest plants in agricultural landscapes. *Biol. Conserv.* **2009**, *142*, 2522–2530. [CrossRef]

129. Londo, M.; Roose, M.; Dekker, J.; de Graaf, H. Willow short-rotation coppice in multiple land-use systems: Evaluation of four combination options in the Dutch context. *Biomass Bioenergy* **2004**, *27*, 205–221. [CrossRef]

130. Homburg, K.; Drees, C.; Boutaud, E.; Nolte, D.; Schuett, W.; Zumstein, P.; Ruschkowski, E.; Assmann, T. Where have all the beetles gone? Long-term study reveals carabid species decline in a nature reserve in Northern Germany. *Insect Conserv. Divers.* **2019**, icad.12348. [CrossRef]

131. Maes, D.; Titeux, N.; Hortal, J.; Anselin, A.; Decleer, K.; De Knijf, G.; Fichefet, V.; Luoto, M. Predicted insect diversity declines under climate change in an already impoverished region. *J. Insect Conserv.* **2010**, *14*, 485–498. [CrossRef]

132. Wilson, R.J.; Maclean, I.M.D. Recent evidence for the climate change threat to Lepidoptera and other insects. *J. Insect Conserv.* **2011**, *15*, 259–268. [CrossRef]

133. Wagner, D.L. Insect Declines in the Anthropocene. *Annu. Rev. Entomol.* **2020**, *65*, 457–480. [CrossRef]

134. Didham, R.K.; Basset, Y.; Collins, C.M.; Leather, S.R.; Littlewood, N.A.; Menz, M.H.M.; Müller, J.; Packer, L.; Saunders, M.E.; Schönrogge, K.; et al. Interpreting insect declines: Seven challenges and a way forward. *Insect Conserv Divers.* **2020**, *13*, 103–114. [CrossRef]

135. Forister, M.L.; Pelton, E.M.; Black, S.H. Declines in insect abundance and diversity: We know enough to act now. *Conserv. Sci Pr.* **2019**, *1*. [CrossRef]

136. Eyre, M.D.; McMillan, S.D.; Critchley, C.N.R. Ground beetles (Coleoptera, Carabidae) as indicators of change and pattern in the agroecosystem: Longer surveys improve understanding. *Ecol. Indic.* **2016**, *68*, 82–88. [CrossRef]

137. Jowett, K.; Milne, A.E.; Metcalfe, H.; Hassall, K.L.; Potts, S.G.; Senapathi, D.; Storkey, J. Species matter when considering landscape effects on carabid distributions. *Agric. Ecosyst. Environ.* **2019**, *285*, 106631. [CrossRef]

138. Zaller, J.G.; Moser, D.; Drapela, T.; Frank, T. Ground-dwelling predators can affect within-field pest insect emergence in winter oilseed rape fields. *BioControl* **2009**, *54*, 247–253. [CrossRef]

139. Suenaga, H.; Hamamura, T. Occurrence of carabid beetles (Coleoptera: Carabidae) in cabbage fields and their possible impact on lepidopteran pests. *Appl. Entomol. Zool.* **2001**, *36*, 151–160. [CrossRef]

140. Vichitbandha, P.; Wise, D.H. A field experiment on the effectiveness of spiders and carabid beetles as biocontrol agents in soybean. *Agric. For. Ent* **2002**, *4*, 31–38. [CrossRef]

141. Warner, D.J.; Allen-Williams, L.J.; Warrington, S.; Ferguson, A.W.; Williams, I.H. Implications for conservation biocontrol of spatio-temporal relationships between carabid beetles and coleopterous pests in winter oilseed rape. *Agric. For. Entomol.* **2008**, *10*, 375–387. [CrossRef]

142. Schlein, O.; Buechs, W. Approaches to assess the importance of carnivorous beetles as predators of oilseed rape pests. *Iobc Wprs Bull.* **2004**, *27*, 289–292.

143. Schmidt, M.H.; Lauer, A.; Purtauf, T.; Thies, C.; Schaefer, M.; Tscharntke, T. Relative importance of predators and parasitoids for cereal aphid control. *Proc. R. Soc. Lond. B* **2003**, *270*, 1905–1909. [CrossRef]

144. Woodcock, B.A.; Harrower, C.; Redhead, J.; Edwards, M.; Vanbergen, A.J.; Heard, M.S.; Roy, D.B.; Pywell, R.F. National patterns of functional diversity and redundancy in predatory ground beetles and bees associated with key UK arable crops. *J. Appl. Ecol.* **2014**, *51*, 142–151. [CrossRef]

145. Honek, A.; Martinkova, Z.; Jarosik, V. Ground beetles (Carabidae) as seed predators. *Eur. J. Entomol.* **2003**, *100*, 531–544. [CrossRef]

146. Menalled, F.D.; Smith, R.G.; Dauer, J.T.; Fox, T.B. Impact of agricultural management on carabid communities and weed seed predation. *Agric. Ecosyst. Environ.* **2007**, *118*, 49–54. [CrossRef]

147. Sasakawa, K. Field Observations of Climbing Behavior and Seed Predation by Adult Ground Beetles (Coleoptera: Carabidae) in a Lowland Area of the Temperate Zone: Table 1. *Environ. Entomol.* **2010**, *39*, 1554–1560. [CrossRef] [PubMed]

148. Pannwitt, H.; Westerman, P.R.; de Mol, F.; Selig, C.; Gerowitt, B. Biological control of weed patches by seed predators; responses to seed density and exposure time. *Biol. Control.* **2017**, *108*, 1–8. [CrossRef]

Article

Could Supercritical Extracts from the Aerial Parts of *Helianthus salicifolius* A. Dietr. and *Helianthus tuberosus* L. Be Regarded as Potential Raw Materials for Biocidal Purposes?

Anna Malm [1], Agnieszka Grzegorczyk [1], Anna Biernasiuk [1,*], Tomasz Baj [2], Edward Rój [3], Katarzyna Tyśkiewicz [3], Agnieszka Dębczak [3], Mariusz Jerzy Stolarski [4], Michał Krzyżaniak [4] and Ewelina Olba-Zięty [4]

1   Department of Pharmaceutical Microbiology, Faculty of Pharmacy, Medical University of Lublin, Chodźki 1, 20-093 Lublin, Poland; anna.malm@umlub.pl (A.M.); agnieszka.grzegorczyk@umlub.pl (A.G.)
2   Department of Pharmacognosy, Faculty of Pharmacy, Medical University of Lublin, Chodźki 1, 20-093 Lublin, Poland; tomasz.baj@umlub.pl
3   Supercritical Extraction Department, Łukasiewicz Research Network—New Chemical Syntheses Institute, Al. Tysiąclecia Państwa Polskiego 13A, 24-110 Puławy, Poland; edward.roj@ins.lukasiewicz.gov.pl (E.R.); katarzyna.tyskiewicz@ins.lukasiewicz.gov.pl (K.T.); agnieszka.debczak@ins.lukasiewicz.gov.pl (A.D.)
4   Department of Plant Breeding and Seed Production, Faculty of Environmental Management and Agriculture, Centre for Bioeconomy and Renewable Energies, University of Warmia and Mazury in Olsztyn, Plac Łódzki 3, 10-724 Olsztyn, Poland; mariusz.stolarski@uwm.edu.pl (M.J.S.); michal.krzyzaniak@uwm.edu.pl (M.K.); e.olba-ziety@uwm.edu.pl (E.O.-Z.)
*   Correspondence: anna.biernasiuk@umlub.pl

Citation: Malm, A.; Grzegorczyk, A.; Biernasiuk, A.; Baj, T.; Rój, E.; Tyśkiewicz, K.; Dębczak, A.; Stolarski, M.J.; Krzyżaniak, M.; Olba-Zięty, E. Could Supercritical Extracts from the Aerial Parts of *Helianthus salicifolius* A. Dietr. and *Helianthus tuberosus* L. Be Regarded as Potential Raw Materials for Biocidal Purposes? *Agriculture* **2021**, *11*, 10. https://dx.doi.org/10.3390/agriculture11010010

Received: 31 October 2020
Accepted: 23 December 2020
Published: 26 December 2020

Publisher's Note: MDPI stays neutral with regard to jurisdictional claims in published maps and institutional affiliations.

Abstract: Extracts from the June collection of aerial parts of *Helianthus salicifolius* A. Dietr and *Helianthus tuberosus* L. were obtained using carbon dioxide supercritical fluid extraction with water as co-solvent. The antimicrobial effect in vitro of these extracts was then determined against reference species of bacteria, as well as against fungi (represented by *Candida* spp.). Both extracts were found to possess antimicrobial activity, with MIC = 0.62–5 mg mL^{-1} for bacteria and MIC = 5–10 mg mL^{-1} for yeasts, and both extracts demonstrated suitable bactericidal and fungicidal effect. The highest activity was observed against *S. aureus* ATCC 29213 (MIC = 0.62 mg mL^{-1} for *H. salicifolius* extract; MIC = 2.5 mg mL^{-1} for *H. tuberosus* extract) as confirmed by time–kill assay. Higher antioxidant activity was found for *H. tuberosus* extract (EC$_{50}$ = 0.332 mg mL^{-1}) as compared to that of *H. salicifolius* (EC$_{50}$ = 0.609 mg mL^{-1}). The total polyphenol content (TPC) expressed as gallic acid equivalents (GAE) was 13.75 ± 0.50 mg GAE g^{-1} of *H. salicifolius* extract and 33.06 ± 0.80 mg GAE g^{-1} of *H. tuberosus* extract. There was a relationship between the antioxidant potential of both extracts and TPC, but not between antistaphylococcal activity and TPC. The ATIR–FTIR spectra of both extracts showed similar main vibrations of the functional groups typical for phytoconstituents possessing bioactivity. The obtained data suggest potential application of these extracts as natural antioxidants and preparations with biocidal activity. Additionally, both extracts may be regarded as potential natural conservants in cosmetics, as well as natural preservatives in food.

Keywords: willow-leaf sunflower; Jerusalem artichoke; supercritical extraction; water as co-solvent; antimicrobial activity; biocidal effect

1. Introduction

Perennial herbaceous crops, including *Helianthus tuberosus* L. (also called Jerusalem artichoke or topinambur) and *Helianthus salicifolius* A. Dietr, (willow-leaf sunflower) belong to the group of plants of potentially high importance for energy use [1,2]. This is due to high biomass production and limited cultivation requirements. It should be added that these species are resistant to frost and possible infestation by diseases and pests. However, when harvesting the aerial parts biomass of these species, there may be periodic lodging problems, which may make harvesting difficult, but this occurs mainly at the end of the

229

vegetation period [2]. While the biomass of these species can be a raw material for the production of biogas, liquid biofuels, and solid biofuels [3–6], it should be emphasized that the energetic use of biomass of these species maybe of the least value regarding their production purposes.

Recently, much attention has been paid to various plants as a source of alternative antimicrobial compounds and strategies. It is well-known that plants are valuable and rich sources of a wide range of secondary metabolites that possess multidirectional biological activity [7]. Studies suggest that *H. tuberosus* exerts antioxidant, anticancer, antidiabetic, and α-glucosidase inhibitory activity and produces inulin. Indeed, studies have noted the antimicrobial activity of *H. tuberosus*, e.g., against several fungal phytopathogens such as *Rhizoctonia solani*, *Gibberella zeae*, *Alternaria solani*, *Botrytis cinerea*, *Colletotrichum gloeosporioides*, and *Phytophthora capsici* Leonian [8–10]. However, no data concerning antimicrobial (antifungal) properties of *H. salicifolius* are available. The possibility, therefore, exists that Jerusalem artichoke and willow-leaf sunflower can be used as functional food with many medical benefits [11,12]. Thus, new possibilities of using the biomass of these species should be searched for in order to obtain bioactive substances from them and ascertain their further application in the production of high-value bioproducts.

The aim of this study was to determine the antimicrobial properties of extracts obtained from the aerial parts of *H. salicifolius* and *H. tuberosus* using carbon dioxide supercritical fluid extraction with water as co-solvent. The biomass for the extraction purposes was collected at the end of June. The extracts were assayed for their activity together with the mode of action (bactericidal/fungicidal vs. bacteriostatic/fungistatic) against Gram-positive (*Staphylococcus aureus*) and Gram-negative (*Escherichia coli*) bacteria and against fungi (yeasts from *Candida* spp.)—these being the component of human skin, oral, and gut microbiota, which under predisposing conditions can be considered human pathogens [13–15]. The antimicrobial activity of these extracts was analyzed in a correlation with their total polyphenol content (TPC) and antioxidant properties. Attenuated total reflection–Fourier transform infrared (ATR–FTIR) spectra analyses of the extracts were also performed in order to obtain preliminary data on the presence of functional groups characteristic for bioactive phytoconstituents.

2. Materials and Methods

2.1. Plant Material

The green aerial parts biomass of *H. salicifolius* and *H. tuberosus* were collected on June 24, 2019, from the experimental plantation owned by the University of Warmia and Mazury in Olsztyn (Figure 1). This was biomass from the beginning of the growing season (April–June) obtained from nine-year-old stools. The plot size was 20 m^2. The plants were harvested with a lawn trimmer and weighed on an electronic scales. During the harvest, biomass samples were obtained in order to determine its moisture content. On the basis of the yield of fresh biomass and its moisture content during harvest, the yield of dry biomass was calculated and expressed in Mg ha^{-1}. The number of replications was three. After harvesting, the biomass of *H. salicifolius* and *H. tuberosus* was transported to the drying plant. This biomass was then dried at 40 °C to a moisture level below 10%. After drying, it was ground using a mill with 6 mm mesh sieves. Subsequently, the biomass was packed in bags and transported to the supercritical extraction plant.

(a)   (b)

Figure 1. (**a**) *H. salicifolius* and (**b**) *H. tuberosus* plants during the harvest period June 24, 2019, (photo: M.J.S.).

2.2. Extraction Method

The ground material was extracted with supercritical carbon dioxide with the addition of water as co-solvent in the amount of 1 wt%. The supercritical fluid extraction was performed on a pilot plant produced by Natex, Austria, with two extractors of 40 dm^3 each, working under the pressure of up to 1000 bar and temperature up to 90 °C. Each raw material (5 kg for *H. salicifolius* and approx. 2 kg for *H. tuberosus*) was extracted with co-solvent under parameters that were set as follows: temperature at 40 °C and pressure at 330 bar. The extract obtained was in the form of an aqueous mixture. The water was evaporated using a Buchi R-220SE vacuum evaporator. The extraction yield expressed in % was determined for the dried extract in relation to the raw material as the ratio of the amount of dried extract to the mass of raw material. These supercritical extracts were named further as CO_2 + H_2O extracts.

2.3. Determination of Total Polyphenol Content (TPC)

The total phenolic content in CO_2 + H_2O extracts from *H. salicifolius* and *H. tuberosus* was determined spectrophotometrically by a modified method previously described by Clarke et al. [16] and Nickavar and Esbati [17]. A 20 µL of extract dissolved in DMSO (conc. 10 mg mL^{-1}) and 100 µL of freshly prepared Folin–Ciocalteu reagent (diluted 1/10 with redistilled water) were added to the wells of a 96-well plate. After 5 min, 100 µL of a 7.5% Na_2CO_3 solution was added. The plates with the mixtures were incubated for 60 min. at room temperature; the absorbance was then measured using an EPOCH spectrophotometer (Biotek, USA, Software ver. 3.08.01) at a wavelength of 760 nm. The same method was used to establish a calibration curve for the standard gallic acid in the concentration ranges 7.5–120.0 µg mL^{-1} (y = 0.054 x + 0.029, R^2 = 0.996). The analysis was performed in triplicate using DMSO as the blank. The content of total phenolic content expressed in equivalents as mg GAE/g of extract was calculated according to the formula described elsewhere [18].

2.4. Determination of Antibacterial and Antifungal Activity

The assay of antibacterial and antifungal activity of CO_2 + H_2O extracts from *H. salicifolius* and *H. tuberosus* was performed by the broth microdilution method according to EUCAST (the European Committee on Antimicrobial Susceptibility Testing) recommendations [19]. The following reference strains were used in the study: *Staphylococcus aureus* ATCC 29213 (representative of Gram-positive bacteria), *Escherichia coli* ATCC 25922 (representative of Gram-negative bacteria), *Candida albicans* ATCC 10231, and *Candida glabrata* ATCC 90030 (representatives of yeast fungi). All the used microbial strains were first subcultured on Mueller—Hinton Agar (MHA for bacteria) or Mueller–Hinton Agar with 2% glucose (MHA + 2% glucose for fungi) and incubated at 35 °C for 24 h. Microbial colonies were collected and suspended in sterile physiological saline to obtain inoculum of 0.5 McFarland standard, corresponding to 1.5×10^8 CFU (colony forming units) mL^{-1}

for bacteria and 5×10^6 CFU mL^{-1} for fungi. The $CO_2 + H_2O$ extracts were dissolved in DMSO to obtain the final concentration 100 mg mL^{-1}.

The two-fold dilutions of the extracts in Mueller–Hinton Broth (MHB for bacteria) or by Mueller–Hinton Broth with 2% glucose (MHB + 2% glucose for fungi) were prepared in 96-well polystyrene plates. The final concentrations of the extracts ranged from 40 to 0.155 mg mL^{-1}. Next, 2 µL of each bacterial or fungal inoculum was added to each well containing 200 µL of the serial dilution of the extracts in the appropriate culture medium. After incubation at 35 °C for 24 h, the MIC (Minimum Inhibitory Concentration) was assessed spectrophotometrically as the lowest concentration of the extract showing complete bacterial or fungal growth inhibition. Appropriate DMSO, growth, and sterile controls were carried out. Vancomycin (range of 0.03–10 µg mL^{-1}), ciprofloxacin (range of 0.007–10 µg mL^{-1}), and fluconazole (range of 0.03–10 µg mL^{-1}) were included as the reference antimicrobial substances active against Gram-positive bacteria, Gram-negative bacteria, and yeasts. The MBC (minimum bactericidal concentration) or MFC (minimum fungicidal concentration) were determined by removing 20 µL of the bacterial or fungal culture used for MIC determinations from each well and spotting this onto appropriate agar medium. The plates were incubated at 35 °C for 24 h. The lowest extracts concentrations with no visible bacterial or fungal growth were assessed as MBC or MFC, respectively. The experiments were performed in triplicate. Of the three MIC, MBC, and MFC values, the most common representative value, i.e., mode was presented.

2.5. Determination of Antioxidant Activity

The antioxidant activity of $CO_2 + H_2O$ extracts from *H. salicifolius* and *H. tuberosus* extracts was determined using the method described by Gai et al. [20] with modifications. Briefly, a starting solution was prepared by dissolving 10 mg of extract in 1 mL of DMSO solution. A series of dilutions were then prepared in the same solvent at a concentration of 0.16–10 mg mL^{-1}. Subsequently, 0.05 mL of each concentration was mixed with 0.15 mL of DPPH methanol solution (0.078 mg mL^{-1}). The 96-well plate with the mixtures was incubated in the dark for 30 min in room temperature. Absorbance was measured at 515 nm (Biotek, Epoch, Software Version 3.08.01). The extract concentration needed to capture 50% of the initial DPPH (EC_{50}) was determined automatically using four parameter logistic regression (4LP) from the plate reader software Gen5. The experiments were performed in triplicate.

2.6. Attenuated Total Reflection-Fourier Transform Infrared (ATR–FTIR) Spectra Analysis

ATR–FTIR spectra of $CO_2 + H_2O$ extracts obtained from *H. salicifolius* and *H. tuberosus* were recorded on a Bruker Tensor 27 FTIR spectrometer (Bruker Optic GmbH, Aetlingen, Germany) equipped with single-bounce diamond ATR (Platinum ATR, Bruker Optic GmbH, Aetlingen, Germany). The spectrometer was controlled with the software OPUS 6.5 (Bruker Optic). The scan number of the spectra was 16 recorded at 4 cm^{-1} resolution in the wavenumber range from 4000 to 400 cm^{-1}. A small amount of each extract (5 mg) was placed on the ATR surface that was cleaned using ethanol to eliminate any contamination by the previous sample. A new background was recorded between each replicate, and the scans were run in triplicates.

3. Results

The height of the nearly three-month-old plants of both species was 1.1 m (Table 1). *H. salicifolius* produced slightly thicker shoots, and therefore the yield of fresh biomass for this species was higher and amounted to 12.9 Mg ha^{-1}. Moisture of *H. salicifolius* biomass was 79% and was almost two percentage points lower compared to that of *H. tuberosus*. Therefore, the dry matter yield of *H. salicifolius* harvested at the end of June was 2.7 Mg ha^{-1} and was higher by 0.5 Mg ha^{-1} compared to the yield of *H. tuberosus*.

Table 1. Biometric features, moisture content, and yield of fresh and dry biomass of *H. salicifolius* and *H. tuberosus*.

Species	Plant Height (m) *	Shoot Diameter (mm) *	Fresh Biomass Yield (Mg ha^{-1}) **	Moisture Content (%) **	Dry Biomass Yield (Mg ha^{-1}) **
H. salicifolius	1.1 ± 0.9	7.0 ± 1.0	12.9 ± 2.2	79.1 ± 0.4	2.7 ± 0.4
H. tuberosus	1.1 ± 0.4	5.9 ± 0.6	11.6 ± 3.3	81.0 ± 1.1	2.2 ± 0.8

Mean values $\pm$ standard deviation were presented; * $n = 30$; ** $n = 3$.

The results presented in Table 2 show that the extraction efficiency of *H. salicifolius* (4.97%) was much higher than that of *H. tuberosus*. Due to the fact that the yield of *H. salicifolius* aerial parts biomass was also higher, the amount of extract that could be obtained from the cultivation of this species was approximately 134 kg ha^{-1}. On the other hand, the production potential of *H. tuberosus* extract was almost 20 times lower.

Table 2. Extraction efficiency and extract potential yield of *H. salicifolius* and *H. tuberosus* from dry biomass under supercritical conditions with the participation of water as a co-solvent ($CO_2 + H_2O$ extracts).

Plant Material	Extraction Efficiency (%)	Extract Potential Yield (kg ha^{-1})
H. salicifolius	4.97	134.19
H. tuberosus	0.31	6.82

As revealed in Table 3, the $CO_2 + H_2O$ extracts obtained from *H. salicifolius* and *H. tuberosus* showed differential activity against bacteria (MIC = 0.62–5 mg mL^{-1}) and yeasts (MIC = 5–10 mg mL^{-1}). The highest activity of both extracts was observed against *S. aureus* ATCC 29213 with MIC = 0.62 mg mL^{-1} for *H. salicifolius* extract and MIC = 2.5 mg mL^{-1} for *H. tuberosus* extract. MIC for the reference antimicrobial substances were as the following: MIC of vancomycin for *S. aureus* ATCC 29213 was 1 µg mL^{-1}, MIC of ciprofloxacin for *E. coli*, ATCC 25922 was 0.015 µg mL^{-1} and MIC of fluconazole for *C. albicans*, and ATCC was 1 µg mL^{-1}. As presented in Table 3, both extracts possessed bactericidal (MBC/MIC = 1–4) and fungicidal effect (MFC/MIC = 1–2). It is generally accepted that antimicrobials are usually regarded as bactericidal or fungicidal if the MBC/MIC or MFC/MIC ratio is $\leq$4 [21].

Table 3. Antimicrobial activity of supercritical extracts obtained from *H. salicifolius* and *H. tuberosus* with water as a co-solvent ($CO_2 + H_2O$ extract).

Microorganisms	Extracts					
	H. salicifolius			*H. tuberosus*		
Bacterial Strains	MIC [mg mL^{-1}]	MBC [mg mL^{-1}]	MBC/ MIC	MIC [mg mL^{-1}]	MBC [mg mL^{-1}]	MBC/ MIC
Staphylococcus aureus ATCC 29213	0.62	2.5	4	2.5	5	2
Escherichia coli ATCC 25922	5	10	2	5	5	1
Fungal (Yeasts) Strains	MIC [mg mL^{-1}]	MFC [mg mL^{-1}]	MFC/ MIC	MIC [mg mL^{-1}]	MFC [mg mL^{-1}]	MFC/ MIC
Candida albicans ATCC 10231	5	10	2	5	10	2
Candida glabrata ATCC 90030	10	10	1	10	20	2

The representative data (mode) are presented.

Time–kill assays were performed exposing *S. aureus* ATCC 29213 to various concentrations of the $CO_2 + H_2O$ extracts obtained from *H. salicifolius* and *H. tuberosus* in order to confirm their bactericidal activity. *S. aureus* ATCC 29213 was chosen for this experiment due to its higher sensitivity to both extracts in comparison to *E. coli* ATCC 25922 and the yeast species. Moreover, it is a common human pathogen that causes a wide range of clinical infections. It is assumed that bactericidal effect is defined as greater than 3 $\log_{10}$ fold decrease in CFU mL^{-1} in the presence of antimicrobials as compared to the initial inoculum [22]. As presented in Figure 2, bacterial killing by both extracts was found to be a concentration-dependent process; some biocidal effect occurred even at sub-inhibitory concentrations of both extracts, that was 0.1 mg mL^{-1} for *H. salicifolius* extract and 1 mg mL^{-1} for *H. tuberosus* extract. Moreover, *H. salicifolius* extract was more active than that of *H. tuberosus*.

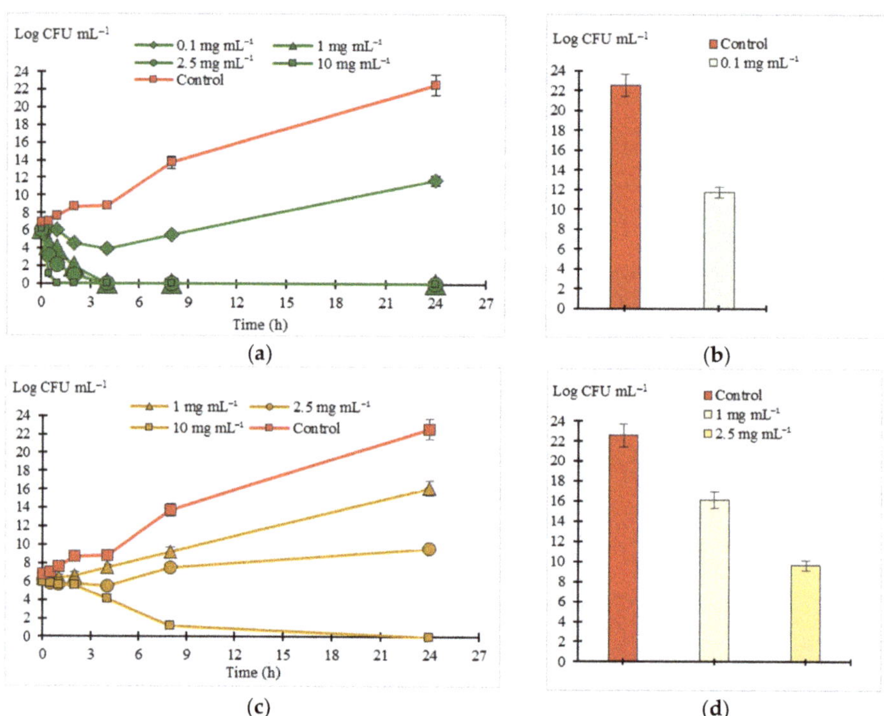

Figure 2. Time–kill curves for *S. aureus* ATCC 29213 at various concentrations of supercritical extracts obtained from: (**a**) *H. salicifolius* and (**c**) *H. tuberosus* with water as a co-solvent ($CO_2 + H_2O$ extracts). Bacterial population density after 24 h exposure to various concentrations of the $CO_2 + H_2O$ extracts obtained from: (**b**) *H. salicifolius* and (**d**) *H. tuberosus*. Mean values ± standard deviations are presented.

As presented in Figure 3, the $CO_2 + H_2O$ extracts obtained from *H. salicifolius* and *H. tuberosus* differed in the total polyphenol content (TPC) expressed as gallic acid equivalents (GAE). This was 13.75 ± 0.50 mg GAE g^{-1} of *H. salicifolius* extract and 33.06 ± 0.80 mg GAE g^{-1} of *H. tuberosus* extract. Both extracts showed different antioxidant activity. *H. tuberosus* extract exhibited almost two-fold higher activity ($EC_{50} = 0.332 ± 0.05$ mg mL^{-1}) as compared to that of *H. salicifolius* ($EC_{50} = 0.609 ± 0.29$ mg mL^{-1}). However, a relationship was observed between the antioxidant potential of both extracts and TPC. It should be noted that despite two-fold lower TPC in the *H. salicifolius* extract than that in the *H. tuberosus* extract (Figure 3), activity of *H. salicifolius* extract against *S. aureus* ATCC 29213 was four-fold higher compared to *H. tuberosus* extract with MIC of 0.62 mg mL^{-1} or 2.5 mg mL^{-1}

(Table 3), respectively. In contrast, activity of both extracts against *E. coli* ATCC 25922 and *Candida* spp. strains was comparable (Table 3), irrespective of TPC (Figure 3).

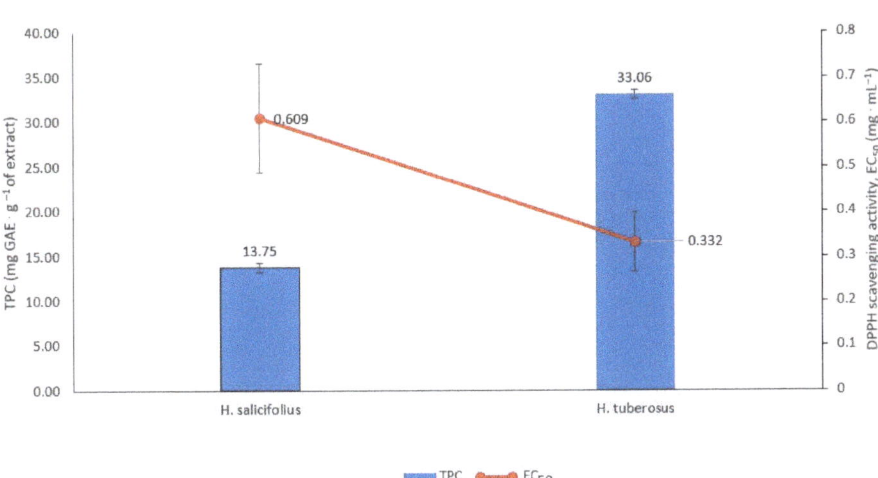

Figure 3. The total polyphenol content (TPC) in supercritical extracts obtained from *H. salicifolius* and *H. tuberosus* with water as a co-solvent (CO_2 + H_2O extracts) together with their antioxidant activity. Mean values ± standard deviation were presented.

The ATR–FTIR spectra of CO_2 + H_2O extracts from *H. salicifolius* and *H. tuberosus* are shown in Figure 4. Interpretation of these spectra was performed according to other authors [23–25]. The main vibrations of the characteristic groups were similar in both extracts; only slight diferrences were observed. At a wave-number of 3392 cm^{-1} for *H. salicifolius* extract and 3401 cm^{-1} for *H. tuberosus* extract, characteristic stretching vibrations were indicated that may have been induced from the OH group derived from carbohydrates proteins and polyphenols. In both analyzed spectra, cis C=C stretching was observed for both extracts at a wavenumbers of 3009 and 3008 cm^{-1}. The absorption bands around 2918 and 2849 cm^{-1} in the *H. salicifolius* extract spectrum and 2921 and 2851 cm^{-1} in the *H. tuberosus* extract spectrum may be due to asymmetric and symmetrical CH_2 stretching vibrations, respectively, while the band at a wavenumber of 1742 cm^{-1} is characteristic of the C=O stretching vibrations of aldehydes, ketones, and carboxylic acids. In the *H. salicifolius* extract, vibrations were recorded at a wavenumber of 1695 cm^{-1}; this band is characteristic of the vibrations of amide I (1600–1700 cm^{-1}), and the effect is related to the stretching vibrations of the C=O and C-N groups. Unconjugated stretching *cis* C=C at 1657 cm^{-1} was also observed. The band at 1462 cm^{-1} wavenumber was probably generated by a CH_2 scissor vibration. In contrast, the presence of the band at wavenumbers of 1369 cm^{-1} (*H. salicifolius* extract) and 1377 cm^{-1} (*H. tuberosus* extract) is characteristic of CH_3 symmetrical bending vibration. At wavenumbers 1239, 1167, 1096, and 1023 cm^{-1} and 1239, 1162, 1096, and 1030 cm^{-1}, respectively, for *H. salicifolius* and *H. tuberosus* extracts, stretching vibrations characteristic for the C-O groups were also seen. Furthermore, in the range of 970–969 cm^{-1} wavenumbers, characteristic vibrations for the groups *trans* double bonds (C=C) and *cis* double bonds (C=C) were revealed. In addition, the bands at 720 cm^{-1} are typical of the CH_2 groups. Moreover, at the wavenumber of 1323 cm^{-1} in the *H. salicifolius* extract, there was an absorption of the band that could be derived from amide III (C-N stretch) with a significant share of CH_2 carbohydrate residue. Table 4 shows major band assignments for the ATR–FTIR spectra of both CO_2 + H_2O extracts.

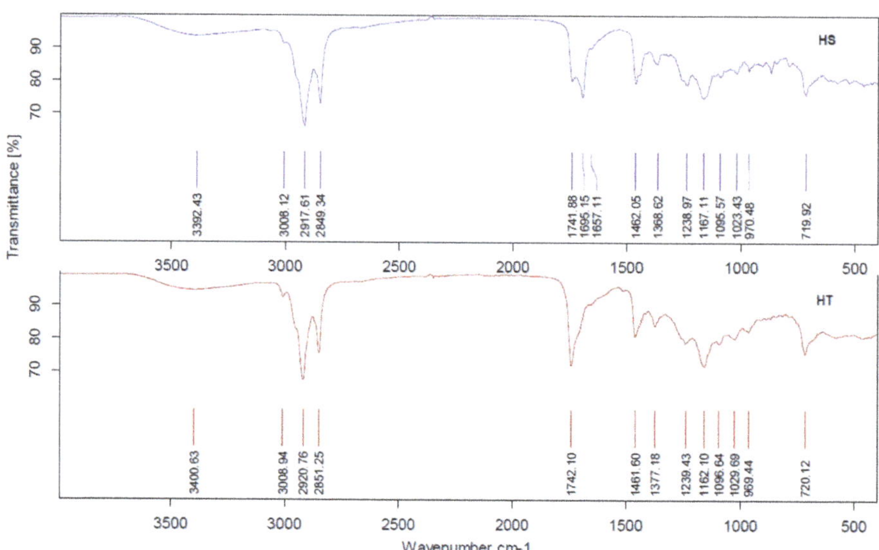

Figure 4. ATR–FTIR spectra of supercritical extracts obtained from *H. salicifolius* (HS) and *H. tuberosus* (HT) with water as a co-solvent (CO_2 + H_2O extracts).

Table 4. Major band assignments for the ATR–FTIR spectra of supercritical extracts obtained from *H. salicifolius* and *H. tuberosus* with water as a co-solvent (CO_2 + H_2O extracts).

Wavenumbers (cm^{-1})		Functional Group Vibration
H. salicifolius	*H. tuberosus*	
3392	3401	OH stretching (carbohydrates, proteins and polyphenols)
3008	3009	cis C=C stretching
2918, 2849	2921, 2851	asymmetric and symmetric stretching vibration of CH_2 group
1742	1742	C=O stretching (aldehydes, ketones, and carboxylic acids)
1695	-	C=O and C-N stretching vibrations
1657	-	unconjugated cis C=C
1462	1462	CH_2 scissor vibration
1369	1377	CH_3 symmetrical bending vibration
1239, 1167, 1096, 1023	1239, 1162, 1096, 1030	C-O stretching vibration
970	969	trans double bonds (C=C) and cis double bonds (C=C)
720	720	bending (rocking) of $-(CH_2)_n-$, -HC-CH-(cis)

4. Discussion

Data presented in this paper showed that CO_2 + H_2O extracts obtained from both *Helianthus* species possessed antimicrobial potential, including activity against Gram-positive bacteria (*S. aureus*) and Gram-negative bacteria (*E. coli*), as well as yeasts (*C. albicans* and *C. glabrata*). Both extracts exerted bactericidal and fungicidal effect. It should be noted that the above microbial species, present within human skin, oral, and gut microbiota, may be regarded as commensals or pathogens, depending on the host–microbe interactions [13–15]. Moreover, these microorganisms may be a cause of cosmetics or food contamination, hence the need to protect these products by substances with antimicrobial activity—conservants or preservatives, respectively [26,27]. According to the literature data [9–11], extracts from

H. tuberosus leaves might be a promising source of natural fungicides active against several phytopathogens, among them caffeoylquinic acid derivatives. However, Liu et al. [9] found that the inhibitory effects of aqueous extracts were significantly less than those of extracts of organic solvents, i.e., petroleum ether, ethyl ether, and ethyl acetate. Of note, there is no literature data on bioactivity and chemical composition of the supercritical extracts obtained from the aerial parts of *H. tuberosus* and *H. salicifolius* using water as co-solvent (CO_2 + H_2O extracts). This is the first report.

The antistaphylococcal activity of CO_2 + H_2O extracts obtained from *H. salicifolius* and from *H. tuberosus* should be underlined. *S. aureus* is known to be an important pathogen related to skin and soft tissue, as well as to food-borne infections [28]. In this paper, we found higher antistaphylococcal effect of *H. salicifolius* CO_2 + H_2O extract, as compared to that of *H. tuberosus* CO_2 + H_2O extract. In contrast, it was found that TPC was higher in *H. tuberosus* extract, in comparison to that in *H. salicifolius* extract. Yuan et al. [29] showed that TPC in aqua ethanolic extracts from *H. tuberosus* leaves was 101.07 mg GAE g^{-1} of dry extract. More detailed studies by Showkat et al. [30] revealed that TPC in *H. tuberosus* was dependent on the plant organ. They determined higher TPC, expressed as mg GAE g^{-1} of dry substance, in ethanolic extracts from leaves (7.9–11.1) than in flower (4.0–5.3), tuber (2.8–3.8), and stem (0.9–1.7) extracts. However, these authors found the overestimation of TPC in the extracts from various organs ranging from 65% (in flowers) to 94% (in stems) and used the Folin–Ciocalteu assay, applying the correction for ascorbic acid.

It should be noted that polyphenols have been recognized as one of the largest and most widespread group of plant secondary metabolites, responsible for both antimicrobial and antioxidant activity [31]. Data presented in this paper suggest that the antistaphylococcal activity of both CO_2 + H_2O extracts, especially that from *H. salicifolius*, may be due to the content of other plant secondary metabolites such as sesquiterpene lactones [32]. These compounds can be regarded as one of the most prevalent and biologically significant classes of plant secondary metabolites, including that in plants from Asteraceae family, e.g. *H. tuberosus* [29]. Sesquiterpene lactones have been assumed to be potent antimicrobials. Some are also considered to be antioxidants [33,34].

The antimicrobial properties of plant-derived products are generally accompanied by a confirmed antioxidant capacity [29,31–34]. In this paper, we found higher antioxidant effect of *H. tuberosus* CO_2 + H_2O extract together with its higher TPC, as compared to those of *H. salicifolius* CO_2 + H_2O extract. These data suggest that the antioxidant properties of CO_2 + H_2O extracts studied may be related to polyphenols content, which is in agreement with the literature data [35,36]. The DPPH assay included in this study for the determination of antioxidant activities of the CO_2 + H_2O extracts from both *Helianthus* species was also used by other authors [30,37,38] in studying the radical scavenging activity of *H. tuberosus*. Showkat et al. [30], for example, showed significant correlation between TPC and radical scavenging activity of ethanolic extracts from *H. tuberosus* leaves. These authors revealed, similarly to Yuan et al. [37] and Nizioł et al. [38], that aqua-ethanolic or ethanolic extracts from *H. tuberosus* leaves, in comparison to those from other organs of this plant (e.g. tubers), possessed higher radical scavenging activity (with EC_{50} about 0.075–0.25 mg mL^{-1}) and could be a potential source of natural antioxidants.

The ATR–FTIR spectroscopic imaging is a suitable technique, not only as a method of identity confirmation, but also for detecting and identifying molecular components in a complex plant matrix [23]. Preliminary phytochemical analysis of both CO_2 + H_2O extracts by ATR–FTIR indicated the presence of similar main vibrations of the functional groups typical for phytoconsituents possessing bioactivity such as polyphenols, aldehydes, ketones, or carboxylic acids [23–25].

5. Conclusions

The presented data suggest that supercritical extracts with water as a co-solvent obtained from the aerial parts of *H. salicifolius* and *H. tuberosus* collected in summer period appeared to be a promising source of natural compounds with biocidal effect. They possessed antibacterial activity against Gram-positive (*S. aureus*) and Gram-negative (*E. coli*, species (MIC = 0.62–5 mg mL^{-1}), as well as antifungal activity against yeasts from *Candida* genus (MIC = 5–10 mg mL^{-1}). It is worth notifying their antistaphylococcal activity (MIC = 0.62–2.5 mg mL^{-1}). These extracts may be also regarded as natural potential antioxidants (EC$_{50}$ = 0.332–0.609 mg mL^{-1}). The ATR–FTIR spectra of both extracts showed similar main vibrations of the functional groups typical for phytoconstituents possessing bioactivity. The obtained data, together with those from literature, suggest that these extracts and their isolated bioactive compounds may be used as conservants in cosmetics and/or natural preservatives in food. However, further studies are needed to confirm the obtained results to define and to quantify constituents present in both extracts, as well as to identify specific applications of supercritical extracts and their phytoconstituents from biomass of these two plant species.

Author Contributions: Conceptualization: A.M., E.R., and M.J.S.; data curation, A.M., E.R., and M.J.S.; formal analysis, A.M., E.R., and M.J.S.; funding acquisition: M.J.S., A.M., and E.R.; investigation: A.M., E.R., M.J.S., A.G., A.B., T.B., K.T., and A.D.; methodology: A.M., E.R., M.J.S., A.G., A.B., T.B., K.T., A.D., M.K., and E.O.-Z.; project administration: A.M. and M.J.S.; validation: A.M., E.R., and M.J.S.; visualization: A.M., E.R., and M.J.S.; writing—original draft: A.M., E.R., and M.J.S.; writing—review and editing: A.M., E.R., M.J.S., A.G., A.B., T.B., K.T., A.D., M.K., and E.O.-Z. All authors have read and agreed to the published version of the manuscript.

Funding: This research was funded by the National (Polish) Centre for Research and Development (NCBiR), entitled "Environment, agriculture and forestry", project: BIOproducts from lignocellulosic biomass derived from Marginal land to fill the Gap In Current national bioeconomy, No BIOSTRATEG3/344253/2/NCBR/2017.

Institutional Review Board Statement: This study did not require ethical approval.

Informed Consent Statement: Not applicable.

Data Availability Statement: Not applicable.

Acknowledgments: We would also like to thank the staff of the Department of Plant Breeding and Seed Production, Didactic and Research Station in Bałdy and Quercus Sp. z o.o. for their technical support during the experiment.

Conflicts of Interest: The authors declare no conflict of interest.

References

1. Rossini, F.; Provenzano, M.E.; Kuzmanović, L.; Ruggeri, R. Jerusalem artichoke (*Helianthus tuberosus* L.): A versatile and sustainable crop for renewable energy production in Europe. *Agronomy* **2019**, *9*, 528. [CrossRef]
2. Stolarski, M.J.; Śnieg, M.; Krzyżaniak, M.; Tworkowski, J.; Szczukowski, S. Short rotation coppices, grasses and other herbaceous crops: Productivity and yield energy value versus 26 genotypes. *Biomass Bioenergy* **2018**, *119*, 109–120. [CrossRef]
3. Volk, G.M.; Richards, K. Preservation methods for Jerusalem artichoke cultivars. *HortScience* **2006**, *41*, 80–83. [CrossRef]
4. Matias, J.; Gonzalez, J.; Cabanillas, J.; Royano, L. Influence of NPK fertilization and harvest date on agronomic performance of Jerusalem artichoke crop in the Guadiana Basin (Southwestern Spain). *Ind. Crop. Prod.* **2013**, *48*, 191–197. [CrossRef]
5. Yang, L.; He, Q.S.; Corscadden, K.; Udenigwe, C.C. The prospects of Jerusalem artichoke in functional food ingredients and bioenergy production. *Biotechnol. Rep.* **2015**, *5*, 77–88. [CrossRef]
6. Stolarski, M.J.; Śnieg, M.; Krzyżaniak, M.; Tworkowski, J.; Szczukowski, S.; Graban, Ł.; Lajszner, W. Short rotation coppices, grasses and other herbaceous crops: Biomass properties versus 26 genotypes and harvest time. *Ind. Crop. Prod.* **2018**, *119*, 22–32. [CrossRef]
7. Kokoska, L.; Kloucek, P.; Leuner, O.; Novy, P. Plant-derived products as antibacterial and antifungal agents in human health care. *Curr. Med. Chem.* **2019**, *26*, 5501–5541. [CrossRef]
8. Szewczyk, A.; Zagaja, M.; Bryda, J.; Kosikowska, U.; Stępień-Pyśniak, D.; Winiarczyk, S.; Andres-Mach, M. Topinambur - new possibilities for use in a supplementation diet. *Ann. Agric. Environ. Med.* **2019**, *26*, 24–28. [CrossRef]
9. Liu, H.W.; Liu, Z.P.; Liu, L.; Zhao, G.M. Studies on the antifungal activities and chemical components of extracts from *Helianthus tuberosus* leaves. *Nat. Prod. Res. Dev.* **2007**, *19*, 405–409.
10. Chen, F.; Long, X.; Yu, M.; Liu, Z.; Liu, L.; Shao, H. Phenolics and antifungal activities analysis in industrial crop Jerusalem artichoke (*Helianthus tuberosus* L.) leaves. *Ind. Crop. Prod.* **2013**, *47*, 339–345. [CrossRef]
11. Chen, F.J.; Long, X.H.; Li, E.Z. Evaluation of antifungal phenolics from *Helianthus tuberosus* L. leaves against *Phytophthora capsici* Leonian by chemometric analysis. *Molecules* **2019**, *24*, 4300. [CrossRef] [PubMed]

12. Kaszás, L.; Alshaal, T.; Kovács, Z.; Koroknai, J.; Elhawat, N.; Nagy, E.; El-Ramady, H.; Fári, M.; Domokos-Szabolcsy, É. Refining high-quality leaf protein and valuable co-products from green biomass of Jerusalem artichoke (*Helianthus tuberosus* L.) for sustainable protein supply. *Biomass Conv. Bioref.* **2020**. [CrossRef]

13. Cogen, A.L.; Nizet, V.; Gallo, R.L. Skin microbiota: A source of disease or defence? *Br. J. Dermatol.* **2008**, *158*, 442–455. [CrossRef] [PubMed]

14. Pickard, J.M.; Zeng, M.Y.; Caruso, R.; Núñez, G. Gut microbiota: Role in pathogen colonization, immune responses, and inflammatory disease. *Immunol. Rev.* **2017**, *279*, 70–89. [CrossRef]

15. Arweiler, N.B.; Netuschil, L. The oral microbiota. *Adv. Exp. Med. Biol.* **2016**, *902*, 45–60. [CrossRef]

16. Clarke, G.; Ting, K.N.; Wiart, C.; Fry, J. High correlation of 2,2-diphenyl-1-picrylhydrazyl (DPPH) radical scavenging, ferric reducing activity potential and total phenolics content indicates redundancy in use of all three assays to screen for antioxidant activity of extracts of plants from the Malaysian rainforest. *Antioxidants* **2013**, *2*, 1–10. [CrossRef]

17. Nickavar, B.; Esbati, N. Evaluation of the antioxidant capacity and phenolic content of three *Thymus* species. *J. Acupunct. Meridian Stud.* **2012**, *5*, 119–125. [CrossRef]

18. Alara, O.R.; Abdurahman, N.H.; Mudalip, S.K.A.; Olalere, O.A. Characterization and effect of extraction solvents on the yield and total phenolic content from *Vernonia amygdalina* leaves. *J. Food Meas. Charact.* **2018**, *12*, 311–316. [CrossRef]

19. European Committee for Antimicrobial Susceptibility Testing (EUCAST) of the European Society of Clinical Microbiology and Infectious Diseases (ESCMID): Determination of minimum inhibitory concentrations (MICs) of antibacterial agents by broth dilution. EUCAST Discussion Document E.Dis 5.1; The European Committee on Antimicrobial Susceptibility Testing. *Clin. Microbiol. Inf. Dis.* **2003**, *9*, 1–7.

20. Gai, F.; Karamać, M.; Janiak, M.A.; Amarowicz, R.; Peiretti, P.G. Sunflower (*Helianthus annuus* L.) plants at various growth stages subjected to extraction-comparison of the antioxidant activity and phenolic profile. *Antioxidants* **2020**, *9*, 535. [CrossRef]

21. Pankey, G.A.; Sabath, L.D. Clinical relevance of bacteriostatic versus bactericidal mechanisms of action in the treatment of gram-positive bacterial infections. *Clin. Infect. Dis.* **2004**, *38*, 864–870. [CrossRef]

22. Silva, F.; Lourenço, O.; Queiroz, J.A.; Domingues, F.C. Bacteriostatic versus bactericidal activity of ciprofloxacin in *Escherichia coli* assessed by flow cytometry using a novel far-red dye. *J. Antibiot. (Tokyo)* **2011**, *64*, 321–325. [CrossRef] [PubMed]

23. Huck, C.W. Advances of infrared spectroscopy in natural product research. *Phytochem. Lett.* **2015**, *11*, 384–393. [CrossRef]

24. Jiménez-Sotelo, P.; Hernández-Martínez, M.; Osorio-Revilla, G.; Meza-Márquez, O.G.; García-Ochoa, F.; Gallardo-Velázquez, T. Use of ATR-FTIR spectroscopy coupled with chemometrics for the authentication of avocado oil in ternary mixtures with sunflower and soybean oils. *Food Addit. Contam Part A* **2016**, *33*, 1105–1115. [CrossRef] [PubMed]

25. Rai, A.; Mohanty, B.; Bhargava, R. Supercritical extraction of sunflower oil: A central composite design for extraction variables. *Food Chem.* **2016**, *192*, 647–659. [CrossRef]

26. Halla, N.; Fernandes, I.P.; Heleno, S.A.; Costa, P.; Boucherit-Otmani, Z.; Boucherit, K.; Rodrigues, A.E.; Ferreira, I.C.F.R.; Barreiro, M.F. Cosmetics preservation: A review on present strategies. *Molecules* **2018**, *23*, 1571. [CrossRef]

27. Pisoschi, A.M.; Pop, A.; Georgescu, C.; Turcuş, V.; Olah, N.K.; Mathe, E. An overview of natural antimicrobials role in food. *Eur. J. Med. Chem.* **2018**, *143*, 922–935. [CrossRef]

28. Tong, S.Y.C.; Davis, J.S.; Eichenberger, E.; Holland, T.L.; Fowler, V.G., Jr. *Staphylococcus aureus* infections: Epidemiology, pathophysiology, clinical manifestations, and management. *Clin. Microbiol. Rev.* **2015**, *28*, 603–661. [CrossRef]

29. Yuan, X.; Yang, Q. Simultaneous quantitative determination of 11 sesquiterpene lactones in Jerusalem artichoke (*Helianthus tuberosus* L.) leaves by ultrahigh performance liquid chromatography with quadrupole time-of-flight mass spectrometry. *J. Sep. Sci.* **2017**, *40*, 1457–1464. [CrossRef]

30. Showkat, M.M.; Falck-Ytter, A.B.; Strætkvern, K.O. Phenolic acids in Jerusalem artichoke (*Helianthus tuberosus* L.): Plant organ dependent antioxidant activity and optimized extraction from leaves. *Molecules* **2019**, *24*, 3296. [CrossRef]

31. Mojzer, E.B.; Hrnčič, M.K.; Škerget, M.; Knez, Ž.; Bren, U. Polyphenols: Extraction methods, antioxidative action, bioavailability and anticarcinogenic effects. *Molecules* **2016**, *21*, 901. [CrossRef]

32. Cartagena, E.; Alva, M.; Montanaro, S.; Bardón, A. Natural sesquiterpene lactones enhance oxacillin and gentamicin effectiveness against pathogenic bacteria without antibacterial effects on beneficial lactobacilli. *Phytother. Res.* **2015**, *29*, 695–700. [CrossRef] [PubMed]

33. Chadwick, M.; Trewin, H.; Gawthrop, F.; Wagstaff, C. Sesquiterpenoids lactones: Benefits to plants and people. *Int. J. Mol. Sci.* **2013**, *14*, 12780–12805. [CrossRef] [PubMed]

34. Fraga, B.M. Natural sesquiterpenoids. *Nat. Prod. Rep.* **2001**, *18*, 650–673. [PubMed]

35. Xu, D.P.; Li, Y.; Meng, X.; Zhou, T.; Zhou, Y.; Zheng, J.; Zhang, J.J.; Li, H.B. Natural antioxidants in foods and medicinal plants: Extraction, assessment and resources. *Int. J. Mol. Sci.* **2017**, *18*, 96. [CrossRef] [PubMed]

36. Dontha, S. A review on antioxidant methods. *Asian J. Pharm. Clin. Res.* **2016**, *9*, 14–32. [CrossRef]

37. Yuan, X.; Gao, M.; Xiao, H.; Tan, C.; Du, Y. Free radical scavenging activities and bioactive substances of Jerusalem artichoke (*Helianthus tuberosus* L.) leaves. *Food Chem.* **2012**, *133*, 10–14. [CrossRef]

38. Nizioł-Łukaszewska, Z.; Furman-Toczek, D.; Zagórska-Dziok, M. Antioxidant activity and cytotoxicity of Jerusalem artichoke tubers and leaves extract on HaCaT and BJ fibroblast cells. *Lipids Health Dis.* **2018**, *17*, 280–292. [CrossRef]

 *agriculture*

Article

Possibly Invasive New Bioenergy Crop *Silphium perfoliatum*: Growth and Reproduction Are Promoted in Moist Soil

L. Marie Ende *, Katja Knöllinger, Moritz Keil, Angelika J. Fiedler and Marianne Lauerer

Ecological Botanical Gardens, Bayreuth Center for Ecology and Environmental Research (BayCEER), University of Bayreuth, 95447 Bayreuth, Germany; katja.knoellinger@uni-bayreuth.de (K.K.); moritz.keil@uni-bayreuth.de (M.K.); angelika.j.fiedler@uni-bayreuth.de (A.J.F.); marianne.lauerer@uni-bayreuth.de (M.L.)
* Correspondence: marie.ende@uni-bayreuth.de

Abstract: The cup plant (*Silphium perfoliatum*) is a new and promising bioenergy crop in Central Europe. Native to North America, its cultivation in Europe has increased in recent years. Cup plant is said to be highly productive, reproductive, and strongly competitive, which could encourage invasiveness. Spontaneous spread has already been documented. Knowledge about habitat requirements is low but necessary, in order to predict sites where it could spontaneously colonize. The present experimental study investigates the growth and reproductive potential of cup plant depending on soil moisture, given as water table distance (WTD). In moist soil conditions, the growth and reproductive potential of cup plant were the highest, with about 3 m plant height, 1.5 kg dry biomass, and about 350 capitula per plant in the second growing season. These parameters decreased significantly in wetter, and especially in drier conditions. The number of shoots per plant and number of fruits per capitulum were independent of WTD. In conclusion, valuable moist ecosystems could be at risk for becoming invaded by cup plant. Hence, fields for cultivating cup plant should be carefully chosen, and distances to such ecosystems should be held. Spontaneous colonization by cup plant must be strictly monitored in order to be able to combat this species where necessary.

Keywords: bioenergy crop; cup plant; groundwater; growth; invasive potential; reproductive potential; *Silphium perfoliatum*; soil moisture; water table distance

Citation: Ende, L.M.; Knöllinger, K.; Keil, M.; Fiedler, A.J.; Lauerer, M. Possibly Invasive New Bioenergy Crop *Silphium perfoliatum*: Growth and Reproduction Are Promoted in Moist Soil. *Agriculture* **2021**, *11*, 24. https://doi.org/10.3390/agriculture11010024

Received: 18 November 2020
Accepted: 28 December 2020
Published: 1 January 2021

Publisher's Note: MDPI stays neutral with regard to jurisdictional clai-ms in published maps and institutio-nal affiliations.

1. Introduction

In Europe, biogas is being increasingly produced as a renewable energy source to replace fossil fuels [1]. Currently maize (*Zea mays* L.) is the most dominant biogas crop, though its cultivation goes along with great ecological damage from the high application of machines, fertilizer, and pesticides. Therefore, alternative bioenergy crops are being sought that are more ecologically agreeable [2–5].

One promising alternative crop in this context is the cup plant (*Silphium perfoliatum* L.) [2]. This perennial, yellow-flowering C3-plant belongs to the Asteraceae family. It develops stems and flowers from the second year onwards and persists many years [6]. Native in the prairies of eastern North America, cup plant was introduced to Europe in the 18th century as an ornamental plant [6]. Since 2004 it has been used as a bioenergy crop in Germany [2], and as of 2019 about 4500 ha have been cultivated there [7]. Many other European countries are cultivating this crop for bioenergy as well [3].

Cup plant has many ecological advantages over maize [2]: It can be harvested profitably for more than 15 years [8], and the application of machines and pesticides is much lower compared to maize, an annual plant [2]. In soil, higher portions of microbial biomass, higher microbial diversity, and higher biological activity comparative to maize have been proven [4]. Benefits for many pollinator species have also been detected: Insects are strongly attracted to the flowers of cup plant, which have a long flowering period relatively late in the year when most other floral resources have already finished blooming [9–11].

Furthermore, cup plant is easy to cultivate, highly productive with a high biogas yield, and it is competitive and very reproductive [2,6]. These traits make it an attractive bioenergy crop. However, in combination with the fact that harvest in agriculture usually takes place after flowering, they carry the risk of spontaneous spreading and settlement out from the fields. Spontaneous colonization was already documented in Upper Franconia (Germany) [12] and other parts of Germany as well as in other European countries (e.g., Austria, Switzerland, Poland) [13–16]. In the Netherlands and Russia, cup plant has already been graded as "potentially invasive" [17,18]. Studies on the invasive potential of this species are essential and should be of interest for all involved stakeholders before cup plant is cultivated on a large scale.

For this purpose, comprehensive knowledge about site requirements is necessary to allow predictions about where cup plant might establish itself. However, little is known about its site preferences in Europe, especially regarding soil moisture. It is assumed that cup plant prefers soils with good moisture provision but is also fairly drought tolerant [3,5,19,20]. So far, there are mainly empirical data or assumptions and only few experimental studies about the yield of cup plant in Central Europe depending on soil moisture. In its native range in North America, cup plant colonizes moist bottomlands and floodplains near streambeds [6]. Assuming that cup plant grows and reproduces in Central Europe in the same way that it does in its native range, it carries a special risk of invasion on moist sites. It is known that these are often ecosystems of high value for nature conservation in Central Europe. To assess the risk of these ecosystems becoming colonized by cup plant, we executed a growth experiment with cup plant over two years in tanks similar to those of Ellenberg's Hohenheimer groundwater table experiment [21–23] at the Ecological Botanical Gardens of the University of Bayreuth, Germany. The question was: How do growth and reproductive potential of cup plant differ depending on groundwater level? This study will not only provide insights into the demands and autecology of cup plant for the first time; the approach is also innovative because the findings are of great interest for nature conservation as well for agriculture.

2. Materials and Methods

2.1. Experimental Setup

The experiment was carried out from May 2018 to September 2019 in tanks similar to those of Ellenberg's Hohenheimer groundwater table experiment [21–23] at the Ecological Botanical Gardens of the University of Bayreuth (Germany, Bavaria). Temperature in the first growing season (May to August 2018) ranged from 1 °C to 35 °C (mean 18 °C) and precipitation sum was 151 mm, and in the second growing season (May to August 2019) between −3 °C and 37 °C (mean 17 °C) and 195 mm, respectively. Seeds of cup plant (Metzler & Brodmann Saaten GmbH, Ostrach, Germany, harvested 2016, pretreated) were sown on 5 March 2018. Seedlings were pricked out three weeks later and cultivated in a greenhouse. On 7 May 2018 the experiment started by planting the saplings into four tanks. For pricking and planting we chose vital plants of equal and mean size.

Each of the four tanks was a south-exposed, 6.4° inclined concrete tank (8 m × 4 m), with a constant soil depth of 90 cm (Figure 1). Substrate was a homogeneous mixture of 40% native soil, 40% compost and 20% quartz sand. In the lower part of each tank water was supplied via a garden hose and a perforated plastic pipe. Excess water could drain through a hole in the tank wall (Figures 1 and 2). The water table was held constant by hand in the first season and automatically by a float switch in the second growing season. Therefore, the plants in the tanks had different water table distances (WTD). In each tank, plants were arranged in nine rows indicating different WTD and in each row, there were nine plants. Distance between rows was 90 cm and between plants in a row 30 cm. For data collection we excluded all margin plants, resulting in seven rows of 28 plants each, divided across the four tanks. After the first growing season, we harvested in each row and in each tank the aboveground biomass of the second, the fourth and the sixth plant (seen from west), resulting in $n = 12$ per treatment (= row) (Figure 2). Afterwards, we removed the central part of the rootstock of these three and of the eighth plant. Consequently, in the second growing season five plants per row were left with distances of 60 cm between the plants. Excluding the margin plants, we had $n = 12$ per treatment (=row) again.

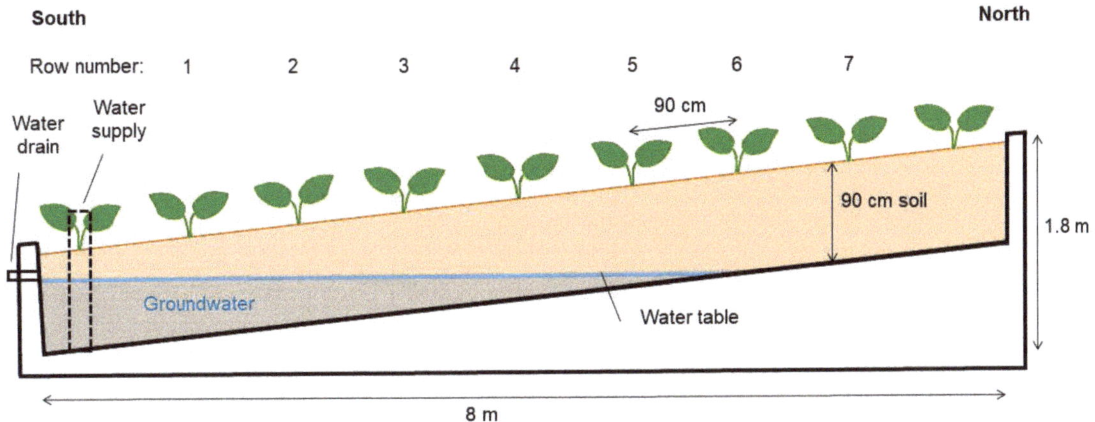

Figure 1. Scheme of a groundwater tank in longitudinal section.

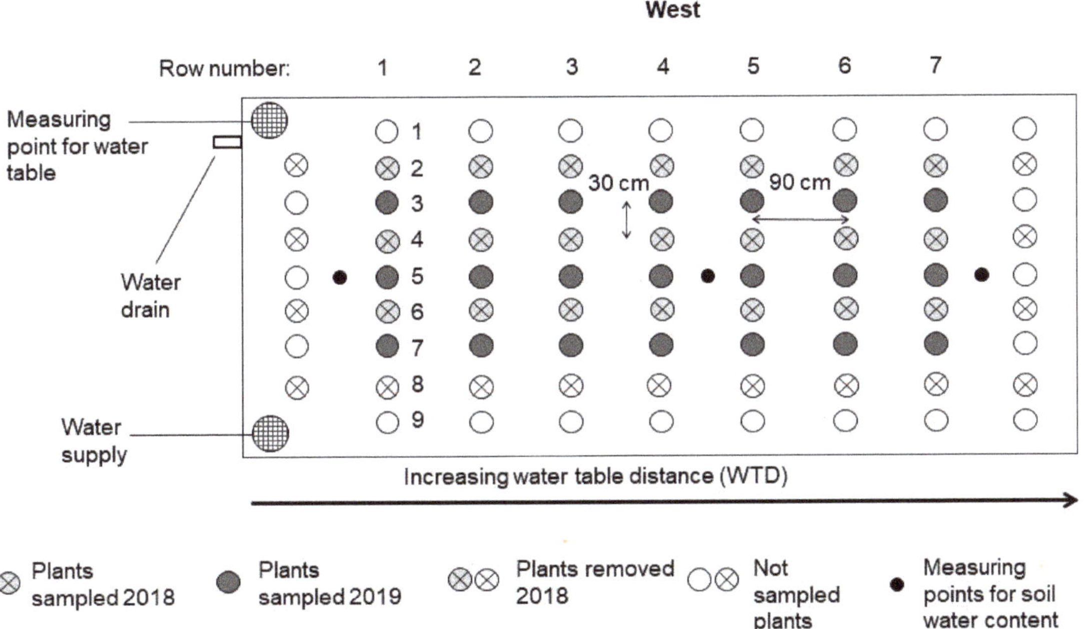

Figure 2. Top view of a groundwater tank indicating the plant arrangement, the harvesting scheme, and the measuring points for soil water content and water table.

2.2. Data Collection

In the first year, we harvested the living aboveground biomass on 9 August 2018, dried it in an oven at 70 °C until weight was constant, and measured the biomass with scales (PM 4600 Delta Range, Mettler-Toledo GmbH, Greifensee, Switzerland, same scales for all further weight measurement unless otherwise noted). Two samples of two different treatments had to be discarded because of plant material loss; consequently, the sample number for biomass of the first growing season was 82 instead of 84. In November 2018, the number of shoots higher than 15 cm of the individuals left in the tanks was counted.

In the second year, sampling and harvesting were carried out between 10 and 13 September 2019. We measured plant height by calculating the mean of the five highest shoots. The number of shoots higher than 50 cm was counted for each plant and was assigned to one of the three stages (1) dead, when more than 50% of biomass was brown, (2) vegetative, without capitula or with buds less than 1 cm in diameter, and (3) generative, with buds of at least 1 cm diameter or flowering or fruiting capitula, respectively. We harvested a representative subsample of the shoots, noting the shoot stages of the subsample, which included at a minimum one-third of the shoots from the original sample. Rosette leaves and shoots lower than 50 cm were harvested completely. The subsamples of the shoots higher than 50 cm were split in compartments of dead and living biomass whose fresh weights were measured. If necessary, again a subsample was taken and its fresh weight was measured before the biomass was dried in an oven at 90 °C until constant weight. Dry weight of biomass was measured and extrapolated to total biomass per plant. Before drying, capitula of the subsets were counted, assigned to the three phenological stages (1) budding, from a diameter of 1 cm on, (2) flowering, when ray florets were visible, and (3) fruiting, when ray florets were fallen off (comprising beginning of fruit development until fallen-off fruits) and extrapolated to the whole plant.

Additionally, we harvested three ripe capitula of the remaining shoots of each plant and dried them separately in paper bags at room temperature. We counted their number of fruits and weighed them with scales (AE 240, Mettler-Toledo GmbH, Greifensee, Switzerland).

Since 12 July 2018, water table was automatically recorded in a perforated pipe in the lower part of each of the four tanks (Figure 2) every 10 minutes by a pressure sensor (BayEOS HX711 Board, BayCEER, Bayreuth, Germany). For data analyses, we averaged water table for the time from the beginning of the water table measurement to particular data sampling for each tank. WTD was calculated using the mean water table, the inclination of the tanks (6.4°), and the distances to water table sensor for each row of each tank. Because soil depth was only 90 cm, plants with a calculated WTD larger than 90 cm had no vertical water access.

Soil water content was measured weekly in the second growing season (May to September 2019) with a TDR probe (TRIME®-FM3, IMKO Micromodultechnik GmbH, Ettlingen, Germany) in plastic pipes on three positions (Figure 2) and two depths (5–25 cm and 40–60 cm) in each tank. In November 2019, we took soil samples in 30 cm soil depth on two positions of each tank to assess the relationship between soil water content and soil water tension. We took an undisturbed soil sample with a core cutter of 100 cm^3 and a disturbed soil sample of about 200 cm^3. The undisturbed soil samples were saturated with water over five days and afterwards dried in an oven at 105 °C until constant weight. The disturbed soil samples were filled each in two sampling rings of 20 cm^3, placed in a pressure pot for 26 days at −15,000 hPa (pF = 4.2), and dried until constant weight. Before and between these steps, all soil samples were weighed each time (PB 3002 DeltaRange, Mettler-Toledo GmbH, Greifensee, Switzerland). After these steps, soil samples were weighed again (PG 503-S DeltaRange, Mettler-Toledo GmbH, Greifensee, Switzerland).

Meteorological data were obtained by a weather station in the Ecological Botanical Gardens 310 m away from the experimental site operated by the Micrometeorology group, Prof. Dr. Thomas, BayCEER, University of Bayreuth.

The data on which calculations in this study are based are available in the supplementary materials (Tables S1–S7).

2.3. Statistics

Data analysis and plot presentation were executed with R version 3.6.1 [24]. Calculating means of data by treatment, we used the function "ddply" from the R package "plyr" version 1.8.4 [25]. To read climate data and groundwater level data we used the R package "bayeos" version 1.4.6 [26]. We used linear models (LM) and checked the diagnostic plots. In case of non-normal distribution or heteroscedasticity of residuals we tried generalized linear models (GLM). If both LM and GLM were not possible, we executed Spearman's rank correlation analysis or the Kruskal–Wallis rank sum test (Kruskal test) with the post-hoc test multiple comparison test after Kruskal–Wallis (KruskalMC) of the R package "pgirmess" version 1.6.9 [27]. The four tanks were considered as four blocks in a block design. We checked the influence of block (tank) with an LM respectively a GLM. In case of non-significance we eliminated the block for the final model. In case of a significant effect of block we exerted a mixed effect model with block as random factor using the R package "lme4" version 1.1–21 [28]. Fits of mixed effect models were built using the mean of intercepts. Level of significance was always 0.05.

3. Results

3.1. Soil Water Conditions

The treatments of the experiment created by the rows in the tanks with increasing water table distance (WTD) described a wide range of soil water conditions (Table 1). Because soil depth was only 90 cm the two driest rows (6 and 7, Figure 1) had no direct vertical access to water table. Logically, soil water content decreased with increasing WTD, as well near soil surface as in deep soil layer. Water content of waterlogged soil was $50 \pm 2\%$ vol (mean $\pm$ standard deviation). Permanent wilting point (pF-value = 4.2) was reached at $9 \pm 2\%$vol water content.

Table 1. Soil water conditions depending on the treatments. Row number in tank is counted from the bottom up (Figure 2). Water table distance (WTD) is given as mean $\pm$ standard deviation for both years separately. Soil depth was 90 cm. Soil water content was measured weekly only in the second growing season (year 2019) at three positions in the tanks (see Figure 2). Given values for each row were calculated by the models described in Figure 3.

Row Number in Tank	First Year (2018)	Second Year (2019)			Classification
	WTD (cm)	WTD (cm)	Soil Water Content (%vol) in Depths		
			5–25 cm	40–60 cm	
1	41 ± 7	40 ± 11	38	57	wet
2	51 ± 7	50 ± 11	31	48	very moist
3	61 ± 7	60 ± 11	26	40	slightly moist
4	71 ± 7	70 ± 11	22	33	fresh
5	81 ± 7	80 ± 11	18	28	slightly dry
6	91 ± 7	90 ± 11	15	23	medium dry
7	101 ± 7	101 ± 11	12	19	rather dry

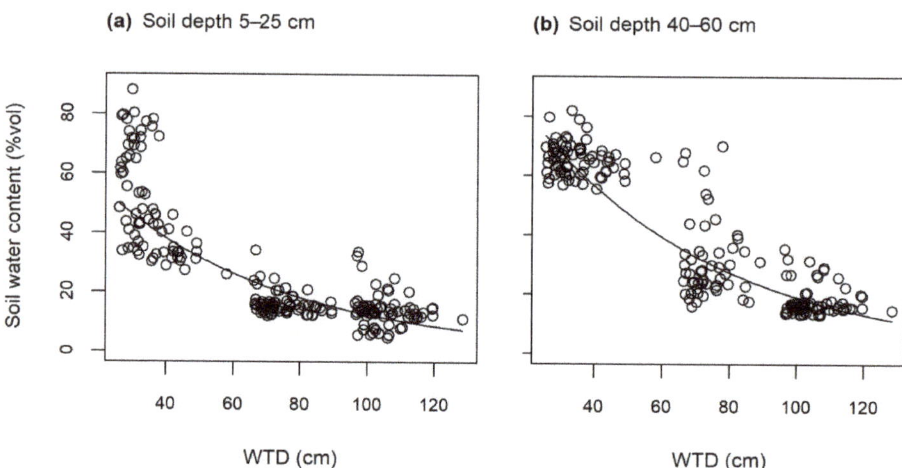

Figure 3. Soil water content in 5–25 cm (**a**) and 40–60 cm (**b**) soil depths depending on water table distance (WTD). Data were collected weekly in the second growing season from May to September 2019 at three positions in each tank (Figure 2). Lines are fitted by mixed effect LM: random effect = tank number (**a**): $\ln(y) = 4.39 - 0.02\,x$, $p < 0.001$, $n = 216$; (**b**): $\ln(y) = 4.77 - 0.02\,x$, $p < 0.001$, $n = 197$.

3.2. Growth and Aboveground Biomass

WTD had a significant effect on living aboveground biomass per plant in both years, although this effect was much smaller in the first than in the second year (Figure 4). Biomass was the highest at a WTD of around 50 to 60 cm (very to slightly moist soil) and achieved 167 ± 49 g in the first and 1491 ± 410 g in the second year (means $\pm$ standard deviation, dry weight). The latter was more than three times as high as in the driest treatment where only 458 g (mean, dry weight) living aboveground biomass per plant was measured. Living aboveground biomass was significantly determined by plant height (Spearman's rho = 0.75, $p < 0.001$) and not by number of shoots (Spearman's rho = 0.14, $p = 0.209$), considering the second year. Therefore, plant height was similarly affected by WTD as the living aboveground biomass and was between 135 and 335 cm (Figure 5). Plant height also reached its maximum at a WTD of around 50 to 60 cm (very to slightly moist soil) with 299 ± 18 cm in mean. Under wetter and drier soil conditions, plant height decreased.

Usually cup plant does not develop shoots before the second year [6]. However, in our experiment some individuals (34 of 84) had already developed one or more shoots (mean 1.6 ± 1.0) in the first growing season. This mainly occurred under moist soil conditions. Indeed, there was a significant correlation between number of shoot-developing individuals and row number of tank (Spearman's rho = -0.92, $p = 0.003$). In the second growing season, each individual independent of WTD developed from 8 to 32 shoots per plant (mean 18 ± 5). There was no significant effect of WTD on shoot number per plant (LM, $p = 0.714$).

All plants of all treatments grew and survived the two years of investigation in the experiment. However, at the end of the second growing season, a high portion of dead biomass in the three dry treatments was evident. There was in mean 23% and up to 73% (maximum) dead biomass in contrast to 6% (mean) in the wet, moist, and fresh treatments (Figure 6). There was a significant effect of WTD on the percentage of dead biomass.

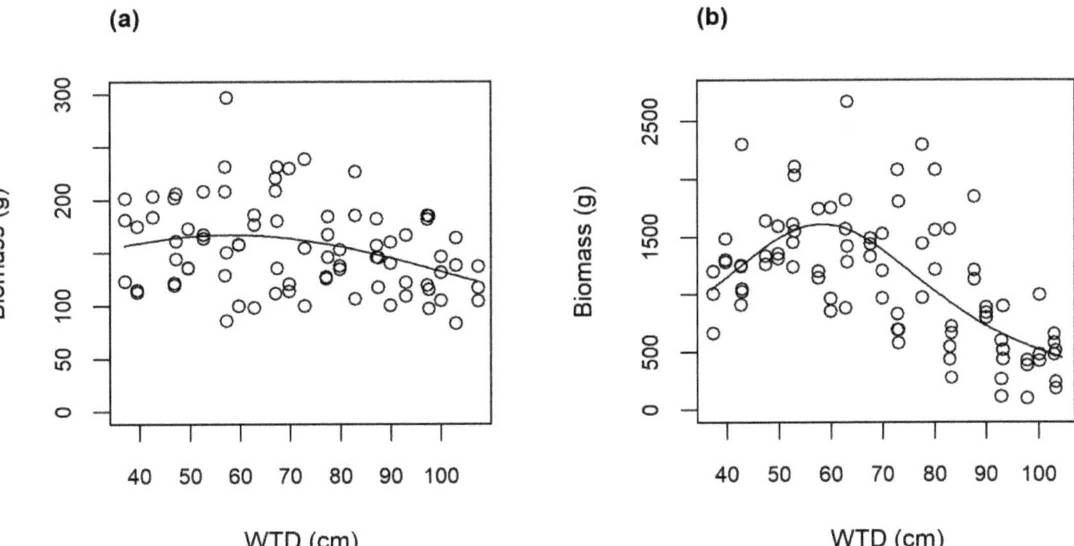

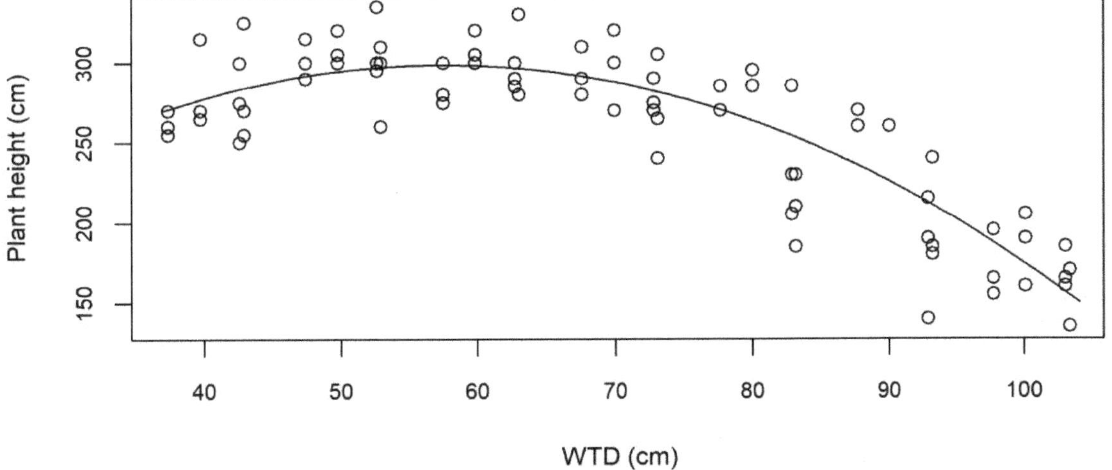

Figure 4. Living aboveground biomass (dry mass) per plant of cup plant, harvested (**a**): at the end of the first growing season (August 2018) and (**b**): at the end of the second growing season (September 2019) depending on water table distance (WTD). GLM: Gamma-distributed residuals, square function, (**a**): $p = 0.007$, $n = 82$; (**b**): $p < 0.001$, $n = 84$.

Figure 5. Plant height of cup plant at the end of the second growing season (September 2019) depending on water table distance (WTD). Mixed effect LM: random effect = tank number, square function, $n = 84$.

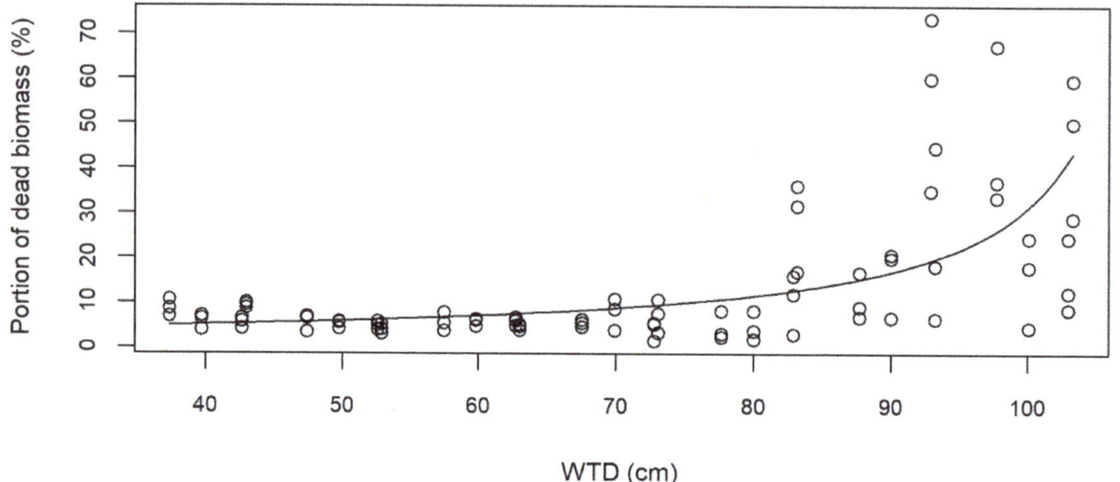

Figure 6. Portion of dead biomass of cup plant at the end of the second growing season (September 2019) depending on water table distance (WTD). GLM: Gamma- distributed residuals, *p* < 0.001, *n* = 84.

3.3. Reproductive Potential

There was a significant effect of WTD on the number of capitula at the end of the second growing season (Figure 7). The plants grown in very moist soil conditions (WTD around 50 cm) had the most capitula (mean ± standard deviation was 349 ± 156), whereas in wet soil conditions they developed slightly less (322 ± 143) and in rather dry soil conditions only a fifth (66 ± 115). In maximum one single plant developed 841 capitula (Figure 7). The number of fruits per capitulum was not affected by WTD (LM, *p* = 0.734) and was in mean 27 ± 4. The thousand grain weight was also not affected by WTD (mixed effect LM, *p* = 0.115) and was in mean 18.1 ± 3.9 g. Summing up, the plants grown in very moist to fresh soil conditions had a higher reproductive potential than those in dry or wet soil conditions because of a higher number of capitula. The number of capitula was significantly correlated with plant height (Spearman's rho = 0.64, *p* < 0.001) and not with number of shoots (Spearman's rho = −0.02, *p* = 0.849).

In September 2019, more than 90% of capitula of the plants grown in wet to fresh soil conditions had completed their flowering period and were already developing fruits (Figure 8). There was no significant difference between these four treatments regarding developmental stages of capitula (KruskalMC, *p* > 0.05). With increasing WTD the development slowed down. The drier the soil, the lower was the portion of fruiting capitula at the time of harvest and the higher was the portion of budding and flowering capitula. Regarding all treatments, there were significant correlations between row in tank and the portion of the three developmental stages of capitula (Spearman's rho for budding = 0.40, flowering = 0.68, fruiting = −0.76, *p* always < 0.001). Thus, plants on drier soil conditions not only produced less capitula (Figure 7) but took also longer to develop them.

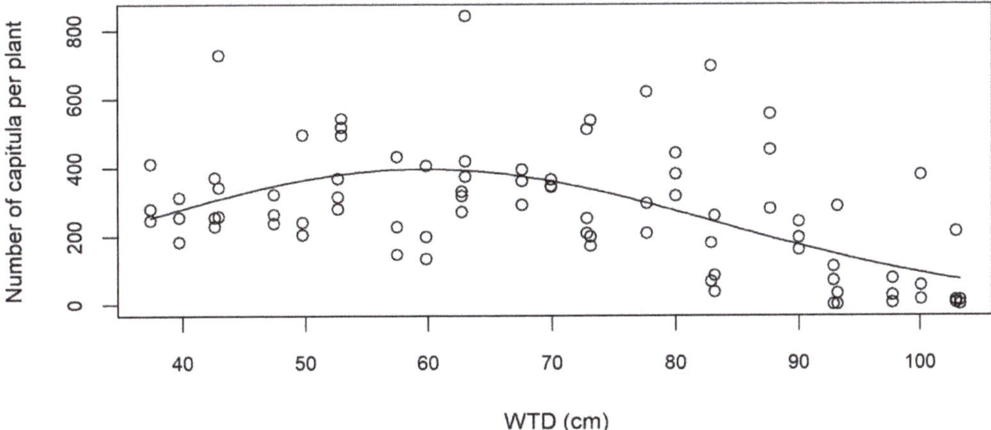

Figure 7. Number of capitula per plant of cup plant, regardless of their developmental stage, depending on water table distance (WTD). Data were collected at the end of the second growing season (September 2019). GLM: Poisson-distributed residuals, square function, $p < 0.001$, $n = 84$.

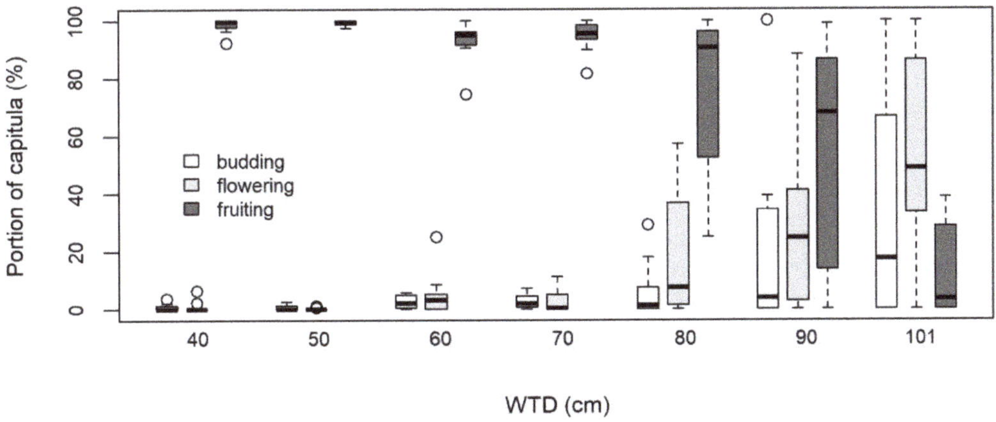

Figure 8. Portion of budding, flowering and fruiting capitula per plant of cup plant depending on water table distance (WTD, given as mean for each row of the four tanks). Data were collected at the end of the second growing season (September 2019). Note that the three different phenological stages of a treatment are always shown next to each other with a slight offset. $n = 12$ (in 90 cm WTD $n = 11$; in 101 cm WTD $n = 10$, because these three plants did not develop any capitulum at that time).

4. Discussion

4.1. Highest Yield on Moist Soil

Soil moisture conditions determined by water table distance (WTD) had a significant impact on growth and development of cup plant. Plant height and living aboveground biomass were the highest on moist soil with about 3 m and 1500 g dry weight per plant and decreased on wetter and especially on drier soil. Therefore, the plant height measured in our study reached the maximum published values for this species [19,29,30], indicating that optimal growth conditions were included in our study.

Several field studies and one pot experiment in Central Europe confirm our results that highest growth is achieved in periodically waterlogged or well-irrigated soil conditions [5,31,32]. In general, highest yields of cup plant are described on soils with good soil moisture; hydromorphic soils are unsuitable [3,6]. Cup plant is able to reach deep water resources with its roots; therefore it is considered as certainly drought-tolerant [3] Because of the limited soil depth in the present study, deep rooting was prevented and cup plant suffered considerable damage in the dry treatments. The number of shoots per plant was in mean 18 and not affected by soil moisture. This value is quite high compared to other studies, which indicated 3.5 to 6.6 shoots per plant in the second growing season [33,34]. Essential for this parameter are stand density and age of the plants [33,34] The fact that cup plant develops shoots already in the first year, as shown in our study has not been published so far. We assume that reasons are the sowing early in spring and the precultivation under optimal conditions in the greenhouse before planting into the experimental tanks.

A high yield of cup plant under moist soil conditions, as demonstrated in our study, is desirable from the farmers' point of view. However, from the perspective of invasion biology, this might carry the risk that spontaneously grown and established cup plants could also become such vigorous plants and might compete with native species. Studies are lacking but necessary to assess the competitiveness of cup plant and its possible risk of suppression of native species in case of spontaneous settlement. A species could be classified as invasive if its spread threatens biodiversity (Article 3 No. 2 EU-Regulation No. 1 143/2014).

4.2. High Reproduction and Rapid Development on Moist Soil

In the present study, cup plant produced the most capitula with about 350 on moist soil, and their development there was faster in comparison to drier soil conditions. Another study in Germany confirms our results, where the number of flowering capitula of cup plant was higher under irrigated than under rainfed conditions [35]. This study also agrees with ours concerning an independence of number of disc florets per capitulum in respect to watering. Although fruits of cup plant are developed from ray florets and not from disc florets [6], this agreement of results confirms that the composition of capitula is independent of soil moisture conditions. The number of fruits per capitulum was about 27 in the present study and therefore in the upper range or even above the values of other studies (10–20 or 20–30) [3,6]. Thousand grain weight varies widely in the literature (14 to 21.5 g [3] and up to 23 g [6]). Our values are with a mean of about 18.1 g rather in the middle.

The rapid development of fruits on moist soils leads to a high proportion of ripe fruits at harvesting time. Together with the high fruit production under these conditions, there is a higher risk of cup plant spreading from the fields—presupposed germination and saplings' establishment are likewise successful.

4.3. Consequences and Recommendations for Nature Conservation

Spontaneous occurrences of cup plant have already been documented in seven federal states of Germany and in other European countries as well [13–16]. From the view of nature conservation, the indication of colonized sites is important to assess the risk for protected or otherwise valuable ecosystems. In its native range in eastern North America, cup plant colonizes moist bottomlands, river valleys, and lakesides [3,6]. This is in line with our results and confirms a possible risk that cup plant could colonize moist habitats in Germany, too. So far, observations of spontaneous occurrences of cup plants in Germany have shown a broader range of habitats. In addition to ruderal places and woody structures, however, even moist ecosystems as perennial fields on river banks as well as bottomland woods are colonized [12,36,37]. This circumstance holds together with the high growth and reproductive potential on moist soils, as shown in our study a particular risk for nature conservation. Moist ecosystems—such as riparian fringes, alluvial forests, fens, and swamps—are valuable for nature conservation, because they are endangered according to the German red list of threatened habitats [38] and protected according to §30 BNatSchG.

Thus, an adequate distance of cup plant fields to moist ecosystems should be kept strictly to prevent their spontaneous colonization by cup plant. Dispersal distance of cup plant is 6 m in median but could be more than 10 m [12]. Therefore, we recommend for cup plant fields distances of several 10 m from valuable ecosystems to preclude fruit dispersal even under extreme wind events. However, dispersal vectors and distances of cup plant fruits are not investigated, and studies are urgently required to be able to give more precise recommendations for minimizing the risk of spreading. So far, it is also unknown whether cup plant fruits can be spread by watercourses and remain viable. As long as this is not examined, it is important to keep a sufficiently large distance to streams, even if they are strongly anthropogenic shaped and not valuable for nature conservation. In order to prevent fruit dispersal by agricultural machines, they should be cleaned before leaving the field and the crop should be covered during transport.

Additionally, the number and size of cup plant fields play a decisive role for the invasion potential because each newly cultivated field enhances the risk for further spontaneous spreading [12,39]. In Germany, more than 1000 ha are newly cultivated with cup plant in each of the recent years, while the older fields remain cultivated [40]. Consequently, further spreading of cup plant is to be expected and needs to be observed. The areas surrounding the cup plant fields and the roads from the fields to the farms should be continuously screened for spontaneous occurrences of cup plant to be able to combat this species where necessary.

5. Conclusions

In Central Europe, cup plant is a promising bioenergy crop that can achieve high yields, especially on moist soils. Wetter and drier soils are less suitable, but cup plant is able to survive on a wide range of soil moisture conditions.

However, a caution in respect to a possible invasiveness of cup plant is advised. The highest risk for spontaneous colonization by cup plant is—similar to the highest yield— supposed for ecosystems with moist soils, which are often valuable for nature conservation in Germany. To assess the actual invasive potential of cup plant, more studies about habitat requirements, competitiveness, and dispersal vectors of cup plant are urgently needed. If precautionary measures are observed, cup plant can take a place in the Central European agricultural landscape and make a valuable contribution to the conservation of biodiversity.

Supplementary Materials: The following tables contain the data on which calculations and figures of the present study are based. They are available online at https://www.mdpi.com/2077-0472/11/1/24/s1: Table S1: Weather conditions during the experiment; Table S2: Soil water content over time; Table S3: Soil water content of soil samples; Table S4: Growth of cup plant at the end of the first growing season (August 2018); Table S5: Shoot development of cup plant at the end of the first growing season (November 2018); Table S6: Growth of cup plant at the end of the second growing season (September 2019); Table S7: Explanation of column names of the Tables S1–S6.

Author Contributions: Conceptualization, M.L. and L.M.E.; methodology, L.M.E. and M.L.; validation, L.M.E. and M.L.; formal analysis, L.M.E.; investigation, K.K., M.K., A.J.F., and L.M.E.; data curation, L.M.E., M.K., K.K., and A.J.F.; writing—original draft preparation, L.M.E.; writing—review and editing, L.M.E., M.L., K.K., A.J.F., and M.K.; visualization, L.M.E.; supervision, L.M.E. and M.L.; project administration, L.M.E.; funding acquisition, L.M.E. and M.L. All authors have read and agreed to the published version of the manuscript.

Funding: This research was funded by the Oberfrankenstiftung and the District Government of Upper Franconia as well as the Studienstiftung des deutschen Volkes. The APC was funded by the German Research Foundation (DFG) and the University of Bayreuth in the funding program Open Access Publishing.

Institutional Review Board Statement: Not applicable.

Informed Consent Statement: Not applicable.

Data Availability Statement: The data presented in this study are available in the supplementary materials (Tables S1–S7).

Acknowledgments: We thank the Oberfrankenstiftung and the District Government of Upper Franconia for financial support as well as the Studienstiftung des deutschen Volkes for scholarship of the first author. Ralf Brodmann (Metzler & Brodmann Saaten GmbH) is thanked for free provision of cup plant seeds. We thank Andreas Schweiger (Plant Ecology, University of Hohenheim) for the support in developing the experimental design, Andreas Kolb (Soil Physics, University of Bayreuth) for the introduction to measurement technologies for soil water content and Oliver Archner as well as Stefan Holzheu (both Bayreuth Center of Ecology and Environmental Research BayCEER) for the installation of the measurement technology for water table. Special thanks are given to Frederik Werner, who supported us generously with plant cultivation and data collection as well as to the gardeners of the Ecological Botanical Gardens, who always assisted us with their expertise and manpower. Ursula Bundschuh (Soil Physics, University of Bayreuth) supported us with laboratory work of soil sampling. Micrometeorology group, University of Bayreuth and Bayreuth Center of Ecology and Environmental Research BayCEER is given thanks for providing meteorological data. Elisabeth Schaefer is thanked for English proof reading.

Conflicts of Interest: The authors declare no conflict of interest.

References

1. IRENA. *Renewable Energy Statistics 2019*; The International Renewable Energy Agency: Abu Dhabi, UAE, 2019; ISBN 978-92-9260-137-9.
2. Frölich, W.; Brodmann, R.; Metzler, T. The cup plant (*Silphium perfoliatum*)—A story of success from agricultural practice. *J. Cultiv. Plants* **2016**, *68*, 351–355. [CrossRef]
3. Gansberger, M.; Montgomery, L.F.R.; Liebhard, P. Botanical characteristics, crop management and potential of *Silphium perfoliatum* L. as a renewable resource for biogas production: A review. *Ind. Crops Prod.* **2015**, *63*, 362–372. [CrossRef]
4. Emmerling, C. Soil quality through the cultivation of perennial bioenergy crops by example of *Silphium perfoliatum*—An innovative agro-ecosystem in future. *J. Cultiv. Plants* **2016**, *68*, 399–406. [CrossRef]
5. Ruf, T.; Audu, V.; Holzhauser, K.; Emmerling, C. Bioenergy from Periodically Waterlogged Cropland in Europe: A First Assessment of the Potential of Five Perennial Energy Crops to Provide Biomass and Their Interactions with Soil. *Agronomy* **2019**, *9*, 374. [CrossRef]
6. Stanford, G. *Silphium perfoliatum* (cup-plant) as a new forage. In Proceedings of the Twelfth North American Prairie Conference, Cedar Falls, IA, USA, 5–9 August 1990; pp. 33–37.
7. Südwestrundfunk. Anbaufläche von Silphie ist größer geworden. *SWR Aktuell.* 5 September 2019. Available online: https://www.swr.de/swraktuell/baden-wuerttemberg/friedrichshafen/groessere-anbauflaeche-silphie-100.html (accessed on 15 May 2020).
8. Hartmann, A.; Lunenberg, T. Yield potential of cup plant under Bavarian cultivation conditions. *J. Cultiv. Plants* **2016**, *68*, 385–388. [CrossRef]
9. Burmeister, J.; Walter, R. Studies on the ecological effect of *Silphium perfoliatum* in Bavaria. *J. Cultiv. Plants* **2016**, *68*, 407–411. [CrossRef]
10. Dauber, J.; Müller, A.L.; Schittenhelm, S.; Schoo, B.; Schorpp, Q.; Schrader, S.; Schroetter, S. Schlussbericht zum Vorhaben: Agrarökologische Bewertung der Durchwachsenen Silphie (*Silphium perfoliatum* L.) als eine Biomassepflanze der Zukunft. 2016. Available online: https://literatur.thuenen.de/digbib_extern/dn056633.pdf (accessed on 13 March 2019).
11. Mueller, A.L.; Dauber, J. Hoverflies (Diptera: Syrphidae) benefit from a cultivation of the bioenergy crop *Silphium perfoliatum* L. (Asteraceae) depending on larval feeding type, landscape composition and crop management. *Agric. For. Entomol.* **2016**, *18*, 419–431. [CrossRef]
12. Ende, L.M.; Lauerer, M. Spontaneous occurrences of the cup plant in the Bayreuth region: Does this new bioenergy crop have invasive potential? *Nat. Landsch.* **2020**, *95*, 310–315. [CrossRef]
13. Roskov, Y.; Ower, G.; Orrell, T.; Nicolson, D.; Bailly, N.; Kirk, P.M.; Bourgoin, T.; DeWalt, R.E.; Decock, W.; van Nieukerken, E.; et al. Species 2000 & ITIS Catalogue of Life, 2019 Annual Checklist. Available online: http://www.catalogueoflife.org/annual-checklist/2019/details/species/id/d1c5f933225a3fa3a7974259458811ac (accessed on 7 February 2020).
14. Brennenstuhl, G. Beobachtungen zur Einbürgerung von Gartenflüchtlingen im Raum Salzwedel (Altmark). *Mitt. Florist. Kart. Sachs. Anhalt.* **2010**, *15*, 121–134.
15. Wölfel, U. Zur Flora von Bitterfeld und Umgebung (11. Beitrag). *Mitt. Florist. Kart. Sachs. Anhalt* **2013**, *18*, 47–53.
16. Verbreitung der Farn- und Blütenpflanzen in Deutschland; aggregiert im Raster der Topographischen Karte 1:25000—*Silphium perfoliatum*. Available online: http://www.floraweb.de/webkarten/karte.html?taxnr=5635 (accessed on 12 March 2019).
17. Matthews, J.; Beringen, R.; Huijbregts, M.A.J.; van der Mheen, H.J.; Odé, B.; Trindade, L.; van Valkenburg, J.L.C.H.; van der Velde, G.; Leuven, R.S.E.W. *Horizon Scanning and Environmental Risk Analyses of Non-Native Biomass Crops in the Netherlands*; Department of Environmental Science, Institute for Water and Wetland Research, Radboud University Nijmegen: Nijmegen, The Netherlands, 2015.
18. Vinogradova, Y.K.; Mayorov, S.R.; Bochkin, V.D. Changes in the spontaneous flora of the Main Botanic Garden, Moscow, over 65 years. *Skvortsovia* **2015**, *2*, 45–95.

19. Bufe, C.; Korevaar, H. *Evaluation of Additional Crops for Dutch List of Ecological Focus Area*; Wageningen Research Foundation (WR) business unit Agrosystems Research: Wageningen, The Netherlands, 2018. [CrossRef]

20. Schoo, B.; Wittich, K.P.; Böttcher, U.; Kage, H.; Schittenhelm, S. Drought Tolerance and Water-Use Efficiency of Biogas Crops: A Comparison of Cup Plant, Maize and Lucerne-Grass. *J. Agron. Crop Sci.* **2017**, *203*, 117–130. [CrossRef]

21. Ellenberg, H. Physiologisches und ökologisches Verhalten derselben Pflanzenarten. *Ber. Deutsch. Bot. Ges.* **1952**, *65*, 350–361.

22. Ellenberg, H. Über einige Fortschritte der kausalen Vegetationskunde. *Plant Ecol.* **1954**, *5*, 199–211. [CrossRef]

23. Hector, A.; von Felten, S.; Hautier, Y.; Weilenmann, M.; Bruelheide, H. Effects of dominance and diversity on productivity along Ellenberg's experimental water table gradients. *PLoS ONE* **2012**, *7*, e43358. [CrossRef] [PubMed]

24. R Core Team. *R: A Language and Environment for Statistical Computing*; R Foundation for Statistical Computing: Vienna, Austria, 2019.

25. Wickham, H. The Split-Apply-Combine Strategy for Data Analysis. *J. Stat. Softw.* **2011**, *40*, 1–29. [CrossRef]

26. Holzheu, S.; Archner, O. *BayEOS: BayEOS Server Access*; R package version 1.4.6.: Bayreuth, Germany, 2017.

27. Giraudoux, P. *Pgirmess: Spatial Analyses and Data Mining for Field Ecologists*; R Package version 1.6.9.: Besançon, France, 2018.

28. Bates, D.; Mächler, M.; Bolker, B.; Walker, S. Fitting Linear Mixed-Effects Models Using lme4. *J. Stat. Soft.* **2015**, *67*. [CrossRef]

29. Blüthner, W.-D.; Krähmer, A.; Hänsch, K.-T. Breeding progress in cup plant-first steps. *J. Cultiv. Plants* **2016**, *68*, 392–398. [CrossRef]

30. Wrobel, M.; Fraczek, J.; Francik, S.; Slipek, Z.; Krzysztof, M. Influence of degree of fragmentation on chosen quality parameters of briquette made from biomass of cup plant *Silphium perfoliatum* L. *Eng. Rural Dev.* **2013**, *12*, 653–657.

31. Schittenhelm, S.; Schoo, B.; Schroetter, S. Yield physiology of biogas crops: Comparison of cup plant, maize, and lucerne-grass. *J. Cultiv. Plants* **2016**, *68*, 378–384. [CrossRef]

32. Figas, A.; Siwik-Ziomek, A.; Rolbiecki, R. Effect of irrigation on some growth parameters of cup plant and dehydrogenase activity in soil. *Ann. Warsaw Univ. Life Sci. SGGW Land Reclam.* **2015**, *47*, 279–288. [CrossRef]

33. Boe, A.; Albrecht, K.A.; Johnson, P.J.; Wu, J. Biomass Production of Cup Plant (*Silphium perfoliatum* L.) in Response to Variation in Plant Population Density in the North Central USA. *Am. J. Plant Sci.* **2019**, *10*, 904–910. [CrossRef]

34. Bury, M.; Możdżer, E.; Kitczak, T.; Siwek, H.; Włodarczyk, M. Yields, Caloric Value and Chemical Properties of Cup Plant *Silphium perfoliatum* L. Biomass, Depending on the Method of Establishing the Plantation. *Agronomy* **2020**, *10*, 851. [CrossRef]

35. Mueller, A.L.; Berger, C.A.; Schittenhelm, S.; Stever-Schoo, B.; Dauber, J. Water availability affects nectar sugar production and insect visitation of the cup plant *Silphium perfoliatum* L. (Asteraceae). *J. Agron. Crop Sci.* **2020**. [CrossRef]

36. *Rothmaler—Exkursionsflora von Deutschland. Gefäßpflanzen: Grundband*, 21st ed.; Jäger, E.J. (Ed.) Springer Spektrum: Berlin, Germany, 2017; ISBN 978-3-662-49707-4.

37. *Schmeil—Fitschen: Die Flora Deutschlands und Angrenzender Länder*, 96th ed.; Parolly, G.; Rohwer, J.G. (Eds.) Quelle & Meyer Verlag QM: Wiebelsheim, Germany, 2016; ISBN 9783494017303.

38. Finck, P.; Heinze, S.; Raths, U.; Riecken, U.; Ssymank, A. Rote Liste der gefährdeten Biotoptypen Deutschlands. *Natursch. Biol. Vielf.* **2017**, *156*, 637.

39. Kowarik, I. Humann agency in biological invasions: Secondary releases foster naturalisation and population expansion of alien plant species. *Biol. Invasions* **2003**, *5*, 293–312. [CrossRef]

40. Donau-Silphie. Available online: https://www.donau-silphie.de/ (accessed on 9 September 2020).

Article

Can *Miscanthus* Fulfill Its Expectations as an Energy Biomass Source in the Current Conditions of the Czech Republic?—Potentials and Barriers

Jan Weger [1,*], Jaroslav Knápek [2], Jaroslav Bubeník [1], Kamila Vávrová [1] and Zdeněk Strašil [3]

1   Silva Tarouca Research Institute for Landscape and Ornamental Gardening, Public Research Institute (VÚKOZ), 252 43 Průhonice, Czech Republic; bubenik@vukoz.cz (J.B.); vavrova@vukoz.cz (K.V.)
2   Faculty of Electrical Engineering (ČVUL FEL), Department of Economics, Management and Humanities, Czech Technical University in Prague, 166 27 Prague, Czech Republic; knapek@fel.cvut.cz
3   Crop Research Institute, Public Research Institute (VÚRV), 161 00 Prague-Ruzyně, Czech Republic; pavelstrasil@mpsprojekt.cz
*   Correspondence: weger@vukoz.cz; Tel.: +420-605-205-995

Abstract: Our article analyzes the main biological potentials and economic barriers of using *Miscanthus* as a new energy crop in agricultural practice in the Czech Republic and the Central-Eastern European region. We have used primary data from long-term field experiments and commercial plantations to create production and economic models that also include an analysis of competitive ability with conventional crops. Our results showed that current economic conditions favor annual crops over *Miscanthus* (for energy biomass) and that this new crop shows very good adaptation to the effects of climate change. Selected clones of *Miscanthus* × *giganteus* reached high biomass yields between 15–17 t DM ha^{-1} y^{-1} despite very dry and warm periods and low-input agrotechnology, and they have good potential to become important biomass crops for future bioenergy and the bioeconomy. Key barriers and factors are identified, including gene pool and agronomy improvement, change of subsidy policy (Common agriculture policy-CAP), climate change trends, and further development of the bioeconomy.

Keywords: *Miscanthus*; energy biomass; yields; invasive behavior; economics

Citation: Weger, J.; Knápek, J.; Bubeník, J.; Vávrová, K.; Strašil, Z. Can *Miscanthus* Fulfill Its Expectations as an Energy Biomass Source in the Current Conditions of the Czech Republic?—Potentials and Barriers. *Agriculture* **2021**, *11*, 40. https://doi.org/10.3390/agriculture11010040

Received: 30 October 2020
Accepted: 29 December 2020
Published: 8 January 2021

Publisher's Note: MDPI stays neutral with regard to jurisdictional clai-ms in published maps and institutio-nal affiliations.

1. Introduction

From the enlargement of the European Union (EU) in 2004 and 2007 to include ten Central and Eastern European (CEE) countries, the share of renewable energy sources in final gross consumption in the EU27 countries has almost doubled from 113 to 220 Mtoe, reaching 18.9% in 2018. Biomass continues to be the most important type of renewable energy source (RES); it contributed 59.2% to the total share of RES in final gross energy consumption in 2018 [1,2]. The total biomass utilization (for solid, liquid, and gaseous biofuels) has been growing in absolute terms since the EU enlargement (from 69 to 135 Mtoe) and is expected to continue to play a significant role in the gradual replacement of fossil fuels [3,4]. For instance, the supply of biomass from agricultural and perennial energy crops would need to increase by 29% to fulfill the ambitious goals of National Renewable Energy Action Plans [2]. These trends are expected to continue under The European Green Deal proposed by the current European Commission [5].

Of all the biomass types, solid biomass is currently most frequently used; in 2017, for the EU as a whole, 69% of the total contribution of biomass for energy from RES was solid biomass. One example of biomass being expected to contribute significantly to meeting 2030 targets is the Czech Republic, whose National Energy Action Plan foresees an increase in the share of RES in final gross consumption from about 15% (2019) to 22% in 2030 and the use of solid biomass for heating to increase by 26% (or 27 PJ) [3].

In CEE countries, of all the renewables, the one with the highest potential is biomass, especially when both residual and intentionally grown biomass sources on agricultural land have been included in the calculations [6]. Second-generation energy crops are an especially promising source for the future. These include selected clones or varieties of fast-growing trees (poplar, willow), perennial and some annual grasses (reed canary grass, *Miscanthus* ssp., triticale), and perennial rhizomatous plants (Schavnat—a spinach-sorrel hybrid, *Sida*—Virginia mallow). These non-food and high yielding crops have a much better energy input/output ratio (five to seven times) than first-generation crops like rape or cereals in regards to how much energy biomass is produced per hectare [7–9].

1.1. Experience with Miscanthus and Energy Crops

In previous analyses, it was expected that second-generation energy crops would be planted on larger areas of agricultural land [10–12] because of the growing demand for energy biomass, and the available political and financial support, although mostly indirect, e.g., subsidized price of electricity from intentionally grown biomass. It was expected that sufficient agricultural land for these plantations, including less productive lands for food crops [13], but also some better quality soils would be available because trends have shown that there is an overproduction of food crops (especially cereals) in the Czech Republic and other CEE countries [14,15].

Nevertheless, the potential of second-generation crops has not been realized as these crops are currently grown only on 0.14% of agricultural land in the Czech Republic. A similar situation can be found in other neighboring countries (Poland, Germany, Austria, Slovakia). For instance, in Germany, the land area suitable for *Miscanthus* has been assessed to be about 4 million ha [16], but the growing area is about 4000 ha [17], e.g., 0.03% of agricultural land. In Poland, according to the Polskie Towarzystwo Biomasy (POLBIOM, Warsaw) association a member of Bioenergy Europe (Brussels, Belgium), *Miscanthus* was grown on approximately 2000 ha from 2009–2011, but this number decreased significantly in the following years and was 733 ha in 2013 [18]. The current area could be even smaller (below 500 ha) because many farmers decided to cease *Miscanthus* plantations for energy biomass due to economic reasons [19]. From the experience of several European regions, it seems that the less favorable economic profitability of growing perennial *Miscanthus* for energy (direct burning) has been the crucial reason why many farmers have ceased planting it and returned to growing annual food crops.

Since the 1990s, many bioenergy projects have focused on planting second-generation crops for energy biomass (direct combustion or pyrolysis)—first, in Western and later, Eastern European regions [20–24]. In the Czech Republic, planting first started in the early 2000s with energy crops that can be easily planted, respectively seeded with standard agricultural mechanization, e.g., Schavnat and triticale (Figure 1) supported by relatively high national subsidies for the establishment of non-woody energy crops. By 2007, there were about 1200 hectares of these crops, but after full adoption of the EU's financial support scheme of the Common Agricultural Policy, when national subsidies no longer applied, the area of planted energy crops decreased. There were also often high losses in newly established plantations partially caused by the low quality of planting material and poor agrotechnology, unsuitable site selection, and sometimes, by extreme climate conditions.

Development of fast-growing trees and *Miscanthus* started in CEE countries after they entered the EU when legislative and 'agro-logistical' conditions (planting material, mechanization) have improved for farmers interested in new biomass crops. While dynamic planting of fast-growing trees (mainly poplar clone Max-4) continued until 2015 when new legislation on soil protection and decreased biomass co-firing has stopped this development, *Miscanthus* never became a 'new biomass crop' attractive enough for pioneering farmers (see Figure 1). The only large *Miscanthus* project in the Czech Republic was carried out by a joint-venture of *Miscanthus* Ltd., of the Pilsen energy company (Pilsen, Czech Republic) and regional forest enterprise, which was terminated after 3–4 years for various reasons, including failures in establishing *Miscanthus* plantations and slower growth than expected.

The Czech Republic's current planting area with *Miscanthus* is around 50 ha and is used mostly to produce animal bedding material, including pellets [25].

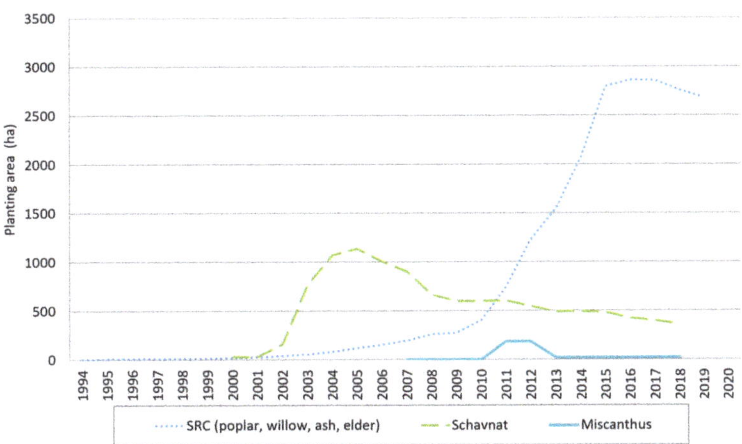

Figure 1. Development of planting area of selected second-generation energy crops in the Czech Republic (different sources compiled by Weger: primary data from annual reports of crops cultivated on soil blocks [MS Excel] of The State Agricultural Intervention Fund, Prague, Czech Republic.

1.2. Miscanthus Resources and Services

The species and varieties of genus *Miscanthus* Andersson consist of 15 taxa of C4 rhizomatous grasses [26]. It naturally occurs in tropical and moderate climate regions of eastern Asia and southeastern Africa. *Miscanthus* ssp. has been planted in many different world regions for ornamental, material, and energy purposes. Commercial and research organizations have produced many genotypes for these purposes. Present-day clones and varieties yield over 55 tons of raw biomass per hectare per year in favorable conditions [13,27]. Currently, the most commonly planted clone in Europe is *M.* × *giganteus* Greef et. Deu ex Hodkinson et Renvoize. It is a sterile triploid hybrid between diploid *M. sinensis* and tetraploid *M. sacchariflorus* [28,29].

Miscanthus straw contains about 41–45% cellulose, 20.6–33% hemicellulose and 19–23.4% lignin [30,31] and has a good heating value (17.6 GJ t^{-1} at 0% moisture respectively 13.60 GJ t^{-1} at average harvesting moisture of 20%). It is also a suitable source for the production of pulp [32]. *Miscanthus* biomass can also be used to produce construction materials [21,33,34] or composted together with cattle or pig slurry [35].

Due to *Miscanthus'* environmental benefits (soil protection, crop diversification), it is especially a suitable alternative in places where food crops are not productive or planted. *Miscanthus*, similar to fast-growing trees [36], does not require high doses of industrial fertilizers [37–39], nor does it tend to be susceptible to diseases or harmful organisms [40,41]. Lower levels of applied fertilizers and pesticides decrease the risk of soil and groundwater contamination. The risk of soil erosion from water or wind in *Miscanthus* plantations is serious only during the first year or two after establishment when the root system has not fully developed, and production of phytomass, especially leaf litter, covering the soil surface is limited. *Miscanthus* can also be planted on grasslands with no-tillage agronomy, thus preventing the loss of biodiversity and soil carbon during establishment [42].

There are also ecological reasons for growing *Miscanthus*, e.g., decreasing the difference between the lowest and highest soil temperatures; facilitating better soil-water management; reducing soil erosion from water and wind; possibly improving the content of the soil's organic matter [43–48].

The main risks for the successful establishment of plantations include low quality of rhizomes, insufficient weeding, site selection (dry or waterlogged) and extreme winter temperatures below minus 3.5 °C in the rhizome zone of the soil [49].

1.3. Invasive Risks of Miscanthus

Miscanthus sinensis has been registered as a weedy or invasive plant in many regions worldwide, e.g., the USA, Canada and Australia [50,51]. An example of its invasiveness can be found in the results of [52], who state that in all experimental field plots, nearly all species and cultivars in the Trinity College Botanic Garden collection in Dublin (Ireland) create viable seeds. Seed viability was also confirmed in genotypes that were considered to be sterile. Seed germination tests of selected decorative *Miscanthus* clones proved that not all genotypes of *Miscanthus sinensis* [29] and other species have the same invasive potential [53]. For example, in eastern parts of the USA, the relatively high invasive potential of self-established local populations of *Miscanthus sinensis* was thoroughly studied; at least four of six populations have origins in nearby growths of ornamental *Miscanthus* plants that were established 20 to 120 years before [54]. From the invasiveness perspective, *Miscanthus* × *giganteus*, a sterile triploid hybrid, is not considered a risky crop [55,56]. Only a few publications mention the uncontrolled spreading or escape of *Miscanthus* ssp. plants, especially *M. sinensis* and *M. sacchariflorus* from field growths in the Czech Republic and Central Europe [57–60], but more articles can be found in other mild-climate countries.

1.4. Objectives of the Article—Potentials and Barriers of Miscanthus as an Energy Crop

Despite the positive production and environmental characteristics partially described above, several barriers and limitations have prevented *Miscanthus* from becoming suitable as a (still) new crop for energy biomass in the Czech and European conditions.

Therefore, in the article, we have focused on the key aspects of *Miscanthus* plantations—production, invasive risk, and economy—to identify the main potentials, barriers, and recommendations for *Miscanthus* to develop as a standard agriculture crop in future farming practice for energy or possibly other sectors of the bioeconomy.

The countries of the CEE region have similar economic conditions—especially the cost of human labor, land costs, and at the same time, have similar conditions in terms of agricultural subsidies. If we assume similar climatic conditions and the use of similar agrotechnical procedures, then similar conclusions can be expected in terms of the economic efficiency of *Miscanthus* cultivation.

2. Materials and Methods

Different methodologies, as described in the following chapters, were used to evaluate the main potentials and risks of producing *Miscanthus*. We have mainly analyzed primary data from our field experiments to evaluate production potential and invasive risk. These data were then compared and confirmed with data from commercial plantations and scientific literature from areas with similar growing conditions.

To analyze the economic efficiency of *Miscanthus* cultivation, we use a modeling approach based on the real economic conditions in the Czech Republic at the price level of 2019. We have also added the aspect of competition with conventional agricultural crops into our economic model.

Since 1989, numerous field experiments have been established to evaluate yield and growth of *Miscanthus* species and genotypes in EU countries within European international programs and projects (JOULE, FAIR) [49]. The European *Miscanthus* Improvement Project (EMI, 1996–1999) also started a breeding program to improve yield and growth of this 'novel crop' for European conditions. Based on the EMI results and with support of the consortium members, we have established experiments to evaluate *Miscanthus* potential in Czech environmental and economic conditions—first, the genotype collection at the Průhonice-Zelinářská zahrada location in 2002 and five years later, the clonal test at two locations: Průhonice-Michovky and Lukavec in 2007 (Figure 2). The distance between the genotype collection and the clonal field experiment in Průhonice is 1.4 km.

Figure 2. Pictures of experimental sites with *Miscanthus*: (**a**) Genotype collection Průhonice-Zelinářská zahrada with marked seedlings of *Miscanthus sinensis* invading the neighboring nursery in 2008; (**b**) Clonal experiment in Lukavec before spring harvest 2019; (**c**) Clonal experiment in Lukavec—lodging of *Miscanthus sinensis* (clone M4 GOFAL) under early snow in October 2009; (**d**) Clonal experiment in Průhonice-Michovky in October 2014.

2.1. Climatic and Soil Conditions of Experimental Sites

Soil and climatic conditions at both locations, Lukavec and Průhonice, are described in Table 1. Experiments in Průhonice are located between 310–330 m above sea level at a flat area and without exposition. Soil is Cambisol [61]. The mean year temperature is 8.8 °C. Mean sum of precipitation is 580 mm. The Průhonice sites are in a region where cereals are produced, while Lukavec is a potato-producing region. Changes in weather parameters during the experiment and their comparison using long-term temperature and precipitation averages (1961–1990) [62] are shown in Table 2.

Table 1. Site conditions at locations of experiments (before planting, 2007, respectively 2002).

Factor	Průhonice Michovky	Průhonice Zelinářská Zahrada	Lukavec
Latitude	49°59′	49°59′	49°34′
Longitude	14°34′	14°34′	14°59′
Altitude (m)	332	310	570
Soil texture	clay-loess	clay-loess	sandy-loam
Soil type	Cambisol	Cambisol	Cambisol
Mean year temperature (°C)	8.8	8.8	7.3
Mean year sum of precipitation (mm)	580	580	682
Agrochemical properties of soil (before establishment):			
Content of humus (%)	1.0	1.3	3.4
pH (H_2O)	6.22	7.11	6.14
Content of P (Mehlich III, mg.kg^{-1})	54	395	112
Content of Mg (Mehlich III, mg.kg^{-1})	179	190	114
Content of K (Mehlich III, mg.kg^{-1})	143	354	337

Table 2. Annual temperatures and precipitations in Průhonice and Lukavec between 2007–2017.

No.	Year	Average Annual Temperature		Annual Sum of Precipitation	
		Průhonice	Lukavec	Průhonice	Lukavec
-		(°C)	(°C)	(mm)	(mm)
1	2007	10.2	8.5	517	777
2	2008	9.8	8.6	502	604
3	2009	9.4	8.4	599	788
4	2010	8.0	7.2	764	940
5	2011	9.6	8.4	563	670
6	2012	9.5	8.1	553	724
7	2013	9.0	7.3	667	876
8	2014	10.6	8.6	548	707
9	2015	10.7	8.7	427	576
10	2016	10.0	7.9	499	601
11	2017	10.0	7.9	553	777
12	2018	11.0	8.8	354	509
13	2019	10.7	9.1	521	680
	Average	9.9	8.3	544	710

2.2. Assortment and Design of the Clonal Experiment

The clonal experiment was established with four clones of *Miscanthus sinensis* and two clones of the triploid hybrid *Miscanthus × giganteus*. The clones come from the national collection (Table 3). The experiment was planted using *Miscanthus* rhizomes that were at least 70 mm long. As shown in Figure 3, the experiment's design includes 4 random repetitions, a 1 × 1 m planting scheme and a density of one plant per 1 m². There are 18 plants of one clone in an individual plot, respectively 36 plants for clones M1 and M6 (double-sized plots). Additionally, the clone *Miscanthus × giganteus* (M12) from the Crop Research Institute in Praha-Ruzyně was planted as an isolation row around the experiment. Plants were measured every year: number and height of plants, number of stems, and fresh weight of biomass (straw) of each plant. Twice as many rhizomes of clones M1 and M6 were planted (36 plants per plot) to compare autumn and spring harvest.

Table 3. Assortment of *Miscanthus* clones included in the experiments.

Clone No.	Clone Code	Taxonomical Classification	Origin	Number of Individuals	Number of Individuals
-		Clonal field experiment		Průhonice Michovky	Lukavec
M1	M-GigM53-003	*M. × giganteus*	Germany	144	144
M2	M-GigFou-009	*M. × giganteus*	Denmark	72	72
M3	M-sin902-005	*M. sinensis*	Denmark	72	72
M4	M-sinGOF-002	*M. sinensis*	Germany	72	72
M5	M-sin903-006	*M. sinensis*	Denmark	72	72
M6	M-sinM43-004	*M. sinensis*	Germany	144	144
M12 **	M-GigVUR-012	*M. × giganteus*	Czech Rep.	100	100
-		Genotype collection		Průhonice Zelinářská zahrada	-
M7 *	M-sin101-007	*M. sinensis*	Denmark	27	-
M8 *	M-sin906-008	*M. sinensis*	Denmark	27	-
M9 *	M-GigFou-009	*M. × giganteus*	Denmark	27	-
M10 *	M-sacHon-010	*M. sacchariflorus*	Denmark	27	-
M11 *	'Goliath'	*M. sinensis*	Czech Rep.	27	-
M13 *	M-sinJes-001	*M. sinensis*	Germany	27	-

* Additional clones in the genotype collection used for monitoring invasive behavior; ** used only in isolation row of clonal field experiment (Průhonice-Michovky).

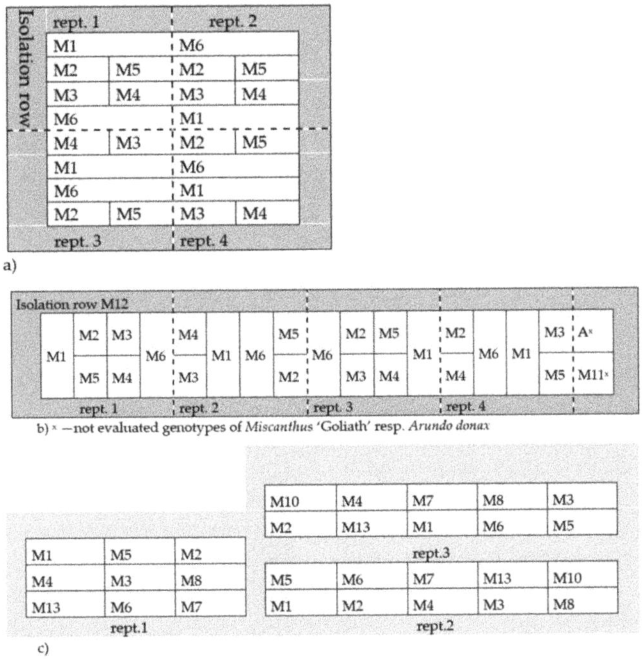

Figure 3 scheme **(a)** — Scheme of the clonal experiment in Lukavec:

Isolation row / Isolation row	rept. 1		rept. 2	
	M1		M6	
	M2	M5	M2	M5
	M3	M4	M3	M4
	M6		M1	
	M4	M3	M2	M5
	M1		M6	
	M6		M1	
	M2	M5	M3	M4
	rept. 3		rept. 4	

a)

Figure 3 scheme **(b)** — Scheme of the clonal experiment in Průhonice-Michovky (Isolation row M12):

	rept. 1			rept. 2			rept. 3			rept. 4			
M1	M2	M3	M6	M4	M1	M6	M5	M6	M2	M5	M1	M6	M1	M3	A×
	M5	M4		M3			M2		M3	M4	M4			M5	M11×

b) × —not evaluated genotypes of *Miscanthus* 'Goliath' resp. *Arundo donax*

Figure 3 scheme **(c)** — Genotype collection Průhonice-Zelinářská zahrada:

rept.1

M1	M5	M2
M4	M3	M8
M13	M6	M7

rept.3

M10	M4	M7	M8	M3
M2	M13	M1	M6	M5

rept.2

M5	M6	M7	M13	M10
M1	M2	M4	M3	M8

c)

Figure 3. Design of field experiments used for data collection: (**a**) Scheme of the clonal experiment in Lukavec; (**b**) Scheme of the clonal experiment in Průhonice-Michovky; (**c**) Genotype collection Průhonice-Zelinářská zahrada; rept.1–4 are repetitions of field experiments.

2.3. Establishment and Maintenance of the Clonal Experiment

The experimental field was plowed and harrowed in autumn 2006 and leveled in spring 2007 to be ready for manual planting of *Miscanthus* rhizomes in May 2007. The design of the experiment was in a semi-randomized block design (Figure 3). Rhizomes were collected from the genotype collection a few days before planting and planted on the 4th May in Lukavec and 15th May 2007 in Průhonice.

For *Miscanthus × giganteus*, we used one standard rhizome (70–80 mm) and for *Miscanthus sinensis*, which has thinner rhizomes, we used two rhizomes in one 3–5 cm deep hole. In the first and second years after planting rhizomes, the plantations were weeded manually. Herbicides were not used. Later, when the *Miscanthus* plants grew fast enough, they did not need any weeding. Potential weeds were also suppressed by rich leaf fall.

As we aimed to study *Miscanthus* production with minimized inputs, no fertilizers were used at the Průhonice sites before establishment and during growth at both sites. The Lukavec site field was fertilized once before establishment using 70 kg ha^{-1} K (potassium salt) and 40 kg ha^{-1} P (superphosphate).

2.4. Evaluation of Yield and Biometric Characteristics

The following biometric parameters were measured in the clonal experiment: number of stems, the height of plants, and volumes of fresh biomass. The number of plants was counted before harvests (autumn and spring). The height of individual plants was measured from the ground surface to its highest point (of straw) in an upright position. All biometric parameters were measured a few days before the autumn and spring harvests.

Harvests were performed at both sites, usually between November–December for autumn harvest and March–April for spring harvest. In autumn and spring, individual plants were first cut down using a brush cutter, and immediately, each plant was weighed on digital scales with an accuracy of ±5 g to obtain fresh weight. One fresh biomass sample was taken from each harvested clone to analyze moisture content and calculate dry mass yield. These samples were dried at 105 °C until constant weight. At this temperature, the energy (fuel) characteristics of biomass are not influenced. In our conditions, drying usually lasted 1.5–2 days. Yields of dry biomass were calculated from the field weight of fresh biomass multiplied by dry matter content in samples from each clone. Harvested biomass was calculated in tons of dry biomass per hectare and year [63,64].

2.5. Modeling the Economic Efficiency of Miscanthus Plantations

The life span of *Miscanthus* plantations is expected to range from 8–14 years. *Miscanthus* plantations characteristically have high costs of establishment and do not reach maximum yields immediately after establishment, but only after several years. Thus simple calculation methods based on annual yields and costs, such as for conventional crops, cannot be used to assess the economic efficiency of *Miscanthus* cultivation [65]. The evaluation must include the whole, relatively long, life cycle of this energy crop—from land preparation, stand establishment, crop maintenance throughout the life of the stand (e.g., fertilization), to harvest of biomass and finally, stand eradication at the end of the stand's life followed by site restoration.

Using a methodology based on simulating net cash flows throughout the planted energy crop's life cycle is appropriate for assessing the economic efficiency of *Miscanthus* cultivation [66]. This standard procedure is based on the calculation of the net present value (NPV) of the project [67], not including the NPV directly, but rather, counting the so-called minimum price of production (here, the minimum price of biomass produced), which will provide the investor with the required return on his investment. Therefore, the minimum price calculation is based on finding a biomass price such that NPV = 0 (see equation (1)). The investor then produces a return on the capital transferred at the discount rate.

$$c_{min} : NPV = \sum_{t=1}^{T_h} CF_t \cdot (1 + r_n)^{-t} = \sum_{t=1}^{T_h} (c_{min,t} \cdot Q_t + S_t - E_t) \cdot (1 + r_n)^{-t} = 0 \qquad (1)$$

where

$c_{min,t}$—minimum price of biomass in year t (EUR/GJ)
Q_t—biomass production measured in heat energy (GJ)
S_t—project subsidies in year t (EUR)
E_t—project expenditure in year t (EUR)
r_n—nominal discount (−)
T_n—evaluation period (here, 10 years)
CF_t—cash flow in year t (EUR).

Note: For practical reasons, the minimum price of biomass is expressed in monetary units per GJ (at moisture levels at harvest). This then also allows a direct comparison of the minimum price of different energy crops, irrespective of their moisture levels.

For analysis purposes, we created a model of a 10 ha *Miscanthus* plantation that reflects the typical conditions of growth of this energy crop in the Czech Republic. The model is based on the assumption of a rigorous valuation of all costs at market prices (2019 price level), an estimation of the scope of individual activities according to the data obtained in field trials on experimental sites, and respecting the time value of money. The model uses long-term average inflation of 2%, a nominal discount of 10%, and an income tax of 19%. All the financing of the project is assumed to be from the investor's own resources.

The model works with 6 different yield curves (Yc) reflecting expected yields at different site conditions (climatic and soil)—see Figure 4.

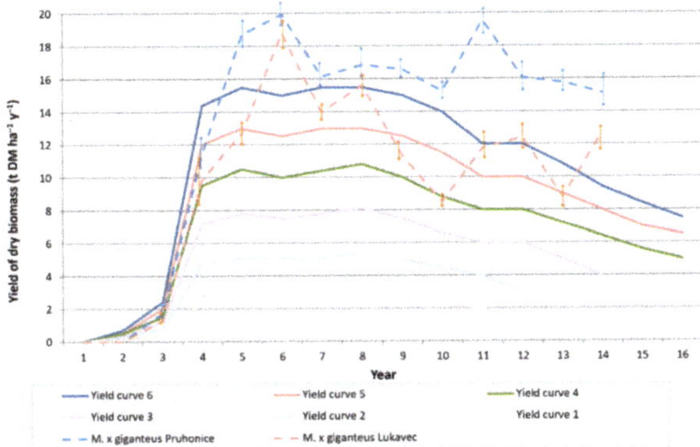

Figure 4. Dry biomass yields of *Miscanthus* × *giganteus* (average with standard error bars of clones M1, M2) from plantations in Průhonice-Michovky and Lukavec, and yield curves (Yc 1–6) of expected yields in commercial practice created for different growing zones for *Miscanthus* in the Czech Republic; growing years (1–16) are considered as financial years (e.g., from January to December).

For modeling purposes, the *Miscanthus* crop processes have been divided into the following:

1. Project and land preparation: Land preparation in the autumn before spring planting includes moderate deep plowing and harrowing, fertilization with NPK (eventually, lime) according to the land's condition. Pre-fertilization in the form of NPK is equivalent to approximately 60 kg N/ha.
2. Costs of establishing a stand: Planting 8000 rhizomes per hectare using a potato planter, post-emergence weeding using a herbicide (eradication of dicotyledonous weeds after one year of *Miscanthus* growth).
3. Planting material: Price used is 0.12 EUR/*Miscanthus* rhizome, which has been typical for a larger amount of purchased material.
4. Harvesting and processes between harvests: Harvesting (bales 80 × 90 cm) takes place in the winter season, with a *Miscanthus* moisture content of 20% at harvest and a calorific value of 13.75 GJ/t (raw biomass). Yield curves already respect the assumed losses of biomass due to the winter harvest. Fertilizer costs (60 kg N/ha in NPK) are estimated from experience with experimental plots once every three years. After the fifth harvest, Ca fertilizing with approximately 2–2.5 tons of dolomitic limestone per hectare is expected.
5. Crop management, subsidies, and land rent: Rent for land is assumed to be 200 EUR/ha/year (approximate median Czech cost of land rent), and overheads are estimated at 40 EUR/ha/year.
6. Costs of stand eradication: After the tenth harvest, the crop is eradicated by deep plowing.
7. Subsidy: Single Area Payment Scheme (SAPS), including a Greening Payment is approximately 210 EUR/ha/year.

3. Results

3.1. Climatic Conditions During Experiments

From 2007 to 2019, the mean annual temperatures in Průhonice ranged between 8.0 °C (2010) and 11.0 °C (2018), while in Lukavec, between 7.2 °C (2010) and 9.1 °C (2019). In Průhonice during the same period, annual sums of precipitation ranged from 354 mm (2018) to 764 mm (2010), while in Lukavec, it was 509 mm (2018) and 940 mm (2010). Compared with the climatological long-term normal (1960–1991) from the Czech Hydrometeorological Institute in Prague [68], temperatures in Průhonice were from normal to extraordinary above normal, while for Lukavec, from normal to above normal. Precipitations ranged from very below normal to above normal in Průhonice, while in Lukavec, from normal to very above long-term normal. The year 2018 was the driest at both localities during the experiment (354 mm respectively 509 mm) and over the last 30 years. In comparison with the long-term normal, the weather in Průhonice was warmer (annual average daily temperatures +1.7 °C) and drier (deficit precipitations minus 661 mm), and the weather in Lukavec was also warmer (annual average daily temperatures +1.1 °C) and precipitation was higher than normal (surplus of precipitations 842 mm) within 13 years of the experiment.

3.2. Survival and Lodging

The survival rate of tested *Miscanthus* clones was quite good on both sites, reaching 90% in Lukavec respectively 89% in Průhonice-Michovky, with only the M5 clone exceeding 20% losses in Lukavec. Survival rates remained stable after the first two years—percentages of living plants are shown in Table 4. Some tested clones have shown insufficient adaptability to winter snow cover, causing serious damage by lodging the whole plant. In the case of clone M6, which has been preselected from the genotype collection as well growing, lodging caused by snow led to reduced yields and loss of biomass quality (Figure 2b).

Table 4. Survival rates (%) of tested *Miscanthus* clones in clonal experiments in Průhonice-Michovky and Lukavec in the year of establishment (2007) and after ten years (March 2017).

Clone	Průhonice Michovky 2007 (XII)	Průhonice Michovky 2017 (III) *	Lukavec 2017 (III)
M1	84	88	85
M2	88	99	85
M3	88	100	96
M4	93	88	79
M5	85	99	99
M6	95	99	94
Average	89	96	90

* After the first year, dead plants were replaced in the experimental plantation in Průhonice-Michovky.

3.3. Biomass Yields

The mean yield of dry biomass of all clones of *Miscanthus* after 13 (spring harvests) from both sites was 10.03 t dry matter (DM) ha^{-1} y^{-1}, respectively 10.80 t DM ha^{-1} y^{-1} at Průhonice-Michovky and 9.27 t DM ha^{-1} y^{-1} at Lukavec. Yields from spring harvests of all clones in the experiment are shown in Table 5. Yields from the establishment year (2007, harvested in spring 2008) were not calculated because the development of *Miscanthus* plants was slow (1–3 stems below 1 m per plant), and biomass harvests are not performed in practice.

The best yielding clones were M2 with 14.7 t DM ha^{-1} y^{-1} and M1 with 13.5 t DM ha^{-1} y^{-1} after 12 spring harvests from the Průhonice-Michovky site. Both clones are *Miscanthus* × *giganteus*. At the Lukavec site, the same clones M1 and M2 reached yields of 10.8 t DM ha^{-1} y^{-1} (M1) and 10.4 t DM ha^{-1} y^{-1} (M2). The best yielding *Miscanthus sinensis* clones were M4 and M5. After the last harvest in spring 2020, M5 is the highest yielding *Miscanthus sinensis* clone at Lukavec (10.2 t DM ha^{-1} y^{-1}), while M4 is the highest yielding clone at the Průhonice site (12.7 t DM ha^{-1} y^{-1}).

At the Průhonice-Michovky site, yields of clones M1, M2, M4 and M5 continuously increased for four years. At the Lukavec site, harvests of clones M5 and M6 increased for three years, while clones M1, M2, M3, M4 increased for four years.

Our statistical examination provided the following results: At the Lukavec site, even though Miscanthus × *giganteus* had higher yields than *Miscanthus sinensis*, no statistical difference was found between yields of these species due to the high variability of data. At the Průhonice-Michovky site, clones of *Miscanthus* × *giganteus* (M1, M2) grew better than those of Miscanthus *sinensis* (M3, M5, M6, M4).

During the experiment's existence at the Lukavec site, the highest harvest yield of dry biomass was reached in autumn (19.1 t DM ha^{-1} y^{-1} in 2011) by clone M1. The highest yield of dry biomass at the Průhonice-Michovky site (21.1 t DM ha^{-1} y^{-1}) was reached in autumn 2011, also by clone M1.

Table 5. Dry biomass yields (t DM ha^{-1} y^{-1}) of *Miscanthus* clones in spring harvest in clonal experiment at Lukavec and Průhonice-Michovky (2008–2017).

Lukavec												
Year [†]	M1	s	M2	s	M3	s	M4	s	M5	s	M6	s
2008 *	-	-	-	-	-	-	-	-	-	-	-	-
2009	1.2 AB	0.45	1.4 ABC	0.86	2.3 C	0.94	1.1 A	0.73	2.3 BC	0.58	1.9 ABC	0.62
2010	9.3 A	3.63	10.4	4.69	7.7	3.18	7.5	5.03	10.5	0.54	10.0 A	1.44
2011	11.2 A	0.75	14.2	1.35	9.4	0.80	10.9	3.33	11.3	0.57	14.0 B	1.62
2012	19.1 D	1.74	18.5 D	2.75	10.9 A	1.45	16.2 CD	3.49	14.3 BC	1.06	12.4 AB	0.54
2013	14.2 B	0.91	13.8	1.73	10.2	1.51	12.8	2.88	17.1	6.72	11.0 A	1.75
2014	16.3 E	1.46	14.9 DE	2.16	8.0 A	1.75	12.4 CD	2.71	11.5 BC	1.84	9.4 AB	1.37
2015	11.6 C	1.32	11.0 C	1.82	6.4 A	1.34	9.6 BC	2.10	10.3 C	1.99	7.3 AB	0.66
2016	8.5 A	2.13	8.6 A	1.57	7.9 A	0.45	11.4 BC	2.42	14.2 D	1.93	10.4 B	0.78
2017	12.4 D	2.5	11.1 C	1.9	6.2 A	1.7	9.8 C	1.0	8.8 BC	4.3	7.7 B	0.9
2018	13.8 B	2.21	11.0 AB	1.03	7.9 A	3.06	10.8 AB	3.62	13.3 B	2.64	9.0 A	0.94
2019	9.3 AB	1.68	8.3 A	1.66	8.0 A	2.02	7.8 A	1.54	11.8 B	3.01	7.8 A	1.42
2020	13.2 B	2.75	12.2 B	0.67	7.6 A	1.23	13.1 B	1.00	11.8 B	1.60	8.2 A	0.95
Average **	10.8		10.4		6.9		9.2		10.2		8.1	
Průhonice-Michovky												
Year	M1	s	M2	s	M3	s	M4	s	M5	s	M6	s
2008 *	-	-	-	-	-	-	-	-	-	-	-	-
2009	2.1 AB	0.93	1.4 A	0.86	1.9 AB	1.14	1.3 A	0.78	1.5 A	0.59	3.5 B	1.59
2010	12.2 B	2.64	10.6 AB	2.47	7.7 A	1.76	9.2 AB	3.09	8.0 A	1.36	11.7 B	1.54
2011	17.5 D	2.10	19.9 D	1.36	6.0 A	0.95	14.1 C	1.77	8.1 AB	0.44	9.9 B	2.33
2012	18.8 B	1.48	21.1 B	1.97	8.6 A	0.87	19.3 B	3.24	10.5 A	1.04	10.8 A	1.63
2013	15.4 B	2.19	16.9 B	1.60	9.1 A	1.16	16.1 B	2.40	9.9 A	1.04	9.2 A	1.41
2014	15.8 CD	1.91	17.9 D	2.22	8.6 AB	0.86	14.0 C	1.28	10.6 B	1.02	8.0 A	1.25
2015	15.9 CD	2.00	17.3 D	1.18	7.9 AB	0.68	14.9 C	1.36	10.0 B	1.18	7.3 A	2.18
2016	14.4 C	0.97	16.1 C	0.86	7.1 A	1.52	14.1 C	1.21	10.3 B	1.26	10.0 B	0.97
2017	18.4 CD	2.1	20.7 D	0.99	9.1 A	0.8	15.5 C	1.39	11.6 B	1.36	9.3 A	1.89
2018	14.5 CD	1.89	17.6 DE	1.60	7.5 A	1.59	18 E	3.25	12.4 BC	2.56	9.7 AB	1.12
2019	14.9 B	1.94	16.7 B	1.54	7.5 A	1.61	14.4 B	2.60	9.7 A	2.06	8.5 A	0.64
2020	15.3 B	2.06	14.9 B	3.46	9.8 A	0.73	14.3 B	1.36	10.0 A	2.24	9.3 A	1.34
Average **	13.5		14.7		7.0		12.7		8.7		8.3	

Table legend: s—standard deviation; * yield was not evaluated due to slow growth in the year of the experiment establishment (2007); [†] considered as financial years from January to December (e.g., yield from spring harvest in 2008 represents biomass from the 2007 vegetative period); [A-E] indexed capital letters show results of statistical analysis of yields of tested clones in individual years (MANOVA–Duncan; homogenous groups); ** average yields from 12 harvests are calculated for 13 years of the experiment.

The expected yields of *Miscanthus*, expressed in yield curves (Figure 4), are linked to the site's soil and climatic conditions. The expected yield can be understood as the long-term average yield at the given age of the crop plantation in areas with similar climatic-soil conditions while respecting proper cultivation conditions.

Yield curves and growing zones for *Miscanthus* as well as other energy crops in the Czech Republic were created by energy crop experts. These experts used field data (yields) from long-term experimental plots and commercial plantations, and the Czech typology of agricultural land for agriculture production [69] that contains climatic, soil and site conditions of each farm field in the Czech Republic, i.e., the Valuation soil ecological unit. By combining these characteristics with empirically measured yields, six land suitability types and growing zones have been deduced for *Miscanthus*. Yield curves for each growing zone were then created using yields from consecutive harvests from long-term experimental plots and commercial plantations. A more detailed description of the methodology used to create yield curves and growing zones for *Miscanthus* can be found in [10,66,70–72].

3.4. Biomass Parameters in Autumn and Spring Harvests

Comparison of dry biomass yields (t DM ha^{-1} y^{-1}) from the autumn and spring harvests for *Miscanthus* clones M1 (*Miscanthus* × *giganteus*) and M6 (*Miscanthus sinensis*) at Průhonice-Michovky and Lukavec in consecutive harvests have shown that spring harvest yields are between 18–31% lower than the autumn harvest yields, depending on the clone and site (see Table 5 for spring harvest and Table 6 for autumn harvest). Differences between autumn (November) and spring (April) yields in individual harvests of both clones have been diverse, predominantly because of the course of winter weather and the occurrence of biotic and abiotic damage (lodging by snow, animals—Figure 2b) in the individual years.

Moisture content in harvested biomass varied between 7–36% (Ø 15%) in the spring harvest and between 20–62% (Ø 43%) in the autumn harvest depending on clone (M1 and M6) and year (weather before harvest) (see Table 6).

Table 6. Dry biomass yields in the autumn harvest and moisture content in biomass from autumn and spring harvests of *Miscanthus* clones M1 and M6 in the clonal experiment (Lukavec, Průhonice-Michovky).

Biomass yields in the Autumn Harvest (t (DM)ha^{-1} year^{-1})							
	Lukavec				**Průhonice-Michovky**		
Year [†]	**M1**	**s**	**M6**	**s**	**M1**	**s**	**M6**	**s**
2008	3.0 [A]	1.18	3.7 [A]	0.73	3.1 [A]	1.53	4.5 [A]	1.74
2009	10.1 [A]	0.87	12.3 [B]	2.69	13.0 [A]	2.74	13.2 [A]	0.19
2010	19.1 [A]	4.81	15.9 [A]	1.19	19.6 [B]	2.57	13.2 [A]	2.76
2011	23.6 [B]	1.34	14.8 [A]	0.84	21.3 [B]	3.68	13.7 [A]	1.68
2012	21.6 [B]	1.07	16.2 [A]	1.18	19.2 [A]	5.62	12.2 [A]	2.21
2013	29.1 [B]	1.26	17.8 [A]	1.52	21.4 [B]	4.29	12.4 [A]	2.51
2014	21.5 [B]	1.85	13.5 [A]	1.86	18.2 [A]	4.30	13.8 [A]	1.28
2015	14.0 [A]	1.20	12.6 [A]	1.00	16.8 [A]	3.45	12.7 [A]	3.05
2016	13.0 [A]	1.47	13.2 [A]	2.37	22.4 [B]	3.30	16.0 [A]	2,52
2017	16.4 [B]	1.23	12.4 [A]	0.84	19.9 [B]	3.93	11.9 [A]	1,51
2018	12.8 [A]	2.25	14.1 [A]	1.30	17.0 [B]	2.09	11.6 [A]	2,52
2019	12.2 [A]	2.07	10.1 [A]	1.57	20.3 [B]	1.23	11.1 [A]	1.79
Average	15.1		12.0		16.4		11.1	

Moisture content in harvested biomass (%)							
	Lukavec				**Průhonice-Michovky**		
	M1		**M6**		**M1**		**M6**	
Year	**Autumn**	**Spring**	**Autumn**	**Spring**	**Autumn**	**Spring**	**Autumn**	**Spring**
2008	38		35		53		59	
2009	51	11	54	10	50	28	44	13
2010	41	9	34	10	56	20	62	17
2011	49	33	47	22	41	25	34	24
2012	43	8	42	5	37	17	51	20
2013	25	7	20	7	26	14	39	19
2014	45	11	41	10	44	21	35	31
2015	42	9	35	9	50	18	42	36
2016	57	8	42	6	35	21	31	12
2017	71	8	68	6	47	16	32	29
2018	32	8	37	6	50	22	51	15
2019	32	9	37	8	41	20	28	29
2020		9		9		10		11
Average *	44	11	41	9	44	19	42	21

Table legend: s—standard deviation; [†] years are considered as financial years (I-XII); * average yields from 12 harvests are calculated for 13 years of the experiment; [A–B] indexed capital letters show results of statistical analysis of yields (MANOVA–Duncan; homogenous groups) in individual years and sites.

3.5. Invasive Behavior

Observations of experimental sites in Průhonice (Michovky, Zelinářská zahrada) resulted in important findings confirming the ability of selected *Miscanthus* clones to spread from the original planting site spontaneously (Figure 2a). We have found relatively large volumes of tens to hundreds of well-growing seedlings of *Miscanthus sinensis* and/or *Miscanthus* ssp. with one individual found ca. 60 m from the experimental site on permanent grassland. By the end of the growing season, around 200 new *Miscanthus* seedlings were found in surrounding growths up to 80 m away. These seedlings were confirmed only on cultivated sites with bare soil. These rooted seedlings were one or two years old, from 5 to 20 cm tall with 1–4 stems. In 2009, samples of 35 individuals of rooted seedlings were taken to find their parental plants. All *Miscanthus sinensis* and *Miscanthus* ssp. clones selected for experimental germination verification could germinate, although in varying volumes depending on the clone. This ability appears more or less risky from the perspective of standard sexual reproduction and potential spreading in neighborhoods. Clones classified as *Miscanthus × giganteus* and *Miscanthus sacchariflorus* were sterile in our climatic conditions. For *Miscanthus sacchariflorus*, sterility can be caused locally by the absence of suitable pollinating plants in the time of flowering or unfavorable temperatures—the tested clone flowers very late. *Miscanthus sacchariflorus* had intensive vegetative reproduction via rhizomes. During our experiment, rhizomes spread 1.5–2 m within a season, presenting a risk in terms of nature protection and field management.

Based on the evaluated experimental results, i.e., the combination of triploidy and verification of the seeds' inability to germinate—the least risky clones, if we consider invasive behavior, are clones M1, M2 (*Miscanthus × giganteus*), and also, the triploid genotype M9 (*Miscanthus sinensis* 'Goliath'). However, when seed germination tests were repeated, the M9 clone showed a slight but certain ability to germinate (similar results were published by [53]).

3.6. Economic Analysis

Results of the calculation of the minimum price with and without subsidies (SAPS-Single Area Payment Scheme) are shown in Table 7. The minimum price of *Miscanthus* biomass was calculated to be between 3.0–15.2 EUR/GJ for the subsidized variant and 4.3 to 23.1 EUR/GJ for the variant without subsidies, in both cases, depending on the yield curve. The project's cost structure for the entire lifespan of the *Miscanthus* stand (in current values) is shown in Figure 5.

Table 7. Results of minimum price modeling of biomass of *Miscanthus × giganteus* for six yield curves (Yc 1–6) with and without subsidies (SAPS).

Yield Curve	Average Yield *	Minimum Price with SAPS **	Minimum Price without SAPS **
	t DM ha^{-1} y^{-1}	EUR/GJ	EUR/GJ
Yc 6	12	3.0	4.3
Yc 5	10	3.4	5.0
Yc 4	8	4.2	6.2
Yc 3	6	5.4	8.0
Yc 2	4	7.8	11.8
Yc 1	2	15.2	23.1

Table legend: * average yield is the average of expected yields during 10 (winter) harvests after establishment; ** the economic model was calculated in Czech Crowns (CZK), and the results recalculated to EUROs using the exchange rate of 1 EUR = 25 CZK.

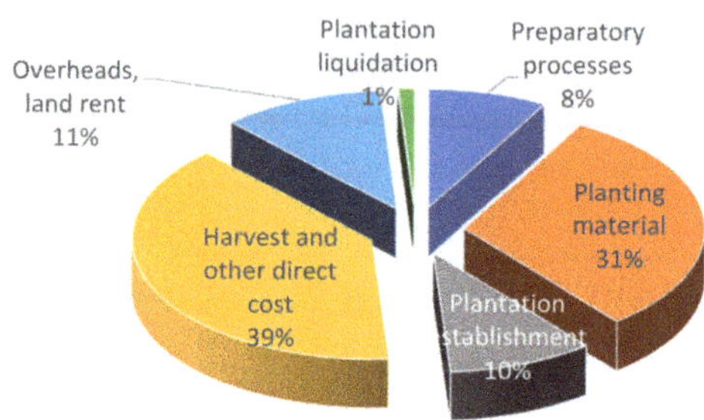

Figure 5. The cost structure of the *Miscanthus* biomass production divided into categories (in current values, with a nominal discount of 10%, Yc 5 = 10 t DM ha^{-1} y^{-1}).

Thus, the minimum price is the price of (raw) biomass after removal from the field to the storage location (distance up to 10 km). Additional costs related to the storage of biomass and delivery to the final consumer (e.g., to the heating plant) or its supply for reprocessing into solid biofuels (pellets, briquettes) are not included. The minimum price also reflects only the losses of biomass in harvesting and transport to the place of storage, but not from storage, or subsequent transport or eventually, processing. Information on the impact of biomass losses and storage of biomass on the minimum price can be found, e.g., in [73].

The costs of establishing a plantation, including planting material, are high and significantly affect the minimum price. Figure 5 shows that this represents 44% of all project expenditure (in present value). Some reduction in the minimum price can be achieved by extending the life of the plantation, which then spreads the cost of establishing the plantation over a longer period. This can be documented with the variant of the calculation, where the lifetime of the stand is assumed to be 15 years and the continuation of a gradual decrease in yields (by about 1/3 in the 15th year compared to 10 years). The minimum biomass price for the Yc-4 yield curve decreases from 4.2 EUR/GJ to 3.9 EUR/GJ. Similarly, in the Yc-5 yield curve, there is a slight decrease from 3.4 EUR/GJ to 3.2 EUR/GJ (both calculations assume SAPS subsidy). This small reduction in the minimum price is mainly because the high costs of establishing a stand are spent at the beginning of the plantation project, while the maximum yields and thus, sales, are achieved only over time. Similarly, extending the life of a plantation generates additional cash flows, but the value of these contributions has little weight on the project's overall effectiveness. The key aspect of improving the economic effectiveness of *Miscanthus* is to reduce the cost of establishing plantations.

The above mentioned process used to determine the minimum price of biomass presents only one of three possible ways to look at the price of biomass.

The second perspective would be to look at it from the producer's point of view (supply side), where, in practice, the producer growing biomass for energy purposes would accept only a price for this biomass that would ensure, at least, a similar economic effect as growing conventional crops. Due to the high subsidies for growing conventional crops and the relatively high prices given for these conventional crops, the price, then, that producers would require for intentionally grown biomass is thus significantly higher. A detailed discussion of this aspect can be found, e.g., in [74]. According to the latest calculations carried out by the authors (price level 2019), high revenues from conventional crops lead to a biomass price increase from the minimum price for biomass from *Miscanthus* crops in comparison to the values given in Table 7, approximately in the range from 12–70%, with an average value of about 42% for all areas.

The third point of view on the price of biomass is that of the buyer (demand). Here, the price that the purchaser is willing to pay for intentionally planted biomass from energy crops does not exceed the price of other fuels (e.g., biomass from other sources such as forest biomass). In many cases, using (burning) raw biomass directly in bale form is not possible and can only be expected for larger heating plants or those equipped with the appropriate technology. For smaller or local heating plants, such biomass must be transformed into solid biofuels—pellets or briquettes. However, manufacturing pellets and briquettes significantly increases the price of biomass, e.g., the cost of pelletization in the Czech Republic is about 4.5–5.0 EUR/GJ, which further significantly increases the (minimum) price of biomass [73].

The limit price of raw biomass from *Miscanthus* plantations in Czech conditions is estimated to range from 6–8 EUR/GJ, assuming price acceptability from the demand side. This is especially valid if the biomass is further processed into solid biofuel, and therefore, the price has been increased due to pelletization. The limit price of biomass in pellet form is, here, the price of wood pellets, i.e., 11.2 EUR/GJ minus the cost of pelletization—for more information, see [75]. If we consider the direct combustion of biomass (straw bales) in large power plants or cogeneration plants, the limit of the biomass price here will be influenced by the amount of support for electricity from direct combustion of biomass that varies according to the year of commissioning and category, see [76]. From the current amount of electricity support in the form of FIP (feed-in-premium) tariffs for intentionally grown biomass combustion, it is possible to estimate a biomass limit price of 6–7 EUR/GJ.

4. Discussion

After twelve consecutive harvest years, results on our two sites show yields similar to those in other experimental plantations in Central-Eastern European conditions and low-input agronomy [9,39,77,78]. For instance, [39] observed a mean yield of dry biomass of 13.7 t DM ha^{-1} y^{-1} in a non-fertilized variant of an 11-year experiment with *Miscanthus* × *giganteus* in southern Germany. The mean biomass yield of M1 and M2 clones (*Miscanthus* × *giganteus*) during spring harvests in Lukavec was 10.8 resp. 10.4 t DM ha^{-1} y^{-1}. In the Průhonice-Michovky site, it was 13.5 respectively 14.7 t DM ha^{-1} y^{-1}. These yields are comparable, if not slightly better than other new lignocellulose energy crops like poplar, willow, reed canary grass, or Schavnat in Czech conditions [79–81].

Even though spring harvests have lower yields than in autumn, they can be recommended because the concentrations of potassium, chlorine, nitrogen, and sulfur in *Miscanthus* biomass decreases significantly due to the translocation of nutrients to the root part and its leaching during winter [82]; a similar result was also recently confirmed by [83]. In comparison with woody energy crops (poplar, willow), *Miscanthus'* spring harvest biomass is less suitable for direct burning in some, especially smaller boilers, where it can create slagging in the heat exchanger. The effectiveness of *Miscanthus* biomass can be improved by mixing it with woodchips to produce pellets [84].

Miscanthus, however, can also be important as an effective source of commodities and materials, e.g., chipboard, pellets for animal (pet) bedding, cement particle boards, biocomposite automotive component, or biogas production from autumn (green) harvest, that have higher added value than energy biomass [85–87].

Knowledge about the invasiveness of some genotypes, resp. species of *Miscanthus* in European conditions, have been taken into account [88] by breeders, and it can be expected that new varieties will be 'minimum or zero invasive' for both generative and vegetative ways of reproduction and dispersal into the surrounding fields and countryside. In the Czech Republic, only clones of *Miscanthus* × *giganteus*, a non-invasive triploid, can be used in agriculture practice [56]. Since 2010, all clones of *Miscanthus sinensis* have been excluded from the "List of plants suitable for cultivation of energy biomass from the point of view of minimizing risks to nature and landscape protection" [89], which is a methodological support tool for decision making in nature protection regarding the use of non-native energy crops in the landscape.

At present, there are economic barriers to the faster development of *Miscanthus* cultivation. Competition with conventional (annual) crops is the main barrier that has the following aspects:

(1) In contrast to conventional crops, *Miscanthus* plantations have high one-off costs for stand establishment. These one-off costs represent around 1/3 of the total cost for the *Miscanthus* stand (in present value) over its entire life cycle (10 years). In this way, the grower must, at the outset, invest significantly more money per unit of area than in the case of conventional agricultural production.

(2) The maximum production of biomass is reached up to 2–3 years after establishment, which, from the producer's point of view, means that cash flow is initially worse.

(3) Having multiyear plantations of energy crops is significantly riskier for producers, both in terms of the higher one-off costs of establishing the stands and losses after establishment due to crop damage or possible changes in the biomass market. An investor or farmer of perennial energy crops cannot react as quickly to market changes as someone who has invested in conventional crops with a one-year production cycle. One reason for this is that most agricultural land in the Czech Republic is still farmed on leased land (about 70%—see [90]), and rental periods are generally shorter than the life cycle of the energy crop plantation, thus further increasing the risk.

Another significant economic barrier is the relatively high costs related to growing biomass in a *Miscanthus* plantation. The minimum price of produced biomass (with a 10% nominal discount) assumed using average to less fertile soils ranges from 5.4–7.8 EUR/GJ, i.e., 57–73 EUR t^{-1} of fresh matter [87] have calculated prices of 35–47 EUR t^{-1} for *Miscanthus* biomass for direct combustion in German conditions, but for higher yields (15–25 t DM y^{-1}), intensive agronomy (fertilization, density), and much longer plantation lifetime (20 years). Farmers, however, in practice, would demand an even higher price that would at least give them the same economic effect that they would have from growing conventional crops, thus increasing the price of raw biomass by, approximately, an additional 43% (on average).

The price of raw *Miscanthus* biomass (without transport, storage, or processing costs (into pellets or briquettes) significantly exceeds the limit of the competitive price of raw biomass estimated at 6 (max. 8) EUR/GJ in the Czech Republic. This limit is important, as can be seen from the results of the authors' analyses, which show that because the minimum price of biomass increases due to the competition from conventional energy crops, there is no land on which any farmer would want to establish *Miscanthus* stands and accept 6 EUR/GJ or less. If the limit price would be 8 EUR/GJ, then producers would consider establishing *Miscanthus* stands on approximately 27% of the Czech Republic's agricultural land.

Another barrier is on the consumers' side and their technological limitations. To date, it has been technologically easier for consumers to focus on woody biomass rather than straw biomass that would need further investment into a suitable boiler using straw fuel. Otherwise, straw biomass would need to be made into pellets or briquettes, which significantly increases the price of the produced biofuel.

Economic barriers to the development of *Miscanthus* plantations (or other perennial crops) can be reduced by the following:

- Providing targeted subsidies for plantation establishment to decrease the investor's risk.
- Supporting long-term contracts to purchase biomass for energy crops using a price formula.
- Using plantations of perennial energy crops for additional benefits, i.e., non-production functions (e.g., decreasing soil erosion, phytoremediation, increasing the soil's humus content and water capacity).

Another measure that would significantly increase, albeit indirectly, the competitive ability of intentionally grown biomass against conventional fuels is to increase markedly the carbon costs (e.g., in the form of an emission allowance or carbon tax) included in the price of fossil fuels.

5. Conclusions

Average yields of *Miscanthus* $\times$ *giganteus* clones tested in our experiment (M1, M2 $\geq$ 10–15 t DM ha^{-1} y^{-1}) are comparable, if not slightly better than other new lignocellulose energy crops (poplar, willow, or Schavnat) in Czech conditions. *Miscanthus × giganteus* clones have good potential for commercial production of energy biomass, especially in warmer regions of Central and Eastern Europe (average annual daily temperature $°t \geq 9\text{–}10\,°C$) with an annual sum of precipitation above 500–550 mm.

Results of monitoring *Miscanthus × giganteus* yields and the course of weather during our experiment (13 years) have shown that *Miscanthus × giganteus* adapts well to dry years (or its parts) characterized by low precipitation ($\sum P = 350\text{–}450$ mm y^{-1}) and increasing annual daily temperatures (average annual daily temperature $°t \geq 10.5\,°C$).

Clones of *Miscanthus sinensis* tested in our experiment could not be recommended for energy biomass production due to their strong invasive ability. The sterile triploid clones of *Miscanthus × giganteus*, however, have been recommended with minimum risks for nature and landscape. Some clones of *M. sinensis* have shown the potential to be bred for colder conditions.

Results of economic modeling have shown that there are significant economic barriers to the development of perennial energy crops, especially those resulting in straw biomass. These barriers not only include the current and relatively high profitability of conventional annual crops, which in turn increases the expected price of biomass from energy crops, but also the economic risk associated with the large portion of one-off initial establishment costs. The competitive ability of straw biomass is significantly lower because of the consumers' technological limitations that usually do not enable them to burn straw biomass directly. Burning straw biomass then is taken into consideration only by larger heating or cogeneration plants. Smaller or local plants need biomass in pellet or briquette form, which means an increase in price and a decrease in competitive ability. At these smaller plants, biomass (processed into pellets or briquettes) can be competitive if natural gas is not available or where using a heat pump instead of a coal furnace is not relevant due to the high costs of reconstructing the heating system.

Regarding the article's question, "Can *Miscanthus* fulfill its potential as a new biomass crop—for energy and material in the Czech Republic (and CEE countries)?", our team would answer positively, but only if the following conditions and steps would materialize in the upcoming years:

- Improvement of *Miscanthus × giganteus* gene pool (new varieties) and agrotechnology (to lower establishment cost, prolong production period to 15–20 years, improve the precision of fertilization, minimize the invasive risk) continues.
- Climate change trends continue with growing effects of weather extremes and changes (droughts, temperature growth) in CEE countries, which may improve growing conditions for *Miscanthus* (C4 plant) over conventional crops (mostly C3 plants).
- A new approach of EC or member states to current agriculture subsidy policy (CAP), which would evaluate environmental services of *Miscanthus* and other new biomass crops, is implemented.
- Further development of the bioeconomy in the EU occurs, thus increasing demand for *Miscanthus* biomass for utilization in products with higher additional value, e.g., construction materials, industrial products, and second-generation biofuels.

Author Contributions: Conceptualization, J.W., J.B., K.V. and J.K.; methodology, J.W., Z.S. (field research) and J.K. (economy); validation, J.W., J.K. and K.V.; formal analysis, J.B.; investigation, J.W., J.B., K.V. and Z.S.; resources, J.B., K.V. and Z.S.; data curation, J.W., J.B. and K.V.; writing—original draft preparation, J.B., J.W. (field research) and K.V. (economic analysis); writing—review and editing, J.W. and J.K.; visualization, J.W.; supervision, J.W., J.K., K.V.; project administration, K.V.; funding acquisition, K.V. All authors have read and agreed to the published version of the manuscript.

Funding: This research was performed as part of project no. TD03000039 supported by TACR—the Technology Agency of the Czech Republic.

Institutional Review Board Statement: Not applicable.

Informed Consent Statement: Not applicable.

Data Availability Statement: The data and genotypes presented in this study are available on request from the corresponding author, Jan Weger, weger@vukoz.cz.

Acknowledgments: We would like to acknowledge Václav Veleta from the field research station in Lukavec (experimental site) for his long-term cooperation and provision of field data from the experimental *Miscanthus* plantation and Michal Severa for carrying out germination tests on *Miscanthus* seeds. Jana Jobbiková, Tereza Humešová, and Aleš Tobyška provided technical and administrative support of the field research; Janice Forry provided proofreading, translations, and language finalization of the text. We would also like to thank Iris Lewandowski, Uffe Jørgensen, and Jens Bonderup Kjeldsen for providing planting material of the original *Miscanthus* genotypes.

Conflicts of Interest: The authors declare no conflict of interest.

References

1. Eurostat. Renewable Energy Statistics—Statistics Explained. 2020. Available online: https://ec.europa.eu/eurostat/statistics-explained/index.php?title=Renewable_energy_statistics_ (accessed on 25 September 2020).
2. Scarlat, N.; Dallemand, J.-F.; Taylor, N.; Banja, M. *Brief on Biomass for Energy in the European Union*; Publications Office of the European Union: Luxembourg, 2019.
3. European Commission. National Energy and Climate Plans (NECPs), European Commission. 2019. Available online: https://ec.europa.eu/info/energy-climate-change-environment/overall-targets/national-energy-and-climate-plans-necps_en (accessed on 4 October 2020).
4. International Renewable Energy Agency. IRENA Global Bioenergy Supply and Demand Projections A Working Paper for REmap 2030. 2014. Available online: www.irena.org/remap (accessed on 4 October 2020).
5. European Commission. The European Green Deal, Brussels. 2019. Available online: https://ec.europa.eu/info/sites/info/files/european-green-deal-communication_en.pdf (accessed on 4 October 2020).
6. Knápek, J.; Králík, T.; Vávrová, K.; Weger, J. Dynamic biomass potential from agricultural land. *Renew. Sustain. Energy Rev.* **2020**, *134*, 110319. [CrossRef]
7. Laurent, A.; Pelzer, E.; Loyce, C.; Makowski, D. Ranking Yields of Energy Crops: A Meta-Analysis Using Direct and Indirect Comparisons. *Renew. Sustain. Energy Rev.* **2015**, *46*, 41–50. [CrossRef]
8. Stolarski, M.J.; Krzyżaniak, M.; Tworkowski, J.; Szczukowski, S.; Gołaszewski, J. Energy intensity and energy ratio in producing willow chips as feedstock for an integrated biorefinery. *Biosyst. Eng.* **2014**, *123*, 19–28. [CrossRef]
9. Stolarski, M.J.; Krzyżaniak, M.; Warmiński, K.; Olba-Zięty, E.; Penni, D.; Bordiean, A. Energy efficiency indices for lignocellulosic biomass production: Short rotation coppices versus grasses and other herbaceous crops. *Ind. Crops Prod.* **2019**, *135*, 10–20. [CrossRef]
10. Havlíčková, K.; Suchý, J.; Weger, J.; Šedivá, J.; Táborová, M.; Bureš, M.; Hána, J.; Nikl, M.; Jirásková, L.; Petruchová, J.; et al. *Analýza Potenciálu Biomasy v České Republice (Analysis of Biomass Potential in the Czech Republic)*; Výzkumný Ústav Silva Taroucy Pro Krajinu a Okrasné Zahradnictví, v.v.i.: Průhonice, Czech Republic, 2010; p. 498.
11. Heděnec, P.; Novotný, D.; Ustak, S.; Cajthaml, T.; Slejška, A.; Šimáčková, H.; Honzík, R.; Kovářová, M.; Jan, F. The effect of native and introduced biofuel crops on the composition of soil biota communities. *Biomass Bioenergy* **2014**, *60*, 137–146. [CrossRef]
12. Rahman, M.M.; Mostafiz, S.B.; Paatero, J.V.; Lahdelma, R. Extension of energy crops on surplus agricultural lands: A potentially viable option in developing countries while fossil fuel reserves are diminishing. *Renew. Sustain. Energy Rev.* **2014**, *29*, 108–119. [CrossRef]
13. Lewandowski, I.; Clifton-Brown, J.; Trindade, L.M.; Van Der Linden, G.C.; Schwarz, K.-U.; Müller-Sämann, K.; Anisimov, A.; Chen, C.-L.; Dolstra, O.; Donnison, I.S.; et al. Progress on Optimizing *Miscanthus* Biomass Production for the European Bioeconomy: Results of the EU FP7 Project OPTIMISC. *Front. Plant Sci.* **2016**, *7*, 1620. [CrossRef]
14. Biedermann, J. *Zemědělství 2013 (Agriculture 2013)*; Ministry of Agriculture: Praha, Czech Republic, 2014.
15. European Commission. Cereals, Oilseeds, Protein Crops and Rice, European Commission. 2016. Available online: https://ec.europa.eu/info/food-farming-fisheries/plants-and-plant-products/plant-products/cereals (accessed on 25 September 2020).
16. Schorling, M.; Enders, C.; Voigt, C.A. Assessing the cultivation potential of the energy crop *Miscanthus × giganteus* for Germany. *GCB Bioenergy.* **2015**, *7*, 763–773. [CrossRef]
17. Lewandowski, I.; Clifton-Brown, J.; Kiesel, A.; Hastings, A.; Iqbal, Y. *Perennial Grasses for Bioenergy and Bioproducts: Production, Uses Sustainability and Markets for Giant Reed, Miscanthus, Switchgrass, Reed Canary Grass and Bamboo*; Efthymia, A., Ed.; Academic Press: London, UK, 2018; pp. 35–60.
18. Bioenergy Europe. Bioenergy Europe Statistical Report. 2018. Available online: https://bioenergyeurope.org/statistical-report-2018 (accessed on 26 October 2020).
19. Stolarski, M.J.; Průhonice, Czech Republic. Personal communication, 2018.
20. Lewandowski, I.; Kicherer, A.; Vonier, P. CO_2-balance for the cultivation and combustion of *Miscanthus*. *Biomass Bioenergy* **1995**, *8*, 81–90. [CrossRef]

21. Venturi, P.; Huisman, W.; Atzema, A. The effect of harvest methods of *Miscanthus × giganteus* on available harvest time, Sustainable Agriculture for Food Energy and Industry. In Proceedings of the International Conference, Braunschweig, Germany, 17–20 June 1997; p. 45.

22. Clifton-Brown, J.C.; Stampfl, P.F.; Jones, M.B. *Miscanthus* biomass production for energy in Europe and its potential contribution to decreasing fossil fuel carbon emissions. *Glob. Chang. Biol.* **2004**, *10*, 509–518. [CrossRef]

23. Everard, C.D.; Schmidt, M.; McDonnell, K.P.; Finnan, J. Heating processes during storage of *Miscanthus* chip piles and numerical simulations to predict self-ignition. *J. Loss Prev. Process Ind.* **2014**, *30*, 188–196. [CrossRef]

24. Moudrý, J.; Strašil, Z. *Alternativní Plodiny (Alternative Crops)*; University of South Bohemia—Faculty of Agriculture: České Budějovice, Czech Republic, 1996; p. 90.

25. Weger, J.; Průhonice, Czech Republic. Personal communication, 2020.

26. Hodkinson, T.R.; Chase, M.W.; Lledó, M.D.; Salamin, N.; Renvoize, S.A. Phylogenetics of *Miscanthus*, Saccharum and related genera (Saccharinae, Andropogoneae, Poaceae) based on DNA sequences from ITS nuclear ribosomal DNA and plastid trnL intron and trnL-F intergenic spacers. *J. Plant Res.* **2002**, *115*, 381–392. [CrossRef] [PubMed]

27. Maucieri, C.; Camarotto, C.; Florio, G.; Albergo, R.; Ambrico, A.; Trupo, M.; Borin, M. Bioethanol and biomethane potential production of thirteen pluri-annual herbaceous species. *Ind. Crops Prod.* **2019**, *129*, 694–701. [CrossRef]

28. Szulczewski, W.; Żyromski, A.; Jakubowski, W.; Biniak-Pieróg, M. A new method for the estimation of biomass yield of giant *Miscanthus* (*Miscanthus giganteus*) in the course of vegetation. *Renew. Sustain. Energy Rev.* **2018**, *82*, 1787–1795. [CrossRef]

29. Stewart, J.R.; Toma, Y.; Fernández, F.G.; Nishiwaki, A.; Yamada, T.; Bollero, G. The ecology and agronomy of *Miscanthus sinensis*, a species important to bioenergy crop development, in its native range in Japan: A review. *GCB Bioenergy* **2009**, *1*, 126–153. [CrossRef]

30. Cerazy-Waliszewska, J.; Jeżowski, S.; Łysakowski, P.; Waliszews, B.; Zborowska, M.; Sobańska, K.; Ślusarkiewicz-Jarzina, A.; Białas, W.; Pniewski, T. Potential of bioethanol production from biomass of various *Miscanthus* genotypes cultivated in three-year plantations in west-central Poland. *Ind. Crops Prod.* **2019**, *141*, 111790. [CrossRef]

31. Grams, J.; Kwapińska, M.; Jędrzejczyk, M.; Rzeźnicka, I.; Leahy, J.J.; Ruppert, A.M. Surface characterization of *Miscanthus × giganteus* and Willow subjected to torrefaction. *J. Anal. Appl. Pyrolysis* **2018**, *138*, 231–241. [CrossRef]

32. McCarthy, S.; Mooney, M. European *Miscanthus* network. In *Biomass for Energy, Environment, Agriculture and Industry*; Guildford and King's Lynn: Biddles Ltd.: Pergamon, MI, USA, 1995; pp. 380–388. ISBN 0080421350.

33. Harvey, J.J.; Hutchens, M. Progress in commercial development of *Miscanthus* in England. In *Biomass for Energy, Environment, Agriculture and Industry*; Guildford and King's Lynn: Biddles Ltd.: Pergamon, MI, USA, 1995; pp. 587–594.

34. Walsh, M.; McCarthy, S. *Miscanthus* handbook. In Proceedings of the 10th European Conference and Technology Exhibition Biomass for Energy and Industry, Würzburg, Germany, 8–11 June 1998.

35. Eiland, F.; Leth, M.; Klamer, M.; Lind, A.-M.; Jensen, H.; Iversen, J. C and N Turnover and Lignocellulose Degradation During Composting of *Miscanthus* Straw and Liquid Pig Manure. *Compost. Sci. Util.* **2001**, *9*, 186–196. [CrossRef]

36. Weger, J.; Bubeník, J.; Dubský, M. Hodnocení vlivu hnojení na růst a výnos klonů vrb a topolů v prvních čtyřech letech pěstování (Evaluation of influence of fertilization on growth and yield of willow and poplar clones in first four years). *Acta Pruhoniciana* **2010**, *94*, 13–20. Available online: https://www.vukoz.cz/acta/dokumenty/acta_94/Acta-94_komplet-cz.pdf (accessed on 25 August 2020).

37. Schwarz, H.; Liebhard, P.; Ehrendorfer, K.; Ruckenbauer, P. The effect of fertilization on yield and quality of *Miscanthus* sinensis 'Giganteus'. *Ind. Crop. Prod.* **1994**, *2*, 153–159. [CrossRef]

38. Nazli, R.I.; Tansi, V.; Öztürk, H.H.; Kusvuran, A. *Miscanthus*, switchgrass, giant reed, and bulbous canary grass as potential bioenergy crops in a semi-arid Mediterranean environment. *Ind. Crop. Prod.* **2018**, *125*, 9–23. [CrossRef]

39. Iqbal, Y.; Gauder, M.; Claupein, W.; Graeff-Hönninger, S.; Lewandowski, I. Yield and quality development comparison between *Miscanthus* and switchgrass over a period of 10 years. *Energy* **2015**, *89*, 268–276. [CrossRef]

40. Loeffel, K.; Nentwig, W. *Ökologische Beurteilung des Anbaus Von Chinaschilf (Miscanthus Sinensis) Anhand Faunistischer Untersuchung*; Agrarökologie Verlag Agrarökologie: Hannover, Germany, 1997.

41. Witzel, C.-P.; Finger, R. Economic evaluation of *Miscanthus* production—A review. *Renew. Sustain. Energy Rev.* **2016**, *53*, 681–696. [CrossRef]

42. Xue, S.; Lewandowski, I.; Kalinina, O. *Miscanthus* establishment and management on permanent grassland in southwest Germany. *Ind. Crop. Prod.* **2017**, *108*, 572–582. [CrossRef]

43. Hůla, J.; Procházková, B. *Vliv Minimalizačních a Půdoochranných Technologií na Plodiny, Půdní Prostředí a Ekonomiku (The Impact of Minimizing and Soil Protection Technologies on Crops, Soil Environment and Economy)*; Ústav Zemědělských a Potravinářských Informací: Praha, Czech Republic, 2002.

44. Raus, A.; Čechová, V.; Šabatka, J. The Effect of Soil Protection Treatment on Soil Organic Matter. *Úroda* **1999**, *47*, 16–17.

45. Strašil, Z. Study of *Miscanthus sinensis*—Source for energy utilization. In Proceedings of the 15th European Biomass Conference and Exhibition, From Research to Market Deployment, Berlin, Germany, 7–11 May 2007; pp. 824–827.

46. McCalmont, J.P.; Hastings, A.; McNamara, N.P.; Richter, G.M.; Robson, P.R.H.; Donnison, I.S.; Clifton-Brown, J. Environmental costs and benefits of growing *Miscanthus* for bioenergy in the UK. *GCB Bioenergy* **2017**, *9*, 489–507. [CrossRef] [PubMed]

47. Sage, R.B.; Cunningham, M.; Haughton, A.J.; Mallott, M.D.; Bohan, D.A.; Riche, A.; Karp, A. The environmental impacts of biomass crops: Use by birds of *Miscanthus* in summer and winter in southwestern England. *IBIS* **2010**, *152*, 487–499. [CrossRef]

48. Kalinina, O. Potential Ecological Impacts of *Miscanthus*, a Prospective Bioenergy Crop, in Agro- and Natural-Ecosystems. In Proceedings of the Book of Abstracts of the 43rd Annual Meeting of the Ecological Society of Germany, Austria and Switzerland, Potsdam, Germany, 9–13 September 2013. Available online: https://www.researchgate.net/publication/258225927_Potential_ecological_impacts_of_Miscanthus_a_prospective_bioenergy_crop_in_agro--_and_natural_ecosystems (accessed on 29 September 2020).

49. Lewandowski, I.; Clifton-Brown, J.; Scurlock, J.; Huisman, W. *Miscanthus*: European experience with a novel energy crop. *Biomass Bioenergy* **2000**, *19*, 209–227. [CrossRef]

50. Global Invasive Species Database Species Profile: Miscanthus Sinensis. 2011. Available online: http://www.iucngisd.org/gisd/speciesname/Miscanthus+sinensis (accessed on 29 September 2020).

51. Global Compendium of Weeds Miscanthus Sinensis (Poaceae). 2007. Available online: http://www.hear.org/gcw/species/Miscanthus_sinensis/ (accessed on 29 September 2020).

52. Scally, L.; Hodkinson, T.; Jones, M.B. Origins and Taxonomy of *Miscanthus*. In *Miscanthus for Energy and Fibre*; Jones, M., Walsh, M., Eds.; Routledge: London, UK, 2007; pp. 1–9.

53. Meyer, M.H.; Tchida, C.L. *Miscanthus* Anderss. Produces Viable Seed in Four USDA Hardiness Zones. *J. Environ. Hortic.* **1999**, *17*, 137–140. [CrossRef]

54. Quinn, L.D.; Allen, D.J.; Stewart, J.R. Invasiveness potential of *Miscanthus* sinensis: Implications for bioenergy production in the United States. *GCB Bioenergy* **2010**, *2*, 310–320. [CrossRef]

55. Barney, J.N.; Ditomaso, J.M. Nonnative Species and Bioenergy: Are We Cultivating the Next Invader? *Bioscience* **2008**, *58*, 64–70 [CrossRef]

56. Severa, M.; Weger, J. Potvrzení generativního šíření vybraných druhů ozdobnic (*Miscanthus* sp.) v přírodních podmínkách České republiky (Confirmation of generative dispersal of selected species of *Miscanthus* species in environmental conditions of the Czech Republic). *Acta Pruhoniciana* **2010**, *96*, 75–78.

57. Mareček, F. *Zahradnický Slovník Naučný 3, CH-M (Horticultural Dictionary)*, 1st ed.; Ústav Zemědělských a Potravinářských Informací: Praha, Czech Republic, 1997.

58. Kubát, K.; Hrouda, L.; Chrtek, J.J.; Kaplan, Z.; Kirschner, J.; Štěpánek, J. *Klíč ke Květeně České Republiky (Key to the Flora of the Czech Republic)*; Academia: Praha, Czech Republic, 2002.

59. Pyšek, P.; Sádlo, J.; Mandák, B. Catalogue of alien plants of the Czech Republic. *Preslia* **2002**, *74*, 97–186.

60. Pyšek, P.; Danihelka, J.; Sádlo, J.; Chrtek, J., Jr.; Chytrý, M.; Jarošík, V.; Kaplan, Z.; Krahulec, F.; Moravcová, L.; Pergl, J.; et al. Catalogue of alien plants of the Czech Republic: Checklist update, taxonomic diversity and invasion patterns—2nd edition. *Preslia* **2012**, *84*, 155–255.

61. IUSS Working Group WRB. *World Reference Base for Soil Resources 2014, Update 2015 International Soil Classification System for Naming Soils and Creating Legends for Soil Maps*; FAO: Rome, Italy, 2015.

62. Kožnarová, V.; Klabzuba, J. Recommendation of World Meteorological Organization to describing meteorological or climatological conditions—Information. *Plant Soil Environ.* **2011**, *48*, 190–192. [CrossRef]

63. Weger, J.; Vávrová, K.; Kašparová, L.; Bubeník, J.; Komárek, A. The influence of rotation length on the biomass production and diversity of ground beetles (Carabidae) in poplar short rotation coppice. *Biomass Bioenergy* **2013**, *54*, 284–292. [CrossRef]

64. Weger, J.; Strašil, Z. Hodnocení polního pokusu s ozdobnicemi (*Miscanthus* sp.) po dvou letech růstu na různých stanovištích (Evaluation of field experiment with *Miscanthus* clones (*Miscanthus* sp.) on two sites after two years). *Acta Pruhoniciana* **2009**, *92*, 27–34.

65. Vavrova, K.; Knápek, J. Economic Assessment of *Miscanthus* Cultivation for Energy Purposes in the Czech Republic. *J. Jpn. Inst. Energy* **2012**, *91*, 485–494. [CrossRef]

66. Havlíčková, K.; Weger, J.; Knápek, J. Modelling of biomass prices for bio-energy market in the Czech Republic. *Simul. Model. Pract. Theory* **2011**, *19*, 1946–1956. [CrossRef]

67. Stapleton, R.C.; Brealey, R.; Myers, S. Principles of Corporate Finance. *J. Financ.* **1981**, *36*, 982. [CrossRef]

68. CHMI Portal. Historical Data: Weather: Territorial Precipitation. Available online: http://portal.chmi.cz/historicka-data/pocasi/uzemni-srazky?l=en (accessed on 25 September 2020).

69. Němec, J. *Bonitace a Oceňování Zemědělské Půdy České Republiky (Typology and Valuation of Agricultural Land in the Czech Republic)*; VUZE: Praha, Czech Republic, 2002.

70. Strašil, Z.; Weger, J. Preliminary zoning of agricultural land for *Miscanthus* (*M.* × *giganteus*) for the Czech Republic. *Ital. J. Agron.* **2008**, *3*, 559–560.

71. Strašil, Z.; Weger, J.; Hutla, P.; Kára, J. Ozdobnice (*Miscanthus*) jako energetická surovina (*Miscanthus* as a source of energy). *Agri-Tech Sci.* **2015**, *9*, 1–11.

72. Vávrová, K.; Knápek, J.; Weger, J. Modeling of biomass potential from agricultural land for energy utilization using high resolution spatial data with regard to food security scenarios. *Renew. Sustain. Energy Rev.* **2014**, *35*, 436–444. [CrossRef]

73. Vávrová, K.; Knápek, J.; Weger, J.; Králík, T.; Beranovský, J. Model for evaluation of locally available biomass competitiveness for decentralized space heating in villages and small towns. *Renew. Energy* **2018**, *129*, 853–865. [CrossRef]

74. Knápek, J.; Vávrová, K.; Valentová, M.; Vašíček, J.; Králík, T. Energy biomass competitiveness-three different views on biomass price. *Wiley Interdiscip. Rev. Energy Environ.* **2017**, *6*, e261. [CrossRef]

75. Tzbinfo. Porovnání Nákladů na Vytápění Podle Druhu Paliva—TZB-Info (Comparison of Heating Costs by Type of Fuel). 2020. Available online: https://vytapeni.tzb-info.cz/tabulky-a-vypocty/139-porovnani-nakladu-na-vytapeni-podle-druhu-paliva (accessed on 4 October 2020).

76. ERO. Cenové Rozhodnutí Energetického Regulačního Úřadu č. 7/2020 (Price Decision of Energy Regulatory Office of the Czech Republic NO. 7/2020), Jihlava. 2020. Available online: https://www.eru.cz/-/cenove-rozhodnuti-c-7-2020 (accessed on 29 October 2020).

77. Gauder, M.; Graeff-Hönninger, S.; Lewandowski, I.; Claupein, W. Long-term yield and performance of 15 different *Miscanthus* genotypes in southwest Germany. *Ann. Appl. Biol.* **2012**, *160*, 126–136. [CrossRef]

78. Ben Fradj, N.; Rozakis, S.; Borzęcka, M.; Matyka, M. *Miscanthus* in the European bio-economy: A network analysis. *Ind. Crop. Prod.* **2020**, *148*, 112281. [CrossRef]

79. Weger, J.; Havlíčko, K.V. The Evaluation of selected willow and poplar clones for short rotation coppice (SRC) after three harvests. In Proceedings of the 17th European Biomass Conference & Exhibition, Hamburg, Germany, 29 June–3 July 2009; pp. 227–230.

80. Strašil, Z.; Weger, J. Studium kostřavy rákosovité (*Festuca arundinacea* Schreb.) pěstované pro energetické využití (Study of tall fescue (*Festuca arundinacea* Schreb.) grown for energy purposes). *Acta Pruhoniciana* **2010**, *96*, 19–26.

81. Petříková, V.; Weger, J. *Pěstování Rostlin Pro Energetické a Technické Využití (Growing of Plants for Energy and Technical Utilization)*; Profi Press: Praha, Czech Republic, 2015.

82. Kätterer, T.; Andrén, O. Growth dynamics of reed canarygrass (*Phalaris arundinacea* L.) and its allocation of biomass and nitrogen below ground in a field receiving daily irrigation and fertilisation. *Nutr. Cycl. Agroecosyst.* **1999**, *54*, 21–29. [CrossRef]

83. Stolarski, M.J.; Śnieg, M.; Krzyżaniak, M.; Tworkowski, J.; Szczukowski, S.; Graban, Ł.; Lajszner, W. Short rotation coppices, grasses and other herbaceous crops: Biomass properties versus 26 genotypes and harvest time. *Ind. Crop. Prod.* **2018**, *119*, 22–32. [CrossRef]

84. Weger, J.; Hutla, P.; Bubeník, J. Yield and fuel characteristics of willows tested for biomass production on agricultural soil. *Res. Agric. Eng.* **2016**, *62*, 155–161. [CrossRef]

85. Tröger, F.; Barbu, M.C.; Seemann, C. Verstärkung von *Miscanthus* und Holzspanplatten mit Flachsfasermatten. *Holz Roh- Werkst.* **1995**, *53*, 268. [CrossRef]

86. Schubert, B.; Simatupang, M.H. Eignung von *Miscanthus* sinensis zur Herstellung von Zementspanplatten. *Holz Roh- Werkst.* **1992**, *50*, 492. [CrossRef]

87. Winkler, B.; Mangold, A.; Von Cossel, M.; Clifton-Brown, J.; Pogrzeba, M.; Lewandowski, I.; Iqbal, Y.; Kiesel, A. Implementing *Miscanthus* into farming systems: A review of agronomic practices, capital and labour demand. *Renew. Sustain. Energy Rev.* **2020**, *132*, 110053. [CrossRef]

88. Jørgensen, U. Benefits versus risks of growing biofuel crops: The case of *Miscanthus. Curr. Opin. Environ. Sustain.* **2011**, *3*, 24–30. [CrossRef]

89. VUKOZ. Seznam Rostlin Vhodných k Pěstování Za Účelem Využití Biomasy Pro Energetické Účely z Pohledu Minimalizace Rizik Pro Ochranu Přírody a Krajiny (List of Plants Suitable for Cultivation and Production of Energy Biomass and Minimizing Risks to Nature). 2020. Available online: https://www.vukoz.cz/index.php/sluzby/energeticke-plodiny (accessed on 4 October 2020).

90. Hruška, M.; Gimunová, T.; Kohlíček, V.; Novotný, I.; Perglerová, M.; Reininger, D.; Smatanová, M.; Trapl, K.; Voltr, V.; Havelka, J.; et al. *Situační a Výhledová Zpráva Půda (Situational and Outlook Report—Soil)*; Ministry of Agriculture: Praha, Czech Republic, 2018; ISBN 978-80-7434-476-3.

 *agriculture*

Article

The Effect of Mineral and Organic Fertilization on Common Osier (*Salix viminalis* L.) Productivity and Qualitative Parameters of Naturally Acidic *Retisol*

Gintaras Šiaudinis [1,*], Danutė Karčauskienė [1], Jūratė Aleinikovienė [1,2], Regina Repšienė [1] and Regina Skuodienė [1]

1 Vėžaičiai Branch of the Lithuanian Research Centre for Agriculture and Forestry, Gargždų 29, 96216 Klaipėda distr., Vėžaičiai, Lithuania; danute.karcauskiene@lammc.lt (D.K.); jurate.aleinikoviene@vdu.lt (J.A.); regina.repsiene@lammc.lt (R.R.); regina.skuodiene@lammc.lt (R.S.)
2 Institute of Agroecosystems and Soil Science, Vytautas Magnus University Agriculture Academy, Studentų 11, LT-53362 Kaunas distr., Akademija, Lithuania
* Correspondence: gintaras.siaudinis@lammc.lt

Abstract: One of the potential options for sewage sludge as an alternative organic material is the fertilization of energy crops. To evaluate the effect of granulated sewage sludge and mineral fertilization N60P60K60 on common osier's (*Salix viminalis* L.) biomass productivity and soil parameters, field trials were held in Western Lithuania's naturally acidic *Retisol* (WB 2014; pH_{KCl} 4.35–4.58). After four years of cultivation and dependent on fertilization type, common osier dry matter (DM) yield varied from 49.60 to 77.92 t ha^{-1}. Higher DM yield was related to an increased number of stems/plants. The application of a 90 t ha^{-1} sewage sludge rate had a significant and positive impact on common osier productivity, as well as on the increment of soil organic carbon, total N, and mobile P_2O_5 content in the upper 0–30 cm soil layer. The use of both sewage sludge rates (45 and 90 t ha^{-1}) had a similar impact on soil bulk density, water-stable aggregates, and the active soil microbial biomass. Annual mineral fertilization had little effect on the parameters studied. When growing common osier in *Retisol*, 45 t ha^{-1} of a single sewage sludge rate was enough to maintain both plant and soil productivity.

Keywords: common osier; fertilization; dry matter yield; soil chemical parameters; soil bulk density; water-stable aggregates; soil microbial carbon

Citation: Šiaudinis, G.; Karčauskienė, D.; Aleinikovienė, J.; Repšienė, R.; Skuodienė, R. The Effect of Mineral and Organic Fertilization on Common Osier (*Salix viminalis* L.) Productivity and Qualitative Parameters of Naturally Acidic *Retisol*. *Agriculture* **2021**, *11*, 42. https://doi.org/10.3390/agriculture11010042

Received: 9 November 2020
Accepted: 7 January 2021
Published: 9 January 2021

Publisher's Note: MDPI stays neutral with regard to jurisdictional clai-ms in published maps and institutio-nal affiliations.

1. Introduction

Sludge is a byproduct of domestic or industrial wastewater treatments. Recycling sewage sludge has remained a significant problem in many countries worldwide [1–4]. By containing a high concentration of organic carbon, nitrogen, phosphorus, and other nutrients, sewage sludge serves as a good organic matter source for plants, either providing the soil with beneficial physical and microbial properties [5–7]. As a sewage sludge substrate is of organic origin, the use of the substrate might maintain soil productivity for several consecutive years. Börjesson and Kätterer point out that the application of 12 t ha^{-1} sewage sludge every four years represents a valuable resource for improving soil fertility vis-à-vis soil organic matter and soil structure; however, its efficiency for nutrient (particularly phosphorus, nitrogen) cycling is very low within this timeframe [8]. Most improvement for the soil aggregate stability is associated with an increase in soil organic carbon content [9,10]. However, other sewage sludge constituents (e.g., heavy metals, pathogens) could be harmful to the environment and human health [11].

According to Polish research, the cultivation of willow on soil treated with 3 and 9 t ha^{-1} sewage sludge results in a gradual increase of humus fractions, total organic carbon content, and bacterial abundance, and a large increase of willow biomass. Organic compounds and high content of sewage sludge nutrients activate soil microbial activity [12].

Low rates of sewage sludge increase soil microbial activity; however, some other research indicates that the excessive application of sewage sludge can cause the accumulation of both organic and inorganic pollutants, which may cause a negative effect on the soil ecosystem [13,14]. It is important to keep in mind that the application of sewage sludge can also result in environmental problems. Apart from organic content and nutrients, sludge also includes toxic compounds, heavy metals, dissolved inorganic salts, chlorinated lignin, and phenolic derivatives [15,16]. The application of sewage sludge can cause a negative ecological impact on terrestrial ecosystems and pose a human health risk [11,17]. For this reason, the use of sewage sludge as an organic fertilizer is limited in many countries worldwide [2,5,18].

Since sewage sludge contains unwanted pollutants, its utilization for traditional crop fertilization is problematic. Yet, to reduce cultivation costs, sewage sludge might be an appropriate alternative fertilizer for energy crops. Many authors have noted that to avoid competition with traditional food crops, energy crops (as well as other non-food crops) should be grown in less productive soils, polluted soils, brownfields, marginal, set-aside, and abandoned lands, all of which might be appropriate for energy crop cultivation [19–22]. For example, *Retisols* and *Fluvisols* are typically found in Western Lithuania. However, due to these soils' high acidity and worsening physical, chemical, and microbial properties, traditional farming is often unprofitable [23]. With increased biomass being used for alternative energy purposes, a share of such fertile land could be assigned for energy crop cultivation.

Salix species are fast-growing and high-yielding; therefore, they can be widely cultivated worldwide. Many studies have assessed the impact of sewage sludge on both common osier (*Salix viminalis* L.) productivity and qualitative indicators [24,25]. However, before our field experiment, we found scarce information on how a common osier (*Salix viminalis* L.) is suitable for cultivation in highly acidic *Retisol* and how sewage sludge affects soil parameters in the common osier growing site.

By executing the experiment, we set a goal complex to evaluate the impact of single sewage sludge on *Salix viminalis* productivity, and changes of soil chemical, physical, and microbial properties of acid moraine loam soil in *Retisol*.

2. Materials and Methods

2.1. Location of the Experimental Site

The experiment with an energy crop common osier (*Salix viminalis* L.) was established at the Vėžaičiai Branch of the Lithuanian Research Centre for Agriculture and Forestry. The site was located at the eastern edge of a seaside lowland area (Western Lithuania, 55°43′ N, 21°27′ E).

2.2. Soil Characteristics

The soil type was a naturally acid moraine loam Bathygleyic Dystric Glossic Retisol (WRB 2014). The following soil chemical parameters (in upper soil layer) were evaluated prior to the experiment: pH_{KCl} was 4.27–4.59, mobile aluminum was 13.69–34.21 mg kg^{-1}, total nitrogen was 0.11–0.12%, total carbon was 1.19–1.25%, mobile P_2O_5 was 50.3–68.3 mg kg^{-1}, and mobile K_2O was 251–303 mg kg^{-1}.

2.3. Mineral Fertilizers and Granulated Sewage Sludge

Mineral fertilizers. The annual application rates for nitrogen, phosphorus and potassium were equal: 60 kg ha^{-1} N, P_2O_5, and K_2O (N60P60K60) (in active ingredient).

Sewage sludge. The chemical composition of the granulated sewage sludge was as follows: pH—5.56, total nitrogen—33.4 g kg^{-1}, total phosphorus—5.02 g kg^{-1}, total potassium—2.80 g kg^{-1}, organic matter—64.97%. The sewage sludge contained the following heavy metals concentrations: lead (Pb)—14.47 mg kg^{-1}, cadmium (Cd)—0.44 mg kg^{-1}, chromium (Cr)—11.51 mg kg^{-1}, copper (Cu)—47.8 mg kg^{-1}, nickel (Ni)—8.22 mg kg^{-1}, zinc (Zn)—287mg kg^{-1}, and mercury (Hg)—0.96 mg kg^{-1}.

Using 45 t ha^{-1} sewage sludge rate, the following amounts of nutrients were inserted into the soil: 1503 kg of total nitrogen, 230 kg of total phosphorus, and 126 kg of total potassium. Accordingly, by the use of 90 t ha^{-1} sewage sludge rate, 3006 kg of total nitrogen, 460 kg of total phosphorus, and 252 kg of total potassium were inserted into the soil.

2.4. Experimental Design

Field research began in 2013. Common osier's (cv. "Tordis") cuttings of about 0.30 m long were planted in the soil on 5 May 2013. Each treatment was composed of two parallel 10 m long rows. The distance between the rows was 0.75 m, the distance between each plant was 0.50 m, the distance between the rows of different treatments was 1.25 m. Next year, i.e., on 23 April 2014, to increase the branching ability, common osier's stems were cut at ~5 cm height. The weight of the first-year stems was not calculated.

The experiment was composed of four treatments: (1) Control (not fertilized); (2) N60P60K60 (mineral fertilization) (in active ingredient); (3) 45 t ha^{-1}, and (4) 90 t ha^{-1} sewage sludge rates. All four treatments were randomly allocated. The number of replications–3. The fertilization of granulated sewage sludge was done once in the 2nd growing year (on 6 May 2014). The granules were immediately inserted into the soil by tillage implements. NPK fertilization was done each year at the beginning of spring vegetation.

2.5. Sampling and Analytical Methods

Common osier yield was harvested after four years of cultivation (2014–2017) on 19 September 2017. The following structural parameters were evaluated: the number of stems per plant stems height, and biomass yield. To evaluate these parameters, five typical plants were chosen from each treatment from all three replications. The common osier dry mass (DM) mass yield was measured by drying plant samples at 105 °C to the constant weight. Dried plant samples were weighed and recalculated into dry matter (DM) yield (t ha^{-1}).

Soil chemical analyses at the growing site of common osiers were done in 2013 (at the beginning of the experiment) and 2016 (in the 3rd experimental year). In both cases, soil samples (in 0–30 cm upper soil layer) were taken in October. The following parameters were evaluated: pH$_{KCl}$ was measured by a potentiometric method in 1 M KCl (1:2.5, *w/v*) extract (ISO 10390:2005); organic C (C$_{org}$) content-by a spectrophotometric measurement at 590 nm after dichromatic oxidation using glucose as a standard (ISO 10694:1995), total N (N$_{tot}$) content—by the Kjeldahl method (ISO 11261-1995), mobile P$_2$O$_5$, and K$_2$O contents-by extraction (A-L) method (both by LVP D-07:2016).

Soil samples for soil physical analysis were taken from the topsoil in 2015 and 2017. Soil samples for soil physical analysis were taken from the topsoil in 2015 and 2017. Dry aggregates size distribution was determined by the standard dry and wet sieving Savinov method [26]. Briefly, 1000 g of air-dried, soil sample is sieved through a nest of sieves having 10, 5, 3, 2, 1, 0.5, and 0.25 mm square openings so eight aggregate size classes are obtained (>10, 10–5, 5–3, 3–2, 2–1, 1–0.5, 0.5–0.25, and <0.25 mm. The soil of each aggregate size classes is weighed separately, and the percentage of the fraction is calculated from the total soil weight. A sample of 50 g is taken from the aggregate fractions in proportion to their percentage composition for wet sieving analysis. By the wet sieving procedure 6 classes were separated >3, 3–2, 2–1, 1–0.5, 0.5–0.25, and <0.25 mm. The soil bulk density (100 cm^3) was estimated according to the Kachinsky method. Soil moisture content was measured by the weighting method.

Soil samples for microbial analysis were taken twice per 2014–2016 in spring and autumn for three treatments (Control, 45 t ha^{-1}, and 90 t ha^{-1} sewage sludge) in 0–30 cm upper soil layer. The chloroform fumigation-extraction (CFE) method was used to evaluate soil microbial biomass carbon (µg g^{-1} C) [27].

2.6. Statistical Analysis

To evaluate the significance of the obtained biomass productive parameters (i.e. number of stems/plants, stems height, stems diameter, and dry mass yield), a one-way statistical analysis was performed on the fertilization rate, using analysis of variance (ANOVA) at LSD_{05} and LSD_{01} (95% and 99% probability levels).

3. Results

3.1. Common Osier Yield and Structure

The biometric parameters for the common osier yield are presented in Table 1. Biomass yield was harvested after 4 years of cultivation.

Table 1. The structural parameters of the common osier's yield after 4 years of cultivation.

Treatments	Number of Stems/Plants	Stems Height, cm	Stems Diameter, mm	DM Yield, t ha^{-1}
Control	2.34	567	29.98	49.60
N60P60K60	2.80	552	29.12	52.00
45 t ha^{-1} sewage sludge	2.87	554	30.80	65.68 **
90 t ha^{-1} sewage sludge	3.47 *	528	31.60	77.92 **
$LSD_{05/01}$	1.12/ns	ns/ns	ns/ns	6.90/10.45

*,** significant at 95% and 99% probability levels, respectively; ns—not significant.

In comparison with the unfertilized plot (in control treatment), fertilizing 90 t ha^{-1} of sewage sludge caused the number of stems/plants too, on average, increase to 3.47 Irrespective of fertilization type, the stem height varied from 528 to 567 cm. The use of sewage sludge also increased stem diameter, though the increment was not statistically substantial. Both (45 and 90 t ha^{-1}) sewage sludge rates increased total dry matter (DM) yield (accumulated per 4 growing years) to 65.68 and 77.92 t ha^{-1}, respectively (significant at 99% probability level). Thus, compared to the control treatment (when growing without fertilization), DM yield increased by 32.4–57.1%. There was a positive average correlation (+0.66) between the number of stems/plants and DM yield: $Y_{(DM\ yield)} = 24.67 + 0.0008x_{(number\ of\ stems/plants)}$. The other two parameters had a weak correlation with DM yield.

It should be noted that the annual use of mineral NPK fertilizers had a weak impact on common osier's DM yield; the increase was not significant. Earlier studies in Sweden showed that nitrogen fertilizers had a positive effect on common osier's yield during the 2nd and 3rd growing years [28].

3.2. Soil Chemical Properties

During the experimental years, soil pH_{KCl} values remained substantially unchanged (Table 2). In 2013, the average soil pH in the common osier's growing site was 4.40 ± 0.16. After three experimental years (in September 2016), irrespective of fertilization type, soil pH_{KCl} varied from 4.41 to 4.49. Thus, the application of sewage sludge had no impact on soil acidity level.

Table 2. Soil chemical content at the growing site of a common osier, specifically for the 0–30 cm upper soil layer (2013 and 2016).

Treatments	pH_{KCl}	Organic C (%)	Total N (%)	$C_{org}:N_{tot}$	Mobile	
					P_2O_5 (mg kg^{-1})	K_2O (mg kg^{-1})
			2013 (before the experiment)			
	4.40 ± 0.16	1.18 ± 0.06	0.07 ± 0.01	16.51 ± 0.98	59.3 ± 13.6	277 ± 0.02
			2016			
Control	4.44	1.16	0.07	15.69	69.4	296
45 t ha^{-1} sewage sludge	4.41	1.34 *	0.09 *	14.92	332	226
90 t ha^{-1} sewage sludge	4.49	1.52 **	0.11 **	13.74 *	816	251
LSD$_{05/01}$	ns/ns	0.15/0.23	0.02/0.03	2.71/ns	146/222	ns/ns

*,** significant at 95% and 99% probability levels, respectively; ns—not significant.

In 2013, the average organic C (C_{org}) content in the topsoil layer was $1.18 \pm 0.06\%$. Over three years of research, the application of 45 and 90 t ha^{-1} sewage sludge rates caused a substantial increase of organic C content (at 95% and 99% probability levels, respectively).

Before the experiment, the average N_{tot} content was $0.07 \pm 0.01\%$. The application of both sewage sludge rates significantly increased N_{tot} content to 0.09–0.11% (at 99% probability level). In the control treatment, even though the common osier utilized high amounts of N for biomass accumulation, the total N content in the topsoil did not significantly change throughout the experimental period.

The application of both sewage sludge rates decreased $C_{org}:N_{tot}$ ratio from 16.51 ± 0.98 (in 2013) to 13.74 (in 2016).

In 2013, the amounts of phosphorus (P_2O_5) and potassium (K_2O) at the experimental site were 59.3 ± 13.6 and 277 ± 0.02 mg kg^{-1}, respectively, indicating that soil reserves were low. On the contrary, potassium concentration in the upper soil layer was sufficient. In 2016, at the end of the growing rotation, mobile P_2O_5 content in the topsoil sharply increased to 332–816 mg kg^{-1} (significant at the 99% probability level). Meanwhile, at the end of the field experiment study, mobile K_2O content in the upper soil layer remained largely unchanged.

3.3. Soil Aggregate Composition and Aggregate Stability

The obtained research data revealed that during the research period the majority (66–73%) of aggregates in the moraine loam soil were composed of agronomically and ecologically valuable mesoaggregates (0.25–5 mm) (Table 3).

Table 3. Soil aggregate composition and the amount of water-stable aggregates in the growing site of common osier dependent on fertilization type in 2015 and 2017.

Treatment	2015 (One Year after Sewage Sludge Application)					2017 (Three Years after Sewage Sludge Application)				
	macro-aggre-gates > 5 mm	meso-aggrega-tes 5–0.25 mm	mikro-aggrega-tes < 0.25 mm	water-stable aggrega-tes > 1.0 mm	water-stable aggrega-tes > 0.25 mm	macro-aggrega-tes > 5 mm	meso-aggrega-tes 5–0.25 mm	mikro-aggrega-tes <0.25 mm	water-stable aggre-gates > 1.0 mm	water-stable aggre-gates > 0.25mm
Control	11.66	67.48	20.86	13.10	47.53	9.15	70.54	20.31	18.8	55.3
NPK	8.02	68.36	23.62	11.68	47.25	-	-	-	-	-
45 t ha^{-1} sewage sludge	12.35	65.86	21.79	21.42 *	64.55 *	9.53	72.95 *	17.52	16.4	54.6
90 t ha^{-1} sewage sludge	11.58	65.60	22.82	14.45	54.77	14.10 *	69.43	16.47 *	18.8	57.2
LSD$_{05}$	ns	ns	ns	8.64	11.63	1.37	2.36	2.81	ns	ns

*—significant at 95% probability level; ns—not significant.

The amount of these aggregates varied slightly from year to year, depending on climatic conditions and fertilization. In 2017, the share of mesoaggregates was 4% higher than in 2015. Compared to the unfertilized soil (i.e., the control), the most valuable mesoaggregates were formed in the common osier growing site three years after the application of the lower sewage sludge rate (45 t ha^{-1}). Sewage sludge had a positive effect not only on the formation of aggregates of different sizes but also on their stability. One year after 45 t ha^{-1} of sewage sludge application, ecologically valuable water-stable aggregates (>1.0 mm and >0.25 mm) accounted for 21.4 and 64.6%, respectively.

In comparison to the unfertilized soil (i.e., the control treatment), the application of 45 and 90 t ha^{-1} increased the amount of water-stable aggregates by 38 and 26%, respectively; and by 45% and 26% in comparison to NPK application. In 2017, three years after the application of sewage sludge (45 t ha^{-1}), the amount of water-resistant aggregates in the soil (>1.0 mm and >0.25 mm) was 5 and 10% lower, respectively, than those determined one year after the application of sludge. This indicates that sewage sludge did not have a long-lasting effect on the stability of aggregates.

According to the average data of 2015 and 2017, in the soil where sewage sludge was used for fertilization, the water-stable aggregates was 9–16% (>0.25 mm) and 4–18% (>1.0 mm) higher compared to the soil in which the sewage sludge was not applied (Control) (Figure 1).

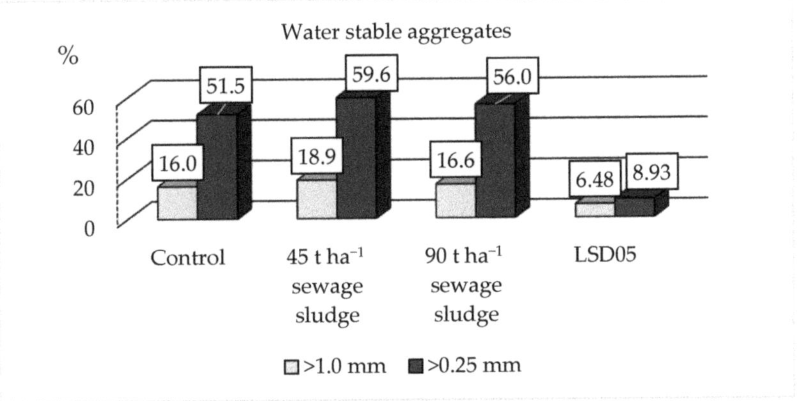

Figure 1. The dependence of soil water-stable aggregates on the fertilization type at the common osier growing site. Average data for 2015 and 2017.

3.4. Soil Bulk Density and Moisture

During the study period, in the common osier growing site, soil bulk density values fluctuated in the range of 1.19–1.32 Mg m^{-3} (Table 4).

The lowest bulk density value was determined in the soil where the lower rate of sewage sludge (45 t ha^{-1}) was applied. Sewage sludge substantially reduced soil bulk density. On average, the soil bulk density was 6.0–7.5% lower compared to unfertilized soil (i.e., the control). The application of sewage sludge (especially its highest 90 t ha^{-1} rate) caused a higher accumulation of soil moisture (by 11.2%) than the control treatment (without fertilization). Compared to soil fertilized with mineral fertilizers, fertilization with organic fertilizers (sewage sludge) accumulated a higher amount of organic carbon in the soil. It was a result of higher organic C content and moisture content in the soil applied by sewage sludge.

Table 4. Soil bulk density and moisture in relation to fertilization type at the common osier growing site in 2015 and 2017.

Treatment	2015 (One Year after Sewage Sludge Application)		2017 (Three Years after Sewage Sludge Application)		Average Per 2015–2017	
	Bulk Density $Mg\ m^{-3}$	Moisture %	Bulk Density $Mg\ m^{-3}$	Moisture %	Bulk Density $Mg\ m^{-3}$	Moisture %
Untreated	1.32	22.13	1. 32	17.87	1.32	20.00
NPK	1.33	21.65	-	-	-	-
45 t ha^{-1} sewage sludge	1.19 *	20.56	1.28	20.42 *	1.24 *	20.49
90 t ha^{-1} sewage sludge	1.22 *	23.29	1.23	21.33 *	1.22 *	22.31
LSD$_{05}$	0.09	ns	0.09	1.90	0.60	ns

*—significant at the 95% probability level; ns—not significant.

3.5. Microbial Activity

The biomass of microorganisms in the soil is expressed as the amount of organic carbon (C) in the biomass, known as microbial biomass carbon [29,30]. The results are presented in Table 5.

Table 5. The changes in soil microbial biomass carbon ($\mu g\ g^{-1}$ C) in humus horizon (0–30 cm depth) at the common osier's site from 2014 to 2016.

Treatments	Soil Microbial Biomass Carbon ($\mu g\ g^{-1}$ C)	
	Spring	Autumn
	2014	
Control	423.1 ± 21.7 a	420.9 ± 14.7 a
45 t ha^{-1} sewage sludge	434.7 ± 21.7 a	432.1 ± 14.5 a
90 t ha^{-1} sewage sludge	457.0 ± 13.5 a	480.0 ± 24.0 b
On average per year	445.8 ± 12.7/	456.1 ± 14.8
	2015	
Control	432.4 ± 18.1 a	434.4 ± 19.2 a
45 t ha^{-1} sewage sludge	451.2 ± 12.9 ab	455.3 ± 20.7 a
90 t ha^{-1} sewage sludge	472.9 ± 16.1 b	492.3 ± 11.0 b
On average per year	462.1 ± 10.3/	473.8 ± 12.2
	2016	
Control	421.6 ± 8.0 a	433.2 ± 19.9 a
45 t ha^{-1} sewage sludge	539.1 ± 31.4 b	625.2 ± 22.1 c
90 t ha^{-1} sewage sludge	625.3 ± 20.6 c	625.6 ± 21.3 c
On average per year	582.2 ± 21.0/	625.4 ± 14.9

Mean values ± standard deviation. The differences between values by different letters are significant.

To evaluate the soil microbial biomass carbon ($\mu g\ C\ g^{-1}$), we sampled soil before the application of sewage sludge (at the beginning of May 2014). In autumn 2014, microbial biomass carbon in the 0–30 cm upper soil layer significantly increased up to 480 $\mu g\ C\ g^{-1}$ only where the 90 t ha^{-1} sewage sludge rate was applied. Similar results were obtained in 2015; in comparison to 2014, soil microbial biomass carbon increased slightly (particularly using the 90 t ha^{-1} sewage sludge rate). Thus, soil microbial biomass carbon during both the 2014 and

2015 seasons (both in spring and autumn) increased only slightly. Nevertheless, during the 3rd year of investigation (in 2016), soil microbial biomass carbon increased significantly in both sewage sludge application sites-in autumn, soil microbial biomass carbon in both sites (applied by 45 t ha^{-1} and 90 t ha^{-1} rates) reached up to 625 μg C g^{-1}. This study showed that the consistent increase in soil microbial biomass during three investigation years could be indicated as a result of the ecophysiological approach of soil microorganisms to adapt to changing soil physical properties and fluctuation of nutrients after sewage sludge application in the common osier's sites.

4. Discussion

Our field and laboratory results revealed that other than a substantial increase in common osier dry matter (DM), the single use of sewage sludge had a positive effect on changing the chemical, physical, and microbiological properties in *Retisol*. Since sewage sludge is an alternative organic matter that contains high amounts of macro- and micronutrients, it might be a successful substitute for mineral fertilization of energy crops and particularly common osier. According to the research data, the use of high 90 t ha^{-1} granulated sewage sludge significantly increased plant biomass. Further, N60P60K60 fertilization had a rather weak impact on DM yield. Irrespective of fertilization type, common osier DM yield varied from 49.60 to 77.92 t ha^{-1} (or 12.4–19.5 t ha^{-1} per year). Canadian authors noted that regarding *Salix* cultivars and the number of growing rotations, the application of sewage sludge caused the dry mass yield to increase from 15 to 22 t ha^{-1} per year [31]. Other research data conducted in Denmark showed that the application of very high sewage sludge rates (i.e., increasing N amount from 120 to 240 kg ha^{-1}) did not have any impact on common osier productivity [32]. Comparing these data with other experimental data in which *Salix viminalis* was annually fertilized with NPK fertilizers, it could be seen that annual mineral fertilization does not have any advantage over organic sewage sludge [33]. We estimated that the productivity of common osier (cv. "Tordis") depended mainly on the number of stems/plants. Neither branch height nor stems diameter had a reliable correlation with DM yield. Based on the literature data of other authors, it can be observed that common osier growth parameters such as the number of stems/plants, stem diameter, stems height, as well as their correlation with DM yield, depending on genotype, growth location, harvest rotation, and their correlation [34].

The application of sewage sludge had no significant impact on soil pH$_{KCl}$. In contrast to our results, other authors emphasized that sewage sludge substantially decreased soil acidity [35,36]. Other authors state that there is a direct relevance between soil pH and calcium carbonate content of sewage sludge [36]. The application of a 90 t ha^{-1} sewage sludge rate significantly increased C$_{org}$, N$_{tot}$, and mobile P$_2$O$_5$ content in the 0–30 cm upper soil layer. Other authors have reported the significant impact of sewage sludge on soil C$_{org}$ content and N$_{tot}$ content, as well as on the fastening of C and N mineralization processes [37–39]. The lower is C:N ratio, the more intense are N mineralization rates in soils [39,40]. Since mineral phosphorus resources are a limited resource, sewage sludge is a promising secondary source containing considerable amounts of phosphorus [41,42].

Water aggregate stability is considered an important indicator of soil physical quality, as it impacts soil functions such as soil aeration, the movement and storage of soil water, soil erodibility, and carbon sequestration. Soil aggregate stability is an important aspect of soil ecological services and health [36,43]. Fertilization with an organic amendment including sewage sludge could potentially alert soil physical properties and thereby affect aggregate stability [8,9]. According to our results, both 45 and 90 t ha^{-1} sewage sludge rates had a positive impact on increasing water-stable aggregates and decreasing soil bulk density. The effect of mineral fertilization on soil quality was insignificant.

Changes in microbial biomass carbon indicated that the increase in microbial carbon was not only due to the providing of the high content of available nutrients in sewage sludge but also due to the intensified rooting system of energy crops that could potentially stimulate microbial biomass increment [44–46].

An increase in soil microbial biomass carbon indicated that microbial activity could indirectly depend on sewage sludge application but either effected by intensified rooting of the energy crops with potential stimulation of microbial biomass increment

Although the 90 t ha^{-1} sewage sludge rate had a significant impact on DM productivity and soil chemical content, the parallel experiments indicated that sewage sludge might be energetically and environmentally inexpedient [7,47]. Thus, to improve soil qualitative parameters and obtain a high common osier DM yield, the application of 45 t ha^{-1} single sewage sludge rate might be sufficient.

Not only from the point of view of plant productivity and soil quality but also from the economic point of view, sewage sludge is a cost-effective organic matter, therefore it is superior to more costly mineral fertilization.

The positive effects of single sewage sludge application on common osier productivity and soil qualitative parameters should remain in the future growing seasons. The most important disadvantage of sewage sludge is its high concentration of heavy metals. We will soon publish another article detailing the dynamics of heavy metal concentration in soil (or its decontamination process) in *Salix viminalis* biomass during the experimental period. Further, field and laboratory experiments are continuing until 2022.

5. Conclusions

The studies conducted in naturally acid *Retisol* revealed that the single application of sewage sludge had a significant impact on plant and soil productivity. The use of 90 t ha^{-1} sewage sludge rate had the highest impact on common osier (*Salix viminalis* L.) dry matter (DM) yield per four years growing rotation. By contrast, the effect of annual mineral fertilizers on DM yield was significantly inferior. As concerning soil parameters, the use of sewage sludge did not change soil pH$_{KCl}$ level, whereas the application of 90 t ha^{-1} rate significantly increased organic C, total N, and mobile P$_2$O$_5$ content in the upper 0–30 cm soil layer over three years of research. Irrespective of sewage sludge application rate, the amount of water-stable aggregates increased, while soil bunk density tended to decrease. It was estimated that the significantly higher microbial biomass carbon content in soil was indicated only in the third year after sewage sludge application. This alteration showed that sewage sludge amendment effect on soil microbial biomass was prolonged and positive with stimulated soil microbial adaptation.

Author Contributions: Conceptualization, G.Š., D.K., and J.A.; methodology, G.Š., D.K., J.A., and R.S.; software, G.Š., D.K., J.A., and R.R.; data curation, G.Š. and D.K.; writing—review and editing, G.Š., D.K., J.A., R.R. and R.S.; supervision, G.Š. and D.K. All authors have read and agreed to the published version of the manuscript.

Funding: This research received no external funding.

Informed Consent Statement: Informed consent was obtained from all subjects involved in the study.

Data Availability Statement: The data presented in this study are available within the article.

Acknowledgments: The study was conducted in compliance with the long-term program "Plant biopotential and quality for multifunctional practice".

Conflicts of Interest: The authors declare no conflict of interest.

References

1. Kocik, A.; Truchan, M.; Rozen, A. Application of willows (*Salix viminalis*) and earthworms (Eisenia fetida) in sewage sludge treatment. *Eur. J. Soil Biol.* **2007**, *43*, 327–331. [CrossRef]
2. Charlton, A.; Sakrabani, R.; Tyrrel, S.; Casado, M.R.; McGrath, S.P.; Crooks, B.; Campbell, C.D. Long-term impact of sewage sludge application on soil microbial biomass: An evaluation using meta-analysis. *Environ. Pollut.* **2016**, *219*, 1021–1035. [CrossRef] [PubMed]
3. Kirchmann, H.; Börjesson, G.; Kätterer, T.; Cohen, Y. From agricultural use of sewage sludge to nutrient extraction: A soil science outlook. *Ambio* **2017**, *46*, 143–154. [CrossRef] [PubMed]

5. Leila, S.; Mhamed, M.; Hermann, H.; Mykola, K.; Oliver, W.; Christin, M.; Onyshchenko, E.; Bouchenaha, N. Fertilization value of municipal sewage sludge for Eucalyptus camaldulensis plants. *Biotechnol. Rep.* **2017**, *13*, 8–12.

6. Singh, R.P.; Agrawal, M. Potential benefits and risks of land application of sewage sludge. *Waste Manag.* **2008**, *28*, 347–358. [CrossRef]

7. Lederer, J.; Rechberger, H. Comparative goal-oriented assessment of conventional and alternative sewage sludge treatment options. *Waste Manag.* **2010**, *30*, 1043–1056.

8. Šiaudinis, G.; Karčauskienė, D.; Aleinikovienė, J. Assessment of a single application of sewage sludge on the biomass yield of Silphium perfoliatum and changes in naturally acid soil properties. *Zemdirb. Agric. Akad.* **2019**, *106*, 213–218.

9. Börjesson, G.; Kätterer, T. Soil fertility effects of repeated application of sewage sludge in two 30-year-old field experiments. *Nutr. Cycl. Agroecosyst.* **2018**, *112*, 369–385. [CrossRef]

10. Haydu-Haudeshell, C.A.; Graham, R.C.; Hendrix, P.F.; Peterson, A.C. Soil aggregate stability under chaparral species in Southen California. *Geoderma* **2018**, *310*, 201–208. [CrossRef]

11. Zoghlami, R.I.; Hamdi, H.; Mokni-Tlili, S.; Hechmi, S.; Khelil, M.N.; Aissa, N.B.; Moussa, M.; Bousnina, H.; Benzarti, S.; Jedidi, N. Monitoring the variation of soil quality with sewage sludge application rates in absence of rhizosphere effect. *Int. Soil Water Conserv. Res.* **2020**, *8*, 245–252. [CrossRef]

12. Buonocore, E.; Mellino, S.; De Angelis, G.; Liu, G.; Cooper, P.; Ulgiati, S. Life cycle assessment indicators of urban wastewater and sewage sludge treatment. *Ecol. Indic.* **2018**, *94*, 13–23. [CrossRef]

13. Urbaniak, M.; Toloczko, W.; Serwecinska, L.; Wyrwicka, A. The effect of sewage sludge application on soil properties and wilow (Salix sp.) cultivation. *Sci. Total Environ.* **2017**, *586*, 66–75. [CrossRef] [PubMed]

14. Usman, K.; Khan, S.; Ghulam, S.; Khan, M.U.; Khan, N.; Khan, M.A.; Khalil, S.K. Sewage sludge: An important biological resource for sustainable agriculture and its environmental implications. *Am. J. Plant Sci.* **2012**, *3*, 1708–1721. [CrossRef]

15. Iglesias, M.; Marguí, E.; Camps, F.; Hidalgo, M. Extractability and crop transfer of potentially toxic elements from Mediterranean agricultural soils following long-term sewage sludge applications as a fertilizer replacement to barley and maize crops. *Waste Manag.* **2018**, *75*, 312–318. [CrossRef] [PubMed]

16. Jarausch-Wehrheim, B.; Mocquot, B.; Mench, M. Absorption and translocation of sludge-borne zinc in field-grown maize (*Zea mays* L.). *Eur. J. Agron.* **1999**, *11*, 23–33. [CrossRef]

17. McGrath, S.P.; Zhao, F.J.; Dunham, S.J.; Crosland, A.R.; Coleman, K. Long-term changes in extractability and bioavailability of zinc and cadmium after sludge application. *J. Environ. Qual.* **2000**, *29*, 875–883. [CrossRef]

18. Žaltauskaitė, J.; Judeikytė, S.; Sujetovienė, G.; Dagiliūtė, R. Sewage sludge application effects to first year willows (*Salix viminalis* L.) growth and heavy metal bioaccumulation. *Waste Biomass Valoriz.* **2017**, *8*, 1813–1818. [CrossRef]

18. Dickinson, N.M.; Pulford, I.D. Cadmium phytoextraction using short-rotation coppice Salix: The evidence trail. *Environ. Int.* **2005**, *31*, 609–613. [CrossRef]

19. Brandao, M.; i Canals, L.M.; Clift, R. Soil organic carbon changes in the cultivation of energy crops: Implications for GHG balances and soil quality for use in LCA. *Biomass Bioenergy* **2010**, *35*, 2323–2336.

20. Richards, B.K.; Stoof, C.R.; Cary, I.J.; Woodbury, P.B. Reporting on marginal lands for bioenergy feedstock production: A modest proposal. *Bioenergy Res.* **2014**, *7*, 1060–1062. [CrossRef]

21. Nilsson, D.; Rosenqvist, H.; Bernesson, S. Profitability of the production of energy grasses on marginal agricultural land in Sweden. *Biomass Bioenergy* **2015**, *83*, 159–168. [CrossRef]

22. Kuzovkina, Y.A.; Schulthess, C.P.; Zheng, D. Influence of soil chemical and physical characteristics on willow yield in Connecticut. *Biomass Bioenergy* **2018**, *108*, 297–306.

23. Mažvila, J.; Adomaitis, T.; Eitmanavičius, L. Changes in the acidity of Lithuania's soils as affected of not liming. *Agriculture* **2004**, *4*, 3–20. (In Lithuanian)

24. Herr, J.R. Bioenergy from trees. *New Phytol.* **2011**, *192*, 313–315. [CrossRef] [PubMed]

25. Karp, A.; Hanley, S.J.; Trybush, S.O.; Macalpine, W.; Pei, M.; Shield, I. Genetic improvement of willow for bioenergy and biofuels free access. *J. Integr. Plant Biol.* **2011**, *53*, 151–165. [PubMed]

26. Vadiunina, A.P.; Korchagina, Z.A. *Methods for Determining the Physical Properties of the Soil*, 3rd ed.; Mosc. Agropromizdat: Moscow, Russian, 1986; pp. 53–79. (In Russian)

27. Vance, E.D.; Brookes, P.C.; Jenkinson, D.S. An extraction method for measuring soil microbial biomass C. *Soil Biol. Biochem.* **1987**, *19*, 703–707. [CrossRef]

28. Alriksson, B.; Ledin, S.; Seeger, P. Effect of nitrogen fertilization on growth in a *Salix viminalis* stand using a response surface experimental design. *Scand. J. For. Res.* **1997**, *12*, 321–327. [CrossRef]

29. Kasel, S.; Bennett, L.T. Land-use history, forest conversion, and soil organic carbon in pine plantations and native forests of south eastern Australia. *Geoderma* **2007**, *137*, 401–413. [CrossRef]

30. Zhang, J.; Bei, S.; Li, B.; Zhang, J.; Christie, P.; Li, X. Organic fertilizer, but not heavy liming, enhances banana biomass, increases soil organic carbon and modifies soil microbiota. *Appl. Soil Ecol.* **2019**, *136*, 67–79. [CrossRef]

31. Nissim, W.G.; Pitre, F.E.; Teodorescu, T.I.; Labrecque, M. Long-term biomass productivity of willow bioenergy plantations maintained in southern Quebec, Canada. *Biomass Bioenergy* **2013**, *56*, 361–369. [CrossRef]

32. Sevel, L.; Nord-Larsen, T.; Ingerslev, M.; Jørgensen, U.; Raulund-Rasmussen, K. Fertilization of SRC willow, I: Biomass production response. *BioEnergy Res.* **2014**, *7*, 319–328. [CrossRef]

33. Šiaudinis, G.; Jasinskas, A.; Karčauskienė, D.; Repšienė, R. The effect of liming and nitrogen application on common osier and black poplar biomass productivity and determination of biofuel quality indicators. *Renew. Energy* **2020**, *152*, 1035–1040.

34. Stolarski, M.J.; Krzyżaniak, M.; Załuski, D.; Tworkowski, J.; Szczukowski, S. Effects of Site, Genotype and Subsequent Harvest Rotation on Willow Productivity. *Agriculture* **2020**, *10*, 412. [CrossRef]

35. Neilsen, G.H.; Hogue, E.J.; Neilsen, D.; Zebarth, B.J. Evaluation of organic wastes as soil amendments for cultivation of carrot and chard on irrigated sandy soils. *Can. J. Soil Sci.* **1998**, *78*, 217–225. [CrossRef]

36. Eitminavičiūtė, I.; Matusevičiūtė, A.; Gasiūnas, V.; Radžiūtė, M.; Grendienė, N. Ecotoxicological assessment of arable field soils fertilized with sewage sludge. *Ekologija* **2009**, *55*, 142–152. [CrossRef]

37. Sommers, L.E. Chemical composition of sewage sludges and analysis of their potential use as fertilizers. *J. Environ. Qual.* **1977**, *62*, 225–232.

38. Hemmat, A.; Aghilinategh, N.; Rezainejad, Y.; Sadeghi, M. Long-term impacts of municipal solid waste compost, sewage sludge and farmyard manure application on organic carbon, bulk density and consistency limits of a calcareous soil in central Iran. *Soil Tillage Res.* **2010**, *108*, 43–50. [CrossRef]

39. Huang, C.C.; Chen, Z.S. Carbon and nitrogen mineralization of sewage sludge compost in soils with a different initial pH. *Soil Sci Plant Nutr.* **2009**, *55*, 715–724.

40. Roig, N.; Sierra, J.; Martí, E.; Nadal, M.; Schuhmacher, M.; Domingo, J.L. Long-term amendment of Spanish soils with sewage sludge: Effects on soil functioning. *Agric. Ecosyst. Environ.* **2012**, *158*, 41–48. [CrossRef]

41. Andriamananjara, A.; Rabeharisoa, L.; Prud'homme, L.; Morel, C. Drivers of plant-availability of phosphorus from thermally conditioned sewage sludge as assessed by isotopic labeling. *Front. Nutr.* **2016**, *3*, 19. [CrossRef]

42. Semerci, N.; Kunt, B.; Calli, B. Phosphorus recovery from sewage sludge ash with bioleaching and electrodialysis. *Int. Biodeterior Biodegrad.* **2019**, *144*, 104739. [CrossRef]

43. Toosi, E.R.; Kravchenko, A.N.; Mao, J.; Quigley, M.Y.; Rivers, M.L. Effects of management and pore characteristics on organic matter composition of macroaggregates: Evidence from characterization of organic matter and imaging. *Eur. J. Soil Sci.* **2017**, *68*, 200–211. [CrossRef]

44. Martins, T.; Saab, S.D.C.; Milori, D.M.B.P.; Brinatti, A.M.; Rosa, J.A.; Cassaro, F.A.M.; Pires, L.F. Soil organic matter humification under different tillage managements evaluated by Laser Induced Fluorescence (LIF) and C/N ratio. *Soil Tillage Res.* **2011**, *111*, 231–235. [CrossRef]

45. Shahzad, T.; Rashid, M.I.; Maire, V.; Barot, S.; Perveen, N.; Alvarez, G.; Mougin, C.; Fontaine, S. Root penetration in deep soil layers stimulates mineralization of millennia-old organic carbon. *Soil Biol. Biochem.* **2018**, *124*, 150–160. [CrossRef]

46. Brust, G.E. Management strategies for organic vegetable fertility. In *Safety and Practice for Organic Food*; Academic Press: Cambridge, MA, USA, 2019; pp. 193–212.

47. Šiaudinis, G.; Karčauskienė, D. The effect of sewage sludge on and cup plant's (*Silphium perfoliatum* L.) biomass productivity under Western Lithuania's Retisol. In Proceedings of the International Scientific Conference "Rural Development", Kaunas, Lithuania, 23–24 November 2017; pp. 148–152.

Article

Could the Content of Soluble Carbohydrates in the Young Shoots of Selected Willow Cultivars Be a Determinant of the Plants' Attractiveness to Cervids (Cervidae, Mammalia)?

Maciej Budny [1], Kazimierz Zalewski [2,*], Lesław Bernard Lahuta [3], Mariusz Jerzy Stolarski [4], Robert Stryiński [2], Adam Okorski [5]

[1] Polish Hunting Association, Research Station, Sokolnicza 12, 64-020 Czempiń, Poland; m.budny@pzlow.pl
[2] Department of Biochemistry, Faculty of Biology and Biotechnology, University of Warmia and Mazury in Olsztyn, Oczapowskiego 1a, 10-719 Olsztyn, Poland; robert.stryinski@uwm.edu.pl
[3] Department of Plant Physiology, Genetics and Biotechnology, University of Warmia and Mazury in Olsztyn, Oczapowskiego 1a, 10-719 Olsztyn, Poland; lahuta@uwm.edu.pl
[4] Department of Genetics, Plant Breeding and Bioresource Engineering, University of Warmia and Mazury, in Olsztyn, Plac Łódzki 3, 10-719 Olsztyn, Poland; mariusz.stolarski@uwm.edu.pl
[5] Department of Entomology, Phytopathology and Molecular Diagnostics, University of Warmia and Mazury in Olsztyn, Prawocheńskiego 17, 10-719 Olsztyn, Poland; adam.okorski@uwm.edu.pl
* Correspondence: k.zalewski@uwm.edu.pl

Citation: Budny, M.; Zalewski, K.; Lahuta, L.B.; Stolarski, M.J.; Stryiński, R.; Okorski, A. Could the Content of Soluble Carbohydrates in the Young Shoots of Selected Willow Cultivars Be a Determinant of the Plants' Attractiveness to Cervids (Cervidae, Mammalia)? *Agriculture* **2021**, *11*, 67. https://doi.org/10.3390/agriculture11010067

Received: 20 November 2020
Accepted: 12 January 2021
Published: 15 January 2021

Abstract: Ten willow cultivars grown in experimental plots were evaluated for performance, attractiveness to foragers, and the content and composition of soluble carbohydrates. The survival of willow cuttings in a thicket and in browse plots differed subject to cultivar, soil quality, and soil moisture content. The number of stump sprouts varied considerably, from 1.1 shoots in the weakest soils in Słonin, Poland, to 3.43 in the plot in Czempin, Poland. Browse plots were established in 2017. They were cut, and fencing was removed in early spring of 2019. Young shoots (10 cm shoot tip with buds, preferably eaten by animals) were sampled for analyses of soluble carbohydrates as potential attractors for foraging cervids. All willow cultivars contained the same soluble carbohydrates: glucose, fructose, galactose, sucrose, *myo*-inositol, galactinol, and raffinose. Total carbohydrate content ranged from 21.31 (*S. amygdalina* 1045) to 69.37 mg/g^{-1} DM (dry matter) (*S. purpurea*). Glucose was the predominant soluble sugar in the shoots of all willow cultivars, excluding *S. viminalis*. The fructose content of the shoots was approximately twice lower than their glucose content in all willow cultivars. Smaller differences were observed in the content of *myo*-inositol, which ranged from 4.61 (*S. amygdalina* 1045) to 8.26 mg/g^{-1} DM (*S. fragilis* cv. Kamon/Resko). The phloem of all willow species contained small quantities of galactinol and trace amounts of raffinose. Weak negative correlations were noted between total carbohydrate content, the content of glucose, fructose, and galactose vs. the attractiveness of willow shoots to foraging cervids. The remaining carbohydrates that occurred in smaller quantities in willow shoots were not correlated with their attractiveness to cervids.

Keywords: willow browse; soluble carbohydrates; browsing damage; cervids; gas chromatography

1. Introduction

Animals, including wild animals, are guided by the senses of smell and taste when selecting food [1]. The taste and aroma of potential food sources play a secondary role only in extreme situations, such as drought, deep snow cover, or the risk of hunger. Such extreme conditions persisted in Poland in the winters of 1962/63, 1978/79, 1986/87, and 2005/06, when roe deer, red deer, elk, as well as hares foraged on the young and green plant parts protruding above the deep snow cover. In those years, animals caused considerable damage to young forests and nurseries of forests, orchards, and ornamental trees and shrubs. Cervids also cause considerable browsing damage in years with less severe weather conditions, which forces forest managers to implement costly protective measures. In 2018, the total area affected by the foraging and browsing behavior of game animals, including red deer, fallow deer, roe deer, wild boars, and hares, in Poland was estimated at 56,300 ha, including 27,500 ha of agricultural land, 21,500 ha of young forests and 7300 ha of mature forests. According to State Forests data [2], around €44.8 million was spent on protecting forests against foraging animals in 2019. Chemical and mechanical protection of seedlings and fencing generate the highest costs. In 2018, seedlings were protected in a total area of 91,500 ha, and 22,600 ha of forest nurseries were fenced. In Germany, €11 million was spent on forest nurseries and forest protection, of which €3.2 was dedicated to combating woodworm infestations. Considerable funds are also allocated to the protection of young forests against wild animals [3].

European foresters and hunters have long expressed an interest in experiments where alternative food sources for cervids, including mixed willow, sycamore, beech, oak, and linden stands, as well as pure willow stands, are grown in food plots, unused forest enclaves, browse plots, and on fallow land by forest and orchard nurseries [4]. The cultivation of willow browse is relatively problem-free. Willows (genus *Salix*) can be planted in a wide variety of soils, but only selected cultivars, such as *S. alba* cv. Tristis, *S. integra* cv. Hakuro Nishiki, *S. purpurea* cv. Gracilis, *S. purpurea* cv. Nana, or other cultivars of *S. purpurea* perform well on sandy and dry soils [5,6]. *S. acutifolia* is the most drought-resistant cultivar that is used to stabilize sand dunes on the Baltic coast and in the Błędów desert in Poland [7,8]. However, its attractiveness to foraging animals has never been studied. Some of the above willow cultivars (*S. acutifolia*) proved to be highly invasive, and their cultivation was discontinued.

The attractiveness of willow browse is a complex problem that is influenced by various factors. In a study performed by Drogoszewski and Wlazełko [9], only 6 out of 133 tested willow cultivars were not browsed by roe deer. Four of those were *S. purpurea* cultivars (out of the analyzed 18). Willow cultivars that are not attractive to foraging animals usually contain bitter inorganic salts, tannins, salicin, alkaloids, and other phenols in the leaves and bark. It is also possible that wild animals are able to detect the presence of substances with medicinal properties in willow shoots [10,11]. On the other hand, plants contain attractants such as complex carbohydrates—starches, soluble sugars, vitamins, mineral salts with essential macronutrients and micronutrients, proteins, fat, and fiber. Not all animal species have a preference for the same plant substances. For example, aspartame, an artificial sweetener composed of two amino acids, tastes sweet to humans, but not to mice [1]. Bears, monkeys, dogs, wild boars, squirrels, ants, and other animals have a preference for sweet foods. Cervids' preferences for sweet-tasting foods have never been investigated, and the extent to which the chemical composition of trees and shrubs contributes to their attractiveness to wild animals remains unknown. On the other hand, getting to know the preferences of cervids with regard to the feeding attractiveness to various species, varieties, or clones of willow might be for them an indicator of the quantity and quality of substances contained in these plants. In view of the above, the aim of this study was to determine the content and composition of soluble carbohydrates in ten willow cultivars, and to evaluate their attractiveness to cervids.

2. Materials and Methods

2.1. Experimental Plots

In the spring of 2015, a willow thicket was established at the research station of the Polish Hunting Association in Czempiń near Poznań, Poland. Willow cuttings were obtained from experimental nurseries in the Resko Forest Inspectorate (Region of Western Pomerania), and field-plots established at experimental stations (owned by the University of Warmia and Mazury in Olsztyn) in Obory, Poland (Region of Pomerania), and Bałdy, Poland (Region of Warmia and Mazury) (Table 1). Weeds were removed only mechanically and manually in July of 2015 and 2016. The survival of willow cuttings and shoot growth were evaluated in September 2015. The willow thicket in Czempiń, Poland, was cut in the spring of 2016 and in early March 2017, and one-year-old shoots were used to establish three experimental browse plots near Grzybno, Poland (X: 482950.30; Y: 351700.42), Bieczyny, Poland (X: 484298.21; Y: 345303.04), and Słonin, Poland (X: 475485.18; Y: 344974.35) (Figure 1).

Willow varieties, breeding lines, and cultivars (collectively referred to as cultivars in the study) for the establishment of the willow thicket at the research station in Czempiń, Poland, were selected based on their availability and the results of previous research on willow browse [3,8,9]. Based on the numerous studies investigating willow browse and the medicinal properties of willows, *S. purpurea* 1126, which was completely or partly ignored by wild animals during previous observations, was introduced as the control cultivar on account of its high salicin content [8,9,12–15].

Table 1. Cultivar designations used later in the work in the tables and figures.

Name of the Cultivar	Willow Species	Cultivar No. Used in Tables and Figures
Resko Forest Inspektorate	*S. amygdalina* cv. Dunajec	1
Resko Forest Inspektorate	*S. amygdalina* cv. Krakowianka	2
Experimental Station of University of Warmia and Mazury in Obory	*S. amygdalina* 1045	3
Experimental Station of University of Warmia and Mazury in Obory	*S. amygdalina* 1102	4
Experimental Station of University of Warmia and Mazury in Obory	*S. amygdalina* 1036	5
Experimental Station of University of Warmia and Mazury in Obory	*S. fragilis* cv. Kamon	6
Resko Forest Inspektorate	*S. fragilis* cv. Kamon	7
Experimental Station of University of Warmia and Mazury in Bałdy	*S. laurina* 220/225	8
Experimental Station of University of Warmia and Mazury in Bałdy	*S. pentederana*	9
Experimental Station of University of Warmia and Mazury in Obory	*S. purpurea* 1126	10

Figure 1. Location of the experimental willow plots.

Willows were planted in 2017, with inter-row spacing of 75 cm and intra-row spacing of 30 cm. The area of the plot established in 2015 in Czempiń was 2500 m^2, where 5045 cuttings were planted. Respectively, the plots established in spring 2018: Bieczyny—800 m^2 and 2800 cuttings; Grzybno—870 m^2 and 3080 cuttings; and Słonin—920 m^2 and 3430 cuttings. The number of individual cultivars planted in a plot depended on their availability on the market in 2015 and was as follows (according to the numbering from Table 1): Cultivar No 1-50; 2-1000; 3-1000; 4-1000; 5-1000; 6-110; 7-400; 8-45; 9-40; and 10-400 cuttings. Each of the cultivars numbered 2, 3, 4, 5, 7, and 10 were randomly planted in several rows. Thus, e.g. *S. amygdalina* cv. Krakowianka was planted in 5 rows separated by other cultivars. The planting of cuttings on the mother plot was completely random, as well as on the occlusal plots established two years later.

Browse plots were fenced with wire mesh to a height of 2.5 m, for two years, and weeds were removed mechanically each year in July. The performance of the willow thicket was evaluated in September of 2017 and 2019, and the survival of the willow cuttings in browse plots was evaluated in September 2019.

2.2. Browsing Damage Assessment

In the fall of 2017 and 2019, the damage caused by wild animals in browse plots after fencing had been removed was assessed on a 5-point scale proposed by Bukiewicz [12], with some modifications, where 1 point—absence of damage, no traces of shoot, bud or leaf browsing; coefficient for calculating browsing damage—0.00 (0%); 2 points—minor damage to selected shrubs, with full recovery; coefficient for calculating browsing damage—0.05 (5%); 3 points—moderate damage to the top part of the main shoot or side shoots; signs of foliage browsing; bark stripping on short stem segments; absence of terminal buds, which can inhibit shoot growth and weaken willow shrubs; coefficient for calculating browsing damage—0.15 (15%); 4 points—considerable damage to the main shoot, side shoots and buds; complete or partial loss of foliage; damage to stump sprouts; bark stripping on more than half of stem length, which can significantly inhibit shrub growth and lead to tissue necrosis in stumps under unfavorable conditions; coefficient for calculating browsing damage—0.50 (50%); and 5 points—absence of shoots, damaged and necrotized shoots; coefficient for calculating browsing damage—1.00 (100%).

When assessing the browsing damage, five repetitions were made, each with 20 consecutive seedlings in a row. For cultivars with less than 200 plants per plot (cultivars No. 1, 6, 7, and 8), all seedlings were assessed.

2.3. Analysis of the Content of Soluble Carbohydrates

At the end of July 2017, ten one-year old shoots with leaves grown from two-year-old stool were obtained from each willow cultivar (three replicates) in the thicket in Czempiń, Poland, and the harvested material was analyzed for the content of soluble carbohydrates. A total of 30 samples of 10 cm tops of one-year-old shoots were analyzed. Each sample of 10 stems was analyzed separately. Since *S. pentederana* and *S. amygdalina* cv. Krakowianka did not survive in the three-year-old thicket in Czempiń, Poland, shoot samples from these cultivars were obtained from the experimental plots in Bałdy, Poland.

Shoot samples (10 cm tips with the buds of one-year old shoots) were weighed, frozen in liquid nitrogen, and lyophilized. Dry tissues were crushed in a mixer mill MM200 (Retsch, Katowice, Poland). Soluble carbohydrates were extracted from 40 mg of dry material using 800 μL of 50% ethanol (containing 100 μg of xylitol, as an internal standard) and analyzed by gas chromatography method on the ZEBRON ZB-1 capillary column (Phenomenex, CA, USA), as described previously [16]. The carbohydrate content was calculated using the internal standard method. Standards of carbohydrates (xylitol, fructose, galactose, glucose, *myo*-inositol, sucrose, galactinol, raffinose, 1-kestose, stachyose, and verbascose) were purchased from Sigma (USA). The amount of unknown carbohydrates with retention times (Rt) of 6.52 and 7.39 was calculated based on the nearest known standards (1-kestose and stachyose, respectively).

2.4. Experiment with Farmed Red Deer and Fallow Deer

In January 2020, twenty one-year-old shoots grown from leafless three-year-old stool were harvested from each willow cultivar (in three replicates) and fed to European red deer at the research station in Czempiń, Poland, and red deer and fallow deer at the research station of the Polish Academy of Sciences in Kosewo Górne, Poland. Both deer species were kept separately. The shoots came from the plot located in Bieczyny, Poland. One hundred and twenty shoots were cut from each cultivar and divided into 6 bunches, three for each town (Czempiń and Kosewo Górne). All bunches from all cultivars were collected at the same time and their fresh weight was determined. All bunches were tied to the fence where the animals were kept. The decrease in shoot mass caused by browsing was determined after 3 days. The degree of browsing damage was determined after the loss of fresh weight. All tests were subjected to the same atmospheric conditions.

The analysis of correlations between the relative sweetness of the total carbohydrates and severity of browsing damage on willow shoots was performed. The sweetness of the individual sugars and total carbohydrates was determined by multiplying the sugar content of the willow tissues (mg g^{-1} DM) by the relative sweetness (RS) indicators (Table 2) given in the literature. Galactinol and sucrose are non-reducing disaccharides because both galactose and *myo*-inositol have α-(1-1)-glycosidic bonds; therefore, an RS index of 0.6 was adopted for galactinol [17–19].

Table 2. Sweetness of the selected carbohydrates relative to sucrose (see the source in the reference list).

Sugar	[20]	[21]	Average
Glucose–fructose syrup/relative sweetness	100	100	100
Sucrose	100	100	100
Fructose	173	180	176.50
Glucose	74	75	74.50
Galactose	32.1	32	32.05
Maltose	32	30	31
Mannose	-	30	30
Lactose	16	25	20.5
Raffinose	10	-	10
Myo-inositol	50	-	50
Galactinol	60	-	60

2.5. Statistical Analysis

The results were processed statistically in the Statistica program (v. 13.1, Dell Inc., Tulsa, OK, USA). The least significant difference was calculated at $p \leq 0.01$, with the use of Tukey's test. The relationships between the carbohydrate content of the shoots and the severity of browsing damage caused by cervids were determined based on the calculated values of Pearson's linear correlation coefficient R ($p \leq 0.05$; $p \leq 0.01$).

3. Results

3.1. Soil Analysis

Soil samples for laboratory analyses were collected from the experimental plots where the evaluated willow cultivars were grown. The soil acidity (KCl), content of organic carbon, total nitrogen, and major macronutrients (P, K, and Mg) were determined in 100 g soil samples (Table 3). Soil samples from the willow thicket in Czempiń, Poland, were acidic, moderately abundant in phosphorus, highly abundant in potassium, and moderately abundant in magnesium. The content of total organic carbon was determined at 1.15%, which, when converted to humus content (1.15 × 1.7 = 1.96), is a satisfactory (average) result for slightly loamy sand. The experimental plot in Słonin, Poland, was established on slightly alkaline soil with a moderate phosphorus and potassium content, which was highly permeable due to an 80 cm-deep layer of loose sand. The experimental plots were established in the following types of habitats: Czempiń, Poland—fresh forest established on former agricultural land (Category 2); Bieczyny, Poland—dehydrated humid forest established on former agricultural land (Category 2); Grzybno, Poland—fresh mixed coniferous forest with low groundwater levels (Category 1); Słonin, Poland—fresh forest with very low groundwater levels (Category 1). Categories 1 and 2 characterize the degree of moisture in the soil in which the forest grows, where Category 1 indicates poor soil irrigation and Category 2 good soil irrigation.

Table 3. Soil parameters in the experimental willow plots.

Plot	Soil Type	pH (in KCl)	Soil Content (Average of Three Replicates)				
			P_2O_5 (mg/100 g of Soil)	K_2O (mg/100 g of Soil)	Mg (mg/100 g of Soil)	N Total (%)	C Organic (%)
Czempin	Loamy sand	4.98	14.8	14.0	3.6	0.086	1.98
Bieczyny	Sandy loam	7.60	7.3	7.44	4.6	0.35	1.62
Grzybno	Slightly loamy sand	6.42	12.0	3.1	1.9	0.07	1.02
Słońsk	Slightly loamy sand	7.31	13.8	8.2	2.1	0.098	0.75

3.2. Willow Performance (Survival)

Willow survival in the first and third year of growth were high (above 60%) in the thicket and in the experimental plots in Bieczyny and Grzybno, Poland (Figure 2). In Słonin, Poland (Figure 2D), the survival exceeded 80% in three cultivars only. Two years after establishment, numerous stumps were lost in all plots due to drought, and the highest losses were noted in Słonin, Poland (Figure 2D). Losses ranged from 80% to 95%, depending on the cultivar (applies to Cultivars No. 1, 2, 4, and 8). In 2017, 2018, and 2019, the Wielkopolska Region in Poland was affected by soil drought, which led to a massive loss of willow stumps in browse plots and partial losses in the thicket. Willow survival in the plots was evaluated for the second time in September in the third year of growth, whereas the thicket was assessed three times during the experiment. Browsing damage was most extensive in the experimental plot in Słonin, Poland, where willows were grown on sandy and highly permeable soil. *S. laurina* 220/225 (cv. No. 8) was nearly completely eliminated, and less than 5% of the *S. amygdalina* 1102 (cv. No. 4), *S. amygdalina* cv. Dunajec (cv. No. 1), and *S. amygdalina* cv. Krakowianka (cv. No. 2) shrubs survived (Figure 2). In the browse plot in Grzybno, Poland, *S. amygdalina* cv. Dunajec (cv. No. 1) was infested with dock bugs (*Coreus marginatus* L.), which damaged nearly 80% of the leaves and developing buds. The above cultivar survived the infestation.

3.3. Analysis of Shoot Browsing

The quality of willow cultivars and the survival differed across the plots after fencing had been removed (Figure 3). Willow performance was largely influenced by the type of habitat, in particular in dry years. The plot in Słonin, Poland, was established on highly permeable soil of the lowest quality (classes IV–VI). In the north, the plot was bound by a fresh mixed forest and a fresh forest composed of pines, birches, and, sporadically, oaks, with a dense undergrowth of black cherry and hazel. Signs of foraging by fallow deer and roe deer (which permanently inhabited the region), as well as mouflons (which were periodically observed in the region) were noted (Figure 3).

In comparison with the remaining two plots, the plot in Bieczyny, Poland offered the optimal habitat for red deer and roe deer. The plot in Grzybno, Poland, was characterized by medium-quality soils, and it was visited mainly by roe deer. Signs of foraging and tracks of other animal species were observed sporadically after fencing had been removed. In the willow thicket, two cultivars (*S. amygdalina* cv. Krakowianka and *S. pentederana*) were also lost due to drought, but only in the fourth and fifth year of growth (Figure 2).

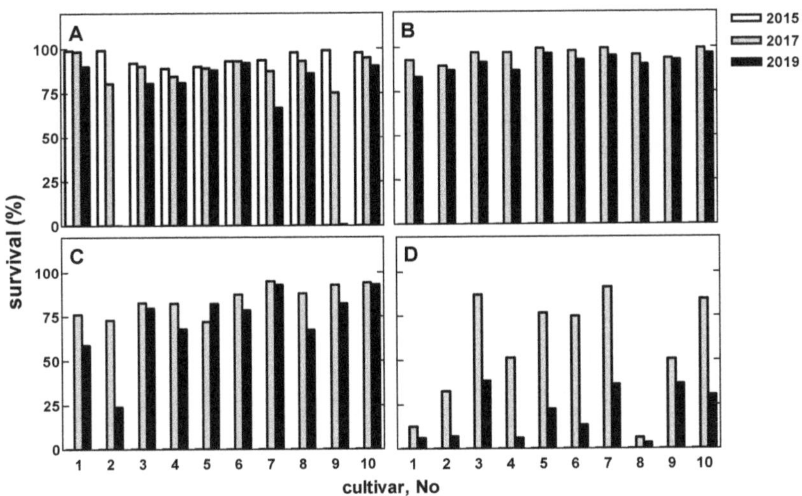

Figure 2. Willow survival (%) in the experimental plots in the first (2015), second (2017), and third (2019) year of vegetation in (**A**) Czempiń-backwoods; (**B**) Bieczyny; (**C**) Grzybno; and (**D**) Słonin. All stumps were counted. For the cultivar explanation, see Table 1.

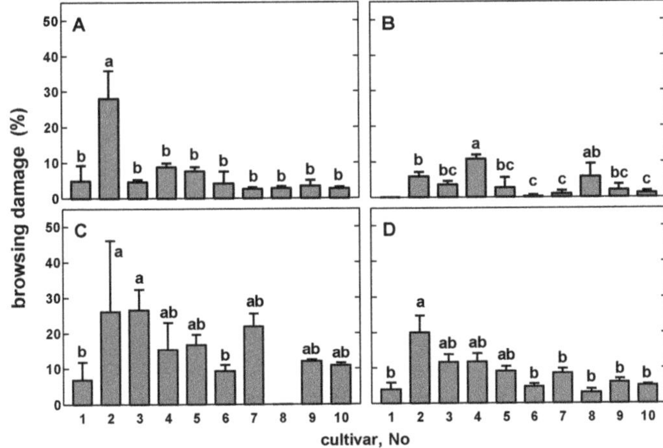

Figure 3. Severity of browsing damage to the analyzed willows cultivated in (**A**) Bieczyny; (**B**) Grzybno; and (**C**) Słonin. The mean values for three localization (**A**–**C**) are shown in box (**D**). Data were processed by analysis of variance. Bars with the same letters did not differ significantly ($p \leq 0.01$) in Tukey's test. For the cultivar explanation, see Table 1.

3.4. Content of Soluble Carbohydrates

The shoots of all analyzed willow cultivars contained the same soluble carbohydrates: glucose, fructose, sucrose, *myo*-inositol, galactinol, and raffinose. Total carbohydrate content ranged from 21.31 (*S. amygdalina* 1045) to 69.37 mg/g^{-1} DM (*S. purpurea*).

Glucose was the predominant sugar in most cultivars, accounting for more than 50% of the identified soluble carbohydrates (Figure 4). Glucose levels were highest in *S. purpurea*

(45.93 mg/g^{-1} DM), *S. pentaderana* (35.08 mg/g^{-1} DM), and *S. fragilis* (29.19 mg/g^{-1} DM on average), and lowest in *S. amygdalina* 1045 (11.61 mg/g^{-1} DM on average). The fructose content of the shoots was approximately two-fold lower than their glucose content in all willow cultivars, whereas the galactose content was approximately 10-fold lower than the fructose content. Most cultivars contained even less sucrose than galactose, excluding *S. purpurea* 1126 (0.55 mg/g^{-1} DM). Smaller differences were noted in the content of *myo*-inositol, which ranged from 4.61 (*S. amigdalina* 1045) to 8.26 mg/g^{-1} DM (*S. fragilis* cv. Kamon/Resko). The phloem of all willow species also contained small quantities of galactinol (0.07–0.55 mg/g^{-1} DM) and trace amounts of raffinose (0–0.13 mg/g^{-1} DM). Raffinose was not detected in Cultivars No. 2, 3, 4, and 6 (Figure 4).

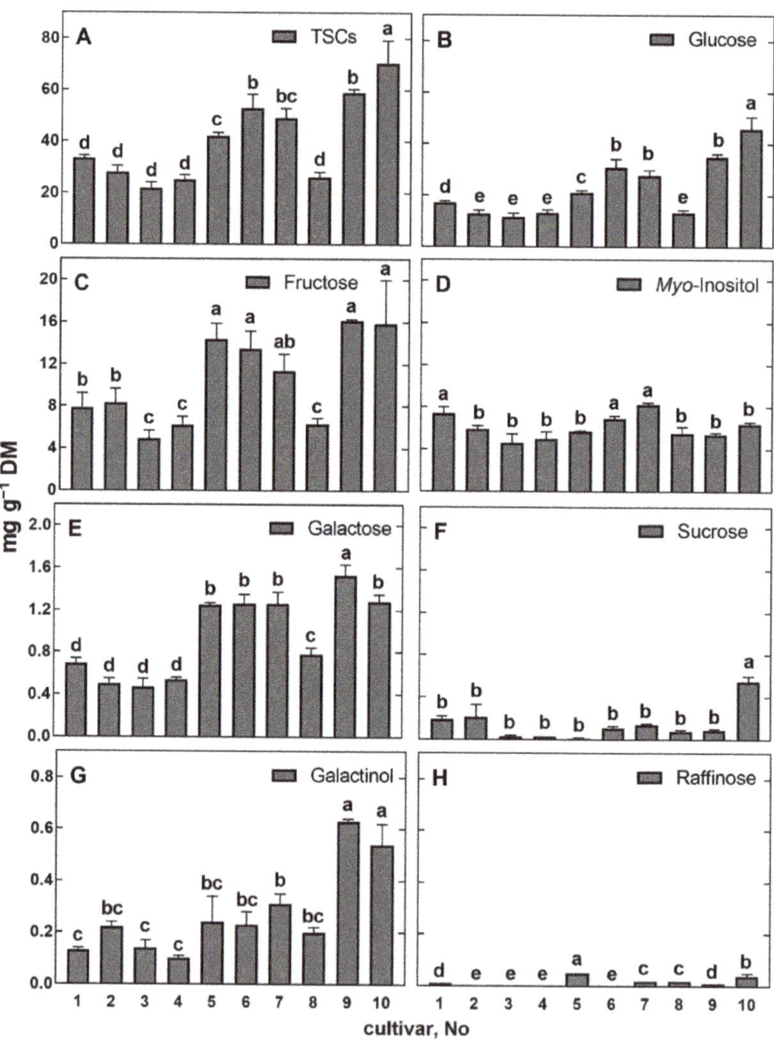

Figure 4. The content of soluble carbohydrates in the shoots of the analyzed willow cultivars: (**A**) total soluble carbohydrates; (**B**) glucose; (**C**) fructose; (**D**) *myo*-inositol; (**E**) galactose; (**F**) sucrose; (**G**) galactinol; and (**H**) raffinose. Values marked with different letters (a to e) are significantly different at $p \leq 0.01$. For the cultivar explanation, see Table 1.

3.5. Shoot Sweetness and Browsing Damage

The extent to which browsing damage was correlated with the sweetness of the evaluated willow cultivars was evaluated in a statistical analysis using Pearson's linear correlation coefficient R ($p \leq 0.05$; $p \leq 0.01$).

The analysis revealed weak negative correlations between the RS of total carbohydrates (R = −0.28) (Figure 5A), the RS of glucose (R = −0.26) (Figure 5B), and the RS of fructose (R = −0.27) (Figure 5C) vs. the severity of browsing damage on willow shoots. No such correlations were observed for the remaining carbohydrates that were less abundant in willow tissues (Figure 5D–H).

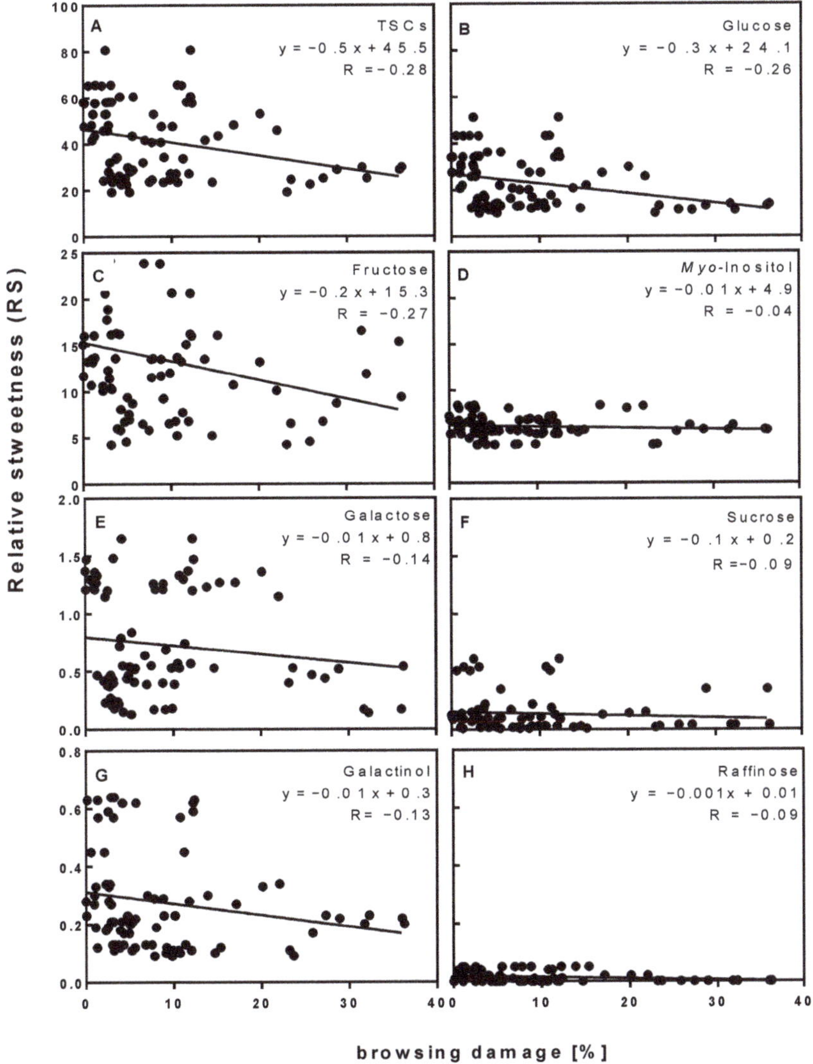

Figure 5. The analysis of correlations between the relative sweetness of the total carbohydrates and severity of browsing damage on willow shoots: (**A**) total soluble carbohydrates; (**B**) glucose; (**C**) fructose; (**D**) *myo*-inositol; (**E**) galactose; (**F**) sucrose; (**G**) galactinol; and (**H**) raffinose. R—Pearson's linear correlation coefficient.

Farmed red deer and fallow deer did not express particular interest in one-year-old shoots harvested from the studied willow cultivars (Table 4). The animals had the greatest preference for *S. cordata* and *S. laurina* (red deer), as well as *S. cordata*, *S. amygdalina* cv Dunajec, *S. amygdalina* cv. Krakowianka, and *S. laurina* (fallow deer), but intensive foraging behavior was not observed. It should also be noted that the content of soluble sugars in overwintered shoots increased by 52.3% relative to the shoots harvested in the middle of the growing season (July). An analysis of *S. cordata* tissue samples also demonstrated changes in the proportions of soluble carbohydrates. The content of fructose decreased, whereas sucrose levels increased between July and January (Table 5).

Red deer expressed varied interest in willow shoots, which generally differed between the studied locations. The animals in both stations had a greater preference for *S. cordata*, but fallow deer were less discriminating in their choice of willow cultivar than red deer (Table 4).

Table 4. The attractiveness of the analyzed willow cultivars to red deer and fallow deer at the research station of the Polish Hunting Association in Czempiń and the Cervid Farm of the Polish Academy of Sciences in Kosewo Górne, based on the average decrease in shoot mass (%) of duplicate measurements.

Cultivar No.	Deer Czempiń	Deer Kosewo G.	Fallow Deer Kosewo G.
1	10.76	0	5.94
2	3.91	0	6.44
3	4.48	0	3.00
4	18.50	1.02	4.25
5	14.49	13.54	8.90
6	8.58	0	6.39
7	8.99	0	2.18
8	18.32	6.92	4.80
9	16.76	0	0.07
10	4.49	0	4.72

Table 5. Average sugar content (mg/g^{-1} DM) of *Salix cordata* shoots harvested in July and January ($n = 3$ for each species and sugar) with the standard deviation.

	Cultivar	
	S. cordata **UWM 1036** (July 2019)	*S. cordata* **UWM 1036** (January 2020)
Fructose	14.37 ± 1.50	2.92 ± 0.03
Galactose	1.25 ± 0.02	0.00
Glucose	21.00 ± 1.02	2.04 ± 0.01
Myo-inositol	5.70 ± 0.07	0.43 ± 0.01
Sucrose	0.01 ± 0.01	37.85 ± 0.48
Galactionol	0.24 ± 0.10	0.00
Raffinose	0.05 ± 0.00	14.18 ± 0.68
Stachyose	0.00	5.28 ± 0.20
Verbascose	0.00	0.34 ± 0.02
Total sugars	41.37 ± 2.40	63.03 ± 1.25

4. Discussion

Human interventions in reducing damage to forests, such as the use of fencing and chemicals, should be minimized or completely eliminated in order to preserve the biodiversity of natural habitats. Artificial structures and chemicals should be replaced with trees and shrubs that, at least partially, protect valuable crops from damage and reduce crop losses to an economically acceptable level.

The attractiveness of willow shoots to herbivores is determined by various factors, in particular the chemical composition of willow cultivars and their growth stage. The survival and growth rates of willows and, to a certain extent, the chemical composition of the shoots is correlated with the soil quality and soil moisture content (Figures 2 and 4; Table 3). In the present study, the willow thicket and the experimental plots differed significantly in soil parameters. The soil in Słonin, Poland, was susceptible to dehydration due to low water-holding capacity and the absence of capillary action. In this experimental plot, the performance of willows was completely dependent on precipitation, in particular in the dry years 2017–2019. The survival was very low in the first year after planting, and only 21.7% of the planted cuttings survived the dry years 2018 and 2019. *S. laurina* was the least resistant to drought (only 3.2% of stumps survived), whereas *S. amygdalina* 1045, *S. amygdalina* cv. Kamon/Resko, and *S. pantederana* were least susceptible to water deficit (50.1% of stumps survived). The results of the study clearly demonstrate that soil parameters and habitat type significantly influence root establishment and sprouting.

The content and composition of soluble carbohydrates in willows remain insufficiently investigated. Aliferis et al. [22] identified glucose, fructose, galactose, *myo*-inositol, trehalose, and sugar alcohols—arabitol, mannitol, and glycosides—in fully developed leaves of 6- to 8-week-old *S. purpurea* L. (cv Fish Creek) plants. Sucrose, galactinol, and raffinose were not detected. Similar to this study, the predominant sugars in the leaves of hydroponically grown *S. viminalis* were sucrose, glucose, and fructose (Figure 4) [23]. However, the sucrose, glucose, and fructose concentrations were more than twice higher in the cited experiment than in this study. In the leaves harvested from the top shoots of two *S. viminalis* × *S. schwerinii* hybrid cultivars grown in a field, the glucose and sucrose concentrations were similar and significantly higher than the fructose levels [24]. The sugar concentrations were lower in older than in younger plants, and similar observations were made in the hybrid poplars *Populus deltoides* × *P. nigra* [25]. Corol et al. [24] also identified small quantities of raffinose and stachyose in the leaves of *S. viminalis* × *S. schwerinie*. In the present study, raffinose (but not stachyose) was detected in most willow cultivars (excluding *S. amigdalina*), and it was most abundant in *S. cordata* (0.05 mg g^{-1} DM). The presence of galactinol (a precursor of raffinose and its higher homologs) in all willow species indicates this oligosaccharide is synthesized in tissues in response to water and temperature stress (drought) [26]. Unlike glucose and fructose, galactinol does not appear to increase the sweetness of willow shoots, due to its low content. However, galactinol acts as a donor of the galactosyl groups for other sugars, and therefore its presence can enhance sweetness [27]. The statistical analysis revealed significant differences in the concentrations of individual soluble carbohydrates and in the total sugar content between the evaluated willow cultivars. The differences in the content of soluble carbohydrates between the cultivars studied here could be associated with the rate of the plant's growth, and/or rate of shoot tip elongation. In our study, the much higher concentration of monosaccharides (glucose and fructose) (Figure 5B,C) than that of sucrose, found for all cultivars, could be an indirect confirmation of the role of the shoot tips as the major sink tissues of willow plants, as in other species [28]. Sucrose is the major photoassimilate distributed from the leaves (sources) to the sinks [29], where it is hydrolyzed (presumably by invertases) into monosaccharides and serving as the carbon and energy sources for fast-growing tissues.

According to research, sweet taste perception differs considerably among animal species [30]. Foods rich in soluble carbohydrates are attractive to many animals and promote foraging behavior. The presence of sugars is responsible for sweet taste perception, but perceived sweetness can also be enhanced or reduced by phenols and other compounds. These compounds can exert a greater effect on sweetness in combination than alone.

In the current study, an experiment involving farmed red deer and fallow deer was carried out to determine whether the concentration of soluble sugars influences the attractiveness of willow shoots to cervids. One-year-old shoots of the analyzed willow cultivars harvested in mid-January were less attractive to the animals than the shoots harvested in spring and summer. Red deer and fallow deer had the greatest preference for *S. cordata* and *S. laurina* (signs of browsing damage were observed on up to 30% of the shoots). Sugar content could play a role in the foraging preferences of cervids because shoots harvested in January contained significantly more soluble carbohydrates (approx. 52% in *S. cordata*) than those harvested in late July. However, the statistical analysis revealed no significant correlations between the severity of browsing damage in spring and summer and the content of soluble sugars in the evaluated willow cultivars. These findings suggest that soluble carbohydrates do not act as typical attractants for red deer and fallow deer. Further research is needed to determine whether a sweet taste is not perceived by fallow deer and red deer due to the absence of the respective receptors [31], too low sugar concentrations in willow shoots, or the presence of other compounds, such as phenols, that mask the sweet taste of willow shoots.

5. Conclusions

The data presented in this work show that success in the cultivation of willow depends on the chosen cultivar and soil quality, especially its hydration. Young shoots of the analyzed cultivars contained from 2.1% to 6.9% soluble sugars, among which quantitatively dominated glucose. These sugars did not have a significant impact on the attractiveness of willows for cervids, which means that other chemical compounds are such a factor. It is possible that there are quantitative relationships between soluble sugars, phenolics, and other compounds that determine food attractiveness for cervids. Roe deer, fallow deer, and European deer do not have identical preferences for all willow cultivars, but they are those that are eaten very willingly by all of these animal species. This fact should be taken into account when selecting species for established plantations.

Author Contributions: Conceptualization, K.Z. and M.B.; methodology, K.Z., L.B.L. and M.B.; software, M.B. and A.O.; validation, M.B. and K.Z.; formal analysis, M.B. and K.Z.; investigation, M.B.; resources, M.J.S.; data curation, M.B., A.O., M.J.S. and K.Z.; writing—original draft preparation, M.B.; writing—review and editing, K.Z., R.S.; visualization, A.O.; supervision, K.Z.; project administration, K.Z. and M.B.; funding acquisition, K.Z. and M.B. All authors have read and agreed to the published version of the manuscript.

Funding: This research was funded by the University of Warmia and Mazury in Olsztyn, grant number 12.610.012-110; and the Main Board of the Polish Hunting Association, grant number 1/PZŁ/2016.

Institutional Review Board Statement: Ethical review and approval were waived for this study because in the test system used, animals were fed with no additives or chemical modifications. The animals received food that occurs naturally in the environment, and the care of the animals was provided under the status of the Research Station of Polish Hunting Association in Czempin, Poland, where these animals were kept.

Informed Consent Statement: Not applicable.

Conflicts of Interest: The authors declare no conflict of interest. The funders had no role in the design of the study; in the collection, analyses, or interpretation of data; in the writing of the manuscript, or in the decision to publish the results.

References

1. Scott, K. Taste Recognition: Food for Thought. *Neuron* **2005**, *48*, 455–464. [CrossRef] [PubMed]
2. *State Forests Poland Forests in Poland 2018*; Milewski, W. (Ed.) The State Forests Information Centre: Warsaw, Poland, 2018; ISBN 978-83-65659-40-8.
3. Kokocki, C.; Kowalski, A. *Growth and Survival of Some Species of "Forage" Willows in Forest and Non-Forest Habitats*; [Wzrost i przeżywalność niektórych gatunków wierzb "paszowych" na siedliskach leśnych i nieleśnych.]; University of Life Sciences: Poznań, Poland, 2012.
4. Klocek, A.; Hycza, T.; Jodłowski, K.; Kalinowski, M.; Kołakowski, B.; Szmidla, H.; Zachara, T. Forest Research Institute Bulletin No 1/2018. *From the Forest World [Z leśnego świata]* **2018**, *102*, 1–4.
5. Willow—Cultivation, Requirements, Varieties [Wierzba—Uprawa, Wymagania, Odmiany]. Available online: https://www.sadowniczy.pl/Wierzba-uprawa-wymagania-pielegnacja-cinfo-pol-624.html (accessed on 10 July 2020).

5. Szczukowski, S.; Tworkowski, J.; Wiwart, M.; Przyborowski, J. *Wicker (Salix sp.). Cultivation and Possibilities of Use*; [Wiklina (Salix sp.). Uprawa i możliwości wykorzystania.]; Szczukowski, S., Ed.; Wydawnictwo ART: Olsztyn, Poland, 2002; ISBN 8387443409.

7. Rahmonov, O.; Szczypek, T.; Wach, J. The Blendow desert (Pustynia Błędowska) a unique phenomenon of the Polish landscape. *Ann. Geogr.* **2006**, *39*, 34–36.

8. Wiaderek, I.; Waliszewska, B. Selected mechanical properties of one-year old twigs of Salix acutifolia. *Ann. Warsaw Univ. Life Sci. SGGW. For. Wood Technol.* **2010**, *72*, 427–432.

9. Drogoszewski, B.; Wlazełko, M. *The Trial of Determining the Food Preference of Roe Deer (Capreolus capreolus L.) in Relation of Some Willow Varieties*; [Próba określania wybiórczości żerowej sarny (*Caprcolus capreolus* L.) w odniesieniu do nickiórych odmian wierzby]; PTPN: Poznań, Poland, 1980.

10. Krauze-Baranowska, M.; Szumowicz, E. Willow—The source of antiinflammatory and analgesic medicinal plants. *Postępy Fitoter.* **2004**, *2*, 77–86.

11. Niemiec, P.; Dubas, J.W. Forest Self-Service Pharmacies for Animals. [Leśne samoobsługowe apteki dla zwierzyny]. *Brać Łow.* **2015**, *10*, 50–51.

12. Bukiewicz, H. Willow species useful for occlusal plots in forest fisheries. [Gatunki wierzb przydatne na poletka ogryzowo-zgryzowe w łowiskach leśnych.]. *Zach. Porad. Łow.* **1963**, *3*, 7–8.

13. Kenstaviciene, P.; Nenortiene, P.; Kiliuviene, G.; Zevzikovas, A.; Lukosius, A.; Kazlauskiene, D. Application of high-performance liquid chromatography for research of salicin in bark of different varieties of Salix. *Medicina* **2009**, *45*, 644–651. [CrossRef]

14. Förster, N.; Ulrichs, C.; Zander, M.; Kätzel, R.; Mewis, I. Factors Influencing the Variability of Antioxidative Phenolic Glycosides in Salix Species. *J. Agric. Food Chem.* **2010**, *58*, 8205–8210. [CrossRef]

15. Sulima, P.; Krauze-Baranowska, M.; Przyborowski, J.A. Variations in the chemical composition and content of salicylic glycosides in the bark of Salix purpurea from natural locations and their significance for breeding. *Fitoterapia* **2017**, *118*, 118–125. [CrossRef]

16. Zalewski, K.; Lahuta, L.B.; Martysiak-Żurowska, D.; Okorski, A.; Nitkiewicz, B.; Zielonka, Ł. Effect of Exogenous Application of Methyl Jasmonate on the Lipid and Carbohydrate Content and Composition of Winter Triticale (Triticosecale Wittm.) Grain and the Severity of Fungal Infections in Triticale Plants and Grain. *J. Agric. Food Chem.* **2019**, *67*, 5932–5939. [CrossRef] [PubMed]

17. Martínez-Villaluenga, C.; Frias, J.; Vidal-Valverde, C. Alpha-Galactosides: Antinutritional Factors or Functional Ingredients? *Crit. Rev. Food Sci. Nutr.* **2008**, *48*, 301–316. [CrossRef] [PubMed]

18. Crittenden, R.G.; Playne, M.J. Production, properties and applications of food-grade oligosaccharides. *Trends Food Sci. Technol.* **1996**, *7*, 353–361. [CrossRef]

19. The Importance of Myo-Inositol in Plants. Available online: https://www.lebanonturf.com/education-center/plant-biostimulants/the-importance-of-myo-inositol-in-plants (accessed on 10 June 2020).

20. Cichon, R.; Wądołowska, L. Carbohydrates. In *Human Nutrition. Introduction to Nutrition Science*; [Żywienie człowieka. Podstawy nauki o żywieniu]; Gawęcki, J., Ed.; Polish Scientific Publishers PWN: Warsaw, Poland, 2010; pp. 149–150.

21. Sweet Life of a Chemist [Słodkie Życie Chemika]. Available online: https://mlodytechnik.pl./eksperymenty-i-zadania-szkolne/chemia/29068-slodkie-zyciechemika?highlight=WyJzXHUwMTQyb2R5Y3oiXQ (accessed on 10 July 2020).

22. Aliferis, K.A.; Chamoun, R.; Jabaji, S. Metabolic responses of willow (*Salix purpurea* L.) leaves to mycorrhization as revealed by mass spectrometry and 1H NMR spectroscopy metabolite profiling. *Front. Plant Sci.* **2015**, *6*. [CrossRef]

23. Gąsecka, M.; Mleczek, M.; Drzewiceka, K.; Magdziak, Z.; Rissmann, I.; Chadzinikolau, T.; Golinski, P. Physiological and morphological changes in Salix viminalis L. as a result of plant exposure to copper. *J. Environ. Sci. Health Part A* **2012**, *47*, 548–557. [CrossRef]

24. Corol, D.; Harflett, C.; Beale, M.; Ward, J. An Efficient High Throughput Metabotyping Platform for Screening of Biomass Willows. *Metabolites* **2014**, *4*, 946–976. [CrossRef]

25. Regier, N.; Streb, S.; Zeeman, S.C.; Frey, B. Seasonal changes in starch and sugar content of poplar (Populus deltoides x nigra cv. Dorskamp) and the impact of stem girdling on carbohydrate allocation to roots. *Tree Physiol.* **2010**, *30*, 979–987. [CrossRef]

26. ElSayed, A.I.; Rafudeen, M.S.; Golldack, D. Physiological aspects of raffinose family oligosaccharides in plants: Protection against abiotic stress. *Plant Biol.* **2014**, *16*, 1–8. [CrossRef]

27. Peterbauer, T.; Richter, A. Biochemistry and physiology of raffinose family oligosaccharides and galactosyl cyclitols in seeds. *Seed Sci. Res.* **2001**, *11*, 185–197. [CrossRef]

28. Slewinski, T.L.; Braun, D.M. Current perspectives on the regulation of whole-plant carbohydrate partitioning. *Plant Sci.* **2010**, *178*, 341–349. [CrossRef]

29. Kozlowski, T.T. Carbohydrate sources and sinks in woody plants. *Bot. Rev.* **1992**, *58*, 107–222. [CrossRef]

30. Yarmolinsky, D.A.; Zuker, C.S.; Ryba, N.J.P. Common Sense about Taste: From Mammals to Insects. *Cell* **2009**, *139*, 234–244. [CrossRef] [PubMed]

31. Lee, A.; Owyang, C. Sugars, Sweet Taste Receptors, and Brain Responses. *Nutrients* **2017**, *9*, 653. [CrossRef] [PubMed]

 *agriculture*

Article

Content and Uptake of Ash and Selected Nutrients (K, Ca, S) with Biomass of *Miscanthus* × *giganteus* Depending on Nitrogen Fertilization

Izabela Gołąb-Bogacz [1], Waldemar Helios [2], Andrzej Kotecki [2], Marcin Kozak [2] and Anna Jama-Rodzeńska [2,*]

[1] Bugaj Sp. z o.o, Bugaj Zakrzewski 5, 97-512 Kodrąb, Poland; iza.golab@o2.pl
[2] Institute of Agroecology and Plant Production, Wroclaw University of Environmental and Life Sciences, Pl. Grunwaldzki 24A, 50-363 Wrocław, Poland; waldemar.helios@upwr.edu.pl (W.H.); andrzej.kotecki@upwr.edu.pl (A.K.); marcin.kozak@upwr.edu.pl (M.K.)
* Correspondence: anna.jama@upwr.edu.pl; Tel.: +48-320-1627

Abstract: Fertilisation has a significant impact not only on the yielding, but also on the quality of the harvested biomass. Among energy crops, *Miscanthus* × *giganteus* are some of the most important plants used for combustion process. The chemical composition of biomass has significant impact on the quality of combustion biomass. The effect of nitrogen fertilisation (with dose of 60 kg N ha^{-1}) in different terms of biomass sampling on the content and uptake of crude ash, potassium, calcium and sulphur by rhizomes, stems, leaves and the aboveground part of miscanthus was evaluated in the paper. Nitrogen fertilisation contributed to the increase of ash content in the rhizomes and the aboveground part of plants. Independently of nitrogen fertilisation potassium content decreased in the whole vegetation period; in the case of stems this decrease amounted 60%. Calcium content in various parts of plants was highly differentiated compared to potassium content. Average calcium content in the aboveground parts was 2.68 higher compared to rhizomes. Nitrogen fertilisation affected significantly on potassium, calcium and sulphur uptake in all examined parts of plants (except stems in the case of calcium uptake). Uptake of crude ash under nitrogen fertilisation was significantly higher in all examined parts of plants during the whole vegetation period.

Keywords: aboveground; belowground part of *Miscanthus* × *giganteus*; ash; potassium; calcium; sulphur content; uptake

Citation: Gołąb-Bogacz, I.; Helios, W.; Kotecki, A.; Kozak, M.; Jama-Rodzeńska, A. Content and Uptake of Ash and Selected Nutrients (K, Ca, S) with Biomass of *Miscanthus* × *giganteus* Depending on Nitrogen Fertilization. *Agriculture* **2021**, *11*, 76. https://doi.org/10.3390/agriculture11010076

Received: 20 December 2020
Accepted: 14 January 2021
Published: 18 January 2021

Publisher's Note: MDPI stays neutral with regard to jurisdictional claims in published maps and institutional affiliations.

1. Introduction

The need to counteract and prevent increasingly rapid climate change is leading to the implementation of processes that will reduce greenhouse gas emissions by replacing fossil fuels with renewable energy sources. Besides the continued use of non-renewable fossil fuels, which include hard coal, lignite, natural gas and oil, energy from renewable sources is increasingly used. The acquisition of renewable energy sources is currently directed towards agriculture [1–8].

Energy from plant biomass is mainly obtained by pyrolysis, gasification or direct combustion of appropriately ground or granulated mass [9,10]. Miscanthus (*Miscanthus* × *giganteus* Greef et Deuter) can play a significant role as a source of renewable energy for Europe [11–13]. Obtaining high quality biomass for the combustion process depends on the quality of the raw material (biomass) [14,15], while the quality of the raw material depends on the content of various elements (for example, high lignin content is desirable for thermochemical and undesirable for biochemical processes) [16,17].

The content of elements in the biomass is significantly influenced by genetic properties [14,18] which can be modified by environmental conditions, such as soil properties, pH, weather conditions (precipitation, temperature), as well as agrotechnical treatments—mainly fertilisation [19–23]. Date of harvest (late winter or spring) can also contribute to the

303

reduced content of nutrients that results from their translocation from aboveground part of plant to rhizomes or natural leaching of components from leaves and stems [23–25]. Appropriate chemical composition, especially low content of contaminants in biomass, is desirable during harvest, especially for biomass for thermal combustion, as it contributes to the minimisation of their emissions [23].

Most of the available studies on the content and nutrient uptake of miscanthus concern nitrogen, phosphorus, potassium and magnesium [25–28], while only a few works concern calcium and sulphur content [29,30]. An innovative part of the study was to examine the dynamics of sulphur uptake during the whole vegetation period, taking into account nitrogen fertilisation in various parts of plants.

Crude ash content and examined macroelements have a significant impact on the quality of biomass combustion; therefore, the relevance of these elements is discussed. High ash concentration decreases the heating value [31,32]. Potassium, alongside silicon, is the main component of ash [12]. The potassium content of biomass is very important because its high content can increase the corrosion effect in heating systems and lower the melting point of ash [31], and is regarded as a critical element in ash-related problems [32]. Therefore, the potassium content should be as low as possible [32]. For optimal plant growth, the potassium content should be 10–50 g of DM [31]. Sulphur also plays an important role during the combustion process. Sulphur compounds that are formed during this process lead to corrosion and are emitted into the atmosphere [30]. In turn, calcium can inhibit the occurrence of silicate melt-induced slagging and bed agglomeration, as a result of forming melting calcium potassium phosphates and silicates at high temperatures [30–32].

The work hypothesis assumes that fertilisation in a of dose 60 kg ha^{-1} will contribute to changes in content an uptake of selected macronutrients and ash. It has been estimated that particular parts of the plant (rhizomes, stems, leaves) will be characterised by different ash, Ca, K, and S accumulation. Additionally, fertilisation at a dose 60 kg ha^{-1} N causes the increase in uptake of ash and selected macroelement.

The aim of the study was to determine the effect of nitrogen fertilisation on the content and uptake of ash and selected macroelements in *Miscanthus × giganteus*.

2. Materials and Methods

2.1. Study Site and Materials

The experiment with miscanthus and nitrogen fertilisation started by separating plots on the plantation carried out in 2004. Detailed information is contained in the article by Bogacz et al. 2020 [33]. The study with miscanthus was conducted in the years 2014–2016 at Experimental Station belonging to Wroclaw University of Environmental and Life Sciences, Pawlowice (geographical location 17°7′ E and 51°08′ N in the Lower Silesian Voivodship (Figure 1)). The tested factor was nitrogen fertilisation (0, 60 kg ha^{-1} N). Miscanthus sampling started from the 30th day of the vegetation period and was done every 30 days until the end of the vegetation period (June, July, August, September, October, November and December). At each date of sampling, a plant sample of the aboveground part of the plant and rhizomes was sampled from an area of 0.25 m^2. Samples for chemical analysis were reduced according to the standard requirements of PN-EN 96 ISO 14780:2017-07 [34] (which defines methods for reducing combined samples to laboratory samples and laboratory samples to sub-samples and general analysis samples, and is applicable to solid biofuels). Plant samples were sampled from the area of 0.25 m^2 by gentle extraction of rhizomes from the soil with the whole stems. Dry mass for laboratory samples was determined by air-drying the dry mass at 105 °C for three hours according to Polish standard (PN-R-04013:1988).

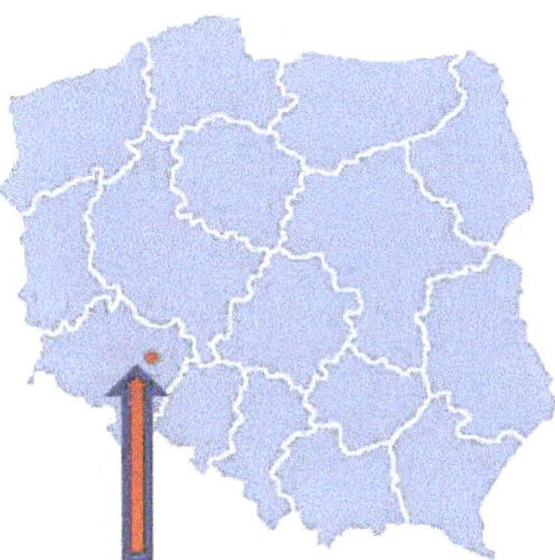 

Figure 1. Location of experiment.

The weather and soil condition, experiment design and agrotechnical treatments are described in research by Bogacz et al. [33].

2.2. Chemical Analysis of Plant Material

The content of ash and macroelements in plant material was determined in the laboratory belonging to the Institute of Agroecology and Plant Production. The content of crude ash and macroelements in the aboveground part was calculated on the basis of the content of these elements in the leaves and stems, taking into account the structure of the dry matter yield.

Chemical analyses comprised:

- crude ash by burning dry plant material at 600 °C in an electric furnace: incineration of plant material and combustion 1/2 g weighing the analytical sample of plant material in the muffle furnace at 600 ± 15 °C and baking the remaining ash;
- potassium and calcium on the flame photometer (BWB Technologies UK LTD), using flame photometry; mineralization of plant material through the use of sulphuric acid and perhydrol and subsequent determination on a flame photometer;
- total sulphur by nephelometric method, after wet mineralisation with concentrated sulphuric acid with 30% perhydrol, by the Bradley–Lancaster nephelometric method.

Uptake of crude ash and selected elements vas calculated based on yield biomass and chemical content of the examined parts of plants.

2.3. Statistical Analysis

The experiment was conducted in four replications in order to test the effects of N fertilisation on the content and uptake of ash and macroelements in *Mischanthus giganteus*. The analysis of variance (ANOVA) and the mixed model with repeated measurements were used. Doses of nitrogen fertilisers were assumed to be a fixed factor, while years was assumed to be random. The results of chemical analysis of the Mischanthus were analysed by ANOVA in the Statistica program (13.1 StatSoft, Kraków, Poland). One-way ANOVA (nitrogen fertilisation, then year of experiment) was performed including post-hoc analysis. The level of significance was determined as $p < 0.05$.

Homogeneous groups were determined on the basis of the Tukey test. The groups were determined from the lowest to the highest value. The correlation of repeated measurements was performed as the average value over the three-year growing season of each month. The p-value concerns the subsequent months.

3. Results

3.1. Crude Ash Content and Uptake

The effect of nitrogen fertilisation on ash content in the rhizomes ($p = 0.0035$), stems ($p = 0.0002$) and aboveground part of *Miscanthus* × *giganteus* ($p < 0.001$) except for the leaves was found. Even though rhizomes are not involved in the combustion process, knowledge of the ash content of rhizomes allowed the ash content to significantly increase from 2014 to 2016 in rhizomes ($p = 0.0156$), whereas the highest content was found in the leaves ($p = 0.0312$) in 2015 (the lowest annual sum of precipitation—392 mm). The highest content of ash was observed in the aboveground part of plants in the first year ($p = 0.0047$). The highest content of this component was found in leaves, which is particularly beneficial as the stem has the greatest share in the process of biomass combustion (Table 1). The highest content of crude ash was found at the beginning of the vegetation period, and as the plants developed (and also as a result of the ageing processes), its content decreased. The decrease in ash content in stems was greater than in leaves at the beginning of the vegetation period (Figure 2). The figures show the significance values of differences (p-values) of ash content in subsequent months of observation for control and dose 60 (Figure 2).

Table 1. Crude ash content in dry matter of miscanthus in g kg^{-1} (average for the years 2014–2016).

Dose kg ha^{-1} N	Rhizomes	Stems	Leaves	Aboveground Part
0	43.6 a	37.6 a	56.7 a	53.6 a
60	46.5 a	42.2 a	58.3 a	57.6 a
p-value	0.0035	0.0002	0.2418	<0.001
2014	43.3 a	39.7 a	58.6 a	57.3 a
2015	45.2 a	39.0 a	59.3 a	54.5 a
2016	46.7 a	41.0 a	54.7 a	55.1 a
p-value	0.0156	0.1980	0.0312	0.0047

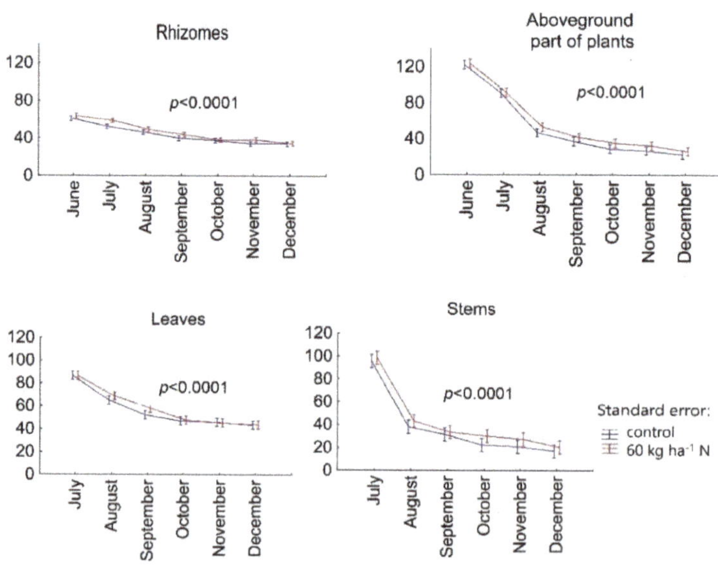

Figure 2. Crude ash content in examined part of miscanthus (g kg^{-1}) (three-year average content from measurements during the growing season every 30 days).

The crude ash uptake through individual elements of the plant was significantly dependent on the nitrogen fertilisation ($p \leq 0.001$). The highest uptake in the rhizomes ($p < 0.001$) was found in the third year, whereas the highest uptake in the stems ($p \leq 0.001$) and aboveground part of plants ($p = 0.0467$) was found in the second year of the experiment (Table 2). Crude ash accumulation by *Miscanthus × giganteus* per 1 m^2 in rhizomes increased throughout the entire vegetation period, while in stems and aboveground parts of the plant, it decreased at the end of the vegetation period. Nitrogen fertilization caused greater uptake of crude ash in all examined parts of plants during the whole vegetation period (Figure 3). The *p*-values presented on the figure concern the date of plant material sampling. The figures show the significance values of differences (*p*-values) of ash uptake in subsequent months of observation for control and dose 60 (Figure 3).

Table 2. Crude ash uptake by g·m^{-2} (average for 2014–2016).

Dose kg ha^{-1} N	Rhizomes	Aboveground Part			Rhizomes and Aboveground Part
		Stems	Leaves	All Together	
0	44.8 a	46.0 a	33.9 a	74.4 a	119.2 a
60	54.3 b	60.1 b	46.3 b	99.0 b	153.3 b
p-value	<0.001	<0.001	<0.001	<0.001	<0.001
2014	48.5 a	49.8 a	40.8 a	85.5 a	134.0 a
2015	47.7 b	57.9 a	39.1 a	89.1 a	136.8 a
2016	52.6 b	51.4 a	40.3 a	85.5 a	138.1 a
p-value	<0.001	<0.001	0.3064	0.0467	0.1679

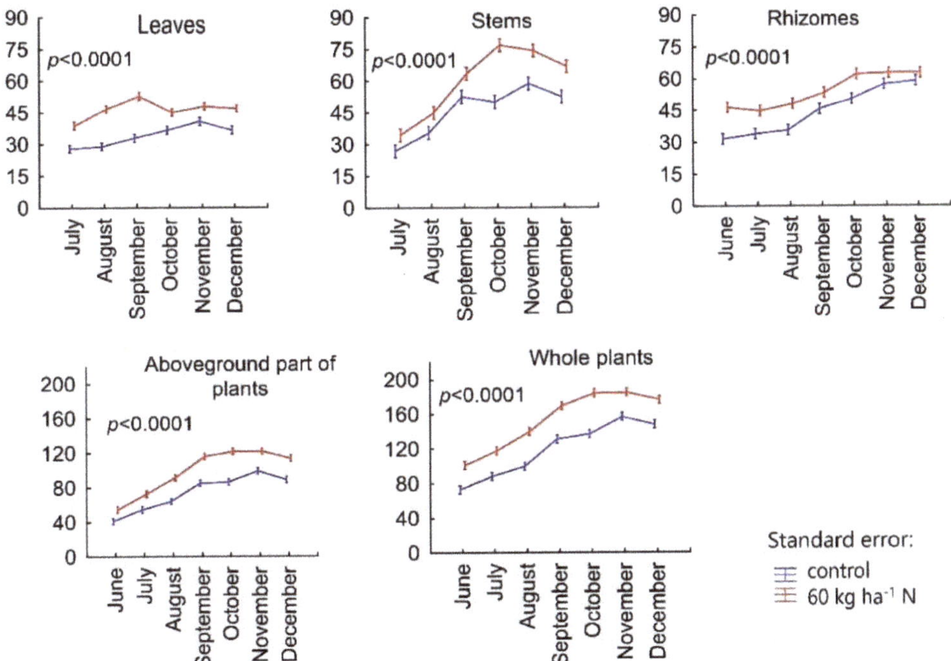

Figure 3. Crude ash uptake in examined part of miscanthus (g m^{-2}) (three-year average content from measurements during the growing season every 30 days).

3.2. Potassium Content and Uptake

The potassium content in leaves ($p = 0.0085$) was significantly dependent on the nitrogen fertilisation. In the stem of *Miscanthus* × *giganteus*, the highest content of potassium was found in the third year of the study ($p = 0.0032$), and in the second year in rhizomes ($p = 0.0219$) and leaves ($p < 0.001$) (Table 3).

Table 3. Potassium content in dry matter of miscanthus g kg^{-1} (average for the years 2014–2016).

Dose kg ha^{-1} N	Rhizomes	Stems	Leaves	Aboveground Part
0	12.7 a	11.6 a	12.3 a	12.0 a
60	11.9 a	11.6 a	13.9 a	12.6 a
p value	0.1455	0.9491	0.0085	0.1643
2014	12.7 a	11.0 a	13.6 a	12.1 a
2015	13.0 a	10.3 a	15.1 ab	12.1 a
2016	11.1 a	13.5 a	10.6 a	12.7 a
p value	0.0219	0.0032	<0.001	0.4601

A decrease was observed in potassium content in the leaves, stems and aboveground part of *Miscanthus* × *giganteus* since August to the December. The lowest level of this element was found in December, when the potassium content in the aerial part of plants was on average about twice as low as in June. In turn, a decrease in potassium content in the rhizomes was found from the beginning of vegetation period until November. The increase in potassium content in the rhizomes from November to the end of the vegetation period (Figure 4) might be the result of translocation of this element from the aboveground part of plants to the rhizomes. The figures show the significance values of differences (*p*-values) of potassium content in subsequent months of observation for control and dose 60 (Figure 4).

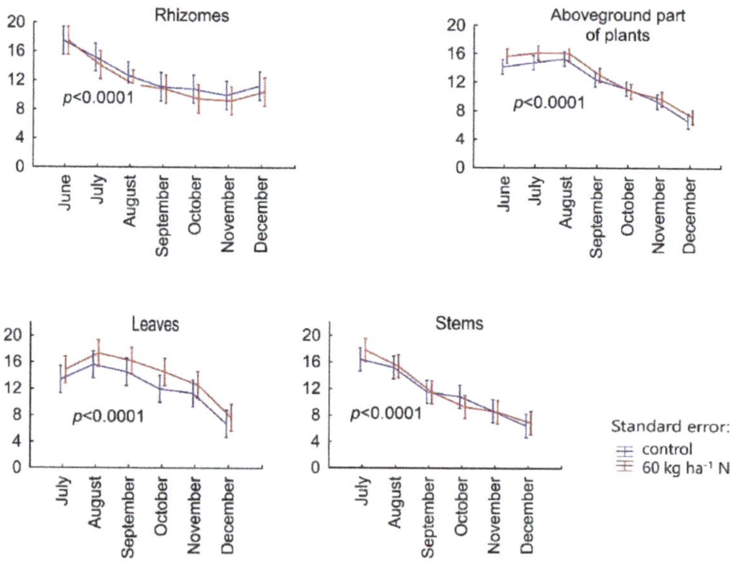

Figure 4. Potassium content in examined part of miscanthus (g kg^{-1}).

Potassium uptake (g m^{-2}) by *Miscanthus* × *giganteus* was dependent on nitrogen fertilisation and the years of the experiment. Nitrogen fertilisation caused an increase in potassium accumulation (g m^{-2}) in all examined parts of plants. The highest potassium uptake was found in the rhizomes ($p < 0.001$) and the aboveground part of plants ($p < 0.001$) in the first year of research (Table 4).

Table 4. Potassium uptake of the giant miscanthus in g m^{-2} (average for 2014–2016).

Dose kg ha^{-1} N	Rhizomes	Aboveground Parts			Rhizomes and Aboveground Part
		Stem	Leaves	Together	
0	13.3 a	17.9 a	7.6 a	22.5 a	35.8 a
60	13.9 a	19.1 a	11.4 b	27.1 b	41.0 a
p-value	0.0064	0.0021	<0.001	<0.001	<0.001
2014	15.1 b	19.7 a	10.2 b	26.5 a	41.6 a
2015	14.3 b	16.9 a	10.4 b	24.1 a	38.4 a
2016	11.4 a	18.9 a	7.8 a	23.8 a	35.1 a
p-value	<0.001	<0.001	<0.001	<0.001	<0.001

Rhizomes accumulated potassium until the end of vegetation (increasing trend). Without nitrogen fertilization in the aboveground part of plants, the peak potassium uptake was observed in November, whereas the highest accumulation was seen earlier on the plots with nitrogen fertilization (Figure 5). The figures show the significance values of differences (*p*-values) of potassium uptake in subsequent months of observation for control and dose 60 (Figure 5).

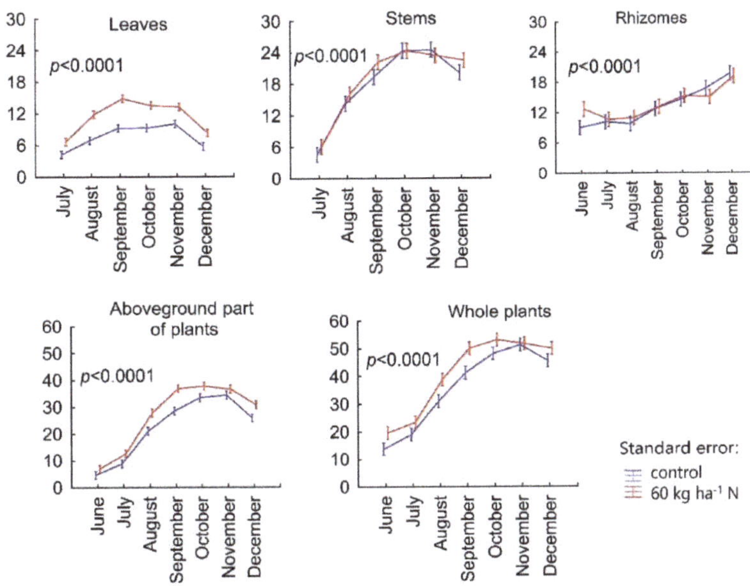

Figure 5. Potassium uptake in examined part of miscanthus (g m^{-2}) (three-year average content from measurements during the growing season every 30 days).

3.3. Calcium Content and Uptake

Nitrogen fertilisation had no significant effect on the calcium content in the examined parts of plants (Table 5). The year of the experiment had a significant effect on calcium content in rhizomes ($p < 0.001$), stems ($p = 0.0036$) and leaves ($p < 0.001$) (Table 5).

Table 5. Calcium content in dry matter of the giant miscanthus in g·kg^{-1} (average for the years 2014–2016).

Dose kg ha^{-1} N	Rhizomes	Stems	Leaves	Aboveground Part
0	0.58 a	1.34 a	1.78 a	1.52 a
60	0.55 a	1.24 a	1.93 a	1.51 a
p-value	0.5401	0.2250	0.1717	0.8787
2014	0.64 b	1.24 a	1.80 a	1.51 a
2015	0.79 b	1.11 a	2.31 a	1.50 a
2016	0.27 a	1.52 a	1.45 b	1.54 a
p-value	0.0000	0.0036	<0.001	0.8886

An increase in the content of this element in rhizomes was found until August and in the stems to the end of the vegetation period (Figure 6). The figures show the significance values of differences (*p*-values) of calcium content in subsequent months of observation for control and dose 60 (Figure 6).

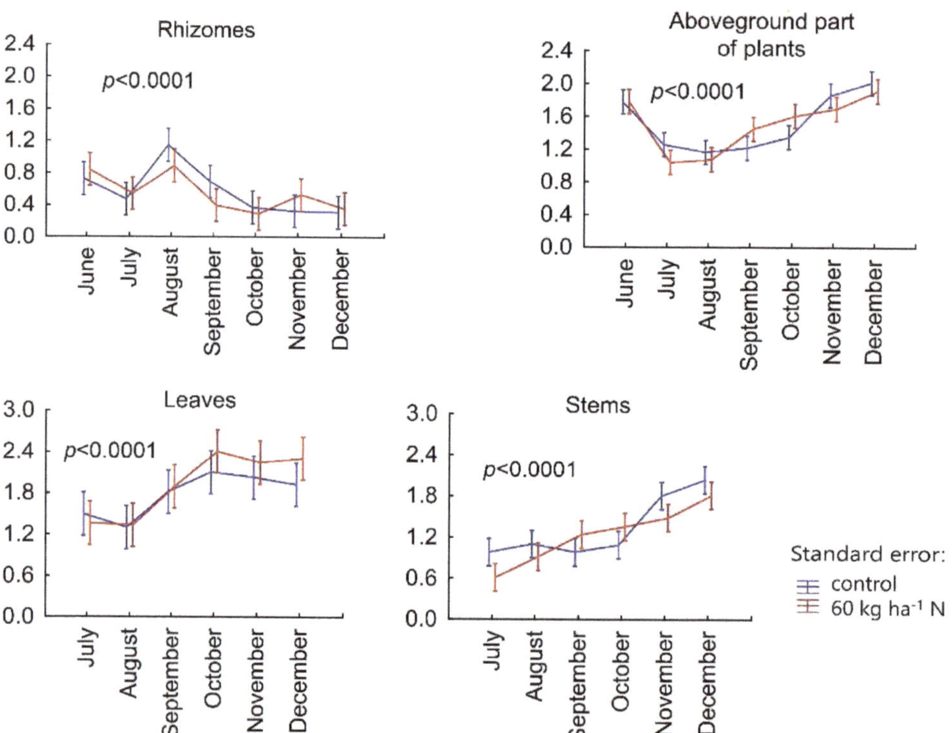

Figure 6. Calcium content in examined part of miscanthus (g kg^{-1}) (three-year average content from measurements during the growing season every 30 days).

Calcium uptake depended on nitrogen fertilisation in all parts of plants ($p < 0.001$) except stems. Significant changes in calcium uptake were found during the years of research in the rhizomes ($p \leq 0.001$), stems ($p \leq 0.001$), leaves ($p \leq 0.001$) and the whole plants ($p \leq 0.001$) (Table 6).

Table 6. Calcium uptake by the giant miscanthus in g·m^{-2} (average for the years 2014–2016).

Dose kg ha^{-1} N	Rhizomes	Aboveground Parts			Rhizomes and Aboveground Parts
		Stems	Leaves	All Together	
0	0.56 a	2.80 a	1.20 a	3.51 a	4.07 a
60	0.61 a	2.80 a	1.71 b	3.98 a	4.59 a
p-value	<0.001	0.9045	<0.001	<0.001	<0.001
2014	0.64 b	2.81 a	1.44 ab	3.76 a	4.40 a
2015	0.83 c	2.41 a	1.73 a	3.64 a	4.47 a
2016	0.28 a	3.18 a	1.19 b	3.84 a	4.12 a
p-value	<0.001	<0.001	<0.001	0.0573	<0.001

An increase in calcium uptake was seen in the stems and aboveground part of plants through the entire vegetation period. Changes in calcium uptake in the rhizomes were lower in this period compared to aerial parts of *Miscanthus × giganteus* (Figure 7). The figures show the significance values of differences (*p*-values) of calcium uptake in subsequent months of observation for control and dose 60 (Figure 7).

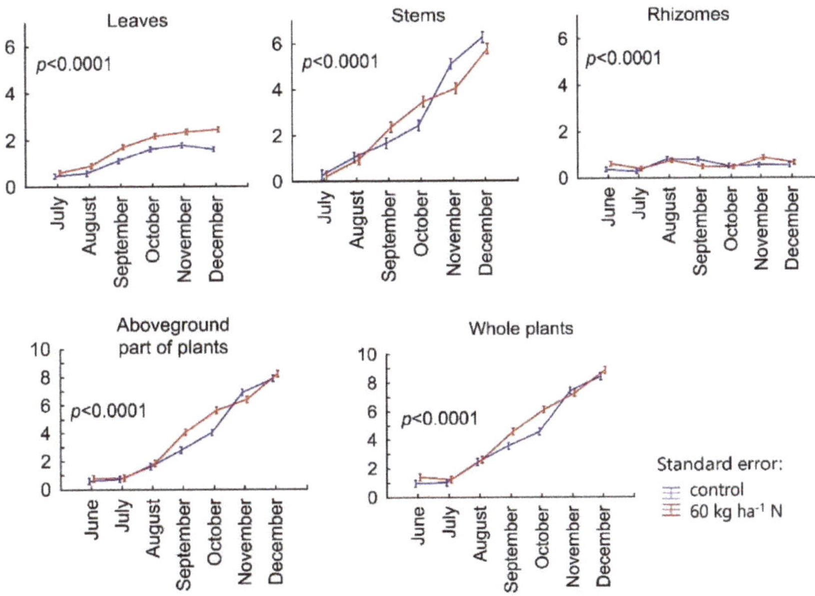

Figure 7. Calcium uptake in examined part of miscanthus (g m^{-2}) (three-year average content from measurements during the growing season every 30 days).

3.4. Sulphur Content and Uptake

Nitrogen fertilisation had a significant impact on sulphur content in the stems ($p = 0.0485$) and aboveground parts ($p = 0.0067$). Significant changes in sulphur content were found in the different years of the experiment for rhizomes ($p = 0.0345$), stems ($p < 0.001$), leaves ($p < 0.001$) and aboveground parts of plants ($p = 0.0219$). The highest sulphur content in the rhizomes and stems was seen in the first year of field experiments and in the leaves and aboveground part of plants in the third year (Table 7).

Table 7. Sulphur content in dry matter of miscanthus in g kg^{-1} (average for years 2014–2016).

Dose kg ha^{-1} N	Rhizomes	Stems	Leaves	Aboveground Parts
0	0.78 a	0.62 a	0.63 a	0.69 a
60	0.81 a	0.67 a	0.64 a	0.75 a
p-value	0.4928	0.0485	0.5357	0.0067
2014	0.90 b	0.71 b	0.56 a	0.71 a
2015	0.77 ab	0.55 a	0.66 ab	0.69 a
2016	0.72 a	0.67 b	0.69 b	0.76 a
p-value	0.0345	<0.001	<0.001	0.0219

The sulphur content in the aboveground parts, stems and leaves decreased with the development of plants. The dynamic changing of sulphur content in the aerial part of *Miscanthus × giganteus* was the highest at the beginning of the vegetation period. The lowest sulphur content in the rhizomes was found in October (Figure 8). The figures show the significance values of differences (*p*-values) of sulphur content in subsequent months of observation for control and dose 60 (Figure 8).

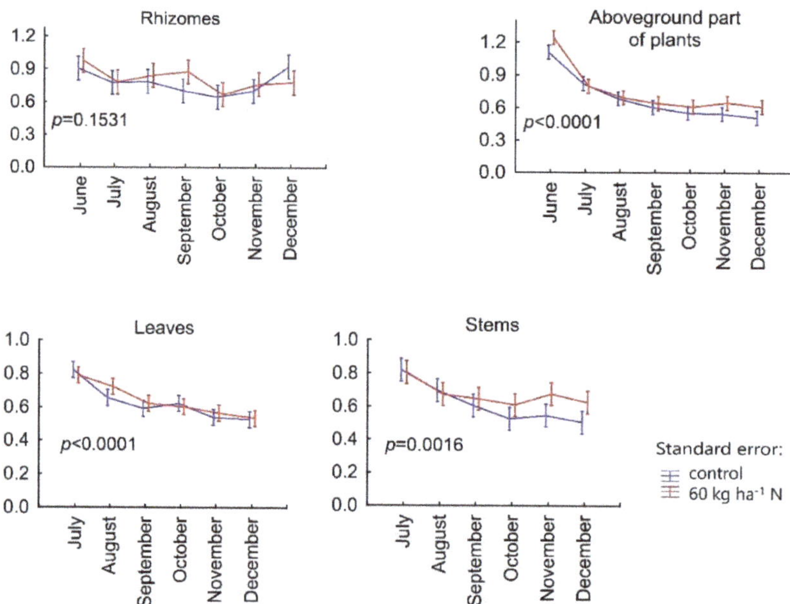

Figure 8. Sulphur content in examined part of miscanthus (g kg^{-1}) (three-year average content from measurements during the growing season every 30 days).

Sulphur uptake (g m^{-2}) by *Miscanthus × giganteus* was significantly dependent on nitrogen fertilisation and year of the experiment (Table 8). The highest sulphur uptake was found on plots with nitrogen fertilisation in all examined parts of plants ($p < 0.001$). The highest sulphur accumulation per m^2 in the rhizomes and aboveground part of plants was observed in the first year of the field experiment (Table 8).

Table 8. Sulphur uptake by the giant miscanthus in g·m^{-2} (average for the years 2014–2016).

Dose kg ha^{-1} N	Number of Days after the Start of the Vegetation	Rhizomes	Aboveground Part			Rhizomes and Aboveground Parts
			Stems	Leaves	All Together	
0		0.87 a	1.03 a	0.39 a	1.27 a	2.14 a
60		0.99 b	1.28 b	0.53 b	1.63 b	2.62 b
p-value		<0.001	<0.001	<0.001	<0.001	<0.001
	2014	1.06 c	1.36 b	0.40 a	1.57 a	2.63 a
	2015	0.89 ab	0.99 a	0.45 a	1.30 a	2.19 a
	2016	0.83 a	1.12 ab	0.52 b	1.48 a	2.31 a
p-value		<0.001	<0.001	<0.001	<0.001	<0.001

The highest sulphur uptake by rhizomes and stems was found in December. It should be noted that stems accumulated over 2–3 times more sulphur than leaves. Sulphur uptake in *Miscanthus × giganteus* increased with the progressing vegetation period in all parts of the field experiment (Figure 9). The figures show the significance values of differences (*p*-values) of Sulphur uptake in subsequent months of observation for control and dose 60 (Figure 9).

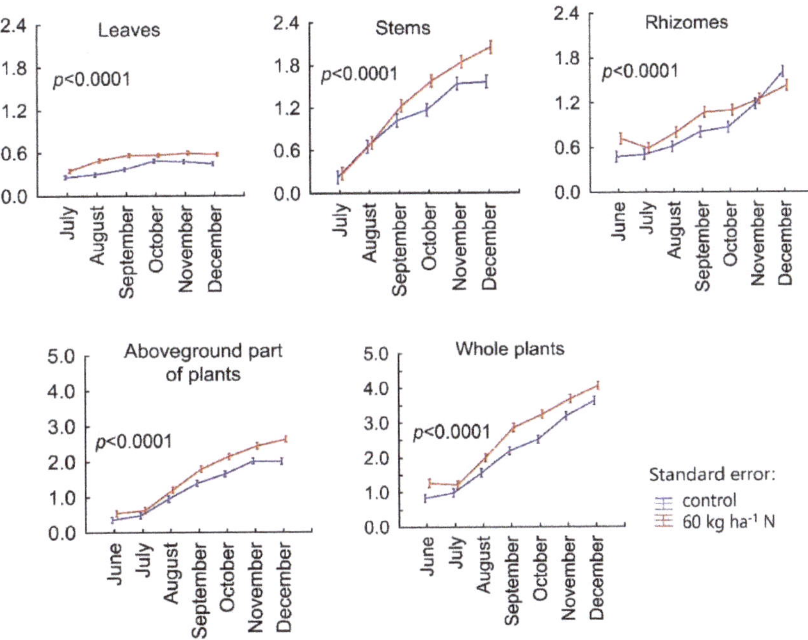

Figure 9. Sulphur uptake in examined part of miscanthus (g m^{-2}) (three-year average content from measurements during the growing season every 30 days).

4. Discussion

Mineral concentration plays an essential role in biomass combustion quality [14,35]. To improve biomass quality of *Miscanthus × giganteus*, cultivation practice should be based on keeping the nitrogen fertilisation rate as low as possible and delaying harvest until the spring following growth, as this will allow nutrient remobilisation and leaching of soluble minerals like K and Cl through rainfall [35,36].

Nutrient remobilisation seems to be a good strategy for perennial rhizomatous grasses [37,38] and represents an environmentally friendly strategy to reduce fertiliser applications [36]. When calculating the nutrient balance and fertiliser recommendations, the remobilisation of nutrients within the plant must be taken into account [25]. September and March are irrelevant for nutrient remobilisation [28]. The increase in nutrient content found in rhizomes in autumn and winter may be caused by remobilisation from the aboveground parts to the underground part [25]. In our research, generally, nutrient concentrations were highest at the beginning of the growing period and decreased clearly during the growing season. There were no significant differences caused by N fertilisation (except for potassium in leaves and sulphur in stems). The large loss of K from shoots between September and harvest can be attributed to leaching from the senescent plant material as K is not organically metabolised [39]. Some leaves fell after the end of the growing period, contributing to improved properties of soil by increasing the contents of elements and organic matter, thereby leading simultaneously to a decrease in ash uptake by plants [24]. The mineral concentration of aerial biomass is at its highest during spring and early summer and then declines, probably as a result of remobilization [24]. These results are also confirmed by own research in the aboveground parts of plants. It is documented that mineral concentration in the aboveground biomass of *Miscanthus × giganteus* decreases gradually from autumn to winter [24,28]. Our results highlight a decline in the concentration of crude ash and macronutrients in aboveground parts of plants from spring to autumn.

The average ash content in Giant Mischanthus according to Borkowska (2007) [31] is about 27.6 g kg^{-1}. In the study Baxter et al. (2014) [15], the average ash content in leaves was between 40 and 60 g kg^{-1} DM, and in stems the mean value was lower, between 10 and 30 g kg^{-1} dm. In our research the highest average ash content was found in leaves (57.5 g kg^{-1}), less in rhizomes (45.1 g kg^{-1}) and the lowest in stems (39.9 g kg^{-1}). In the research by Lewandowski and Heinz (2002) [36], the content of ash in the aboveground part of plants decreased from December to February. Ash content decreased also from autumn to spring in the study of Lewandowski et al. (2003) [40]. Similarly, delayed harvesting in the research of Lewandowski and Heinz (2003) [36] contributed to a reduction in ash content by 28% on average in Portugal and Great Britain, by 42% in Germany, by 50% in Sweden and by 54% in Denmark. Kotecki et al. (2010) [41] found that nutrient and crude ash yields were higher during the autumn harvest compared to the winter harvest, rising from 31 to 69%, while nitrogen fertilisation contributed to an increase in ash content. In our research, the ash content depended on nitrogen fertilisation and years of experiment in the rhizomes and aboveground part of miscanthus. Ash content decreased during the whole vegetation period. Studies by Lewandowski and Kircherer (1997) [42] showed that miscanthus leaves have a higher ash content than stems, which is also confirmed by own research.

The content of potassium in the aboveground part of plants ranges from 4.3 to 10.5 g kg^{-1} DM [22]. According to Borzęcka-Walker's (2010) [31] study, the potassium content in the aerieal parts of miscanthus plants ranged from 2.7 to 9.9 g kg^{-1} DM on heavy black soil and from 1.6 to 9.4 g kg^{-1} DM on medium heavy black soil, depending on the genotype and year of cultivation. In the research of Kalembasa et al. (2019) [22] the mean potassium content in mischanthus grass biomass was 15.66 g kg^{-1} D.M. Furthermore, Lewandowski et al. 2000 [43] presented a review of potassium content obtained in field studies by several authors for some locations in Europe. Potassium concentration was significantly influenced during harvest time. According to Jensen et al. (2017) [38], potassium content decreased over the three harvests from June (2009) to February (2010) with the highest concentration during the summer. As expected, delaying the harvest by three to four months improved the combustion quality by reducing potassium content from 9 to 4 g kg^{-1} DM.

Their experiment indicated that many genotypes of *Mischanthus* are characterised by higher concentrations of potassium in autumn. According to Beale and Long (1997) [24], potassium concentrations in the aboveground dry matter decreased from 32 to 12.0 g kg^{-1} during whole vegetation period. In our study, the potassium content also decreased from summer till the end of vegetation in aboveground part of plants. Kalembasa et al. (2019) [22] proved that potassium was transferred from the aboveground parts of plants to rhizomes at the end of the growing season. According to Christian et al. (2008) [44] transfer of potassium from leaves and stem to rhizomes is 14–30%.

The uptake of macronutrients is strongly dependent on the yield. The higher obtained yield, the higher the uptake of the following element [44]. In the Christian et al. (2008) [44] experiment between 1993 and 1995, the mineral uptake increased when the yield increased rapidly. The translocation of the elements during harvest depends on many external factors, especially weather conditions. While *Mischanthus* is characterised by higher dry yields (about 30 Mg ha^{-1}) from a three-year old crop, Beale and Long (1997) [24] found high potassium uptake gaining 38.0 g m^{-2}. Nassi o Di Nasso et al. (2011) [45] obtained potassium uptake of around 27.0 g m^{-2}. In our research potassium uptake was around 24.8 g·m^{-2}. Greater uptake of potassium in the Roncucci et al. (2015) [28] study was found in autumn (16.0 g·m^{-2}), and uptake was lower during wintertime. In this research, potassium uptake by the aboveground part of miscanthus at wintertime had values corresponding to around 33% of those recorded at autumn harvest. The time of harvest was the most relevant factor influencing miscanthus nutrient uptake in own experiments and those by Roncucci et al. (2015) [28].

Aerial parts of grasses accumulated mostly calcium, potassium and magnesium. The issue of calcium content in the rhizomes was undertaken by Stypczyńska et al. (2017) [21]. The concentration of this element in their study in the rhizomes was 1.5 g kg^{-1} DM. In turn, Nassi di Nasso et al. 2010 [45] studied calcium content in the rhizomes and obtained values of 0.5–1.4 g kg^{-1} DM which are confirmed in our research. The content of calcium was affected by the following factors: genotypes, geographical location of plantation and weather conditions, according to Helios (2018) [46]. In the lack of calcium fertilisation the content of this element relies on age of the plantation. In a 12-year study by Helios (2018) [46], the calcium content ranged from 3.1 g kg^{-1} DM while calcium uptake amounted to 0.45 g m^{-2} in the first year of experiment to 0.5 g kg^{-1} DM, while the calcium uptake was 1.2 g m^{-2} in the tenth year of cultivation. In studies by Baxter et al. (2014) [15] and Stypczyńska et al. (2017) [21], leaves of miscanthus were characterised by higher calcium content compared to the stems, which is confirmed by our research. In conducted experiments by Lewandowski and Kicherer [42], the calcium concentrations in the leaves ranged from 2.3 to 3.7 g kg DM while that in the stems ranged from 0.5 to 1.1 g kg DM.

In our research, the trends of changes in calcium content during vegetation were similar to those of Kotecki et al. (2010) [41] who showed that the content of this element in the aboveground part of the plant was decreasing until summer and then it increased.

Sulphur plays an important role during the combustion process. Sulphur compounds that are formed during this process lead to corrosion and are emitted into the atmosphere [30]. In Lewandowski and Kicherer's [42] research no definite effect of nitrogen fertilisation on sulphur concentration in the leaves and stems was found. In our experiment, the content of this element in the stems was dependent on examined factor ($p = 0.0485$) while the nitrogen fertilisation had no significant impact on sulphur content in the leaves. For the entire vegetation period of miscanthus, Spiak et al. (2012) [29] showed that almost half the sulphur content is present in the stems compared to that in the leaves. In the study by Baxter et al. (2014) [15], on the other hand, the opposite results were obtained. In our study, the highest sulphur content was found in the rhizomes and there was less in the leaves and stems. The sulphur content in the leaves (0.64 g kg^{-1} DM) and stems (0.65 g kg^{-1} DM) was similar. Concentration of this element in aboveground parts of miscanthus amounted to 0.72 g kg^{-1} DM. In research by Kotecki (2010) [41], sulphur content in aerial parts of plants was 0.5–0.8 g kg^{-1} DM.

In our field experiment, the highest content of this component was found in young plants. As the vegetation progressed, the sulphur content decreased in the aboveground part of plants by around 50%. In contrast to the content, the sulphur uptake was significantly higher in stems than in leaves. The uptake of sulphur in the aboveground part and whole plants with an increased trend was observed until the end of the vegetation season. A similar tendency was observed in rhizomes from July to December.

5. Conclusions

Because of the need to reduce emissions, and to avoid worsening the air quality by producing the compounds during combustion of, e.g., hard coal for heating purposes in many Polish cities and other Central and Eastern European countries, the low content of mineral components in *Miscanthus × giganteus* biomass is very desirable and may constitute an alternative source of biomass for energy purposes.

While the research hypothesis was verified, it should be stated that only ash content in rhizomes and aboveground part of plants depended significantly on nitrogen fertilisation while potassium (except in leaves), calcium and sulphur content (except in stems and aboveground parts) were not significantly influenced by this factor. The uptake of the studied elements was significantly dependent on nitrogen fertilisation in the case of ash, potassium, sulphur and calcium (except for stems). K and S concentrations were highest at the beginning of the growing period and decreased clearly during the growing season.

The ash content was significantly higher under the influence of nitrogen fertilisation in leaves at 58.3 g m^{-2} and the lowest in stems at 42.2 g m^{-2}, and the highest intake by stems at 60.1 g m^{-2} and the lowest in leaves at 46.3 g m^{-2}. Significantly higher sulphur uptake was found in stems under the influence of nitrogen fertilisation at the amount of 1.28 g m^{-2}.

Author Contributions: Conceptualization: A.J.-R., I.G.-B., W.H.; Methodology: A.K., M.K., W.H.; Software: I.G.-B., W.H., A.J.-R.; Validation: A.K., M.K.; Investigation: I.G.-B., W.H., A.K.; Formal Analysis: I.G.-B., W.H., A.J.-R.; Writing: I.G.-B., W.H., A.J.-R.; Review: M.K., A.K., A.J.-R. All authors have read and agreed to the published version of the manuscript.

Funding: This research received no external funding.

Conflicts of Interest: The authors declare no conflict of interest.

References

1. Hager, H.A.; Sinasac, S.E.; Gedalof, Z.E.; Newman, J.A. Predicting potential global distributions of two Miscanthus grasses: Implications for horticulture, biofuel production, and biological invasions. *PLoS ONE* **2014**, *9*, e100032. [CrossRef] [PubMed]
2. McCalmont, J.P.; Hastings, A.; McNamara, N.P.; Richter, G.M.; Robson, P.; Donnison, I.S.; Clifton-Brown, J. Environmental costs and benefits of growing Miscanthus for bioenergy in the UK. *Glob. Chang. Biol. Bioenergy* **2017**, *9*, 489–507. [CrossRef] [PubMed]
3. Schou, P. Polluting non-renewable resources and growth. *Environ. Resour. Econ.* **2000**, *16*, 211–227. [CrossRef]
4. Quinn, L.D.; Allen, D.J.; Stewart, J.R. Invasiveness potential of Miscanthus sinensis: Implications for bioenergy production in the United States. *Glob. Chang. Biol. Bioenergy* **2010**, *2*, 310–320. [CrossRef]
5. Mann, J.J.; Barney, J.N.; Kyser, G.B.; DiTomaso, J.M. Root System Dynamics of *Miscanthus × giganteus* and Panicum virgatum in Response to Rainfed and Irrigated Conditions in California. *Bioenergy Res.* **2013**, *6*, 678–687. [CrossRef]
6. Barney, J.N.; DiTomaso, J.M. Nonnative Species and Bioenergy: Are We Cultivating the Next Invader? *BioScience* **2008**, *58*, 64–70. [CrossRef]
7. Hastings, A.; Clifton-Brown, J.; Wattenbach, M.; Mitchell, C.P.; Stampfl, P.; Smith, P. Future energy potential of Miscanthus in Europe. *Glob. Chang. Biol. Bioenergy* **2009**, *1*, 180–196. [CrossRef]
8. Hastings, A.; Mos, M.; Yesufu, J.A.; McCalmont, J.; Schwarz, K.; Shafei, R.; Ashman, C.; Nunn, C.; Schuele, H.; Cosentino, S.; et al. Economic and Environmental Assessment of Seed and Rhizome Propagated Miscanthus in the UK. *Front. Plant Sci.* **2017**, *8*, 1058. [CrossRef]
9. Collura, S.; Azambre, B.; Finqueneisel, G.; Zimny, T.; Victor Weber, J. Miscanthus × Giganteus straw and pellets as sustainable fuels. *Environ. Chem. Lett.* **2006**, *4*, 75–78. [CrossRef]
10. Bilandžija, N.; Krička, T.; Matin, A.; Leto, J.; Grubor, M. Effect of Harvest Season on the Fuel Properties of Sida hermaphrodita (L.) Rusby Biomass as Solid Biofuel. *Energies* **2018**, *11*, 3398. [CrossRef]
11. Lanzerstorfer, C. Chemical composition and properties of ashes from combustion plants using *Miscanthus* as fuel. *J. Environ. Sci. (China)* **2017**, *54*, 178–183. [CrossRef] [PubMed]

12. Baxter, X.C.; Darvell, L.I.; Jones, J.M.; Barraclough, T.; Yates, N.E.; Shield, I. Study of *Miscanthus × giganteus* ash composition—Variation with agronomy and assessment method. *Fuel* **2012**, *95*, 50–62. [CrossRef]

13. Chou, C.-H. Miscanthus plants used as an alternative biofuel material: The basic studies on ecology and molecular evolution. *Renew. Energy* **2009**, *34*, 1908–1912. [CrossRef]

14. Van der Weijde, T.; Kiesel, A.; Iqbal, Y.; Muylle, H.; Dolstra, O.; Visser, R.G.F.; Lewandowski, I.; Trindade, L.M. Evaluation of Miscanthus sinensis biomass quality as feedstock for conversion into different bioenergy products. *Glob. Chang. Biol. Bioenergy* **2017**, *9*, 176–190. [CrossRef]

15. Baxter, X.C.; Darvell, L.I.; Jones, J.M.; Barraclough, T.; Yates, N.E.; Shield, I. *Miscanthus* combustion properties and variations with Miscanthus agronomy. *Fuel* **2014**, *117*, 851–869. [CrossRef]

16. Kadžiulienė, Ž.; Jasinskas, A.; Zinkevičius, R.; Makarevičienė, V.; Šarūnaitė, L.; Tilvikienė, V.; Šlepetys, J. Miscanthus biomass quality composition and methods of feedstock preparation for conversion into synthetic diesel fuel. *Zemdirbyste-Agric.* **2014**, *101*, 27–34. [CrossRef]

17. Arnoult, S.; Brancourt-Hulmel, M. A Review on Miscanthus Biomass Production and Composition for Bioenergy Use: Genotypic and Environmental Variability and Implications for Breeding. *Bioenergy Res.* **2015**, *8*, 502–526. [CrossRef]

18. Kering, M.K.; Butler, T.J.; Biermacher, J.T.; Guretzky, J.A. Biomass Yield and Nutrient Removal Rates of Perennial Grasses under Nitrogen Fertilization. *Bioenergy Res.* **2012**, *5*, 61–70. [CrossRef]

19. Pedroso, G.M.; Hutmacher, R.B.; Putnam, D.; Six, J.; van Kessel, C.; Linquist, B.A. Biomass yield and nitrogen use of potential C4 and C3 dedicated energy crops in a Mediterranean climate. *Field Crops Res.* **2014**, *161*, 149–157. [CrossRef]

20. Cunniff, J.; Purdy, S.J.; Barraclough, T.J.P.; Castle, M.; Maddison, A.L.; Jones, L.E.; Shield, I.F.; Gregory, A.S.; Karp, A. High yielding biomass genotypes of willow (*Salix* spp.) show differences in below ground biomass allocation. *Biomass Bioenergy* **2015**, *80*, 114–127. [CrossRef]

21. Stypczyńska, Z.; Jaworska, H.; Dziamski, A.; Majtkowski, W. Content of selected macroelements in the aerial and underground biomass of plants from old stands of the genus *Miscanthus*. *J. Elem.* **2012**. [CrossRef]

22. Kalembasa, D.; Jeżowski, S.; Symanowicz, B.; Głowacka, K.; Kaczmarek, Z.; Kalembasa, S.; Cerazy-Waliszewska, J. Changes in the content, uptake and bioaccumulation of macronutrients in genotypes of Miscanthus grass. possibilities of using ash from *Miscanthus* biomass. *Fresenius Environ. Bull.* **2019**, *28*, 1933–1942.

23. McKendry, P. Energy production from biomass (part 1): Overview of biomass. *Bioresour. Technol.* **2002**, *83*, 37–46. [CrossRef]

24. Beale, C.V.; Long, S.P. Seasonal dynamics of nutrient accumulation and partitioning in the perennial C4-grasses *Miscanthus × giganteus* and Spartina cynosuroides. *Biomass Bioenergy* **1997**, *12*, 419–428. [CrossRef]

25. Himken, M.; Lammel, J.; Neukirchen, D.; Czypionka-Krause, U.; Olfs, H.-W. Cultivation of *Miscanthus* under West European conditions: Seasonal changes in dry matter production, nutrient uptake and remobilization. *Plant Soil* **1997**, *189*, 117–126. [CrossRef]

26. Neukirchen, D.; Himken, M.; Lammel, J.; Czypionka-Krause, U.; Olfs, H.-W. Spatial and temporal distribution of the root system and root nutrient content of an established *Miscanthus* crop. *Eur. J. Agron.* **1999**, *11*, 301–309. [CrossRef]

27. Borkowska, H.; Lipiński, W. Content of selected elements in biomass of several species of energy plants. *Acta Agrophys.* **2007**, 287–292.

28. Roncucci, N.; Di Nassi O Nasso, N.; Tozzini, C.; Bonari, E.; Ragaglini, G. *Miscanthus × giganteus* nutrient concentrations and uptakes in autumn and winter harvests as influenced by soil texture, irrigation and nitrogen fertilization in the Mediterranean. *Glob. Chang. Biol. Bioenergy* **2015**, *7*, 1009–1018. [CrossRef]

29. Spiak, Z.; Piszcz, U.; Zbroszczyk, T. Modification of sulfur content in *Miscanthus × giganteus* under different nitrogen and potassium fertilization. *Ecol. Chem. Eng. A Chemia Inzynieria Ekologiczna A* **2012**, 213–221. [CrossRef]

30. Atienza, S.G.; Satovic, Z.; Petersen, K.K.; Dolstra, O.; Martin, A. Influencing combustion quality in *Miscanthus* sinensis Anderss: Identification of QTLs for calcium, phosphorus and sulphur content. *Plant Breed.* **2003**, *122*, 141–145. [CrossRef]

31. Borzęcka-Walker, M. Nutrient content and uptake by *Miscanthus* plants. *Electron. J. Pol. Agric. Univ.* **2010**, *13*. Available online: http://www.ejpau.media.pl/volume13/issue3/art-10.html (accessed on 14 January 2021).

32. Demirbas, A. Relationships between Heating Value and Lignin, Moisture, Ash and Extractive Contents of Biomass Fuels. *Energy Explor. Exploit.* **2016**, *20*, 105–111. [CrossRef]

33. Gołąb-Bogacz, I.; Helios, W.; Kotecki, A.; Kozak, M.; Jama-Rodzeńska, A. The Influence of Three Years of Supplemental Nitrogen on Above- and Belowground Biomass Partitioning in a Decade-Old *Miscanthus × giganteus* in the Lower Silesian Voivodeship (Poland). *Agriculture* **2020**, *10*, 473. [CrossRef]

34. *PN-EN ISO 14780:2017-07. Solid Biofuels—Sample Preparation*; ISO: Geneva, Switzerland, 2017.

35. Iqbal, Y.; Lewandowski, I. Inter-annual variation in biomass combustion quality traits over five years in fifteen Miscanthus genotypes in south Germany. *Fuel Process. Technol.* **2014**, *121*, 47–55. [CrossRef]

36. Lewandowski, I.; Heinz, A. Delayed harvest of miscanthus—influences on biomass quantity and quality and environmental impacts of energy production. *Eur. J. Agron.* **2003**, *19*, 45–63. [CrossRef]

37. Scordia, D.; Cosentino, S. Perennial Energy Grasses: Resilient Crops in a Changing European Agriculture. *Agriculture* **2019**, *9*, 169. [CrossRef]

38. Jensen, E.; Robson, P.; Farrar, K.; Thomas Jones, S.; Clifton-Brown, J.; Payne, R.; Donnison, I. Towards Miscanthus combustion quality improvement: The role of flowering and senescence. *Glob. Chang. Biol. Bioenergy* **2017**, *9*, 891–908. [CrossRef]

39. Mengel, K. *Ernahrung und Stoffwechsel der Pflanze*; tav Fischer Verlag: Jena, Germany, 1991.
40. Lewandowski, I.; Clifton-Brown, J.C.; Andersson, B.; Basch, G.; Christian, D.G.; Jørgensen, U.; Jones, M.B.; Riche, A.B.; Schwarz, K.U.; Tayebi, K.; et al. Environment and Harvest Time Affects the Combustion Qualities of Miscanthus Genotypes. *Agron. J.* **2003**, *95*, 1274–1280. [CrossRef]
41. Uprawa Miskanta Olbrzymiego. *Energetyczne i Pozaenergetyczne Możliwości Wykorzystania Słomy*; Kotecki, A., Ed.; Wydawnictwo Uniwersytetu Przyrodniczego we Wrocławiu: Wrocław, Poland, 2010.
42. Lewandowski, I.; Kicherer, A. Combustion quality of biomass: Practical relevance and experiments to modify the biomass quality of *Miscanthus* × *giganteus*. *Eur. J. Agron.* **1997**, *6*, 163–177. [CrossRef]
43. Lewandowski, I.; Clifton-Brown, J.C.; Scurlock, J.M.O.; Huisman, W. Miscanthus: European experience with a novel energy crop. *Biomass Bioenergy* **2000**, *19*, 209–227. [CrossRef]
44. Christian, D.G.; Riche, A.B.; Yates, N.E. Growth, yield and mineral content of *Miscanthus* × *giganteus* grown as a biofuel for 14 successive harvests. *Ind. Crops Prod.* **2008**, *28*, 320–327. [CrossRef]
45. Di Nassi O Nasso, N.; Roncucci, N.; Triana, F.; Tozzini, C.; Bonari, E. Seasonal nutrient dynamics and biomass quality of giant reed (*Arundo donax* L.) and miscanthus (*Miscanthus* × *giganteus* Greef et Deuter) as energy crops. *Ital. J. Agron.* **2011**, *6*, 24. [CrossRef]
46. Helios, W. *Growth and Yielding of the Giant Miscanthus Miscanthus×giganteus Greef et Deu*; Monogr. CCXIV.; Wydaw. Uniw. Przyr. we Wrocławiu: Wrocław, Poland, 2018; ISBN 978-83-7717-302-2.

 *agriculture*

Article

The Agri-Environment-Climate Measure as an Element of the Bioeconomy in Poland—A Spatial Study

Aleksandra Jezierska-Thöle [1,2,*], Roman Rudnicki [3], Łukasz Wiśniewski [3], Marta Gwiaździńska-Goraj [4] and Mirosław Biczkowski [3]

1. Institute of Geography, Kazimierz Wielki University, Plac Kościeleckich 8, 85-033 Bydgoszcz, Poland
2. Department of Anthropogeography, Freie University Berlin, Malteserstr. 74-100, 12 249 Berlin, Germany
3. Department of Spatial Planning and Tourism, Nicolaus Copernicus University in Toruń, Lwowska 1, 87-100 Toruń, Poland; rudnicki.r@umk.pl (R.R.); lukaszwisniewski@umk.pl (Ł.W.); mirbicz@umk.pl (M.B.)
4. Department of Socio-Economic Geography, University of Warmia and Mazury in Olsztyn, Prawochenskiego 15, 10-720 Olsztyn, Poland; marta.gwiazdzinska-goraj@uwm.edu.pl
* Correspondence: alekjez@ukw.edu.pl; Tel.: +48-52-327-71-45

Citation: Jezierska-Thöle, A.; Rudnicki, R.; Wiśniewski, Ł.; Gwiaździńska-Goraj, M.; Biczkowski, M. The Agri-Environment-Climate Measure as an Element of the Bioeconomy in Poland—A Spatial Study. *Agriculture* **2021**, *11*, 110. https://doi.org/10.3390/agriculture11020110

Academic Editor: Mariusz J. Stolarski

Received: 11 December 2020

Accepted: 25 January 2021

Published: 1 February 2021

Abstract: The Polish agricultural economy has a chance to dynamically develop and influence the innovation policy in the EU model of bioeconomy. The research aims to assess the spatial diversification of the level and structure of spending funds for two Rural Development Program (RDP) measures: agri-environment-climate measures (AECM) and organic farming scheme (OFS) aimed at supporting proenvironmental forms of agricultural management in the context of bioeconomy development. The EU financial perspective determined the time range for 2014–2020. The study was conducted on the example of Poland in two spatial scales: regional (province) and local (community). The analysis was based on partial indicators, which were then subjected to the standardisation procedure and included in the total as a synthetic indicator of the utilisation of RDP 2014–2020 funds aimed at supporting proenvironmental forms of farming. The following information was included in the evaluation: the number of farms, the size of utilised agricultural area (UAA) covered by support and the amounts of payments made under the two analysed RDP measures. In the research, the size and distribution of farms benefiting from AECM and OFS were determined. Besides, the relationship between funds absorption and socioeconomic development, as well as natural and non-natural conditions, were identified. The synthetic indicator of AECM/OFS usage showed a strong spatial differentiation, determined by the impact of several conditions: the level of socioeconomic development, the level of agriculture development, natural conditions of agriculture, land with significant natural and ecological values, and proenvironmental forms of land use on farms. Spatial diversification is more often the result of the impact of proenvironmental or natural-ecological factors than of socioeconomic conditions, or the level of agricultural development.

Keywords: bioproduction; CAP payments; sustainable agriculture; Poland

1. Introduction

Poland ranks third in Europe after France and Spain in terms of the share of agricultural land in the total area of the country (56%). In 2018, the global production value (in current prices) of Polish agricultural holdings ranked the country's agriculture 7th in the European Union, behind France, Germany, Italy, Spain, the UK, and the Netherlands [1,2]. Following the definition of the European Commission (EC) bioeconomy is "one of the oldest economic sectors known to humanity, and the life sciences and biotechnology are transforming it into one of the newest" [3,4]. Bioeconomy, i.e., industry based on bio-based raw materials and biotechnology, is concentrated around traditional sectors: agriculture, forestry, and food processing [5–7]. Polish agriculture may become an essential element in the development of the bioeconomy by supplying critical resources [8–10]. Bioeconomy is an important branch of the Polish economy, responsible for about 20% of the employment and 10% of the total production volume [11]. The development of bioeconomy is determined by the depletion of available natural resources, climate change and the need to implement sustainable agriculture [12–14]. The key determinants of bioeconomy development are the adopted legal regulations implementing international obligations in the form of the UN sustainable development goals and climate and energy policy combined with innovation [15–17]. The Polish bioeconomy was formed under the commitments resulting from the membership in the European Union, the communication of the European Commission to the European Parliament and the Council of Europe titled "Innovation in the service of sustainable growth: bioeconomy for Europe" contributed to the development of bioeconomy in Poland [8,18,19]. References to bioeconomy are found in documents such as Strategy for Innovation and Efficiency of the Economy [20], Strategy for Sustainable Rural Development [21], and the Energy Security and Environment Strategy, all of which promote growing efficiency of the use of natural resources and raw materials [22]. The Organization for Economic Cooperation and Development (OECD) has produced a "policy agenda" pushing for biotechnology as a new "bioeconomy" [3,23–26].

Traditional bioeconomy includes primary production, i.e., agriculture, the development of which in Poland follows two tracks. The areas with a favourable agrarian structure are dominated by intensive agriculture with high rates of plant and animal production [23,27,28]. At the same time, it is accompanied by extensive, traditional agriculture and organic farming predominantly located in naturally valuable areas [29]. Bearing in mind that agricultural production is based on natural resources and its durability depends on the state of the natural environment, the type of agricultural production is of great importance (including industrial and bioenergy types), not only due to the quantity and quality of production, but also its impact on the natural environment and climate [30,31]. Bioeconomy, based on biodiversity, is of particular importance in agricultural areas which are protected and financially supported by EU programs [32]. One of the challenges for agriculture is to ensure food security while maintaining the postulates contained in the concept of bioeconomy [8]. Agricultural production has a significant impact on the natural environment, including responsibility for a significant part of greenhouse gas emissions [33] and at the same time is a sector susceptible to climate change [34]. These relationships are two-sided: environmental resources determine the size and directions of agricultural production; at the same time, agriculture changes the existing ecosystems, shapes the landscape and affects the individual components of nature [35–37].

Increased competition on the market and pressure to increase the agricultural production efficiency in the EU contribute to the loss of biodiversity, the disappearance of traditional forms of farming and local varieties of crop plants [38]. Therefore, it becomes essential to reconcile the increase in agricultural productivity and its competitiveness with the simultaneous reduction of its negative impact on the environment. Therefore, agricultural production must use energy, water and soil in a more effective and proenvironmental way, while reducing greenhouse gas emissions [19,39–41].

The answer to the above challenge is more sustainable agriculture, which combines production, economic, social, and ethical priorities with ecological safety [42,43]. The concept of sustainable development postulates the simultaneous implementation of goals relating to three independent but related areas: environmental (ecological), social and economic [44,45]. Sustainable agriculture could become an essential element in the development of the bioeconomy [46–49]. According to Kłodziński [50], it should be remembered that sustainable development of rural areas requires, above all, a compromise between agricultural producers, whose aim is to maximise the effects of their activities, and the interests of society, for which protection and management taking into account the state of the natural environment is becoming more and more critical. In the conditions of the Common Agricultural Policy (CAP), this leads to a redefinition of the concept of agriculture: from a narrow, productive approach, to holistic, sustainable and rational management of natural resources recognised as particularly protected public goods [51].

From the beginning of Poland's membership in the EU, i.e., 2004, the most effective CAP program aimed at minimising the negative impact of agriculture on the environment was the Agro-environmental Program. It had three goals: protection of the environment and landscape, development of organic farming and preservation of biodiversity. Implemented in the first (incomplete) financial period (2004–2006), it continued in 2007–2013, and now functions under the Rural Development Program (RDP, 2014–2020 perspective). In the concept of multifunctional and sustainable agricultural development adopted for implementation, two measures are of particular importance for shaping the relationship between agriculture and the environment—agri-environment-climate measures (AECM) and organic farming scheme (OFS; in previous EU financial periods, these measures were included in one agri-environmental program). It should be emphasised that agri-environment-climate measures of the RDP are one of the financial instruments of the bioeconomy and are part of its development trend [52–54]. These activities are mainly aimed at strengthening two nonmarket functions of agriculture:

- Green—related to the management of land resources in order to maintain their valuable properties, with the creation of conditions for wild animals and plants, protection of animal welfare, maintenance of biodiversity and improvement of the circulation of chemicals in agricultural production systems [11,55];
- blue—related to water resources management, water quality improvement, flood prevention, hydropower and wind energy generation [5].

When considering the multifunctionality of agriculture, support for farms in areas with unfavourable farming conditions is aimed at securing their possibility of further operation. Land management and land use should be based on environmentally friendly principles, while supporting functions other than food production, thus preventing trends of marginalisation and degradation of these areas [43,56]. The answer to the problems related to sustainable development is to improve the methods of managing the environment and natural resources. In this context, social-ecologically sustainable agricultural management is gaining more and more importance [40,57].

The European Bioeconomy Strategy developed by the EC (adopted on 13 February 2012) is based on three pillars [58], that is the support from EU and national funds, providing knowledge for sustainable production growth and the creation of a bioeconomy panel and bioeconomy observatory. Bearing the above in mind, the article presents and describes the first pillar of the strategy, i.e., supporting the bioeconomy with EU funds from the RDP [59].

The main objective of the research is to assess the spatial diversity of farms acquiring CAP funds aimed at supporting sustainable agriculture, i.e., concerning two RDP measures 2014–2020, "Agri-Environment-Climate Measures" and "Organic Farming Scheme". The second aim of the study is an attempt to delimit the determinants affecting the level of use of the researched funds, and thus the possibilities of sustainable agricultural development.It was assumed that the identification of such targeted activity of farmers is a sine qua non condition for broader inclusion of agriculture in the framework of bioeconomy [60].The research goals set in this way will help to answer the question "Can the spatial diversification of the use of proecological CAP funds be the key to a more regionally and locally optimised development of the bioeconomy in Polish agriculture?"

The research used an indicator of the share of completed applications of the measures mentioned above in the total number of farms to assess the level of interest of farmers towards proenvironmental forms of agricultural management along with its spatial diversification. Additionally, the strength of the relationship (correlation) between the level of activity determined in this way and the adopted conditions thus identified i.e., the levels of socioeconomic and agricultural development, and two environmental determinants related to the assessment of the natural conditions of agriculture and the share of proenvironmental forms of land use.

Earlier studies [61–65] indicate territorial disparities in obtaining CAP funds, resulting from the characteristics of a given area, both human (socioeconomic) and environmental.

2. Materials and Methods

2.1. Study Area and Materials

Taking up the above topic was motivated by the need to summarise the effects of two RDP measures of 2014–2020 AECM and OFS, which taken together constituted the basis for recognising the strength of the relationship between the different level of absorption of funds from the measures as mentioned earlier and the level of natural and non-natural conditions. The spatial scope of the research covered the territory of Poland (NUTS 0) in the system of province (16 NUTS 2 units) and communities (2477 units; the third-order administrative division of the country sometimes referred to as "communes" or "municipalities" until 2016—according to Local Administrative Units—LAU level 2).

Data on the implementation of AECM/OFS were obtained from the Agency for Restructuring and Modernisation of Agriculture (ARMA; Warsaw, Poland). They took into account the number of beneficiaries, the area covered by the payment and the amounts of payments made. The second leading source of data was the Local Data Bank of the Central Statistical Office (Warsaw, Poland) [66–68]. The obtained data from the ARiMR related to two RDP 2014–2020 measures (AECM, OFS), including:

- the number of completed applications—97,200, which constituted 10.4% of the total number of farms;
- the surface of the subsidised area—1,259,600 ha, which constituted 10.8% of the total agricultural area of farms;
- the realised payments—EUR 933.8 million (at the rate of PLN 4.295 to EUR 1), which was EUR 1190 per farm.

The data concerned a wide range of issues that allowed for spatial assessment, among others, of the level of socioeconomic development, the level of agricultural development or natural and ecological valorisation, which were adopted as the level of conditions for the sustainable development of bioagriculture (see Table 1).

The main criteria for assessing spatial differentiation were the number of farms, the area of UAA covered by the support and the amounts of AECM/OFS payments made. The research assumptions included analysis in two spatial scales:

- macroscale—comprehensive nationwide analysis;
- microscale—enabling the identification of specific areas in which activities aroused extreme interest and areas in which farmers showed passivity in applying for funds for agri-environment-climate activities. Such an approach is an advantage of the work, as most of the analyses related to the evaluation of the implementation of EU funds are conducted only on a regional scale, without in-depth analysis at the local level (LAU 2 units).

The primary analysis was based on the number of applications completed within the framework of the said measures and the volume of funds obtained. Both elements allow assessing the scale of farmers' interest in activities aimed at diversifying the sources of income. The empirical nature of the article, to a large extent, contributes to the development of the cognitive thread in the field of the impact of EU funds on the diversification of farmers' income sources and the development of entrepreneurship in rural areas towards the development of nonagricultural activities, with particular emphasis on the bioeconomy.

2.2. Methods

The implementation of the set research goal required the adoption of an appropriate research procedure and the construction of a whole set of indicators. The research was conducted in several stages (cf. Figure 1). In the first one, three partial indicators (IAF, ITR, IFSF) were used to assess the spatial level of the use of proenvironmental CAP funds (IUF-RDP). ARMA data and normalisation methods were used. The aim of the second stage was spatial delimitation of selected determinants (LSED, LAO, APS, NEA), which should determine the scale and directions of using the researched funds (IUF-RDP). They were defined on the basis of 12 partial indices. The last stage was a comparative analysis of both planes, which allowed to assess the role of individual determinants in the level of the use of proenvironmental CAP funds.

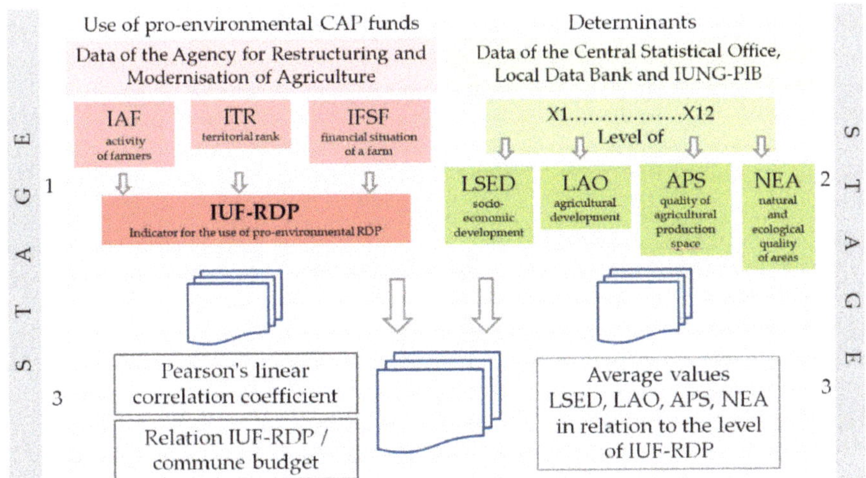

Figure 1. Research procedure.

2.2.1. Stage 1. Indicator of the Use of Proenvironmental CAP Funds

The basis for assessing the use of proenvironmental CAP measures were normalised values of three indicators illustrating:

- (IAF) the activity of farmers in terms of obtaining funds (ratio of the number of applications to the total number of farms expressed in percentages);
- (ITR) territorial rank (the ratio of the number of applications to the area of agricultural land expressed in percentages);
- (IFSF) impact on the financial situation of a farm (ratio of obtained subsidies in EUR per farm). The above indicators were subjected to the normalisation procedure following the formula [69–71].

$$Z_{ij} = \frac{X_{ij} - av.X_i}{\delta_i} \tag{1}$$

where:

Z_{ij}—standardised value of the diagnostic feature 'i' in spatial unit 'j'
X_{ij}—value of diagnostic feature 'i' in spatial unit 'j'
$av.X_i$—average value of diagnostic feature 'i'
δ_i—standard deviation of diagnostic feature 'i'.

The next step was to calculate the synthetic indicator for the use of proenvironmental RDP funds (IUF-RDP), according to the formula:

$$G = \frac{1}{M}\left(Z_{i1} + Z_{i2} + \ldots + Z_{ij}\right) \tag{2}$$

where:

G—average standardised value of selected diagnostic features within the respective group of features
Z_{ij}—standardised value of diagnostic feature 'i' in spatial unit 'j'
M—number of diagnostic features.

The zero values (national means) of the indicators, assuming a standard deviation threshold of $+/-0.5$, were the basis for distinguishing three classes characterised by low (below $-0.50\ \delta$), medium (from -0.50 to $0.50\ \delta$), and high (above $0.50\ \delta$) level of the phenomenon.

2.2.2. Stage 2. Assessment Planes—Determinants of Sustainable Development of (Bio)Agriculture

The spatial distribution of the level of the use of proenvironmental CAP funds was compared to synthetic indicators illustrating the level of socioeconomic development (LSED), the level of agricultural development (LAD), the quality of agricultural production space (APS), and the natural and ecological quality of areas (NEA). The indicators were constructed based on normalisation and then averaging of several partial indicators and division into units with the low, medium and high level of the studied phenomenon (the same as in the case of IUF-RDP). The basis for delimiting selected indicators is presented in Table 1.

Table 1. Assessment planes—determinants of development of sustainable agriculture in Poland characteristics. Source: own elaboration on [66–68,72,73].

Assessment Plane	Delimitation Indicators	Data Source
The level of socioeconomic development (LSED)	(x^1) business entities in the REGON register per 10,000 population	[70]
	(x^2) unemployed per 10,000 population (destimulant)	[70]
	(x^3) population with the access to the sewage network as a percentage of the total population	[70]
	(x^4) own incomes of communities in PLN per capita	[70]
The level of agricultural development (LAD)	(x^5) the average area of a farm in ha (2018, according to ARMA)	[70]
	(x^6) farmers with secondary and higher education as a percentage of the total number of farmers	[66]
	(x^7) young farmers (up to 34 years of age) as a percentage of the total number of farms	[66]
	(x^8) noncereal crops as a percentage of the total sown area	[66]
The quality of agricultural production space (APS)	(x^9) indicator of the quality of agricultural production space	[67]
The natural and ecological quality of areas (NEA)	(x^{10}) forests and wooded and bushy land as a percentage of the total area	[70]
	(x^{11}) grasslands, water bodies, and the legally protected areas as a percentage of the total area (2018, according to CSO BDL)	[70]
	(x^{12}) priority zones of the agri-environmental program delimited in the period 2004–2006 as a percentage of the total area	[68]

2.2.3. Stage 3. Assessment of IUF-RDP in the Context of Conditions

The assessment of IUF-RDP in the context of separate groups of conditions for sustainable agricultural development was carried out in two ways. Firstly, it will support the calculation of the average values of LSED, LAD, APS, NEA indicators for each category (low, medium, high) for the level of use of IUF-RDP funds. In the analysis aimed at determining the strength and direction of the relationship between the synthetic index of the use of proenvironmental RDP measures and the conditions of sustainable agricultural development, the linear Pearson correlation coefficient (according to the product-moment) was used. Correlation factor took numerical values from (-1) to $(+1)$, where the value with zero indicates no statistical relationship. Additionally, in order to better illustrate the rank of the studied group of funds, an indicator of the ratio (IR) of the value of proenvironmental

EU funds to the size of commune budgets (EU payments as % of the commune budget in 2015–2019) was introduced.

3. Results and Discussion

The research questions and adopted methods mean that the discussion of the results is preceded by explanatory background. It is a theoretical framework on the essence and importance of agri-environment-climate action for the development of bioeconomy and implementation of the agri-environmental program in Poland in 2007–2013 against the European Union.

3.1. The Role and Importance of Agri-Environment-Climate Action

One of the objectives of the EU's Common Agricultural Policy is to promote environmentally friendly agricultural practices. This goal is being implemented, among others, by implementing agri-environment-climate measures in all EU member states. These measures are part of the Rural Development Program for 2014–2020 (RDP 2014–2020) and are mostly a continuation of the previous measure, RDP agri-environmental program for 2007–2013 [73]. It was planned as one of the components implementing strategic EU and national environmental goals, taking into account the economic and social importance of agriculture in the context of the growing demand for agricultural raw materials, including for the bioeconomy, and the still high importance of agricultural activity for employment and territorial development in Poland [74,75]. The agri-environmental program is a financial instrument aimed at encouraging farmers to continue or apply agricultural practices leading to the greening of agricultural production. The implementation of the agri-environmental program contributes to the sustainable development of rural areas and the preservation of biodiversity in these areas. The primary assumption of the program is to promote agricultural production based on methods consistent with the requirements of environmental and nature protection [76–78]. An additional goal of the program is to increase the environmental awareness of the rural community. According to research [47,79], the agri-environmental program has become an impulse mainly for the development of multifunctional agriculture.

In the period 2004–2006, the agri-environmental program included seven packages, while in the years 2007–2013, nine packages were implemented (RDP 2004–2006 and RDP 2007–2013). The agri-environment-climate measure (RDP 2014-2020) consists of seven packages: (1) Sustainable agriculture, (2) Soil and water protection, (3) Preservation of orchards with traditional varieties of fruit trees, (4) Valuable habitats and endangered bird species in Natura 2000 areas, (5) Valuable habitats outside Natura 2000 areas, (6) Conservation of endangered plant genetic resources in agriculture, and (7) Preservation of endangered animal genetic resources in agriculture.

The limit of funds for Poland for 2014–2020 for this priority was approximately EUR 4.2 billion [80] and addressed the implementation of the following specific objectives:

- restoring, protecting, and enhancing biodiversity, including in Natura 2000 sites and areas facing natural or other specific constraints, and HNV farming and the state of European landscapes;
- improving water management, including fertilisation and pesticide use,
- preventing soil erosion and improving soil management [81].

The limit of funds allocated to Poland for this measure for 2014–2020 is approximately EUR 1.4 billion, which constitutes 33% of Priority 4. These measures (excluding Package 2. Organic farming under the agri-environment-climate measures) were implemented for a total of 11.6% of the area reported for direct payments (as of 2015). In subsequent campaigns, this area was systematically decreasing, and in 2018 the estimated area amounted to 7.1% of the area reported for direct payments. In the context of specific Priority 4: Preventing soil erosion and improving soil management, the most crucial action is the organic farming scheme (OFS). The total funds allocated for this purpose amount to less than EUR 700 million, which constituted approximately 17% of Priority 4 funds. In total, by the end of 2018, based on issued decisions granting ecological payment, the size of the physical area covered by support was 531,816 ha, which was 3.7% of the area reported for direct payments. Although this is a highly unsatisfactory effect, in the years 2016–2018 a reasonably stable increase (approximately 12% per year) of the agricultural area on which this measure is implemented was observed.

As part of agri-environmental programs, which are followed by specific financial support, farmers are encouraged to act to protect the natural environment, biodiversity and preserve landscape values. The beneficiary may join the measure, first of all, if the farm has at least 1 ha of UAA (3 ha in Package 1. Sustainable agriculture) or at least 1 ha of natural areas and has the producer's identification number assigned by the Agency for Restructuring and Modernisation of Agriculture.

3.2. Implementation of Proenvironmental CAP Funds in Poland in 2014–2020

In Poland, in 2014–2020, 97,200 projects were implemented as AECM/OFS activities, including 67,200 for the operation of AECM and 30,100 of OFS. These ranged from less than 1000 in Opolskie (700) and Śląskie (900) to over 10,000 in Lubelskie (13,300), Podlaskie (11,700), and Warmińsko-Mazurskie (10,400; see Table 2, Figure 2a). The researched group of farms accounted for 10.4% of the total number of farms in Poland (IAF, according to the register of agricultural producers by the ARMA).

Table 2. Proenvironmental instruments of the Rural Development Program (RDP) 2014–2020 in Poland—together agri-environment-climate (AECM) and organic farming (OFS) measures—selected elements. Source: own elaboration on [66–68,72,73].

	Specification	Number of Applications		Subsidised Area		Payments Made		IUF-RDP **
		In Total Thousand Requests *	IAF **	In Thousand ha *	ITR **	In Million Euros	IFSF **	
	POLAND	97.2	10.4	1259.6	10.8	933.8	1.190	0.00
I	Dolnośląskie	5.1	10.0	77.6	9.2	64.7	1.259	−0.03
II	Kujawsko-Pomorskie	3.8	6.3	68.7	6.7	37.2	623	−0.26
III	Lubelskie	13.3	7.9	119.3	8.8	87.8	522	−0.19
IV	Lubuskie	4.7	24.4	98.4	24.1	83.2	4.304	1.05
V	Łódzkie	2.6	2.2	20.4	2.2	12.9	112	−0.52
VI	Małopolskie	3.8	3.3	18.7	3.8	17.1	150	−0.46
VII	Mazowieckie	7.4	3.7	57.0	3.1	40.4	204	−0.46
VIII	Opolskie	0.7	2.6	17.2	3.4	8.1	302	−0.46
IX	Podkarpackie	9.8	8.8	64.1	11.9	66.9	600	−0.10
X	Podlaskie	11.7	14.9	102.9	10.1	82.1	1.051	0.07
XI	Pomorskie	6.1	16.2	109.9	15.2	70.6	1.889	0.33
XII	Śląskie	0.9	2.1	9.9	3.0	6.5	159	−0.50
XIII	Świętokrzyskie	4.1	5.0	26.0	5.3	18.3	223	−0.38
XIV	Warmińsko-Mazurskie	10.4	24.6	195.2	20.1	145.8	3.460	0.85
XV	Wielkopolskie	5.3	4.5	76.2	4.4	48.7	419	−0.38
XVI	Zachodniopomorskie	7.8	28.2	198.4	23.5	143.6	5.173	1.25

* annual average values 2015–2019, (for 2281 communities); ** IAF—the activity of farmers; ITR—territorial rank; IFSF—impact on the financial situation of a farm; IUF-RDP—indicator for the use of proenvironmental RDP (more about indicators: see methodology).

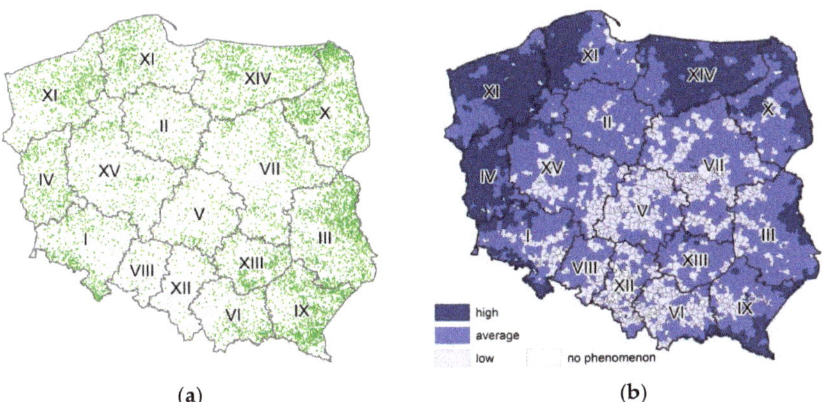

(a) (b)

Figure 2. Number of submitted applications (**a**) and the synthetic indicator of the utilisation of the Rural Development Program (IUF-RDP-indicator for the use of proenvironmental RDP (**b**)). *1 dot = 10 applications. Source: own elaboration on [66–68,70,71].

In the province system, the most significant activity was recorded by Zachodniopomorskie (28.2%), Warmińsko-Mazurskie (24.6%), and Lubuskie (24.4%), the lowest by Śląskie (2.1%), Łódzkie (2.2%), and Opolskie (2.6%). The UAA covered by the cofinancing was 1,284,900 ha, while the ITR was 10.8%. The spatial analysis showed that the highest ITR was recorded in Lubuskie (24.1%) and zachodniopomorskie (23.5%), while the lowest in Łódzkie (2.2%) as well as Mazowieckie (3.1%) and Śląskie (3.0% each).

The AECM and OFS funds are commonly used in communities with natural and ecological values (X^4, X^5). The communities showing high interest in proenvironmental forms of EU support include Gołdap (534 applications), UstrzykiDolne (506), UjścieGorlickie (386), and DrawskoPomorskie (379). Generally, however, the group of the most active communities is relatively small. More than 100 applications were implemented only in 261 communities (10.5% of the total). The group of communities where the beneficiaries showed passivity and distance to the implementation of proecological activities is much more numerous. It is confirmed by this study, as it shows that in 338 communities, only 1–5 applications were implemented, while in another 304 communities, 6–10 applications. In total, these administrative units constitute as much as a quarter of all communities in Poland. Besides, there is a relatively large group of communities (198, i.e., 8%) in which none of the farmers took advantage of the possibility of obtaining EU payments for proecological activities. Thus, the activity rate of beneficiaries is relatively low. Statistically, the leaders among local authorities in implementing solutions supporting the bioeconomy are found in one in 10 communities in Poland.

Bioeconomy, based on biodiversity, is of particular importance in areas used for agriculture, which are covered by EU payments for proenvironmental measures, i.e., in this case from AECM and OFS. Hence, one of the essential measures in the spatial dimension is the UAA to which payments under the instruments above have been granted. This indicator shows extensive spatial spans. From this perspective, the leading communities in Poland include Gołdap (the area covered by AECM and OFS payments is 8885 ha), Szczecinek (8617), DrawskoPomorskie (7035), Słońsk (6388), Bobolice (6322), and BiałyBór (6290). In total, in 387 communities, more than 1000 ha of UAA were qualified for the subsidies. In the next 299 communities it was 500–999 ha, and in as many as 918 from 100 to 499 ha. In total, the area eligible for support for proecological activities in Poland, i.e., AECM and OFS, amounts to 1.259 million ha, i.e., approximately 10.8% of the total UAA. For comparison, in Germany—the most developed EU country in this respect—subsidies to the area covered by the agri-environmental program were granted to nearly 5.3 million ha, i.e., about one-fourth of the total UAA [81].

The total value of cofinancing was EUR 933.8 million (on average, EUR 145 per farm), the majority of which was related to the AECM (68.6%). In terms of the volume of the payments, it was shown that an average farm received almost EUR 1200 (IFSF). This type of impact on the financial situation of farms was the highest in Zachodniopomorskie (EUR 5173). At the community level, the threshold of EUR 10,000 was exceeded in 34 of them, while two communities (Stepnica in Zachodniopomorskie and Lutowiska in Podkarpackie) exceeded EUR 19,000.

It was assumed that IUF-RDP (average IAF, ITR, IFSF) is a determinant of proenvironmental preferences of farms, therefore recognising its spatial diversity is an essential element of bioeconomy development opportunities.

The synthetic indicator of the utilisation of the AECM and OFS resources (normalised average) showed high spatial differentiation. In terms of regions, the leader in the use of funds was Zachodniopomorskie (1.25), followed by Lubuskie (1.05) and Warmińsko-Mazurskie (0.85), which achieved the highest values in the system of partial indicators (i.e., activity and absorption). Low values of the AECM/OFS utilisation rate were noted in Łódzkie (−0.52) and Śląskie (−0.50). Such a territorially oriented implementation of AECM and OFS measures shows that the use of biodiversity in agriculture is pragmatic, which increases the possibilities for sustainable use and good management of environmental resources. On the other hand, it also helps to strengthen the environmental ecosystem, which will help preserve biodiversity at the farm level and beyond. Appropriate (i.e., in line with environmental preferences, as well as agricultural and economic opportunities) territorially oriented AECM and OFS activities should significantly strengthen the desired economic profile, i.e., aimed at the bioeconomy. However, it should be noted that the implementation of proecological activities by farmers is not only motivated by care for the environment. As such, no other behaviours are mainly due to the economic factor associated with the possibility of obtaining subsidies from the CAP funds. It is evidenced by the fact that the activities of AECM and OFS in Poland are used mainly by large farms located in the northern and western parts of the country (Zachodniopomorskie, Warmińsko-Mazurskie, Lubuskie). Thus, it points to the opposite relationship than, for example, in neighbouring Germany, where smaller farms, focused on more extensive forms of management, are more eager to use funds from proecological activities.

In the system of communities, the highest level of activity in the use of EU funds was found in Dziwnów (8.6), Zachodniopomorskie and Górowoławeckie (7.90), Warmińsko-Mazurskie. The remaining clusters of communities are located in the coastal and lake district regions, within the Natura 2000 areas, within the territory of national parks, e.g., Słowiński (NP communities of Łeba 5.62, Smołdzino 5.2) and Bieszczady NP (community of Lutowiska 4.9) (cf. Figure 2b).

The value of IUF-RDP was the basis for the classification of communities into three groups, namely with the low, medium, and high levels of absorption of proenvironmental RDP funds (see Figure 2b).

3.3. An Attempt at Spatial Delimitation of Conditions for the Development of Sustainable Agriculture in Poland

Apart from recognising the spatial differentiation of the use of proenvironmental RDP funds, one of the objectives of the study was an attempt to delimit the spatial conditions influencing the sustainable development of agriculture, and thus the absorption of these funds. The analysis of the values of synthetic indicators (LSED, LAD, APS, NEA) both on a regional (see Table 3) and local scale (see Figures 3 and 4) shows that the potential factors determining the sustainable development of agriculture create an extremely complex system.

Table 3. Selected conditions for the development of sustainable agriculture in Poland (indicators in the form of a normalised value). Source: own elaboration on [66–68,72,73].

	Province	The Level of Socioeconomic Development	The Level of Agricultural Development	The Quality of Agricultural Production Space	The Natural and Ecological Quality of Areas
		LSED *	LAD *	APS *	NEA *
I	Dolnośląskie	0.30	0.21	0.39	−0.12
II	Kujawsko-Pomorskie	−0.35	0.36	0.39	−0.27
III	Lubelskie	−0.66	−0.04	0.97	−0.27
IV	Lubuskie	0.00	0.24	−0.54	0.20
V	Łódzkie	−0.16	−0.11	−0.27	−0.41
VI	Małopolskie	−0.03	−0.24	0.03	0.17
VII	Mazowieckie	0.51	−0.04	−0.59	0.13
VIII	Opolskie	−0.11	0.17	0.72	−0.52
IX	Podkarpackie	−0.61	−0.29	0.27	0.40
X	Podlaskie	−0.43	0.17	−1.09	0.28
XI	Pomorskie	0.30	0.21	0.44	−0.05
XII	Śląskie	0.13	−0.36	0.15	−0.25
XIII	Świętokrzyskie	−0.57	−0.21	−0.03	0.29
XIV	Warmińsko-Mazurskie	−0.46	0.36	0.15	0.17
XV	Wielkopolskie	0.20	0.14	−0.28	−0.10
XVI	Zachodniopomorskie	0.20	0.71	−0.02	0.13

* LSED—the level of socioeconomic development; LAD—the level of agricultural development; APS—the quality of agricultural production space; NEA—the natural and ecological quality of areas.

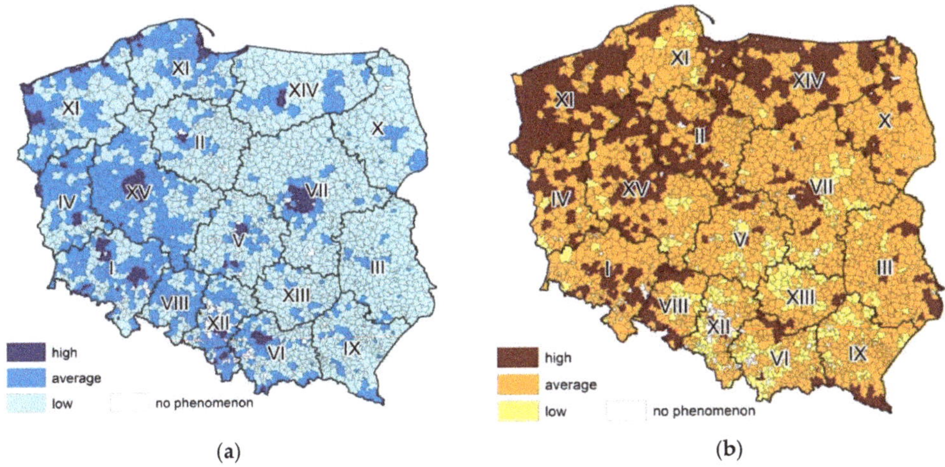

Figure 3. The level of socioeconomic development ((**a**) LSED) and the level of agricultural development ((**b**) LAD). Source: own elaboration on [66–68,70,71].

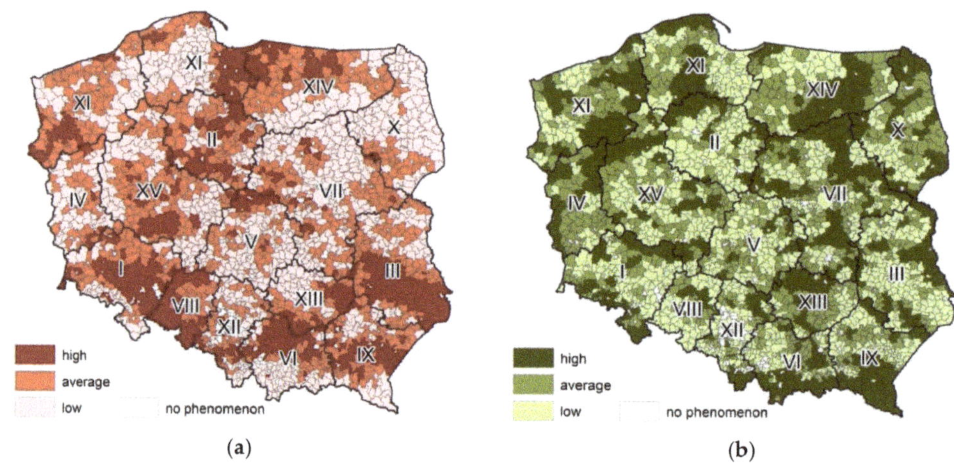

Figure 4. The quality of agricultural production space ((**a**); APS) and the natural and ecological quality of areas ((**b**); NEA). Source: own elaboration on [66–68,72,73].

The analysis showed a considerable spatial differentiation of the level of socioeconomic development at the level of communities, confirming two basic rules indicated in numerous studies dealing with the problem under consideration [82–84]. The first rule results from historical conditions and is related to the differences between communities in eastern and western Poland. The second one stems from the differentiation of the level between the centre (agglomerations) and the periphery (rural areas; cf. Figure 3a). The indicated factors, especially the old political and historical divisions, play a significant role in the case of differentiation in the level of agricultural development (cf. Figure 3b) [66]. It is confirmed by the lowest ratings of this sector in Podkarpackie (−0.29) and Małopolskie (−0.24; Austrian partition). The low rating of Śląskie (−0.36) is associated with the domination of the industry (Upper Silesian Industrial District; cf. Figure 3b).

The natural conditions (APS and NEA) refer primarily to the differentiation of soils, which in turn is the result of soil-forming processes dependent on bedrock, climate, water, living organisms, including humans, as well as topography and the passage of time. The quality and usefulness of soils is the most crucial component in the valorisation of the quality of agricultural production space (cf. Figure 4a) [85–87]. Soil production capacity is later reflected in the land use; therefore, high scores of the APS index inversely correlate with NEA (cf. Figure 4b). The high value of the latter should be associated with the areas which show an above-average share of forests, grasslands, water bodies, and legally protected areas in the total area.

3.4. Spatial Assessment of the Use of Proenvironmental RDP Funds in the Context of the Conditions of Sustainable Agricultural Development

The conducted spatial analysis allows concluding that proenvironmental activities are more willingly implemented in peripheral (border) regions, mainly in northern, as well as western and eastern parts of Poland. These are the provinces with lower urbanisation levels than the more centrally located ones, more often characterised by high natural and ecological values and proenvironmental forms of land use (NEA = 0.44; see Table 4). In terms of the land use structure, significant areas are covered by forests (Lubuskie) or grasslands (Podlaskie). In some of these areas, more extensive forms of farming are preferred. From the point of view of enhancing natural ecosystems and biodiversity, these are, therefore, preferable areas for such projects.

Table 4. Conditions for sustainable agricultural development against the level of use of proenvironmental RDP funds. Source: own elaboration.

Level of IUF-RDP	Conditions			
	LSED	**LAD**	**APS**	**NEA**
Low	−0.62	−0.07	0.08	−0.43
Average	−0.76	0.05	−0.06	−0.13
High	−0.62	0.32	−0.51	0.44

The directions of spatial differentiation of the implemented applications from the AECM and OFS also refer to the areas covered by the NATURA 2000 program. These spatially oriented activities of the beneficiaries help to maintain a balance between environmental resources and the requirements of the economy [7]. Thus, this trend is a strong element of the concept of sustainable development [88] and brings benefits to the natural environment and society. Moreover, the high level of use of proenvironmental RDP measures should be associated with the areas of low natural suitability for agricultural production (APS = −0.51) as well as with a relatively well-developed agricultural sector (LAD = 0.32; see Table 4). On the other hand, the areas with low activity of farmers in introducing proenvironmental management methods show a relatively low average level of agriculture (LAD = −0.07) in the conditions of above-average natural predispositions for agricultural production (APS = 0.08) and very low natural and ecological quality (NEA = −0.43).

Such a territorially oriented implementation of activities shows that the use of biodiversity in agriculture is pragmatic [89], which increases the possibilities of sustainable use and good management of environmental resources [76]. On the other hand, it also contributes to strengthening the environmental ecosystem, which will help preserve biodiversity at the farm level and beyond [90]. Appropriate (i.e., in line with environmental, ecological and agricultural preferences, as well as economic opportunities) territorially oriented activities of AECM and OFS should significantly strengthen the desired economic profile, i.e., aimed at bioeconomy. However, it should be noted that the implementation of proecological activities by farmers is not only motivated by care for the environment. As such, no other behaviours are primarily due to the economic factor related to the possibility of obtaining subsidies from the CAP funds. It is evidenced by the fact that large farms located in the northern and western part of the country (Zachodniopomorskie, Warmińsko-Mazurskie, Lubuskie) benefit to a large extent from the activities of the AECM and OFS in Poland. Thus, this indicates the opposite relationship than, for example, in neighbouring Germany, where smaller farms, focused on more extensive forms of management, are more eager to use funds from proecological activities [39,81].

In the context of bioeconomy development, the economic factor is of crucial importance. One of the three pillars of the European Bioeconomy Strategy is the support of EU funds for proenvironmental forms of farming. The results of the analyses of the authors so far [63,73,77] indicate that it is an important, if not the essential motive that guides Polish farmers in implementing proecological solutions on their farms. The financial factor is also of fundamental importance from the point of view of influencing the broadly understood socioeconomic development of rural areas, mainly since it includes the inflow of funds supporting local economies. From this point of view, funds aimed at the development of proecological activities had an unusually high rank in the communities of Trzcianne—EUR 7.3 million (data for 2015–2019) and Szczecinek—EUR 6.7 million. In both communities, the level of payments was higher than in all communities of Śląskie taken together (a total of EUR 6.5 million). It proves the scale of the inflow of EU funds and the importance of proecological activities, thus supporting the economic development of these communities towards the broadly understood bioeconomy.To better illustrate the rank of this group of funds the ratio (Wr) of the value of proenvironmental EU funds to the size of community budgets (EU payments as percentage of the community budget in 2015–2019) was used. Such targeted analysis confirmed the importance of AECM and OFS measures and their impact on the economic situation of communities. In the Community of Trzcianne, the ratio as mentioned above was Wr = 0.31 (i.e., the amount of EU payments corresponds to 31% of the community's budget). These funds are of a slightly lower rank in Szczecinek community (ratio IR = 0.15), but their importance for the economic situation of the community is still significant. The communities mentioned above are not the only examples of high positioning of proenvironmental activities in the local economy. In four more administrative units of this level, funds directed at proecological forms of agricultural activity exceeded the level of EUR 5 million: Gołdap (EUR 6.2 million, IR = 0.06), Słońsk (EUR 5.9 million, IR = 0.23), Komańcza (EUR 5.8 million, IR = 0.29), and Drawsko Pomorskie (EUR 5.4 million, IR = 0.06). In total, the value of subsidies in 276 communities exceeded EUR 1 million. On the other hand, in 924 communities (37%) the level of payments did not exceed EUR 100,000. From the economic and bioeconomic development perspective, these funds are of crucial importance in 41 communities (IR > 0.10), and in another 134 they play an essential role (IR 0.05 < 0.10). It can be said that those communities mentioned above are avant-garde in the implementation of proecological solutions strengthening the biodiversity of the natural environment and thus directing development towards the bioeconomy. Nevertheless, the majority of beneficiaries still do not see the potential of ecological solutions that can stimulate the development of local economies. It is confirmed by the fact that in 65% of communities the ratio IR < 0.010 (excluding 191 communities where AECM and OFS activities were not implemented).

The indicator of the use of proenvironmental RDP funds is distinguished by a significant spatial differentiation resulting from the impact of several conditions. The spatial distribution of the indicator was compared to the indicators (expressed in the form of the average normalised value) illustrating the general level of socioeconomic development, level of agriculture, quality indicator of agricultural production space and land use, both at the community level (land with significant natural and ecological values as a percentage of the total community area) and a farm (share of proenvironmental forms of land use in the total area of farms).

The analysis of the correlation of the AECM and OFS utilisation rates showed a relatively weak correlation (−0.5 to +0.5) concerning the adopted indicators (see Table 5). The highest value of the correlation was recorded concerning the index (land with significant natural and ecological values − r = 0.282), which proves a low environmental pressure and concerning the index (agricultural development level (0.233). It proves that the interest among farmers in the implementation and use of funds from agri-environment-climate measures is more significant in areas of natural value and those with a high level of agricultural development. Low values of the correlation were recorded concerning the index (level of socioeconomic development − r = 0.046).

Table 5. Dependencies between selected conditions for the development of sustainable agriculture and the level of use of proenvironmental RDP funds. Source: own elaboration.

Specification	LSED	LAD	APS	NEA	IUF-RDP
LSED	1	0.105	−0.025	−0.078	0.046
LAD	0.105	1	0.309	−0.172	0.233
APS	−0.025	0.309	1	−0.514	−0.203
NEA	−0.078	−0.172	−0.514	1	0.282
IUF-RDP	0.046	0.233	−0.203	0.282	1

Nonsignificant correlations are marked in grey ($\alpha = 0.05$).

Therefore, the above results confirm the fact that the socioeconomic conditions do not have such an enormous impact on the territorial orientation of the implementation of proecological activities as environmental factors. In the context of activities aimed at directing the development of local and regional economies towards the bioeconomy profile, this is of great importance. It clearly shows that the territorialisation of AECM and OFS actions should be related to local conditions (mainly environmental). Then, profiling for the bioeconomy has a chance for sustainable development, which will be sustained by grassroots activities of farmers with the support of the CAP funds. More urbanised, industrialised regions with a higher level of socioeconomic development base their economic profile more on other industries (e.g., automotive, electromechanical, IT sector, highly specialised services).

The results of the correlation analysis and the distribution of the values of synthetic indicators (see Tables 4 and 5) show that particular attention should be paid to the regional component, as each province shows its specificity, potential and conditions. Thus, it creates different possibilities from the point of view of bioeconomy development. In order to effectively manage and influence the rationality of spending funds from AECM and OFS activities, in line with the objectives of the EU environmental policy (including enhancing biodiversity and bioeconomy), the funds should include a regional component, as is the case in Germany, where each state has specific autonomy in the field of creating a development policy taking into account the existing conditions [91].

Generally, it should be noted that in the whole EU the rank of proecologically oriented activities is gradually increasing, which is a derivative of the change of priorities and the successive strengthening of this direction of development. As a result, in the EU countries, over the last three decades, public funds allocated to the development of organic farming have been gradually increasing and becoming more available [92]. Despite the change of direction in the ecological policy strengthening the bioeconomy, there is still a large gap between funds aimed at conventional agriculture and expenditure on agri-environmental measures, including organic farming (they accounted for about 7%, i.e., nearly EUR 20 billion, of total EU funding for the CAP 2014–2020; European Commission, 2013). Even in the countries with the highest input rates for organic farming in the EU (Germany), this represents only a small part of the total expenditure on agricultural policy [93,94].

4. Conclusions

The aim of the study was to identify the spatial level of the use of proenvironmental CAP funds, which constitute an important element in the sustainable development of agriculture, and thus the bioeconomy. Analyses carried out in high spatial resolution allowed showing areas where the proenvironmental management system is widely accepted by farmers and applied on the majority of farmland, thus significantly affecting farm income. It was natural to try to look for factors influencing the differentiation of the studied phenomenon. The detected relationships indicate the complexity of the problem, with stronger relationships related to environmental determinants.

Considering strengthening natural ecosystems and biodiversity, such targeted territorialisation of activities has a solid foundation. It is conducive not only to maintaining the balance between environmental resources but also meeting economic needs, and thus building the foundations for the development of the bioeconomy.

However, the adopted set of conditions does not sufficiently explain the complexity of the process of absorption of proenvironmental RDP funds. Therefore, further research explaining the mechanisms of sustainable agriculture development is essential.

In future activities, it is recommended to strengthen the territorialisation of proecological activities (AECM and OFS, but also others), which should be strongly related to local conditions, mostly natural and ecological. It will provide a stable foundation for sustainable development kept and reinforced by bottom-up activities. The cyclical approach, in line with the current fashions, but not based on local resources, does not create opportunities for the stable long-term development of bioeconomy.

Considering Polish conditions and economic profile, i.e., a significant share of the agricultural sector, from the point of view of bioeconomy, they predispose to the development based on the agricultural and natural potential. It applies primarily to development towards bioenergy, organic food production, and ensuring food security, as well as sustainable agricultural production, which combines production, economic, social, and ethical priorities with environmental security.

The empirical nature of the article contributed to the development of a cognitive thread regarding the impact of EU funds on the development of proenvironmental forms of agricultural management as an element of bioeconomy. The conducted research is also of great application value. It enriches works in the field of bioeconomy, spatial planning and strategic agriculture with knowledge about the impact of the CAP instruments on the multifunctional and sustainable development of agriculture.

Author Contributions: Conceptualisation, A.J.-T., R.R., Ł.W. and M.G.-G. methodology, Ł.W. and R.R. validation, A.J.-T., Ł.W., R.R. and M.B.; formal analysis, A.J.-T., Ł.W., R.R., M.G.-G. and M.B. investigation, A.J.-T., Ł.W., R.R. and M.B.; resources, Ł.W. and R.R.; data curation, Ł.W. and R.R. writing—original draft preparation, A.J.-T., R.R., Ł.W. and M.B.; writing—review and editing, A.J.-T. Ł.W., R.R. and M.G.-G.; visualisation, Ł.W.; supervision, A.J.-T., R.R., Ł.W. and M.B.; funding acquisition, R.R. and M.B. All authors have read and agreed to the published version of the manuscript.

Funding: This paper was written in the framework of the research project of the National Science Centre (no. DEC-2012/07/B/HS4/00364).

Conflicts of Interest: The authors declare no conflict of interest.

References

1. Komunikat Komisji Europejskiej. Europejska Strategia I Plan Działania W Kierunku Zrównoważonej Biogospodarki Do 2020 Roku. Available online: Europa.eu/rapid/press-release_IP-12-124_pl.htm (accessed on 7 November 2020).
2. Ratajczak, E. Rolnictwo i leśnictwo w świetle koncepcji biogospodarki. In Proceedings of the IX Kongres Ekonomistów Polskich, Warszawa, Poland, 28–29 September 2013.
3. Chyłek, E.K. Nowe strategie Komisji Europejskiej dotyczące biogospodarki i gospodarki wewnętrznej o obiegu zamkniętym *Pol. J. Agron.* **2016**, *25*, 3–12.
4. European Commission (EC). *New Perspectives on the Knowledge-Based Bio-Economy*; European Commission: Brussels, Belgium, 2005
5. Hanley, N.; Banerjee, S.; Lennox, G.D.; Armsworth, P.R. How should we incentivize private landowners to 'produce' more biodiversity? *Oxf. Rev. Econ. Policy* **2012**, *28*, 93–113. [CrossRef]
6. European Commission (EC). *Innovating for Sustainable Growth: A Bioeconomy for Europe*; Publication Office of the European Office: Luxembourg, 2012.
7. Goldberger, J.R. Conventionalization, civic engagement, and the sustainability of organic agriculture. *J. Rural Stud.* **2011**, *27*, 288–296. [CrossRef]
8. Żmija, J.; Czekaj, M. The diversity of farm production as the basis for the development of bio-economy in the malopolskie province. *Econ. Reg. Stud.* **2014**, *7*, 33–42.
9. Bański, J. *Geografia Polskie Jwsi*; Polskie Wydawnictwo Ekonomiczne: Warszawa, Poland, 2006; p. 220, ISBN 83-208-1596-7.
10. Jezierska-Thöle, A.; Biczkowski, M. The Analysis of Conditions and Level of Sustainable Development of Rural Areas in Poland and Germany. *Eur. J. Sustain. Dev.* **2014**, *3*, 387. [CrossRef]
11. Adamowicz, M. Bioeconomy-concept, application and perspectives. *Probl. Agric. Econ.* **2017**, *1*, 29–49. [CrossRef]
12. Moore, J.W. The end of the road? Agricultural revolutions in the capitalist world-ecology, 1450–2010. *J. Agrar. Chang.* **2010**, *10*, 389–413. [CrossRef]
13. Abler, D.G. Multifunctionality, agricultural policy, and environmental policy. *Agric. Resour. Econ. Rev.* **2004**, *33*, 8–17. [CrossRef]
14. European Commission (EC). *Innovating for Sustainable Growth: A Bioeconomy for Europe*; COM (2012) Final; European Commission: Brussels, Belgium, 2012.
15. Birch, K.; Levidow, L.; Papaioannou, T. Sustainable capital? The neoliberalization of nature and knowledge in the European "knowledge-based bio-economy". *Sustainability* **2010**, *2*, 2898–2918. [CrossRef]
16. Anderson, L.G.; Seijo, J.C. *Bioeconomics of Fisheries Management*; John Wiley & Sons: Hoboken, NJ, USA, 2010; p. 320, ISBN1 0813817323, ISBN2 9780813817323.
17. Kołodziejczak, A.; Rudnicki, R. Spatial diversification of energy crops in Polish agriculture: A study of plantation concentration based on local indicators of spatial association (LISA). *Quaest. Geogr.* **2017**, *36*, 49–56. [CrossRef]
18. COM (2012) 60. Available online: Ec.europa.eu/research/bioeconomy/pdf/201202_innovating_sustainable_growth_pl.pdf (accessed on 7 November 2020).
19. Pajewski, T. Programy rolnośrodowiskowe jako forma wspierania ochrony środowiska na terenach wiejskich. *Zesz. Nauk. Szkoły Głównej Gospod. Wiej. Ekon. I Organ. Gospod. Żywnościowej* **2014**, *107*, 69–80. [CrossRef]
20. Strategia Innowacyjności I Efektywności Gospodarki. Załącznik Do Uchwały Nr 7 Rady Ministrów Z Dnia 15 Stycznia 2013 Roku. Ministerstwo Gospodarki. Available online: https://www.Strategia_Innowacyjnosci_i_Efektywnosci_Gospodarki_2020.pdf (accessed on 7 November 2020).
21. Strategia Zrównoważonego Rozwoju Wsi Rolnictwa I Rybactwa 2030. Available online: https://www.gov.pl/web/rolnictwo/strategia-zrownowazonego-rozwoju-wsi-rolnictwa-i-rybactwa-2030 (accessed on 1 October 2020).

22. Strategia Bezpieczeństwa Narodowego Rzeczypospolitej Polskiej 2020. Warszawa 2020. Available online: https://www.bbn.gov.pl/ftp/dokumenty/Strategia_Bezpieczenstwa_Narodowego_RP_2020.pdf (accessed on 1 October 2020).

23. Organisation for Economic Cooperation and Development (OECD). *The Bioeconomy to 2030: Designing a Policy Agenda, Main Findings*; Organisation for Economic Cooperation and Development: Paris, France, 2006.

24. Aims of the common agricultural policy. Available online: Ec.europa.eu/agriculture/cap-post-2013/.../index_en.htm (accessed on 7 November 2020).

25. European Association for Bioindustries (EuropaBio). *Building a Bio-Based Economy for Europe in 2020*; European Association for Bioindustries: Brussels, Belgium, 2011.

26. European Commission. *Technical Elements of Agri-Environment-Climate Measure in the Programming Period. 2014–2020, (WD 08–18–14)*; European Commission: Brussels, Belgium, 2014.

27. Kluba, M.; Rudnicki, R.; Jezierska-Thöle, A. Poziom i struktura produkcji rolniczej województwa kujawsko-pomorskiego w 2010 roku w warunkach oddziaływania instrumentów WPR Unii Europejskiej. *Rocz. Nauk. Stowarzyszenia Ekon. Rol. I Agrobiz.* **2014**, *16*, 101–108.

28. Rudnicki, R. Natural considerations of the regional diversification of absorption of European Union funds in Polish agriculture. In *Selected Aspects of Transformation in Countries of Central and Central-Eastern Europe*; Michalski, T., Kuczabski, A., Eds.; University of Gdańsk: Gdańsk, Poland, 2010; pp. 63–80.

29. Jezierska-Thöle, A.; Gwiaździńska-Goraj, M.; Wiśniewski, Ł. Current status and prospects for organic agriculture in Poland. *Quaest. Geogr.* **2017**, *36*, 23–36. [CrossRef]

30. Lakner, S.; Zinngrebe, Y.; Koemle, D. Combining management plans and payment schemes for targeted grassland conservation within the Habitats Directive in Saxony, Eastern Germany. *Land Use Policy* **2020**, *97*, 104642. [CrossRef]

31. Adamowicz, M. European concept of bioeconomy and its bearing on practical use. *Econ. Reg. Stud. (Studia Ekon. I Reg.)* **2014**, *7*, 5–21.

32. Financing Natura 2000. Available online: https://ec.europa.eu/environment/nature/natura2000/financing/index_en.htm (accessed on 1 October 2020).

33. Gołębiewski, J. Zrównoważona biogospodarka–potencjał i czynniki rozwoju. *IX Kongr. Ekon. Pol.* **2013**, *2*, 1–13.

34. Chyłek, E.K.; Bielecki, S. Biogospodarka–technologie innowacyjne szansą poprawy konkurencyjności w sektorze rolno-spożywczym i na obszarach wiejskich. In *Bioeconomy–Innovative Technologies as an Opportunity to Improve Competitiveness in the Agri-Food Sector and in Rural Areas*; Chyłek, E.K., Pietras, M., Eds.; III Kongres Nauk Rolniczych "Nauka–Praktyce": Warszawa, Poland, 2015; pp. 9–14.

35. Barberi, P.; Burgio, G.; Dinelli, G.; Moonen, A.C.; Otto, S.; Vazzana, C.; Zanin, G. Functional biodiversity in the agricultural landscape: Relationships between weeds and arthropod fauna. *Weed Res.* **2010**, *50*, 388–401. [CrossRef]

36. Bio-Economy Technology Platforms (BECOTEPS). *The European Bioeconomy in 2030: Delivering Sustainable Growth by Addressing the Grand Societal Challenges*; Bio-Economy Technology Platforms: Brussels, Belgium, 2011.

37. Jezierska-Thöle, A.; Biczkowski, M. Financial Funds from European Union funds as a chance for the development of the sector of organic farms in Poland. *Assoc. Agric. Agribus. Econ.* **2017**, *19*, 95–101. [CrossRef]

38. Król, M. Legal Framework of Environmental law for Agricultural Production in Poland. Available online: http://sj.wne.sggw.pl/pdf/PEFIM_2015_n62_s86.pdf (accessed on 5 December 2020).

39. Zikeli, S.; Gruber, S. Reduced tillage and no-till in organic farming systems, Germany—Status quo, potentials and challenges. *Agriculture* **2017**, *7*, 35. [CrossRef]

40. Zimmermann, A.; Britz, W. European farms' participation in agri-environmental measures. *Land Use Policy* **2016**, *50*, 214–228. [CrossRef]

41. Pondel, H. European Union funds as a tool for creating new functions of rural areas, as illustrated by the example of RDP. *J. Agribus. Rural Dev.* **2017**, *2*, 435–443. [CrossRef]

42. Schmidtner, E.; Lippert, C.; Engler, B.; Häring, A.M.; Aurbacher, J.; Dabbert, S. Spatial distribution of organic farming in Germany: Does neighbourhood matter? *Eur. Rev. Agric. Econ.* **2012**, *39*, 661–683. [CrossRef]

43. Zander, P.; Knierim, U.; Groot, J.C.; Rossing, W.A. Multifunctionality of agriculture: Tools and methods for impact assessment and valuation. *Agric. Ecosyst. Environ.* **2007**, *120*, 1–4. [CrossRef]

44. Birge, T.; Herzon, I. Exploring cultural acceptability of a hypothetical results-based agri-environment payment for grassland biodiversity. *J. Rural Stud.* **2019**, *67*, 1–11. [CrossRef]

45. Desjeux, Y.; Dupraz, P.; Kuhlman, T.; Paracchini, M.L.; Michels, R.; Maigne, E.; Reinhard, S. Evaluating the impact of rural development measures on nature value indicators at different spatial levels: Application to France and The Netherlands. *Ecol. Indic.* **2015**, *59*, 41–61. [CrossRef]

46. Kleijn, D.; Sutherland, W.J. How effective are Europeanagri-environment schemes in conserving and promoting biodiversity? *J. Appl. Ecol.* **2003**, *40*, 947–969. [CrossRef]

47. Kołodziejczak, A.; Rudnicki, R. Instrumenty Wspólnej Polityki Rolnej ukierunkowane na poprawę środowiska przyrodniczego a planowanie przestrzenne rolnictwa. *Acta Sci. Pol. Adm. Locorum* **2012**, *11*, 17–133.

48. Rudnicki, R. *Fundusze Unii Europejskiej jako Czynnik Modernizacji Rolnictwa Polskiego*; Bogucki Wydawnictwo Naukowe: Poznań, Poland, 2010; p. 210.

49. Wilkin, J. Wielofunkcyjność wsi i rolnictwa a rozwój zrównoważony. *Wieś I Rol.* **2011**, *153*, 27–39.

50. Kłodziński, M. Przedsiębiorczość pozarolnicza na wsi w procesie wielofunkcyjnego rozwoju obszarów wiejskich. *Wieś I Rol* **2014**, *162*, 97–112.
51. Treasury, H.M. The Green Book: Central Government Guidance on Appraisal and Evaluation, London. 2018. Available online https://assets.publishing.service.gov.uk/government/uploads/system/uploads/attachment_data/file/685903/The_Green_Book.pdf (accessed on 1 October 2020).
52. Bartolini, F.; Vergamini, D.; Longhitano, D.; Povellato, A. Do differential payments for agri-environment schemes affect the environmental benefits? A case study in the North-Eastern Italy. *Land Use Policy* **2020**. [CrossRef]
53. Batáry, P.; Dicks, L.V.; Kleijn, D.; Sutherland, W.J. The role of agri-environment schemes in conservation and environmental management. *Conserv. Biol.* **2015**, *29*, 1006–1016. [CrossRef] [PubMed]
54. Krajowy Program Badań. Założenia Polityki Naukowo-Technicznej I Innowacyjnej Państwa–Załącznik Do Uchwały Nr 164/201 Rady Ministrów Z Dnia 16 Sierpnia 2011 r. Available online: https://www.bip.nauka.gov.pl/krajowy-program-bada%C5%84$ backslash$h (accessed on 7 November 2020).
55. Ecorys. Mapping and Analysis of the Implementation of the CAP. 2017. Available online: https://op.europa.eu/pl/publication detail/-/publication/65c49958-e138-11e6-ad7c-01aa75ed71a1 (accessed on 22 August 2020).
56. Galler, C.; von Haaren, C.; Albert, C. Optimizing environmental measures for landscape multifunctionality: Effectiveness efficiency, and recommendations for agri-environmental programs. *J. Environ. Manag.* **2015**, *151*, 243–257. [CrossRef] [PubMed]
57. Pe'er, G.; Dicks, L.V.; Visconti, P.; Arlettaz, R.P.R.; Kleijn, D.; Neumann, R.K.; Robijns, T.; Schmidt, J.; Shwartz, A. Sutherland, W.J.; et al. EU agricultural reform fails on biodiversity. *Science* **2014**, *344*, 1090–1092. [CrossRef] [PubMed]
58. European Plant Science Organisation (EPSO). The European Bioeconomy in 2030: Delivering Sustainable Growth by Ad-dressing the Grand Societal Challenges. Available online: http://www.plantetp.org/system/files/publications/files/the_european_bioeconomy_brochure_web_final.pdf (accessed on 1 October 2020).
59. Innovation for Sustainable Growth: Bio-Economy for Europe. Sygnatura COM (2012) 60. Available online: https://wbc-rti.info/object/event/11271/attach/05_Bio_economy_for_Europe.pdffor.org.pl (accessed on 7 October 2020).
60. European Parliament, Council of the European Union Regulation (EU) No 1305/2013 Support for Rural Development by the European Agricultural Fund for Rural Development (EAFRD) and Repealing. Available online: https://www.bip.nauka.gov.pl/krajowy-program-badan/ (accessed on 7 October 2020).
61. Jezierska-Thöle, A.; Rudnicki, R.; Kluba, M. Development of energy crops cultivation for biomass production in Poland *Renew. Sustain. Energy Rev.* **2016**, *62*, 534–545. [CrossRef]
62. Biczkowski, M. *Instrumenty Wspólnej Polityki Rolnej jako Czynnik Wspierający Rozwój Obszarów Wiejskich. Studium na Przykładzie Re-gion. Kujawsko-Pomorskiego*; Wydawnictwo Naukowe Uniwersytetu Mikołaja Kopernika: Toruń, Poland, 2018; p. 306, ISBN 978-83-231-4041-2.
63. Rudnicki, R.; Jezierska-Thole, A.; Biczkowski, M. The impact of the Common Agricultural Policy on socio-economic development in Poland. In Proceedings of the 13th International days of statistics and economics, Prague, Czech Republic, 5–7 September 2019, pp. 1309–1318.
64. Rudnicki, R.; Wiśniewski, Ł.; Kluba, M. Poziom i struktura przestrzenna rolnictwa Polskiego w świetle wyników Powszechnego Spisu Rolnego 2010. *Rocz. Nauk. Stowarzyszenia Ekon. Rol. I Agrobiz.* **2015**, *17*, 337–343.
65. Rudnicki, R. *The Spatial Structure of Polish Agriculture Conditioned by Common Agricultural Policy Instruments*; Wydawnictwo Naukow Uniwersytetu Mikołaja Kopernika: Toruń, Poland, 2016; p. 208.
66. Statistics Poland-Local Data Bank of the Central Statistical Office. Available online: https://bdl.stat.gov.pl/BDL/ (accessed on 3 October 2020).
67. Management Information System of the Agency for Restructuring and Modernisation of Agriculture (ARMA). Available online: https://www.ARiMR-AgencjaRestrukturyzacjiiModernizacjiRolnictwa (accessed on 1 October 2020).
68. Statistics Poland-General Agricultural Census. Available online: https://www.PowszechnySpisRolny2010 (accessed on 1 October 2020).
69. Perdał, R. Zastosowanie analizy skupień i lasów losowych w klasyfikacji gmin w Polsce na skali poziomu rozwoju społeczno-gospodarczego. *Metod. Ilościowe W Bad. Ekon.* **2018**, *19*, 263–273.
70. Sadik-Zada, E.R.; Gatto, A. The puzzle of greenhouse gas footprints of oil abundance. *Socio Econ. Plan. Sci.* **2020**. [CrossRef]
71. Sadik-Zada, E.R.; Ferrari, M. Environmental Policy Stringency, Technical Progress and Pollution Haven Hypothesis. *Sustainability* **2020**, *12*, 3880. [CrossRef]
72. IUNiG-PIB. Available online: https://iung.pl/index (accessed on 1 October 2020).
73. *Plan Rozwoju Obszarów Wiejskich na Lata 2004–2006*; Ministerstwo Rolnictwa I Rozwoju Wsi: Warszawa, Poland, 2004; Available online: http://eko.org.pl/lkp/poradniki/prow_plus.pdf (accessed on 1 October 2020).
74. Früh-Müller, A.; Bach, M.; Breuer, L.; Hotes, S.; Koellner, T.; Krippes, C.; Wolters, V. The use of agri-environmental measures to address environmental pressures in Germany: Spatial mismatches and options for improvement. *Land Use Policy* **2019**, *84*, 347–362.
75. Rudnicki, R. Analiza absorpcji środków WPR i ich wpływu na zmiany strukturalne w rolnictwie polskim. In *Zróżnicowanie Przestrzenne Rolnictwa*; Głębocki, B., Ed.; GUS: Warszawa, Poland, 2014; pp. 441–463.
76. Mijatović, D.; Van Oudenhoven, F.; Eyzaguirre, P.; Hodgkin, T. The role of agricultural biodiversity in strengthening resilience to climate change: Towards an analytical framework. *Int. J. Agric. Sustain.* **2013**, *11*, 95–107. [CrossRef]

77. Niggli, U.; Slabe, A.; Schmid, O.; Halberg, N.; Schlüter, M. *Vision for an Organic Food and Farming Research Agenda to 2025. Organic Knowledge for the Future*; Technology Platform Organics: Brussels, Belgium, 2008. Available online: http://www.tporganics.eu/upload/TPOrganics_VisionResearchAgenda.pdf (accessed on 1 October 2020).

78. Biczkowski, M.; Jezierska-Thöle, A. Impact of EU funds on current status and prospects of organic farming in Poland. *Econ. Sci. Rural Dev.* **2018**, *47*, 123–131. [CrossRef]

79. Rudnicki, R.; Jezierska-Thöle, A.; Wiśniewski, Ł.; Janzen, J.; Kozłowski, L. Former political borders and their impact on the evolution of the present-day spatial structure of agriculture in Poland. *Stud. Agric. Econ.* **2018**, *120*, 8–16. [CrossRef]

80. Przewodnik Po Działaniu Rolno-Środowiskowo-Klimatycznym PROW 2014–2020. Available online: https://www.gov.pl/documents/912055/913531/Przewodnik (accessed on 1 October 2020).

81. Rural Development Programme for Mainland Finland 2014–2020. Available online: https://mmm.fi/en/rural-areas/rural-development-programme (accessed on 1 October 2020).

82. Batáry, P.; Gallé, R.; Riesch, F.; Fischer, C.; Dormann, C.F.; Mußhoff, O.; Császár, P.; Fusaro, S.; Gayer, C.; Happe, A.-K.; et al. The former Iron Curtain still drives biodiversity–profit trade-offs in German agriculture. *Nat. Ecol. Evol.* **2017**, *1*, 1279–1284. [CrossRef] [PubMed]

83. Czyż, T. Poziom rozwoju społeczno-gospodarczego Polski w ujęciu subregionalnym. *Przegląd Geogr.* **2012**, *84*, 219–236.

84. Stanny, M. *Przestrzenne Zróżnicowanie Rozwoju Obszarów Wiejskich W Polsce*; IRWiR PAN: Warszawa, Poland, 2013; p. 329, ISBN 978-83-89900-53-1.

85. Rudnicki, R. Spatial Differences in the Use of European Funds by Agricultural Holdings in Poland in the Years 2002-2010. In *Transformation Processes of Rural Areas*; Kamińska, W., Heffner, K., Eds.; Studia Regionalia KPZK PAN: Warsaw, Poland, 2013; Volume 36, pp. 71–87.

86. Witek, T.; Górski, T. *Przyrodnicza Bonitacja Rolniczej Przestrzeni Produkcyjnej W Polsce*; IUNG: Puławy, Poland, 1977.

87. *Waloryzacja Rolniczej Przestrzeni Produkcyjnej Polski*; IUNG: Puławy, Poland, 2000.

88. Rudnicki, R.; Wiśniewski, Ł. Cechy produkcyjne rolnictwa a poziom absorpcji środków Wspólnej Polityki Rolnej w Polsce. *Studia Obsz. Wiej.* **2016**, *42*, 89–104. [CrossRef]

89. Stawicka, J.; Szymczak-Piątek, M.; Wieczorek, J. *Selected Ecological Issues*; SGGW: Warszawa, Poland, 2004.

90. Moonen, A.C.; Barberi, P. Functional biodiversity: An agroecosystem approach. *Agric. Ecosyst. Environ.* **2008**, *127*, 7–21. [CrossRef]

91. Rahmann, G.; Reza Ardakani, M.; Bàrberi, P.; Boehm, H.; Canali, S.; Chander, M.; David, W.; Dengel, L.; Erisman, J.W.; Galvis-Martinez, A.C.; et al. Organic Agriculture 3.0 is innovation with research. *Org. Agric.* **2017**, *7*, 169–197. [CrossRef]

92. Uthes, S.; Matzdorf, B. Studies on agri-environmental measures: A survey of the literature. *Env. Manag.* **2013**, *51*, 251–266. [CrossRef]

93. Feindt, P.; Proestou, M.; Daedlow, K. Resilience and policy design in the emerging bioeconomy–the RPD framework and the changing role of energy crop systems in Germany. *J. Environ. Policy Plan.* **2020**, *22*, 636–652. [CrossRef]

94. Kirschke, D.; Hager, A.; Jechlitschka, K.; Wegener, S. Distortions in a multi-level co-financing system: The case of the agri-environmental programme of Saxony-Anhalt. *Ger. J. Agric. Econ.* **2007**, *56*, 297–304.

Article

Impact Assessment of the Long-Term Fallowed Land on Agricultural Soils and the Possibility of Their Return to Agriculture

Małgorzata Kozak * and Rafał Pudełko

Department of Bioeconomy and Systems Analysis, Institute of Soil Science and Plant Cultivation—State Research Institute (IUNG-PIB), 24-100 Puławy, Poland; rpudelko@iung.pulawy.pl
* Correspondence: mkozak@iung.pulawy.pl; Tel.: +48-81-47-86-769

Abstract: Agricultural land abandonment is a process observed in most European countries. In Poland and other countries of Central and Eastern Europe, it was initiated with the political transformation of the 1990s. Currently, in Poland, it concerns over 2 million ha of arable land. Such a large acreage constitutes a resource of land that can be directly restored to agricultural production or perform environmental functions. A new concept for management of fallow/abandoned areas is to start producing biomass for the bioeconomy purposes. Production of perennial crops, especially on poorer soils, requires an appropriate assessment of soil conditions. Therefore, it has become crucial to answer the question: What is the real impact of the fallowing process on soil, and is it possible to return it to production at all? For this purpose, on the selected fallowed land that met the marginality criteria defined under the project, physicochemical tests of soil properties were carried out, and subsequently, the results were compared with those of the neighboring agricultural land and with the soil valuation of the fallow land, which was conducted during its past agricultural use. The work was mainly aimed at analyzing the impact of long-term fallowing on soil pH, carbon sequestration and nutrient content, e.g., phosphorus and potassium. The result of the work is a positive assessment of the possibility of restoring fallowed land for agricultural production, including the production of biomass for non-agricultural purposes. Among the studied types of fallow plots, the fields where goldenrod (*Solidago L.*—invasive species) appeared were indicated as the areas most affected by soil degradation.

Keywords: unutilized agricultural areas (uUAA); abandoned areas; land use and land-use change; carbon sequestration; soil properties (physical and chemical)

Citation: Kozak, M.; Pudełko, R. Impact Assessment of the Long-Term Fallowed Land on Agricultural Soils and the Possibility of Their Return to Agriculture. *Agriculture* **2021**, *11*, 148. https://doi.org/10.3390/agriculture11020148

Academic Editor: Mariusz J. Stolarski

Received: 31 December 2020
Accepted: 8 February 2021
Published: 11 February 2021

Publisher's Note: MDPI stays neutral with regard to jurisdictional claims in published maps and institutional affiliations.

1. Introduction

1.1. The Process of Setting Aside/Fallowing Land in the Political and Environmental Context

The time of political transformations that took place in Poland in the 1980s and 1990s significantly contributed to changes in land use [1]. The process of agricultural land abandonment on large scale became visible then. Besides the economic effects resulting from abandoning agricultural activity, also structural and functional transformations of landscape units took place [2,3]. The first visible effect of land abandonment is regrowth of vegetation through natural secondary succession [4]. In many regions of the country, as a result of land use discontinuation and succession of natural vegetation, a significant part of the agricultural plots became permanently covered with trees and bushes. At the same time, regional nature of this process became apparent [5]. It is difficult to assess unequivocally whether it is a negative or a positive phenomenon, it depends mainly on the local environmental, political, or social conditions. In some cases, such a change may result in restoration of the old ecosystem or emergence of completely new landscape or utility functions [6], and in the case of mountain areas, it may also affect the functioning of valley ecosystems [7]. Currently, in order to identify abandoned areas, a number of remote sensing methods are being developed, using satellite photos or Airborne Laser Scanning (ALS), which allows for precise identification of the size of areas covered with high vegetation and the dynamics of changes over time [8]. According to some researchers, tree stands resulting from spontaneous succession are much more abundant in elements of environmental value than conifer monoculture stands [3]. Natural succession also creates ecological corridors, prevents soil erosion thanks to vegetation cover, increases carbon sequestration in the soil, and can be a stronghold of biodiversity in intensively utilized agriculturally areas [9,10]. On the other hand, uncontrolled succession may pose a threat to the biodiversity in open areas and the species occurring there, e.g., by the invasive plants entering those areas, causing the loss of cultural landscapes [6,11]. Invasive plants have a negative effect on native species, not only by displacing them from their growing area, but also by modifying soil conditions [12]. For example, in Poland, we can observe how the goldenrod species (*Solidago L.*) enters after the segetal phase of natural succession as an invasive plant, creating dense fields [13]. Just as it is difficult to unequivocally assess the process of farmland abandonment, so are the decisions on how to manage it. Bell [14] identified three main categories of strategies that are undertaken in the process of abandoned land management in relation to its new functions:

1. Naturality (with or without controlled natural succession),
2. Multi-functionality (e.g., extensive agricultural production, hobby farming (traditional, small-scale food production, agri-tourism),
3. Productivity—sustainable agricultural production or bioenergy and renewable energy sources.

The researchers also admit that implementing the most appropriate post-abandonment strategy will be based on a number of variables. The aforementioned production function related to the acquisition of biomass has recently gained particular importance in formulating bioeconomy development strategies. The research shows that the cultivation of perennial industrial plants on marginal lands can represent a significant potential in obtaining biomass [15–19].

In addition to the many challenges associated with this issue, some opportunities were also recognized for the sustainable management of fallow land to compensate for the consequences of climate change. This is because soils, in addition to the basic functions of food production and ensuring food security, also provide ecosystem services that are necessary for the functioning and resilience of the environment on Earth. The main non-agricultural functions are: (i) Storage of massive amounts of carbon, which helps to regulate CO_2 emissions and shape climate processes; (ii) functioning as the largest water filter and storage tank on Earth, which ensures control of its circulation, retention, and quality of freshwater resources; and (iii) storage of nitrogen, phosphorus, and other essential nutrients [20].

1.2. Testing Conditions of the Fallow Soil

The possibility of carbon sequestration in soil was and still is widely discussed in many publications, also in the context of abandoned agricultural land. Among other things, the effect of converting agricultural land into other forms of land use (arable land to grassland, arable land to abandoned agricultural land, and arable land to afforested land), and the possibility of SOC sequestration in the topsoil were investigated. Kazlauskaite-Jadzevice et al. [21] showed, among others, that carbon sequestration in the Arenosol soil layer was positively influenced by long-term fallowing and transformation into grassland. Abandoned land or fertilized grassland accumulated significantly more CO_2 (48% and 38%—respectively), compared to arable land. Whereas the potential of "mature" forest succession in terms of CO_2 sequestration was confirmed by the studies conducted by Foote and Grogan [22], showing that the total contents of organic carbon and nitrogen in soil at a depth of 10 cm were lower in arable fields compared to forests with secondary succession, by about 32% and 18%, respectively.

Studies on the impact of the natural succession diversity on the storage capacity and rate of C accumulation in soil over a period of two decades were conducted by Yang et al. [23]. The authors concluded that the annual rate of carbon storage was higher in the second period of the study (years 13–22), in the same time suggesting that restoring high plant diversity could significantly increase carbon capture and storage on degraded and abandoned agricultural land. The impact of afforestation and deforestation on soil carbon content was also investigated [24]. The impact of land use change in Mediterranean areas on the processes of co-carbonization, decarbonization, and recarbonization was taken up by researchers Lozano-García et al. [25], anticipating at the same time the possibility of soil regeneration and climate change. The topic of reclamation of degraded agricultural land by changing the variety of vegetation and restoring organic matter was also discussed by researchers Zhang et al. [26].

In publications concerning fallow land, the influence of land use change on physical and chemical properties of soils is often discussed. The negative influence of relatively young (5–10 years) fallow land on the properties of the soil environment was found by Strączyńska et al. [27]. It was manifested in the unfavorable change in pH and reduction of humus resources in silty and clay soils, while in light soils, the humus content slightly increased. Moreover, Tomaszewicz and Chudecka [28] did not find any enrichment of fallow rusty soils with humus either. At the same time, fallow soil was characterized by a lower content of plant-available forms of magnesium, potassium, and phosphorus. On the other hand, the positive effect of setting aside on the properties of the soil environment was observed by Włodek et al. [29]. In their research, the authors observed that after several years of excluding the field from agricultural production, there was a significant increase in the content of carbon, phosphorus, potassium, and magnesium in the soil. In the study of the physicochemical properties of soils, a detailed analysis of changes in the quantitative and qualitative composition of humus compounds is of key importance. Based on their research results, Licznar et al. [30] concluded that the fractional composition of humus compounds in fallow soils shows strong relationships with their physicochemical properties. They also noticed that in set-aside light soils, the process of organic matter accumulation occurs, thus showing a lower degree of humification than in a cultivated soil.

The results presented in this paper were obtained as part of the BioMagic [31] project, whose main objective is to develop bioproducts from lignocellulosic biomass obtained from marginal soils to fill the gap in the national bioeconomy. On the selected fallow land, meeting the marginality criteria defined in the project, physicochemical tests of soil properties were carried out, the results of which were then compared with the neighboring utilized agricultural lands. The main aim of the study is to answer the question, whether it is possible to restore weak and marginal soils to biomass production after a long-term abandonment. The obtained results, within the BioMagic project, were also used to estimate the technical and economic potential of fallow soils to be used for the production of perennial industrial plants.

2. Material and Methods

2.1. Main Assumptions

The following research hypothesis was assumed: Fallowing process does not cause a significant deterioration of soil conditions, which would be a problem when they are returned to agricultural production. An alternative to the research hypothesis is to show that long-term set aside affects the soil in such a negative way that its restoration to agricultural production requires expensive agrotechnical treatments. The following properties were considered essential for soil fertility: Carbon content in the topsoil, soil pH, phosphorus, and potassium content. The nitrogen content in the soil was not analyzed, because this element is not stable (its content changes rapidly during vegetation period, especially in case of intensive agricultural production).

In this study, it was assumed that the necessity to use "expensive treatments" to restore the soil to agricultural production will occur if the tested soil parameters for fallowed land deteriorate so that the ranges adopted in the determination of soil fertility in Poland will be exceeded twice. In the case of soil carbon, it is a 1.7% decrease in its content (compared to the content determined for the sample from arable land). For pH, it is an increase in soil acidity by two units. In the case of phosphorus and potassium, it is a reduction of 10 mg per 100 g of soil (forms: P_2O_5 and K_2O).

In order to assess the impact of fallowing on soil conditions, the following work was carried out:

- A region representative of the country was selected, where fallowing agricultural land is a serious problem for agriculture.
- Materials were collected to track the course and dynamics of the fallowing process.
- Basic types of fallow land were defined with their location on the map of the research area.
- For selected sites, the impact of fallowing on the soil condition was analyzed—by determining its chemical properties and comparing it with soil samples collected in adjacent agricultural fields.

2.2. Area of Research

The research was carried out in the area of Puławy municipality (gmina), Local Administrstive Unit (LAU code:1006061121409) according to the Eurostat nomenclature, located in the north-western part of the Lubelskie Voivodeship (NUTS-2: PL81). This municipality is directly adjacent to the city of Puławy and constitutes its suburban area, which is visible, among others, in employment the structure: In 97% of farms, at least 1 person has an additional non-agricultural source of income [32]. The economic situation and the large fragmentation of farms in this region determine the high percentage of fallow land in the agricultural landscape.

For soil sampling 17 sites were chosen. Research area and locations of sampling plots were shown on the Figure 1. The article's supplement includes a download link of the KLM file, which contains the location of the sampling sites, enabling their identification in the Google Maps/Earth application, which visualizes this region on high-resolution orthophotomaps (like in the right picture on Figure 1).

id	coordinates	
site	Lon.	Lat.
1	21.912	51.417
2	21.911	51.421
3	21.940	51.423
4	21.926	51.432
5	21.903	51.442
6	21.917	51.459
7	21.910	51.460
8	21.879	51.441
9	21.880	51.439
10	21.877	51.414
11	21.902	51.399
12	21.835	51.392
13	21.837	51.388
14	21.800	51.390
15	21.799	51.390
16	21.816	51.387
17	21.801	51.391

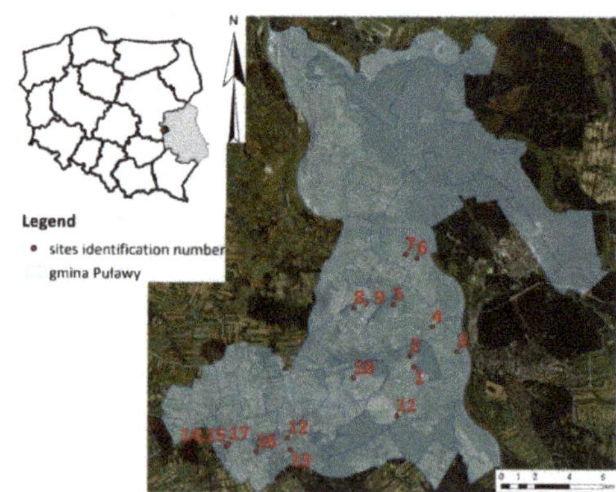

Figure 1. Research area—locations of sampling sites.

2.3. Spatial Data Collection

In the study, the following types of spatial data were used:

- Cadastral maps showing the range of cadastral parcels and classifications of agricultural usefulness of soils (source: GUGiK [33], SIP Pulawy [34]).
- Historical orthophotomaps from the years: 1997, 2006, 2010, 2017–2018—Figure 2 (source: GUGIK [33]).

- Geo-referenced photos taken in-situ during visits at the selected sites (for examples, see Section 2.4).

Figure 2. Time sequence of aerial photographs, showing the approximate moment of the entry of natural succession (1997, 2006, 2010, after 2017). Source: GUGIK [33].

The above data fed the geographic information system, which was built in the QGIS (open source environment).

2.4. Definition of Fallow Types

Due to the diversity of natural succession within the selected plots, the investigated areas of fallow plots have been divided into:

- Grassland (FGL)—mostly newly abandoned agricultural land, possibly fallow land, with a predominance of grassland vegetation, with only a few plantings of later succession, e.g., goldenrod (*Solidago* L.) (Figure 3).

Figure 3. Example of photos on-site No. 2 (left side) and No. 5 (right side).

- Goldenrod (FG)—areas with a predominance of plants of later succession stages, mainly goldenrod (*Solidago* L.), tansy (*Tanacetum vulgare* L.). Criterion: Over 80% share of goldenrod or tansy in land cover (Figure 4).

Figure 4. Example of photos on-site No. 1 (left side) and No. 16 (right side).

- Bushy (FB)—areas where apart from ruderal plants, such as goldenrod (*Solidago* L.) there are bushes, e.g., in the form of blackberries (*Rubus* L.), blackthorn (*Prunus spinosa* there are bushes, e.g., in the form of blackberries (*Rubus* L.), blackthorn (*Prunus spinosa* L.) and single self-seeded trees. Criterion: Over 30% share of bushes in land cover (Figure 5).

Figure 5. Example of photo on-site No. 8.

- Wooded/afforested (FW)—areas with trees, dense shrubs, advanced succession. Criterion: Samples were taken in places where young forest covered at least 0.10 ha (Figure 6).

Figure 6. Example of photos on-site No. 10 (left side) and No. 14 (right side).

2.5. Field Work, Soil Sampling and Laboratory Tests

Selection of 17 sites for the comparison of soil samples collected from fallow land with samples from neighboring utilized agricultural fields was carried out according to the following assumptions:

- Each tested fallowed parcel has to have a fallowing period documented on the aerial photographs, and meet the marginality criteria for agricultural plots, defined by Pudełko et al. [5].
- In the immediate vicinity of the fallow plot, there are agricultural plots with similar site conditions (soil class, topography, water conditions), with no episodes of fallowing.
- One site must consist of at least one fallow plot and one utilized agricultural plot (arable land). However, most often chosen sites offered several types of succession in the unutilized field and both types of land use the utilized field (arable and grassland)— see Figure 7.

Figure 7. Example of soil sampling strategy—site 17, where in the close vicinity were sampled: Two arable plots and three fallowed plots covered by goldenrod, wood, and bushes.

In total, 72 soil sampling points were identified for the 17 selected sites. Soil samples were taken from the top soil layer, 0–20 cm, where the reaction to the change in land use should be clearly visible. For each sample, the following characteristics were determined: Physical and chemical properties of soil, type of land use (arable land, grassland, and fallow type), and site affiliation.

The valuation class of the soil sampling site was preliminary determined based on the utilization class map available in the Spatial Information System of the Puławy poviat (SIS Pulawy) [34] and then verified by laboratory methods. National soil classification used in this work focuses on soil suitability for agricultural purposes, taking into account the morphological features and physicochemical properties of the soil, such as location and structure of the soil profile, water relations/conditions, and pH [35]. For better soil characteristics, each sample was described by its granulometric properties: pl (loose sand), pg (clay sands), ps (weak loamy sand), gp (sandy loam), and pyg (clay dust) [36,37]. Relation between the (Polish) classification used in this study and the international USDA classification was shown on Figure S1 (in the Supplementary part).

The scope of the performed analyses of soil samples included the basic physico-chemical properties, including: pH in a 1-molar KCl solution, granulometric composition according to the norm: BN-78/9180-11 and PTG (2008) [38], which were determined by the laser method. The Egner–Riehm method was used to determine the content of assimilable forms of nutrients (phosphorus, potassium). The Egner–Riehm method is a chemical laboratory method. It involves extraction of the available forms of nutrients from the soil by means of special solutions, usually a buffer one. The same extraction solution is used for phosphorus and potassium. It is lactic acid buffered with calcium lactate, pH = 3.6. This solution is obtained by dissolving calcium lactate with hydrochloric acid. It is well buffered against both hydrogen ions and calcium ions—two factors significantly affecting the solubility of phosphorus compounds in the soil. Organic carbon (C_{org}) and Humus were determined using the modified Tiurin method [39].

2.6. Statistical Analyses

In order to verify the working hypothesis, the results of determination of the chemical properties of soils between the adjacent fallowed and used plots were compared. In the first step, corresponding pairs were selected (classes of pairs):

- 8 pairs of arable (AL) and grassland, former arable (GL),
- 6 pairs of arable and fallow grassland (FGL),
- 17 pairs of arable and fallow goldenrod (FG),
- 15 pairs of arable and fallow bush (FB),
- 13 pairs of arable and fallow wood (FW).

For all classes of pairs, differences in carbon content in soil, pH, and potassium and phosphorus content were assessed in comparison to arable plot. If there were more than one arable plot sampled on the site, then higher values were taken into account. Choosing the highest value in the significance test introduces a more restrictive approach because the working hypothesis is always rejected when the upper-tailed critical region is exceeded. In the first step of statistical analyses, the significance level (α) at which the criteria of the research hypothesis are met was estimated.

Principal component analysis (PCA based on correlations) was also used to capture the correlation between particular parameters in different types of land use [40]. This approach was to test whether the change in land use may affect the relationship between the parameters. Principal components analysis creates new artificial variables (principal components) based on the variables (features) that we analyze. Its main assumption was the possibility of visualizing the relationships of individual variables on a two-dimensional graph, which shows the coordinate system representing the first two principal components. Based on the position of the vectors in space, it can be determined which features are correlated with each other. The smaller the angle between the vectors, the stronger the positive correlation. When vectors are aligned on the same line but in opposite directions, there is a strong negative correlation between the variables. However, when the vectors are at an angle close to 90 degrees, no correlation occurs. Statistical analyses were performed using Statistica software package Statistica v13.1 (TIBCO Software Inc., Palo Alto, CA, USA).

3. Result

The summarized results of field work, remote sensing and laboratory tests are listed on Figure S2 (see Supplementary). For each site, it is specified: Chemical and physical properties for each tested land cover, soil granulometric type and fallow period. The conducted research confirmed the influence of the change in the use of agricultural land on its physicochemical properties. It should be noted that the specific granulometric composition and the valuation class within each of the sites were similar, which made it possible to attempt a comparison of the results of chemical analyses at the sites level. The vast majority of the tested samples belonged to the granulometric group of sandy loam—gp (31 samples) and clay sands—pg (25 samples). Only in sites no. 1, 12, 14, and 15 were the sampled soils classified as lighter groups such as: Loose sand—pl, weak loamy sand—ps and heavy group such as clay dust—pyg in site no. 4 (Figure S2). Detailed characteristics of the granulometric fractions are presented on the Ferret's triangle (see Supplementary, Figure S1).

3.1. Changes in Carbon Content

The conducted research showed differences in the content of organic carbon within the sites in relation to different types of land use. The content of organic carbon in arable land ranged from 0.63% to 1.42%. It can be seen that a relatively high percentage of C_{org} was determined for site No. 12 despite its poor valuation soil class, which is associated with straw management in this field and manure fertilization. Newly abandoned land (FGL) with a predominance of grassy vegetation and several goldenrod instances showed an increased % C_{org} content, compared to arable land, on average by 32%, even up to 46.5%. Similar results were obtained by comparing arable land and permanent grassland (GL), where % of carbon content increased by max. 71.2%. In these soils, similar to meadow soils, the increase in carbon content is associated with the year-round cover of vegetation forming a compacted turf, which promotes binding of carbon from the atmosphere in the process of CO_2 assimilation by these plants [41]. Different tendencies are observed in the case of abandoned land dominated by species of later succession, in this case by goldenrod (*Solidago* L.). FGL species, for in this type of abandoned land one can see a clear decrease in carbon content even by 23.7% compared to arable land. On abandoned lands of the later succession, overgrown with FB bushes or trees, in sites similar to the FW forest, the carbon content in the studied samples is quite diversified. The highest increase, by 103.6%, was recorded for sites no. 14 and no. 7 and 10, by 95.2% and 90.4%, respectively. However, in addition to the increase in carbon content in this type of fallow land, we can also notice a decrease in carbon content in comparison with arable soils, which applies to sites no. 1, 3, and 17. In the case of site no. 1, carbon loss amounted at 52.3%, and the sample from this area was taken six months after the removal of bushy vegetation with single trees.

In general, it can be stated that in all researched sites, no case was observed in which the carbon content determined in the fallow fields, in relation to arable land, fell below the adopted critical value (1.7%). However, in most cases (except for fallow goldenrod—FG) an increase in the content of organic matter was noted. It should also be noted that in each considered case of the fallow type, values of standard deviations in the sample set (SD C%—presented in Table 1) are significantly smaller than the adopted critical value (1.7). A low variance in this case confirms that the observed relationships are not accidental.

Table 1. Standard deviations (SD) of chemical soil properties in various types of land use and various types of fallow land.

Land Use	SD (C%)	SD (pH)	SD (P_2O_5)	SD (K_2O)
arable land (AL)	0.25	0.80	7.45	5.70
grassland (GL)	0.71	0.86	4.54	3.06
grassland fallow (FGL)	0.07	0.75	3.19	5.79
goldenrod fallow (FG)	0.27	0.83	6.28	5.29
bushy fallow (FB)	0.41	1.11	5.92	7.23
wooded (afforested)—FW	0.26	0.86	4.89	16.41

3.2. Changes in pH

The pH indicator in the top layer of the investigated arable lands was mostly very acidic < 4.6 and acidic 4.6–5.5, only in sites 3, 7, 12, and 17 the soil pH was more favorable—slightly acid 5.6–6.5 (Figure S2). Comparing the average values of this parameter in individual types of use (Figure 8), we notice that the most favorable values were found in grasslands and grasslands fallow (pH = 5.3 and 5.1), besides with increasing pH, the percentage of organic carbon also increased.

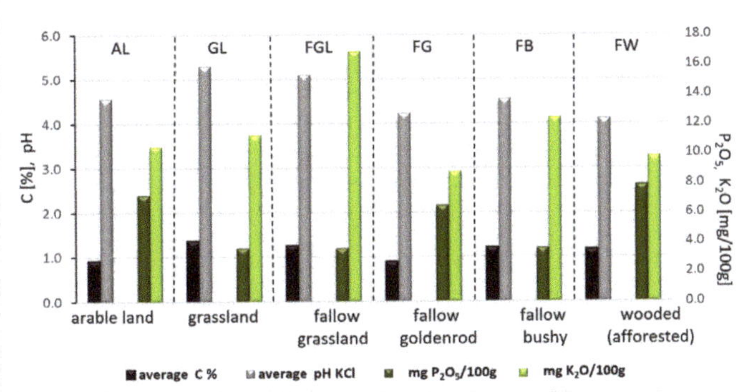

Figure 8. Average chemical soil properties in various types of land use and various types of fallow land.

In the case of the remaining types of fallow land, the mean pH is lower in relation to arable land, and the lowest for perennial fallow land covered with older trees, where the beginning of natural succession is determined for the years 1996, 2006. Despite the lower pH in these lands, the average organic carbon content is about 30% higher than in arable land. The amount of organic carbon was restored through the annual deposition of organic matter from tree leaves.

In general, it can be stated that in all researched sites, four cases were observed in which the pH value determined in the fallow fields, in relation to arable land, fell below the adopted critical value (2). In three cases it was the site no. 17, which may indicate the specificity of this site and its lack of representativeness in relation to other sampling sites. If we reject these outliers, we can conclude that the research hypothesis has been rejected for 1.8% of cases ($\alpha < 0.05$). Similar to the carbon content, in each considered case of the fallow type, values of standard deviations in the sample set (SD pH—presented in Table 1) are smaller than the adopted critical value (2.0). A low variance confirms that the observed relationships are not accidental.

3.3. Changes in K_2O and P_2O_5 Content

Potassium and phosphorus are, besides nitrogen, macronutrients of essential importance in plant nutrition. Phosphorus is one of the compounds building plant cells. Its presence in the soil and absorption by plants determines the absorption of other nutrients, mainly nitrogen. Phosphorus plays an important role in various plant life processes (regulates cell division, root development, flowering processes, seed setting, and maturation processes). The factor that strongly limits phosphorus absorption is the low pH of the soil, and the content of organic matter also plays an important role in this process [35,42,43]. On the other hand, potassium, unlike nitrogen and phosphorus, does not form basic organic substances of the plant. The natural content of potassium in soils depends on their mineralogical structure and granulometry, especially the content of clay minerals, the presence of which is reflected in the share of floated parts in the soil composition. In light soils, the natural potassium content is usually lower than in more compact soils, but with the increase in the share of colloidal parts, the absorption of potassium decreases, because it is strongly bound in the inter-packet spaces of clay minerals. The available forms of potassium are subject to losses due to plant uptake and leaching, especially in light soils. [35,42,43]. Average content of available forms of potassium and phosphorus in relation to the land use type are presented in Figure 8.

In general, it can be stated that in all researched sites, nine cases were observed for phosphorus and six cases for potassium, in which the content of these elements determined in the fallow fields, in relation to arable land, fell below the adopted critical value (10 mg per 100 g of soil). We can conclude that the research hypothesis has been rejected for 15% of cases ($\alpha \sim 0.15$) and 10% ($\alpha \sim 0.1$) respectively for each of these elements.

Contrary to carbon content and pH, potassium and phosphorus are the more labile elements in the soil. In this case, the greater variance in the sample set is expected and natural. Despite this feature, only in one case (Table 1: SD of K_2O for wooded) the values of the standard deviation exceeded the critical value (10), and in others they are close to half of its value. It should be noted, however, that in the case of potassium, the mean values of its content in fallow soils are most often higher than in arable soils (Figure 8).

3.4. Principal Component Analysis (PCA)

The analysis showed different relations between individual parameters depending on the land use type. In the case of arable agricultural land (AL), the positively correlated parameters are: C_{org} content and pH, which are not related to the other group of parameters: content of floatable parts < 0.002 mm and available form of potassium. In case of this group, a negative correlation with the content of available phosphorus can be observed (Figure 9A).

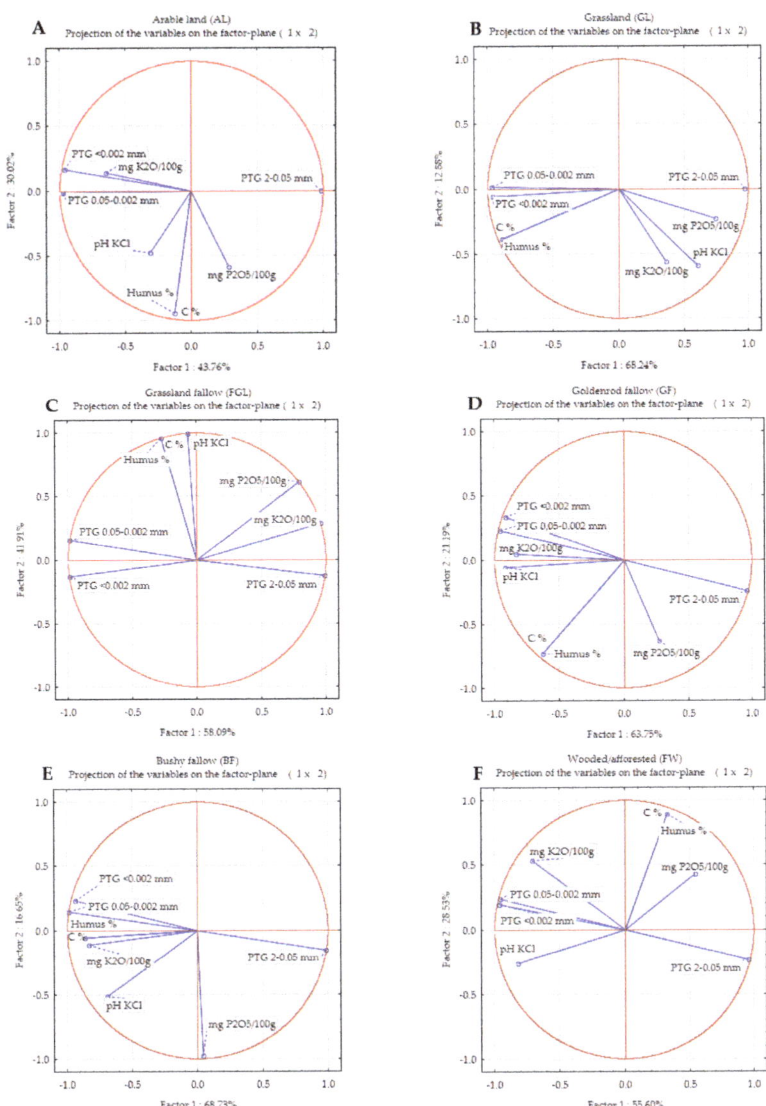

Figure 9. Variables charts showing relationships between different parameters depending on landuse types. Subfigures: (**A**)—arable land, (**B**)—grassland, (**C**)—grassland fallow, (**D**)—goldenrod fallow, (**E**)—bushy fallow, (**F**)—wooded/afforested.

In the case of grassland (GL), the variables form two groups. The first group of correlated parameters is pH, the content of potassium and, to a lesser extent, of phosphorus. The second group, on the other hand, is the C_{org} content and the share of fraction 0.05–0.002 and < 0.002 (Figure 9B). A similar correlation between the C_{org} content and the granulometric composition can be observed in bushy fallow land—FB (Figure 9E), where we can also see a relationship between whose parameters and the content of potassium.

Grassland fallow (FGL) shows correlation between the content of C_{org} and pH, which demonstrates in high values of those parameters compared to the other groups (Figure 9C), as it is in the case of grassland, where the content of potassium and phosphorus is negatively correlated with the content of floatable parts < 0.002 and fraction 0.05–0.002.

Moreover, in this group it can be observed that the first two principal components (PC1 and PC2) explain the largest percentage of the total variation, respectively (58.1% and 41.9%). Goldenrod fallow (FG) shows some similarities in relation of some factors with regard to arable land (AL), and afforested fallow areas (FW), applying mainly to potassium which depends on the content of fraction < 0.002 and fraction 0.05–0.002. The content of C_{org}, on the other hand, is not correlated with any other parameter (Figure 9D).

The last group, which represents advanced succession with trees (FW), shows correlation between the content of C_{org} and phosphorus, which at the same time are negatively correlated with pH (Figure 9F). In this case, the higher content of C_{org} the lower pH value (Figure S1).

4. Discussion

4.1. Impact of Fallowing on Soil

The obtained results generally show that fallowing on weak soils does not lead to such deterioration of chemical properties, which would make it difficult to restore plant production. In the case of fallowing agricultural land as grassland—without allowing secondary succession, even an improvement in the condition of the soil can be observed, i.e., no tendency to soil acidification, high humus and potassium content, which confirms the results published by Kazlauskaite-Jadzevice et al. [21]. Similar consistency with the results described by Foote and Grogan [22] was obtained for the assessment of the impact of mature succession (afforestation), but in this case, the effect of carbon sequestration was not so noticeable, which could have been influenced by the selection of sites for testing (poor and very poor soils).

Another important research result is the negative impact of goldenrod (*Solidago* L.) succession. This invasive species is now common in the agricultural landscape of the country and, as shown in the research of Orczewska [12], Sekutowski and his team [13], as well as this work—it has a negative effect on soil and environmental conditions. For this reason, one of the recommendations for maintaining the soil in good condition should be the requirement to mow fallow land with this type of succession.

The works carried out by the teams of Stolarski [15] and Matyka [10,18] prove that plantations of perennial energy crops can be established even on the weakest soils, classified as marginal soils. This work proves that setting aside agricultural land with poor soils is not an obstacle in restoring it to the production of this type of biomass. This applies in particular to the goldenrod succession sites—in such case, conversion may also contribute to a more sustainable use of agricultural production space.

4.2. Possibility of Returning Fallow Land to Agriculture

Michna et al. [1] wrote about the purposefulness of restoring fallow land to agricultural production. To the conclusion that: "Land of individual farmers may be transformed or used for other purposes only with the consent of the owners. Thus, there is the ownership barrier to transformation and changes in the forms of land use of all, including fallow, soils", we can now add the fact that over the three decades of intensified changes in agricultural production in Poland, which unfortunately show the constantly growing trend of abandoning agricultural land, not much intervention by the administration has taken place. Also, the expected changes after Poland's accession to the EU structures in 2004 did not result in any visible process of returning fallow land to production, but only slowed down the trend of land abandonment [5].

Moreover, the subsequent solutions in the field of bioeconomy, promoting an increase in the share of biomass in the energy mix, did not change the observed situation. Although the RED Directive [44] obligated EU countries to increase the sustainable use of biomass, it was not reflected in the relevant national regulations, which would provide tangible support to farmers and target recipients of the biomass produced [45]. This resulted in a lack of response from industry and energy, which could lead to the creation of a stable biomass market and allow the recovery of fallow land for the production of this raw material, or, alternatively, the resumption of food production.

The new concept of the Green Deal is another attempt to draw attention to the sustainable use of biomass in agriculture and bioeconomy [46]. These activities once again stimulate the interest of the energy and industry sectors in biomass of agricultural origin. This was manifested by the need to estimate the raw material potentials reported to the Ministry of Agriculture and Rural Development and the need to regulate the possibility of non-agricultural use of biomass along with the possibility of reusing waste from biomass processing as fertilizer for soil conservation. The solutions can directly contribute to the greater use of straw, hay, and manure in the bioeconomy, but they can also have a significant impact on the consolidation of fallow land and their recovery to biomass production.

In the BioMagic project, which financed this work, the presented results were used to estimate the theoretical, technical, and economic potentials of producing perennial industrial plants on fallow soils. In subsequent works carried out on this subject, a remote sensing tool will be developed for the remote sensing recognition of the degree of natural succession, and the density and type of biomass on the analyzed land plot.

5. Conclusions

Fallow land in Poland constitutes a significant percentage of agricultural land. Most of it has great potential to be restored for food or biomass production for bioeconomy purposes—such as providing raw materials for the chemical and pharmaceutical industries and energy. The observed effects of fallowing are not an agro-technical problem in this case. Based on the obtained results, it can be concluded that:

- Basic soil parameters, such as the content of organic matter and its acidity—do not deteriorate noticeably during the process of setting aside. In this case, the research hypothesis is accepted at a significance level $\alpha < 0.05$.
- Fallowing shows significant effect on the content of potassium and phosphorus in the soil. In this case, the research hypothesis is satisfied at a significance level $\alpha \sim 0.15$ (P_2O_5) and $\alpha \sim 0.1$ (K_2O). However, fertilization with these components is, in the case of agricultural use, a typical agrotechnical treatment that effectively increases soil fertility.
- Among the examined types of fallow, maintaining it with succession of goldenrod (*Solidago* L.) is the least favorable.
- Changes in the dependence and correlation of the analyzed soil properties are observed in the tested types of use and fallowing. Long-term fallow, which develops mature forms of succession (trees, shrubs), diversifies the carbon content and the acidity of the soil. In this case, a positive aspect is the increase in the content of organic matter in the soil, but at the same time increases also its acidity.
- For agricultural plots where the deterioration of soil conditions has been observed, these conditions can be quickly restored for the selected type of biomass production by applying agro-technical practices in accordance with the recommendations of the Code of Good Agricultural Practices [47].

Supplementary Materials: The following are available online at https://www.mdpi.com/2077-0472/11/2/148/s1, Figure S1: A ternary diagram of the soil texture triangle showing the USDA-based soil texture classifications [S1]., Figure S2: Primary results characterizing each tested site.

Author Contributions: Data preparation, M.K.; calculations and data analysis, M.K.; discussion of the results, M.K. and R.P.; writing of the paper, M.K. and R.P. All authors have read and agreed to the published version of the manuscript.

Funding: This paper is the result of a study carried out at the Institute of Soil Science and Plant Cultivation—State Research Institute, Department of Bioeconomy and Systems Analysis, and it was financed by the National (Polish) Centre for Research and Development (NCBiR), entitled "Environment, agriculture and forestry", project: BIOproducts from lignocellulosic biomass derived from MArginal land to fill the Gap In Current national bioeconomy, No. BIOSTRATEG3/344253/2/NCBR/2017.

Institutional Review Board Statement: Not applicable.

Informed Consent Statement: Not applicable.

Data Availability Statement: Not applicable.

Acknowledgments: We would like to thank Małgorzata Wydra for her help with editing the English manuscript of this paper.

Conflicts of Interest: The authors declare no conflict of interest. The funders had no role in the design of the study; in the collection, analyses, or interpretation of data; in the writing of the manuscript, or in the decision to publish the results.

References

1. Michna, W.; Rokicka, W. Low quality soils, their agricultural use and economic marginalisation. In *Rationalisation of the Use of Marginal Soils*; IERiGŻ: Warsaw, Poland, 1998; pp. 85–92. (In Polish)
2. Kolecka, N.; Kozak, J.; Kaim, D.; Dobosz, M.; Ostafin, K.; Ostapowicz, K.; Wężyk, P.; Prince, B. Understanding Farmland Abandonment in the Polish Carpathians. *Appl. Geogr.* **2017**, *88*, 62–72. [CrossRef]
3. Krysiak, S. Fallow lands in the landscapes of central Poland—Spatial, typological and ecological aspects. *Probl. Landsc. Ecol.* **2011**, *41*, 89–96. (In Polish)
4. Lasanta, T.; Nadal-Romero, E.; Arnáez, J. Managing abandoned farmland to control the impact of re-vegetation on the environment. The state of the art in Europe. *Environ. Sci. Policy* **2015**, *52*, 99–109. Available online: https://linkinghub.elsevier.com/retrieve/pii/S1462901115001094 (accessed on 12 December 2020). [CrossRef]
5. Pudełko, R.; Kozak, M.; Jędrejek, A.; Gałczyńska, M.; Pomianek, B. Regionalisation of unutilised agricultural area in Poland. *Polish J. Soil Sci.* **2018**, *51*, 119–132. [CrossRef]
6. Beilin, R.; Lindborg, R.; Stenseke, M.; Pereira, H.M.; Llausàs, A.; Slätmo, E.; Cerqueira, Y.; Navarro, L.; Rodrigues, P.; Reichelt, N.; et al. Analysing how drivers of agricultural land abandonment affect biodiversity and cultural landscapes using case studies from Scandinavia, Iberia and Oceania. *Land Use Policy* **2014**, *36*, 60–72. [CrossRef]
7. Ortyl, B.; Ćwik, A.; Kasprzyk, I. What Happens in a Carpathian Catchment after the Sudden Abandonment of Cultivation? *Catena* **2018**, *166*, 158–170. [CrossRef]
8. Janus, J.; Bozek, P. Land Abandonment in Poland after the Collapse of Socialism: Over a Quarter of a Century of Increasing Tree Cover on Agricultural Land. *Ecol. Eng.* **2019**, *138*, 106–117. [CrossRef]
9. Novara, A.; Gristina, L.; Sala, G.; Galati, A.; Crescimanno, M.; Cerdà, A.; Badalamenti, E.; La Mantia, T. Agricultural land abandonment in Mediterranean environment provides ecosystem services via soil carbon sequestration. *Sci. Total Environ.* **2017**, *576*, 420–429. [CrossRef]
10. Radzikowski, P.; Matyka, M.; Berbeć, A.K. Biodiversity of Weeds and Arthropods in Five Different Perennial Industrial Crops in Eastern Poland. *Agriculture* **2020**, *10*, 636. [CrossRef]
11. Van der Zanden, E.H.; Verburg, P.H.; Schulp, C.J.E.; Verkerk, P.J. Trade-Offs of European Agricultural Abandonment. *Land Use Policy* **2017**, *62*, 290–301. [CrossRef]
12. Orczewska, A. Who is more dangerous: The alien or the native? The negative impact of Solidago gigantea, Urtica dioica and Galium aparine on the herbaceous woodland species in recent post-agricultural alder woods. *Stud. Mater. CEPL* **2012**, *14*, 217–225. (In Polish)
13. Sekutowski, T.; Włodek, S.; Biskupski, A.; Sienkiewicz-Cholewa, U. Comparison of the content of seeds and plants of the goldenrod (*Solidago* sp.) in the fallow and adjacent field. *Zesz. Nauk. UP Wroclaw* **2012**, *584*, 99–111. (In Polish)
14. Bell, S.; Barriocanal, C.; Terrer, C.; Rosell-Melé, A. Management Opportunities for Soil Carbon Sequestration Following Agricultural Land Abandonment. *Environ. Sci. Policy* **2020**, *108*, 104–111. [CrossRef]
15. Stolarski, M.J.; Szczukowski, S.; Krzyzaniak, M.; Tworkowski, J. Energy Value of Yield and Biomass Quality in a 7-Year Rotation of Willow Cultivated on Marginal Soil. *Energies* **2020**, *13*, 2144. [CrossRef]
16. Stolarski, M.J.; Niksa, D.; Krzyżaniak, M.; Tworkowski, J.; Szczukowski, S. Willow Productivity from Small- and Large-Scale Experimental Plantations in Poland from 2000 to 2017. *Renew. Sustain. Energy Rev.* **2019**, *101*, 461–475. [CrossRef]
17. Berbeć, A.K.; Matyka, M. Planting Density Effects on Grow Rate, Biometric Parameters, and Biomass Calorific Value of Selected Trees Cultivated as SRC. *Agriculture* **2020**, *10*, 583. [CrossRef]
18. Matyka, M.; Radzikowski, P. Productivity and Biometric Characteristics of 11 Varieties of Willow Cultivated on Marginal Soil. *Agriculture* **2020**, *10*, 616. [CrossRef]
19. Von Cossel, M.; Lewandowski, I.; Elbersen, B.; Staritsky, I.; Van Eupen, M.; Iqbal, Y.; Mantel, S.; Scordia, D.; Testa, G.; Cosentino, S.L.; et al. Marginal Agricultural Land Low-Input Systems for Biomass Production. *Energies* **2019**, *12*, 3123. [CrossRef]
20. IPCC. Climate Change and Land Ice. In *IPCC Special Report on Climate Change, Desertification, Land Degradation, Sustainable Land Management, Food Security, and Greenhouse Gas Fluxes in Terrestrial Ecosystems*; Summary for Policymakers; IPPC: Rome, Italy, 2017; pp. 1–15.

21. Kazlauskaite-Jadzevice, A.; Tripolskaja, L.; Volungevicius, J.; Baksiene, E. Impact of Land Use Change on Organic Carbon Sequestration in Arenosol. *Agric. Food Sci.* **2019**, *28*, 9–17. [CrossRef]
22. Foote, R.L.; Grogan, P. Soil Carbon Accumulation during Temperate Forest Succession on Abandoned Low Productivity Agricultural Lands. *Ecosystems* **2010**, *13*, 795–812. [CrossRef]
23. Yang, Y.; Tilman, D.; Furey, G.; Lehman, C. Soil carbon sequestration accelerated by restoration of grassland biodiversity. *Nat. Commun.* **2019**, *10*, 718. [CrossRef]
24. Karhu, K.; Wall, A.; Vanhala, P.; Liski, J.; Esala, M.; Regina, K. Effects of afforestation and aeforestation on boreal soil carbon stocks-Comparison of measured C stocks with Yasso07 model results. *Geoderma* **2011**, *164*, 33–45. [CrossRef]
25. Lozano-García, B.; Francaviglia, R.; Renzi, G.; Doro, L.; Ledda, L.; Benítez, C.; González-Rosado, M.; Parras-Alcántara, L. Land Use Change Effects on Soil Organic Carbon Store. An Opportunity to Soils Regeneration in Mediterranean Areas: Implications in the 4p1000 Notion. *Ecol. Indic.* **2020**, 106831. [CrossRef]
26. Zhang, Z.; Han, X.; Yan, J.; Zou, W.; Wang, E.; Zou, W.; Wang, E.; Lu, X.; Chen, X. Keystone Microbiomes Revealed by 14 Years of Field Restoration of the Degraded Agricultural Soil Under Distinct Vegetation Scenarios. *Front. Microbiol.* **2020**, *11*, 1–13. [CrossRef]
27. Strączyńska, S.; Strączyński, S.; Wojciechowski, W. Effect of different land uses on selected properties of the soils. *Soil Sci. Ann.* **2010**, *61*, 227–232. (In Polish)
28. Tomaszewicz, T.; Chudecka, J. Influence of land use on properties of rusty soils: Fallow and agricultural use in Ginawa. *Agric. Eng.* **2005**, *4*, 311–320. (In Polish)
29. Włodek, S.; Sienkiewicz-Cholewa, U.; Biskupski, A.; Sekutowski, T.R. Comparison of the chosen environmental features of the arable land and fallow. *Ecol. Eng.* **2014**, *38*, 51–59. (In Polish)
30. Licznar, M.; Licznar, S.E.; Walenczak, K.; Brojanowska, M. Humic substances of fallowed soils against a background of their physicochemical properties. *Ann. Soil Sci.* **2009**, *60*, 69–76. (In Polish)
31. Biomagic, BIOproducts from Lignocellulosic Biomass Derived from MArginal Land to Fill the Gap in Current National Bioeconomy, 2017–2020. Available online: http://www.uwm.edu.pl/cbeo/projekty/biomagic (accessed on 30 December 2020).
32. GUS. Central Statistical Office. Natl. Agric. Census. 2010. Available online: https://bdl.stat.gov.pl/BDL/dane/podgrup/temat (accessed on 6 December 2020).
33. GUGiK: The Head Office of Geodesy and Cartography. Available online: https://www.geoportal.gov.pl/#(1) (accessed on 10 December 2020).
34. SIS Pulawy, Spatial Information System of the Puławy poviat. Available online: http://sip.pulawy.powiat.pl/ (accessed on 10 December 2020).
35. Mocek, A. *Soil Science*, 1st ed.; PWN: Warsaw, Poland, 2014; pp. 201–224. (In Polish)
36. Systematyka gleb Polski (Polish soil classification). *Rocz. Glebozn. Soil Sci. Ann.* **2011**, *62*, 1–193.
37. Smreczak, B.; Łachacz, A. Soil Types Specified in the Bonitation Classification and Their Analogues in the Sixth Edition of the Polish Soil Classification. *Soil Sci. Ann.* **2019**, *70*, 115–136. [CrossRef]
38. Stępień, M.; Bodecka, E.; Gozdowski, D.; Wijata, M.; Groszyk, J.; Studnicki, M.; Sobczyński, G.; Rozbicki, J.; Samborski, S. Compatibility of granulometric groups determined based on standard BN-78/9180-11 and granulometric groups according to PTG 2008 and USDA texture classes. *Soil Sci. Ann.* **2018**, *69*, 223–233. [CrossRef]
39. Ostrowska, A.; Gawliński, S.; Szczubiałka, Z. *Methods of Analysis and Evaluation of the Properties of Soils and Plants*; Catalog Institute of Environmental Protection: Warsaw, Poland, 1991; pp. 89–93.
40. Jambu, M. *Exploratory and Multivariate Data Analysis*; Academic Press: New York, NY, USA, 1991; pp. 125–160.
41. Sapek, B. Loss prevention and organic carbon sequestration in meadow soils. *Ecol. Eng.* **2009**, *21*, 48–61. (In Polish)
42. Mengel, K.; Kirkby, E.A. Principles of plant nutrition. *Ann. Bot.* **2004**, *93*, 479–480. [CrossRef]
43. Zawadzki, S. *Soil Sience*, 4th ed.; PWRiL: Warsaw, Poland, 1999; pp. 183–220.
44. Directive 2009/28/EC of the European Parliament and of the Council of 23 April 2009 on the promotion of the use of energy from renewable sources and amending and subsequently repealing Directives 2001/77/EC and 2003/30/EC. *Off. J. Eur. Union* **2009**, *5*, 2009.
45. The Act on Renewable Energy Sources, 20 February 2015. Dz. U. 2015 poz. 478 (Polish Legal Regulation). Available online: https://isap.sejm.gov.pl/isap.nsf/download.xsp/WDU20150000478/U/D20150478Lj.pdf (accessed on 20 December 2020).
46. European Commission: Commission Staff Working Document. Analysis of links between CAP Reform and Green Deal. Brussels SWD 2020. Available online: https://ec.europa.eu/info/sites/info/files/food-farming-fisheries/sustainability_and_natural_resources/documents/analysis-of-links-between-cap-and-green-deal_en.pdf (accessed on 20 December 2020).
47. Duer, I.; Fotyma, M.; Madej, A. *Code of Good Agricultural Practices*; MRiRW MŚ FAPA: Warsaw, Poland, 2004. (In Polish)

Article

Screening of Functional Compounds in Supercritical Carbon Dioxide Extracts from Perennial Herbaceous Crops

Mateusz Ostolski [1], Marek Adamczak [1,*], Bartosz Brzozowski [1] and Mariusz Jerzy Stolarski [2,3]

[1] Department of Process Engineering Equipment and Food Biotechnology, Faculty of Food Science, University of Warmia and Mazury in Olsztyn, Jan Heweliusz St. 1, 10-718 Olsztyn, Poland; mateusz.ostolski@uwm.edu.pl (M.O.); bartosz.brzozowski@uwm.edu.pl (B.B.)

[2] Department of Genetics, Plant Breeding and Bioresource Engineering, Faculty of Agriculture and Forestry, University of Warmia and Mazury in Olsztyn, Plac Łódzki 3, 10-724 Olsztyn, Poland; mariusz.stolarski@uwm.edu.pl

[3] Centre for Bioeconomy and Renewable Energies, University of Warmia and Mazury in Olsztyn, Plac Łódzki 3, 10-724 Olsztyn, Poland

* Correspondence: marek.adamczak@uwm.edu.pl

Citation: Ostolski, M.; Adamczak, M.; Brzozowski, B.; Stolarski, M.J. Screening of Functional Compounds in Supercritical Carbon Dioxide Extracts from Perennial Herbaceous Crops. *Agriculture* **2021**, *11*, 488. https://doi.org/10.3390/agriculture11060488

Academic Editor: Ângela Fernandes

Received: 22 April 2021
Accepted: 21 May 2021
Published: 25 May 2021

Publisher's Note: MDPI stays neutral with regard to jurisdictional claims in published maps and institutional affiliations.

Abstract: The bio-based economy concept requires using biomass not only for energy production but also for bioactive compound extraction, application or biotransformation. This study analyzed the possibility of obtaining bioactive compounds from biomass before its transformation into biofuel. This involved an analysis of the total content of polyphenols (TPC), flavonoids (TFC), and spectral analysis using Fourier transform infrared spectroscopy (QATR- FTIR) as well as analysis of the antioxidant activity of extracts from selected perennial herbaceous crops cultivated on marginal lands in Poland. The extracts were obtained by supercritical carbon dioxide extraction ($scCO_2$) or $scCO_2$ with water as a cosolvent ($scCO_2/H_2O$) from biomass of the following plants: *Helianthus salicifolius*, *Silphium perfoliatum*, *Helianthus tuberosus*, *Miscanthus × giganteus*, *Miscanthus sacchariflorus*, *Miscanthus sinensis* and *Spartina pectinata*. The biomass was harvested twice during the growing period (June and October) and once after the end of the growing period (February). For most of the analyzed extracts obtained from biomass at the growing stage using $scCO_2$ or $scCO_2/H_2O$, a higher TPC was noted than for samples of semi-wood or straw biomass obtained after the end of the growing period. Higher contents of polyphenolic compounds were recorded in extracts obtained using $scCO_2/H_2O$. A positive correlation between TPC and antioxidant activity was noted for the analyzed substrates. Flavonoid contents varied in the analyzed samples, and higher contents were generally obtained in $scCO_2$ extracts from biomass harvested at the beginning of the growing period. A high diversity of extract compositions was confirmed by spectral analysis. The presented data can be used at the initial stage of planning a biorefinery.

Keywords: polyphenols; supercritical CO_2 extraction; perennial industrial crops; antioxidant activity; silvergrass; Jerusalem artichoke; willowleaf sunflower; cup plant; prairie cordgrass

1. Introduction

One of the most attractive renewable sources of energy is plant biomass, particularly perennial plants that are not used in food production [1–3]. However, as the data show, the profitability of producing energy from plant biomass is not always satisfactory. Therefore, consideration should be given to both its multidirectional use and solutions appropriate for sustainable technologies [4]. Perennial plants, including perennial herbaceous crops, offer a wide variety of environmental benefits when compared with annual species. Due to economic reasons resulting from the single-purpose use of perennial plants as solid biofuel, they are currently mainly cultivated on an experimental scale (excluding *Miscanthus × giganteus*). However, the possibility of multidirectional use of these species as substrates for biogas plants, second-generation biofuels or bioproducts (with regard to the content of specific compounds) has been increasingly more frequently indicated.

Agriculture **2021**, *11*, 488. https://doi.org/10.3390/agriculture11060488   https://www.mdpi.com/journal/agriculture

The great diversity of polyphenolic compound structures determines their varied biological activity. Polyphenols are widespread compounds that are safe, cheap and effective for many applications. The contribution of polyphenols to the prevention of cardiovascular diseases, neoplasms, prostate diseases and osteoporosis has been confirmed, and their role in the prevention of diabetes mellitus and neurodegenerative diseases has been suggested [5–7]. Polyphenols can also be used in the food industry as natural antioxidants that delay food spoilage [8]. There has been great interest in the use of plant extracts or selected polyphenols in the synthesis of metal nanoparticles [9]. It is also possible that the poor stability, solubility and bioavailability of polyphenols could be solved by the application of encapsulation, e.g., synthetic polyphenol-loaded nanoparticles [10].

Antioxidants can be synthesized in vivo or taken as dietary antioxidants in the form of mainly phenols and polyphenols from plants, i.e., from the stems, roots, bark, leaves, fruits and seeds [11]. It is estimated that two-thirds of the world's plant species display medicinal importance, and almost all of these have excellent antioxidant potential. The antioxidant activity of polyphenols results from a direct reaction with ROS or the stimulation of natural processes that contribute to an improvement in the cellular resistance to oxidative stress [12].

These compounds reduce the virulence of microorganisms, inhibit biofilm formation [13], inhibit quorum sensing [14] and neutralize bacterial toxins [15].

There is no information on extracts obtained using supercritical carbon dioxide (scCO$_2$) from perennial industrial crops (PIC), particularly from herbaceous crops. These plants can be used for the production of food and pharmaceuticals and as renewable energy sources. However, no analyses are available on the effects of the growing phase on the efficiency of polyphenolic compound extraction using scCO$_2$. The cultivation and application of new plants could influence and protect the natural environment [16]. For example, *H. tuberosus* exhibits varied pharmacological properties. It displays, inter alia, aperient, cholagogue, diuretic, spermatogenic, stomachic and tonic effects and has been used in folk medicine for the treatment of diabetes and rheumatism [17]. Moreover, extracts of *H. tuberosus* L. exhibit antibacterial, antifungal and antineoplastic activities. The leaves and tubers of Jerusalem artichoke are rich sources of phenolic and flavonoid compounds. The leaves contain much higher concentrations of those substances compared to the tubers [18]. Research on the biological activity of *S. perfoliatum* extracts has confirmed their antifungal as well as wound healing accelerating properties. Their antibacterial activity against Gram-positive (*Enterococcus faecalis* and *Staphylococcus aureus*) and Gram-negative (*Escherichia coli* and *Pseudomonas aeruginosa*) bacteria has also been noted [19], along with the positive effect of *S. perfoliatum* extracts on the stability of sunflower oil fatty acid composition [20].

The cultivation conditions and the harvesting period affect the chemical composition of plant biomass, including polyphenolic compounds [21]. The large number of components of plant extracts requires the selection of appropriate extraction techniques, which play a crucial role in the qualitative and quantitative characterization of the product. To the authors' knowledge, there are no studies on the general characteristics of bioactive polyphenolic compounds obtained by green extraction technique using scCO$_2$ from *S. pectinata* and *H. salicifolius*, and there is limited information on polyphenols from other analyzed biomasses. The presented data can be used at the initial stage of planning a biorefinery. Lignocellulosic biomass was delignified during extraction using organic solvents and scCO$_2$ at temperatures above 150 °C. A higher degree of lignin removal was demonstrated in extractions using pure scCO$_2$ than with scCO$_2$/water [22].

The current study aimed to screen the polyphenolic compound content and antioxidant activity in extracts obtained using supercritical carbon dioxide from the biomass of perennial herbaceous crops collected in three different phases of plant growth. Attenuated total reflection Fourier transform infrared (QATR-FTIR) spectral analyses of the selected extracts were performed to obtain data on the general characteristics of functional groups of bioactive compounds.

2. Materials and Methods

2.1. Chemicals

The following chemical reagents were used in the experiments: dimethyl sulfoxide (DMSO, $\geq$99%, Stanlab, Lublin, Poland), methanol ($\geq$99%, Stanlab), Folin–Ciocalteu reagent (Aktyn, Suchy Las, Poland), sodium carbonate ($\geq$99%, Stanlab), gallic acid ($\geq$98%, Sigma Aldrich, Poznan, Poland), 2,2-diphenyl-1-picrylhydrazyl, (DPPH, Sigma Aldrich), quercetin ($\geq$95% Sigma Aldrich) and 6-hydroxy-2,5,7,8-tetramethylchromane-2-carboxylic acid (Trolox, $\geq$98%, Sigma Aldrich).

2.2. Harvesting of the Plant Material

The biomass of the following species of herbaceous (semi-woody) crops was extracted: *Helianthus salicifolius* A. Dietr, *Silphium perfoliatum* L., *Helianthus tuberosus* L., *Miscanthus* × *giganteus* J.M. Greef & M. Deuter, *Miscanthus sacchariflorus* (Maxim.) Hack, *Miscanthus sinensis* (Thunb.) Andersson and *Spartina pectinata* Bosc ex Link. Perennial crops were obtained from the experimental fields of the University of Warmia and Mazury in Olsztyn, located in north-eastern Poland. The details of the sample preparation and extraction were presented by Stolarski et al. [22] and recently by Malm et al. [23] and Ostolski et al. [24]. Entire plants were harvested three times in one year (2018) within one field experiment, where plants regrew following consecutive harvests. One-year-old plants that grew throughout the entire vegetation period of 2017 were harvested in the third decade of February 2018 (biomass C). The collected biomass was in the form of straw or semi-wood featuring a moisture content of approximately 20%. In the third decade of June 2018 (biomass A), a three-month-old biomass in the form of young green shoots was collected from the previously (February 2018) harvested field. Finally, in the third decade of October 2018 (biomass B), crop regrowth in the form of young green shoots was gathered from the plots harvested in June 2018. The moisture content of the biomass obtained in June and October ranged from 60% to 70%. Each biomass was dried in a convection dryer (Memmert, Schwabach, Germany) at 40 °C for 7 consecutive days to achieve a moisture content below 10%, followed by grinding in a hammer mill using a 1 mm sieve.

2.3. Extraction Conditions

The experiments were carried out in a pilot plant extractor with two high-pressure extraction vessels, each with a capacity of approximately 40 dm^3 (NATEX, Ternitz, Austria). Biomass batches of 5 kg were prepared, and supercritical extraction was performed under the following conditions: 40 °C and 33 MPa using scCO$_2$ or scCO$_2$ and 40% (*w/w* based on biomass weight) water (scCO$_2$/H$_2$O) [23,24]. The water from the collected extracts was evaporated using a vacuum evaporator at 50 °C.

To prepare the working solution, 50 mg of the extracts was mixed in 1 cm^3 of DMSO (24 h, 1400 rpm, 22 °C). A 9 cm^3, 5% (*w/v*) aqueous solution of DMSO was added to the obtained solution, which yielded the final extract concentration of 10 mg/cm^3. The samples prepared in this manner were mixed for 1 h to dissolve the extract. Afterwards, the samples were filtered through a cellulose filter and used for analyses or stored at 4 °C until further analyses.

2.4. Determination of Total Polyphenol Content

The total polyphenol content (TPC) was measured using a modified method described by Singleton and Rossi [25]. Briefly, 0.1 cm^3 of extract, 0.5 cm^3 of Folin–Ciocalteu reagent and 0.4 cm^3 of 7.5% (*w/v*) aqueous sodium carbonate solution were added to 5 cm^3 Eppendorf tubes. Samples were degassed by vigorous mixing (IKA MS3 basic). After 30 min incubation in the dark, counted from the moment sodium carbonate solution was added, absorbance was measured at λ = 756 nm (Beckman DU 650 spectrophotometer, Fullerton, CA, USA). The results were expressed as gallic acid equivalent (mg GAE/g d.m. of extract).

2.5. Determination of Total Flavonoid Content

The total flavonoid content (TFC) was measured using the method described by Lamaison and Carnat [26]. Briefly, 0.4 cm^3 of extract and 0.8 cm^3 of 2% (*w/v*) AlCl$_3$ × 6H$_2$O methanol solution were added to 2 cm^3 Eppendorf tubes. After 10 min of incubation in the dark, counted from the moment the complexing reagent (AlCl$_3$ × 6H$_2$O) was added, absorbance was measured at λ = 430 nm (Beckman DU 650 spectrophotometer). The results were determined as quercetin equivalent (QE) and expressed in mg QE/g d.m. of extract.

2.6. Antioxidant Capacity Assay (DPPH Test)

The radical scavenging activity was determined using a modified method described by Blois [27]. To 2 cm^3 Eppendorf tubes, 1 cm^3 of extract solution and 1 cm^3 of DPPH radical solution (0.05 mM) were added. After 30 min incubation in the dark, counted from the moment DPPH was added, absorbance at λ = 517 nm was measured. The results were expressed as Trolox equivalent antioxidant capacity (mg Trolox/g d.m. of extract).

2.7. Fourier Transform Infrared Spectroscopy (QATR-FTIR) Analysis

Spectroscopic measurements were performed by an attenuated total reflectance (ATR) method. The transmittance of plant extract samples (5–6 mg) was measured using a FTIR (IRSpirit-T, Schimadzu, Duisburg, Germany) spectrometer equipped with a QATR-S accessory with a diamond crystal and a DLATGS detector with a germanium-coated KBr beam splitter. Each sample was scanned 25 times within the wavelength range of 4000.0 to 400.0 cm^{-1} at the resolution of 4.0 cm^{-1} using the Happ–Genzel function. Sample spectra obtained with the application of LabSolutions IR software (ver. 2.23, Shimadzu, Duisburg, Germany) were corrected, smoothed with the seven-point Savitzky–Golay algorithm and normalized.

2.8. Statistical Analysis

All analyses were carried out in triplicate. The results were expressed as mean values ± standard deviation (SD). The same letter indicated results that were not statistically different within the parameter under analysis, Bonferroni ANOVA test, $p < 0.05$. A multidimensional principal component analysis (PCA) with varimax rotation was also applied. All analyses were performed using the STATISTICA, ver. 13.3 (TIBCO Software Inc., Palo Alto, CA, USA).

3. Results and Discussion

3.1. Total Polyphenols and Flavonoids

Phenolic compounds are essential for plant functions because they are involved in oxidative stress reactions, defensive systems, growth and development [28]. The problem with phenolic compound definition, classification and chemical characterization was presented in a review by Quideau et al. [29]. It has already been shown that raffinate obtained after extraction of polyphenols using scCO$_2$ could be used as a stable and valuable substrate in a biorefinery or for energy production [22]. The scCO$_2$ extraction removes significant amounts of lipids, terpenes, steroids and resin acids that easily undergo autoxidation, which consequently improves both the quality of biofuel and pyrolysis [30–32]. The results of the current study indicate varied polyphenol content in extracts obtained during a one-year-long experiment depending on both the harvest period and the extraction conditions (Table 1). The aim of crop harvesting at different times was to compare the obtained extract activities in relation to the biomass harvest time and to demonstrate the plant's ability to regrow after harvest at different times of the year (in different growth phases). The study presented in this paper focused only on assessing extract activity. To the best of the authors' knowledge, there is limited information on polyphenols from other analyzed biomasses. The highest polyphenol content ($p < 0.05$) was obtained in the scCO$_2$/H$_2$O extract from *S. pectinata* (168.18 ± 6.26 mg GAE/g d.m.) harvested in June, i.e., at the beginning of the vegetation period, while the lowest content was noted for the extract of *M. giganteus* straw harvested in February, obtained by extraction using only $scCO_2$ (1.24 ± 0.06 mg GAE/g d.m. of the extract). For most of the analyzed extract samples ($scCO_2$ or $scCO_2/H_2O$) obtained from biomass harvested during the growing period, a statistically significantly higher polyphenol content was noted than those for the semi-wood or straw samples collected in February. Only the extracts from *S. perfoliatum* semi-wood biomass ($scCO_2$) and *H. tuberosus*, *M.* × *giganteus* straw ($scCO_2/H_2O$) were characterized by higher polyphenol content than the extracts from biomass harvested during the growing period ($p = 0.05$) (Tables 1 and S1). In another study, the highest polyphenol content in an ethanol extract of Jerusalem artichoke leaves was 4.5–5.7 mg GAE/g d.m. [33], i.e., approximately 2–6-fold lower than the polyphenol content in a supercritical extract of biomass of this species in the current study (Table 1). The contents of polyphenols in $scCO_2/H_2O$ extracts from *H. salicifolius* and *H. tuberosus* obtained by Malm et al. [23] under the same conditions were similar to those reported in the present study, i.e., the values were lower for the former (13.75 mg GAE/g of extract) and higher for the latter (33.06 mg GAE/g of extract) plant.

Table 1. Influence of extraction conditions ($scCO_2$ or $scCO_2/H_2O$) and harvest time ((A) June, (B) October and (C) February) on total polyphenol content (TPC)* in extracts from herbaceous species.

| Plant Material | TPC (mg GAE/g d.m.) | | | | | |
| | $scCO_2$ | | | $scCO_2/H_2O$ | | |
	A	B	C	A	B	C
Helianthus salicifolius	20.22 ± 0.20 [d]	25.02 ± 0.31 [a]	11.92 ± 0.16 [i]	27.51 ± 0.11 [h]	36.08 ± 0.15 [g]	15.2 ± 0.62 [j]
Silphium perfoliatum	11.33 ± 0.07 [j]	13.99 ± 0.24 [h]	17.37 ± 0.54 [f]	58.37 ± 1.11 [e]	21.59 ± 0.15 [i]	18.25 ± 0.32 [i,j]
Helianthus tuberosus	16.50 ± 0.06 [g]	16.01 ± 0.23 [g]	8.47 ± 0.11 [k]	27.57 ± 0.38 [h]	18.34 ± 0.23 [i,j]	26.26 ± 0.36 [h]
Miscanthus × *giganteus*	22.17 ± 0.18 [c]	12.2 ± 0.19 [i]	1.24 ± 0.06 [o]	73.91 ± 0.96 [c]	51.33 ± 0.69 [f]	90.75 ± 0.71 [b]
Miscanthus sinensis	19.01 ± 0.20 [e]	17.11 ± 0.28 [f]	8.84 ± 0.35 [k]	90.98 ± 2.05 [b]	65.36 ± 1.26 [d]	26.61 ± 0.60 [h]
Miscanthus sacchariflorus	19.55 ± 0.17 [d,e]	5.95 ± 0.23 [l]	4.86 ± 0.09 [m]	64.64 ± 0.97 [d]	21.91 ± 0.59 [i]	14.43 ± 0.51 [j]
Spartina pectinata	23.93 ± 0.14 [b]	23.46 ± 0.22 [b]	3.82 ± 0.16 [n]	168.18 ± 6.26 [a]	92.47 ± 1.46 [b]	6.17 ± 0.07 [k]

* Mean values of three different determinations followed by standard deviation are presented. The different letters in the same group indicate statistically significant differences between the samples ($p < 0.05$).

A low content of polyphenols was identified in a water extract of the aboveground parts of *M. sinensis* at the flowering phase ($1.1795 \pm 0.1608\%$ (% d.m. equivalent of pyrogallol)) [34]. A high content of polyphenolic compounds is not a prerequisite for the development of a biorefinery concept. Substrates for biotechnological processes with a low polyphenol content or effective methods of polyphenol extraction preceding the bioprocess are often sought. Based on the analysis of cinnamates in the leaves, stems and flowers of *M. sinensis* and *M. sacchariflorus*, it is suggested that an assay could be useful for selecting plants with low contents of antimicrobial phenols that may be a good feedstock for fermentation [35].

Higher contents of polyphenolic compounds were noted in extracts obtained using $scCO_2/H_2O$ (Figure 1a, marked area, unfilled triangles). The application of this extraction type resulted in an increase in polyphenol content, particularly in the extracts from biomass harvested during the growing period, i.e., in June and October (Table 1). Supercritical fluid extraction is an environment-friendly technology used for the extraction of bioactive compounds from natural products and may provide an alternative to conventional extraction methods. Supercritical fluids exhibit unique characteristics, good solvent properties, a high diffusion coefficient, low viscosity and negligible value of the surface tension coefficient. Modifications of the pressure or temperature produce considerable changes in the fluid properties, including selectivity. These properties enable rapid mass transfer and an increased capability to penetrate the sample matrix, which yields rapid and efficient extraction [36]. The advantage of $scCO_2$ is its ability to produce extracts free of solvent residue. Because CO_2 is a gas at ambient temperature and pressure, it is capable of minimizing the thermal degradation of bioactive compounds. Supercritical CO_2 is safe for human health and the environment and is reusable, inert, nontoxic, nonflammable and noncorrosive [37]. However, CO_2 is a weak solvent for the extraction of polar compounds, including polyphenolic compounds. A frequently proposed solution to this problem is the use of a cosolvent, a modifier that changes the solvent properties. For the extraction of polar compounds such as alkaloids, glycosides, saccharides and cyclitols, water and various organic solvents that are soluble in supercritical fluid have been used as a cosolvent, e.g., methanol, ethanol, acetone and acetonitrile. This modification usually requires the adaptation of extraction conditions to the new solvent composition and its critical parameters. The addition of water as a cosolvent modifies the polarity of the mixture and increases the efficiency of extraction of polar polyphenolic compounds. For the first time, $scCO_2$ was used for the extraction of polyphenols from grape marc, but the effective and selective extraction of proanthocyanidins required a 15% addition of water to the $scCO_2$ [38]. The antioxidant and biocidal properties of extracts obtained from *H. salicifolius* and *H. tuberosus* with $scCO_2/H_2O$ were presented in [23]. Generally, higher contents of polyphenols were obtained in willow and poplar biomass samples after extraction with $scCO_2$ and water compared to extraction with $scCO_2$ alone [24].

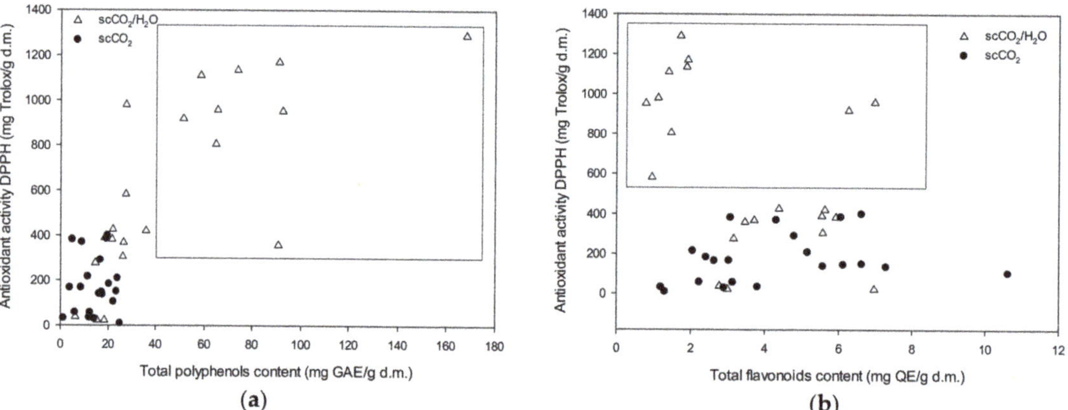

Figure 1. Interaction between (**a**) total polyphenol content (TPC), (**b**) total flavonoid content (TFC) and antioxidant activity determined in $scCO_2/H_2O$ (Δ) and $scCO_2$ (•) extracts from perennial herbaceous crops. DPPH - 2,2-diphenyl-1-picrylhydrazyl.

The highest TFC was obtained in the $scCO_2$ extract from *M. × giganteus* harvested at the beginning of the growing period (June) (10.62 ± 0.09 mg QE/g d.m.) ($p < 0.05$) (Table 2). The application of $scCO_2$ extraction enabled an increase in flavonoid content compared to extraction using $scCO_2/H_2O$ when biomass from the beginning of the growing period (June) was used as the substrate. However, when biomass from October and February was used as the extraction substrate, no correlation was found between the solvent used and the selective flavonoid extraction (Table 2 and Figure 1b). Moreover, it was demonstrated that flavonoids from pomelo peel were preferentially separated by $scCO_2$ at 80 °C and a pressure of 39 MPa with 85% aqueous ethanol as a modifier [39]. The lowest statistically significant ($p < 0.05$) flavonoid content was obtained following extraction using $scCO_2/H_2O$ from all of the plants harvested at the beginning of the growing period except for *S. pectinata* (Tables 2 and S2).

Table 2. Influence of extraction conditions ($scCO_2$ or $scCO_2/H_2O$) and harvest time ((A) June, (B) October and (C) February) on total flavonoid content (TFC)* in extracts from herbaceous species.

Plant Material	TFC (mg QE/g d.m.)					
	$scCO_2$			$scCO_2/H_2O$		
	A	B	C	A	B	C
Helianthus salicifolius	2.41 ± 0.06 [m]	1.29 ± 0.07 [o]	3.80 ± 0.04 [i]	0.95 ± 0.02 [l,m]	5.62 ± 0.01 [d]	2.99 ± 0.09 [g,h]
Silphium perfoliatum	2.05 ± 0.05 [n]	2.90 ± 0.03 [k]	7.28 ± 0.09 [b]	1.39 ± 0.02 [k]	5.92 ± 0.05 [c]	6.95 ± 0.11 [a]
Helianthus tuberosus	4.79 ± 0.02 [g]	5.57 ± 0.05 [e]	3.03 ± 0.11 [j,k]	1.11 ± 0.05 [l]	5.54 ± 0.04 [d]	5.56 ± 0.64 [d]
Miscanthus × *giganteus*	10.62 ± 0.09 [a]	3.14 ± 0.03 [j]	1.19 ± 0.03 [g]	1.87 ± 0.03 [i]	6.25 ± 0.11 [b]	3.46 ± 0.06 [f]
Miscanthus sinensis	6.05 ± 0.05 [d]	6.12 ± 0.03 [d]	4.31 ± 0.01 [h]	1.91 ± 0.03 [i]	6.97 ± 0.11 [a]	3.71 ± 0.12 [f]
Miscanthus sacchariflorus	6.61 ± 0.07 [c]	2.23 ± 0.02 [m,n]	3.07 ± 0.03 [j,k]	1.46 ± 0.02 [j,k]	4.38 ± 0.06 [e]	3.16 ± 0.07 [g]
Spartina pectinata	5.15 ± 0.11 [f]	6.62 ± 0.05 [c]	2.64 ± 0.06 [l]	1.71 ± 0.03 [i,j]	0.78 ± 0.01 [m]	2.78 ± 0.08 [h]

* Mean values of three different determinations followed by standard deviation are presented. The different letters in the same group indicate statistically significant differences between the samples ($p < 0.05$).

The chemical composition of the studied plants depend on the environmental conditions, growing period, etc. [40]. Changes in the chemical composition affect the biological activity of a plant and the possibility of obtaining active compounds. The effect of the season on the chemical composition, and thus bioactivity, can be attributed to climate change, i.e., the temperature, the soil moisture content, precipitation and the plant growth period [41]. A significant seasonal variation of phenolic compounds was observed in the leaves of *Cyclocarya paliurus* (Batal.) Iljinskaja, and the highest phenolic content was recorded in May, July and November [42]. A seasonal variation was observed in the synthesis of secondary metabolites in *Phillyrea angustifolia*, and the highest oleuropein content was confirmed during the winter period [43]. Similarly, Pacifico et al. [44] demonstrated that an extract of *Calamintha nepeta* harvested during the winter period was characterized by a higher acacetin content. The season can affect the chemical composition of plants in different ways, and the possible changes in the chemical composition can occur over several years. A quantitative analysis of phenolic compounds in extracts of *Vaccinium myrtillus* demonstrated relatively small seasonal variations in the composition of phenolic compounds or their absence, while significant changes in the composition were observed over two successive years [45].

The availability of water can also impact the synthesis of polyphenolic compounds by the plants. The relationships between the synthesis of polyphenolic compounds by the plants and the hydration are unclear and are determined by their parameters. High temperatures (30–40 °C) can prevent flavonoid synthesis by inhibiting gene expression and enzyme activity [46], while low temperatures can induce flavonoid biosynthesis (although in the absence of light, the synthesis of these compounds is inhibited) [47].

3.2. Antioxidant Activity

The extracts obtained using a mixture of $scCO_2/H_2O$, in which an increased polyphenolic compound content was also noted, exhibited the highest antioxidant activity (Tables 3 and S3, Figure 1a). The highest antioxidant activity ($p < 0.05$) was noted for the extract from *S. pectinata* (1289.5 ± 34.35 mg Trolox/g d.m.) harvested at the beginning of the growing period, and the lowest was in the extract from *S. perfoliatum* semi-wood biomass (22.85 ± 0.62 mg Trolox/g d.m.). Regardless of the biomass species or the extraction technique applied, the extracts obtained from plants harvested in June (the beginning of the growing period) were characterized by the highest antioxidant activity (Table 3).

Table 3. Influence of extraction conditions ($scCO_2$ or $scCO_2/H_2O$) and harvest time ((A) June, (B) October and (C) February) on antioxidant activity (DPPH assay)* in extracts from herbaceous species.

| Plant Material | Antioxidant Activity Using the DPPH Method (mg Trolox/g d.m.) | | | | | |
| | $scCO_2$ | | | $scCO_2/H_2O$ | | |
	A	B	C	A	B	C
Helianthus salicifolius	183.04 ± 7.77 [f]	9.04 ± 1.95 [n]	34.02 ± 0.21 [m]	582.50 ± 34.76 [h]	420.78 ± 2.94 [i]	23.01 ± 0.13 [n]
Silphium perfoliatum	215.55 ± 4.54 [e]	28.62 ± 1.01 [m]	136.47 ± 0.59 [j]	1110.64 ± 47.27 [c]	383.25 ± 3.17 [j,k]	22.85 ± 0.62 [n]
Helianthus tuberosus	290.43 ± 2.25 [d]	139.48 ± 4.82 [i,j]	167.18 ± 2.25 [g]	980.24 ± 25.15 [d]	389.94 ± 0.51 [j]	305.99 ± 0.67 [l]
Miscanthus × giganteus	104.83 ± 2.97 [k]	56.31 ± 1.52 [i]	31.56 ± 0.24 [m]	1134.86 ± 33.73 [c]	919.02 ± 1.48 [f]	359.16 ± 20.86 [k]
Miscanthus sinensis	384.94 ± 5.52 [b]	146.69 ± 1.20 [h,i]	369.83 ± 3.76 [c]	1170.19 ± 17.96 [b]	958.81 ± 1.87 [d,e]	368.78 ± 5.65 [j,k]
Miscanthus sacchariflorus	399.41 ± 2.67 [a]	57.20 ± 4.56 [i]	382.05 ± 2.67 [a]	805.72 ± 31.51 [g]	426.86 ± 1.94 [i]	276.91 ± 2.10 [m]
Spartina pectinata	208.44 ± 5.21 [e]	150.07 ± 2.74 [h]	166.43 ± 0.85 [g]	1289.50 ± 34.35 [a]	953.15 ± 1.66 [e]	37.85 ± 0.16 [n]

* Mean values of three different determinations followed by standard deviation are presented. The different letters in the same group indicate statistically significant differences between the samples ($p < 0.05$). DPPH - 2,2-diphenyl-1-picrylhydrazyl.

The accumulation, quantity and activity of phenolic compounds in the plant raw material are determined by the previously mentioned factors, primarily the substrate, plant species, the method of biomass preparation and the growth conditions. Antioxidant activity is primarily determined by the content of bioactive phenolic compounds. The current study found a positive correlation between the polyphenolic compound content and the antioxidant activity ($r = 0.80$), regardless of the type of raw material, the harvest date or the extraction conditions. In other studies, the antioxidant activity was closely related to the TPC and TFC in *Nyctanthes arbor-tristis* (Harsingar) leaf extracts used in Indian traditional medicine [48]. Zhang et al. [49] analyzed differences in the contents of polyphenols and flavonoids as well as the antioxidant activity in pomegranate leaves. In the period between April and September, the polyphenol and flavonoid contents decreased and then gradually increased. The antioxidant activity of extracts was also significantly correlated with the polyphenolic compound content ($r = 0.80$).

In the extracts obtained using $scCO_2/H_2O$, a high polyphenol content of over 40 mg GAE/g d.m. was obtained from *S. perfoliatum* and *M. sacchariflorus* harvested in June, from *M. sinensis* and *S. pectinata* harvested during the growing period (June and October) and from *M. × giganteus* harvested at each of the analyzed dates (Figure 1a). No statistical relationship was found between TFC and antioxidant activity (Figure 1b). The antioxidant activity over 600 mg Trolox/g d.m. was obtained for extracts with a flavonoid content of approximately 2, or over 6 mg QE/g d.m. (Figure 1b, marked area, unfilled triangles). The $scCO_2$ extract with the highest flavonoid content (from *M. × giganteus*) from biomass obtained at the beginning of the growing period (June) was characterized by a low antioxidant activity of 104.83 Trolox/g d.m. Therefore, it can be initially concluded that it is the qualitative composition rather than the content that probably determines the antioxidant activity.

The principal components PC1 and PC2 for extracts obtained using $scCO_2$ (Figure 2a) or $scCO_2/H_2O$ (Figure 2b) explained 86.5% (53.8% and 32.7%) and 91.5% (68.3% and 23.2%) of data sets, respectively. However, a great dispersion of the results based on the previous analyses was noticeable. PC1 represents TPC and the PC2 represents antioxidant activity (DPPH). The TPC and TFC values in the extracts obtained using $scCO_2$ were positively correlated with the antioxidant activity (Figure 2a). A negative correlation for TFC and a positive correlation for TPC were demonstrated with the antioxidant activity in the extracts obtained using $scCO_2$ and water.

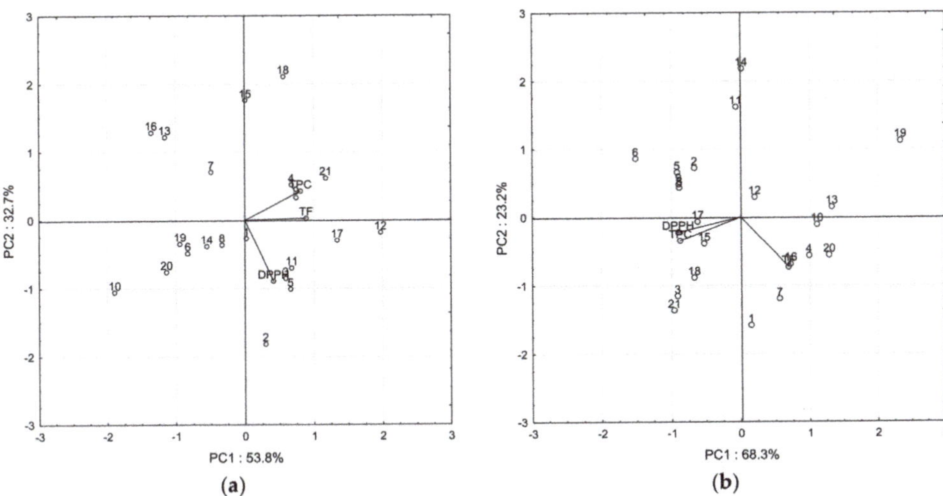

Figure 2. Principal component analysis of total polyphenol content (TPC), total flavonoid content (TFC) and antioxidant activity (DPPH) of extracts obtained with (**a**) scCO$_2$ or (**b**) scCO$_2$/H$_2$O. Points 1–21 present samples obtained from *Helianthus salicifolius, Silphium perfoliatum, Helianthus tuberosus, Miscanthus × giganteus, Miscanthus sacchariflorus, Miscanthus sinensis* and *Spartina pectinate* harvested in June (A), October (B) and February (C).

There are many parameters that influence this assay of antioxidant capacity. Lignin is a polymer constructed of aromatic subunits that account for its radical scavenging activity and was not analyzed in the studied samples [50]. A high content of lignin was determined in an extract from *M. × giganteus* and *Panicum virgatum* L. obtained by different methods in [51]. The presence of residual carbohydrates as well as aliphatic hydroxyls, along with lignin structural monomers, can cause an overestimation of antioxidant capacity.

3.3. FTIR Analysis

Selected extracts (scCO$_2$ and scCO$_2$/H$_2$O) from the plants harvested in June and characterized by a high content of phenolic compounds were used for the identification of functional groups present in extract components. Peak characteristic of functional group bonds in FTIR spectra were identified (Figures S1–S14 and Table S4), classified into nine types groups and assigned a functional group based on data presented in the literature (Table 4) [52–57]. An analysis of FTIR spectra indicated absorption within the wavelength region corresponding to vibrations of bonds present in the principal components of lignocellulose biomass, i.e., saccharides, lignins, proteins and lipids. Absorption within the wavelength region from 3421 to 3335 cm^{-1}, characteristic of O–H and N–H stretching bonds occurring in saccharides, proteins and polyphenols (Table 4) [55–57], was noted in scCO$_2$ and scCO$_2$/H$_2$O extracts from *H. salicifolius, H. tuberosus* and *M. sinensis*, scCO$_2$ extracts from *M. × giganteus* and scCO$_2$/H$_2$O extracts from *S. perfoliatum* and *S. pectinata*. All of the examined extracts displayed absorption within the wavelength region from 3070 to 2849 cm^{-1}, corresponding to vibrations of the C–H stretching bonds present in the functional groups of CH$_2$ and CH$_3$ as well as in aromatic rings and the O–CH$_3$ bonds in lipids, lignins, saccharides and esters (Table 4) [52–57]. The existence of ester bonds in all of the studied extracts was confirmed by the excitation in the wavelength region from 1748 to 1648 cm^{-1}, which corresponds to polyesters and lignins (Table 4) [55–57]. This region is also characteristic of vibrations of the C–N stretching bonds present in proteins and the unconjugated cis C=C bonds in alkenyl groups (Table 4) [52,54,57]. The FTIR spectra of scCO$_2$ extracts from *H. salicifolius*, scCO$_2$ and scCO$_2$/H$_2$O extracts from *H. tuberosus* and scCO$_2$/H$_2$O extracts from *S. pectinata* displayed absorbance within the wavelength region from 1602 to 1514 cm^{-1}, corresponding to vibrations of the C–C and C=C stretching bonds occurring in aromatic rings and the COO$^-$ stretching bonds in phenols, pectins and lignins (Table 4) [53,55,56].

Table 4. Identification of functional groups of supercritical plant extracts (scCO$_2$ and scCO$_2$/H$_2$O) harvested in June (analysis based on literature [52–57]).

Group	Wavenumbers (cm^{-1})	Plant Extract		Bond Type	Functional Group
		scCO$_2$	scCO$_2$/H$_2$O		
1	4000–3100	*H. salicifolius, H. tuberosus, M. × giganteus, M. sinensis*	*H. salicifolius, S. perfoliatum, H. tuberosus, M. sinensis, S. pectinata*	O–H, N–H stretching	Polyphenolic, carbohydrates, proteins
2	3100–2800	*H. salicifolius, S. perfoliatum, H. tuberosus, M. × giganteus, M. sinensis, M. sacchariflorus, S. pectinata*	*H. salicifolius, S. perfoliatum, H. tuberosus, M. × giganteus, M. sinensis, M. sacchariflorus, S. pectinata*	C–H stretching in CH$_2$ and CH$_3$; C–H aromatic stretching; O–CH$_3$	Lipids, lignins, carbohydrates, esters
3	1750–1650	*H. salicifolius, S. perfoliatum, H. tuberosus, M. × giganteus, M. sinensis, M. sacchariflorus, S. pectinata*	*H. salicifolius, S. perfoliatum, H. tuberosus, M. × giganteus, M. sinensis, M. sacchariflorus, S. pectinata*	C=O ester stretching; C–N stretching; *cis* C=C unconjugated	Polyesters, lignins, proteins, alkenyl groups
4	1610–1500	-	*H. salicifolius, S. perfoliatum, H. tuberosus, S. pectinata*	C–C, C=C aromatic stretching; COO$^-$ stretching	Phenolic groups, pectins, lignins
5	1500–1290	*H. salicifolius, S. perfoliatum, H. tuberosus, M. × giganteus, M. sinensis, M. sacchariflorus, S. pectinata*	*H. salicifolius, S. perfoliatum, H. tuberosus, M. × giganteus, M. sinensis, M. sacchariflorus, S. pectinata*	CH$_2$ scissoring, bending, out of plane; C–H in >CH–; C–H in –CH$_3$; C–H, CH$_2$, CH$_3$ deformations; C=C–C aromatic ring stretching; O–H bending; N–H, C–N	triterpenoids, phenyl groups, polysaccharides, pectins, lipids, lignins, tertiary alcohols, proteins
6	1260–1100	*H. salicifolius, S. perfoliatum, H. tuberosus, M. × giganteus, M. sinensis, M. sacchariflorus, S. pectinata*	*H. salicifolius, S. perfoliatum, H. tuberosus, M. × giganteus, M. sinensis, M. sacchariflorus, S. pectinata*	C–O, C–N, C–C, >PO$_2$ stretching; O–H bending; C–O ring vibrations	Phenyl groups, lignins, pectins, triterpenoids, polysaccharides, phospholipids, proteins
7	1100–1000	*H. salicifolius, S. perfoliatum, H. tuberosus, M. × giganteus, M. sinensis, M. sacchariflorus, S. pectinata*	*H. salicifolius, S. perfoliatum, H. tuberosus, M. × giganteus, M. sinensis, M. sacchariflorus, S. pectinata*	C–O, C–N stretching; C–H aromatic stretching; >PO$_2$ stretching	Phenolic groups, phospholipids, polysaccharides, pectins, aliphatic amines
8	1000–510	*H. salicifolius, S. perfoliatum, H. tuberosus, M. × giganteus, M. sinensis, M. sacchariflorus, S. pectinata*	*H. salicifolius, H. tuberosus, M. × giganteus, M. sinensis, M. sacchariflorus, S. pectinata*	C–H aromatic stretching; –CH$_2$–; –HC–; *cis* –CH; *trans* C–H out of plane	Isoprenoids
9	510–400	*H. salicifolius, H. tuberosus*	*H. tuberosus, S. pectinata*	S–S stretching; C–OH$_3$ torsion	Aryl disulfides, polysulfides, methoxy groups

An excitation within the wavelength region from 1464 to 1291 cm^{-1} was recorded for all of the analyzed extracts, which confirms the presence of the functional groups occurring in the compounds, including triterpenoids, polysaccharides, pectins, lignins, lipids, tertiary alcohols, proteins and phenols (Table 4) [52–57]. The largest number of peaks within this wavelength region was noted for scCO$_2$ extracts from *S. perfoliatum* (1464, 1445, 1377, 1310 and 1291 cm^{-1}), while the lowest number of peaks was recorded for scCO$_2$/H$_2$O extracts from *H. tuberosus* (1374 cm^{-1}). Vibrations of the C–O, C–N, C–C and >PO$_2$ stretching bonds, O–H bending bonds and aromatic C–O bonds (Table 4) [52–57] within the wavelength from 1271 to 1117 cm^{-1}, occurring in triterpenoids, phospholipids, polysaccharides, pectins, lignins, proteins and phenolic groups, were detected in all extract samples. However, for extracts obtained from *M. sinensis*, *M. × giganteus* and *M. sacchariflorus*, only one peak was detected (approximately 1170 cm^{-1}) regardless of the solvent used. Excitation within the region from 1104 to 1013 cm^{-1}, caused by vibrations of the C–O, C–N and >PO$_2$ stretching bonds and aromatic C–H stretching bonds characteristic of phospholipids, polysaccharides, pectins, aliphatic amines and phenols (Table 4) [54–57], was recorded in all analyzed extracts. The FTIR spectra indicated that the number of excitations recorded for isoprenoid bonds was greater for scCO$_2$ extracts than for scCO$_2$/H$_2$O extracts, and this was the case for all the analyzed extracts except for extracts from *S. pectinata*. The scCO$_2$/H$_2$O extract from *S. perfoliatum* did not display the presence of isoprenoids. Within the wavelength region from 501 to 427 cm^{-1}, absorbance corresponding to vibrations of the S–S stretching bonds and C–OH$_3$ torsion bonds occurring in polysulfides, aryl disulfides and methoxy groups (Table 4) [52,53] was detected for extracts from *H. salicifolius* (scCO$_2$), *H. tuberosus* (scCO$_2$ or scCO$_2$/H$_2$O) and *S. pectinata* (scCO$_2$/H$_2$O).

4. Conclusions

The presented data indicate varying contents of polyphenolic compounds in the biomass extracts of perennial herbaceous crops depending on the harvest term in the growing period. The varying quality of extract compositions determined by FTIR analysis was noted, which depended on the performance of extraction using scCO$_2$ or the same supercritical solvent combined with water. This information can enable the utilization of the studied biomass for not only the production of bioenergy but also to obtain valuable components of foodstuffs, medicines and cosmetics. However, a comprehensive analysis of the properties and chemical composition (particularly the flavonoid content) of the extracts as well as the relationship between the composition and the properties of the extracts is necessary. Moreover, it is important to study the influence of the extracts on microorganisms and enzymes to determine the cytotoxicity, apoptotic effect and safety of their use. Further development of selective techniques for the extraction of polyphenols, hydrocarbons, fatty acids and acylglycerols from biomass (particularly using scCO$_2$) or of the extract fractioning method is also possible.

Supplementary Materials: The following are available online at https://www.mdpi.com/article/10.3390/agriculture11060488/s1, Table S1. The results of ANOVA ($p < 0.05$) for the influence of substrate, extraction conditions and harvest time (A) June, (B) October, (C) February on total polyphenol concentration (TPC), Table S2. The results of ANOVA ($p < 0.05$) for the influence of substrate, extraction conditions and harvest time (A) June, (B) October, (C) February on total flavonoid concentration (TFC), Table S3. The results of ANOVA ($p < 0.05$) for the influence of substrate, extraction conditions and harvest time (A) June, (B) October, (C) February on antioxidant activity (DPPH assay), Table S4. Wavenumbers of QATR-FTIR peaks of supercritical plant extracts (scCO$_2$ and scCO$_2$/H$_2$O) harvested in June, Figure S1. QATR-FTIR spectrum of the scCO$_2$ extract from *H. salicifolius* harvested in June, Figure S2. QATR-FTIR spectrum of the scCO$_2$/H$_2$O extract from *H. salicifolius* harvested in June, Figure S3. QATR-FTIR spectrum of the scCO$_2$ extract from *S. perfoliatum* harvested in June, Figure S4. QATR-FTIR spectrum of the scCO$_2$/H$_2$O extract from *S. perfoliatum* harvested in June, Figure S5. QATR-FTIR spectrum of the scCO$_2$ extract from *H. tuberosus* harvested in June, Figure S6. QATR-FTIR spectrum of the scCO$_2$/H$_2$O extract from *H. tuberosus* harvested in June, Figure S7. QATR-FTIR spectrum of the scCO$_2$ extract from *M. × giganteus* harvested in June, Figure S8. QATR-FTIR spectrum of the scCO$_2$/H$_2$O extract from *M. × giganteus* harvested in June, Figure S9. QATR-FTIR spectrum of the scCO$_2$ extract from *M. sinensis* harvested in June, Figure S10. QATR-FTIR spectrum of the scCO$_2$/H$_2$O extract from *M. sinensis* harvested in June, Figure S11. QATR-FTIR spectrum of the scCO$_2$ extract from *M. sacchariflorus* harvested in June, Figure S12. QATR-FTIR spectrum of the scCO$_2$/H$_2$O extract from *M. sacchariflorus* harvested in June, Figure S13. QATR-FTIR spectrum of the scCO$_2$ extract from *S. pectinata* harvested in June, Figure S14. QATR-FTIR spectrum of the scCO$_2$/H$_2$O extract from *S. pectinata* harvested in June.

Author Contributions: Conceptualization, M.A.; methodology, M.O., B.B. and M.A.; formal analysis, B.B. and M.A.; investigation, M.O.; resources, M.O., B.B. and M.A.; data curation, M.A.; writing—original draft preparation, M.O. and M.A.; writing—review and editing, B.B., M.J.S. and M.A.; visualization, M.O., B.B. and M.A.; supervision. M.A.; project administration, B.B. and M.A.; funding acquisition. M.A. and M.J.S. All authors have read and agreed to the published version of the manuscript.

Funding: This research was funded by the National (Polish) Centre for Research and Development (NCBiR), titled "Environment, agriculture and forestry", project: BIOproducts from lignocellulosic biomass derived from MArginal land to fill the Gap In the Current national bioeconomy, No. BIOS-TRATEG3/344253/2/NCBR/2017.

Institutional Review Board Statement: Not applicable.

Informed Consent Statement: Not applicable.

Acknowledgments: We would like to thank the University of Warmia and Mazury in Olsztyn, Faculty of Agriculture and Forestry, Department of Genetics, Plant Breeding and Bioresource Engineering (grant No. 30.610.007-110), for the materials and the Supercritical Extraction Department, ŁUKASZEWICZ Research Network—New Chemical Syntheses Institute (Puławy, Poland), for the extracts.

Conflicts of Interest: The authors declare no conflict of interest.

References

1. Keoleian, G.A.; Volk, T.A. Renewable energy from willow biomass crops: Life cycle energy, environmental and economic performance. *Crit. Rev. Plant Sci.* **2005**, *24*, 385–406. [CrossRef]
2. Stolarski, M.J.; Warmiński, K.; Krzyżaniak, M. Energy value of yield and biomass quality of poplar grown in two consecutive 4-year harvest rotations in the north-east of Poland. *Energies* **2020**, *13*, 1495. [CrossRef]
3. Stolarski, M.J.; Niksa, D.; Krzyżaniak, M.; Tworkowski, J.; Szczukowski, S. Willow productivity from small- and large-scale experimental plantations in Poland from 2000 to 2017. *Renew. Sustain. Energy Rev.* **2019**, *101*, 461–475. [CrossRef]
4. James, L.K.; Swinton, S.M.; Thelen, K.D. Profitability analysis of cellulosic energy crops compared with corn. *Agron. J.* **2010**, *102*, 675–687. [CrossRef]
5. Scalbert, A.; Manach, C.; Morand, C.; Rémésy, C.; Jiménez, L. Dietary polyphenols and the prevention of diseases. *Crit. Rev. Food Sci. Nutr.* **2005**, *45*, 287–306. [CrossRef]
6. Madunić, J.; Madunić, I.V.; Gajski, G.; Popić, J.; Garaj-Vrhovac, V. Apigenin: A dietary flavonoid with diverse anticancer properties. *Cancer Lett.* **2018**, *413*, 11–22. [CrossRef]
7. Anhê, F.F.; Desjardins, Y.; Pilon, G.; Dudonné, S.; Genovese, M.I.; Lajolo, F.M.; Marette, A. Polyphenols and type 2 diabetes: A prospective review. *PharmaNutrition* **2013**, *1*, 105–114. [CrossRef]
8. Marranzano, M.; Rosa, R.L.; Malaguarnera, M.; Palmeri, R.; Tessitori, M.; Barbera, A.C. Polyphenols: Plant sources and food industry applications. *Curr. Pharm. Des.* **2018**, *24*, 4125–4130. [CrossRef]
9. Mittal, A.K.; Chisti, Y.; Banerjee, U.C. Synthesis of metallic nanoparticles using plant extracts. *Biotechnol. Adv.* **2013**, *31*, 346–356. [CrossRef] [PubMed]
10. Milinčić, D.D.; Popović, D.A.; Lević, S.M.; Kostić, A.Ž.; Tešić, Ž.L.; Nedović, V.A.; Pešić, M.B. Application of polyphenol-loaded nanoparticles in food industry. *Nanomaterials* **2019**, *9*, 1629. [CrossRef]
11. Krishnaiah, D.; Sarbatly, R.; Nithyanandam, R. A review of the antioxidant potential of medicinal plant species. *Food Bioprod. Process.* **2011**, *89*, 217–233. [CrossRef]
12. Gessner, D.K.; Ringseis, R.; Eder, K. Potential of plant polyphenols to combat oxidative stress and inflammatory processes in farm animals. *J. Anim. Physiol. Anim. Nutr.* **2017**, *101*, 605–628. [CrossRef]
13. González de Llano, D.; Liu, H.; Khoo, C.; Moreno-Arribas, M.V.; Bartolomé, B. Some new findings regarding the antiadhesive activity of cranberry phenolic compounds and their microbial-derived metabolites against uropathogenic bacteria. *J. Agric. Food Chem.* **2019**, *67*, 2166–2174. [CrossRef]
14. Zhu, J.; Huang, X.; Zhang, F.; Feng, L.; Li, J. Inhibition of quorum sensing, biofilm, and spoilage potential in *Shewanella baltica* by green tea polyphenols. *J. Microbiol.* **2015**, *53*, 829–836. [CrossRef]
15. Fleitas Martínez, O.; Cardoso, M.H.; Ribeiro, S.M.; Franco, O.L. Recent advances in anti-virulence therapeutic strategies with a focus on dismantling bacterial membrane microdomains, toxin neutralization, quorum-sensing interference and biofilm inhibition. *Front. Cell. Infect. Microbiol.* **2019**, *9*, 74. [CrossRef]
16. Kowalska, G.; Pankiewicz, U.; Kowalski, R. Evaluation of chemical composition of some *Silphium*, L. species as alternative raw materials. *Agriculture* **2020**, *10*, 132. [CrossRef]
17. Pan, L.; Sinden, M.R.; Kennedy, A.H.; Chai, H.; Watson, L.E.; Graham, T.L.; Kinghorn, A.D. Bioactive constituents of *Helianthus tuberosus* (Jerusalem artichoke). *Phytochem. Lett.* **2009**, *2*, 15–18. [CrossRef]

18. Nizioł-Łukaszewska, Z.; Furman-Toczek, D.; Zagórska-Dziok, M. Antioxidant activity and cytotoxicity of Jerusalem artichoke tubers and leaves extract on HaCaT and BJ fibroblast cells. *Lipids. Health Dis.* **2018**, *17*, 280. [CrossRef] [PubMed]

19. Kowalski, R.; Kędzia, B. Antibacterial activity of *Silphium perfoliatum* extracts. *Pharm. Biol.* **2007**, *45*, 494–500. [CrossRef]

20. Kowalski, R. *Silphium*, L. extracts–composition and protective effect on fatty acids content in sunflower oil subjected to heating and storage. *Food Chem.* **2009**, *112*, 820–830. [CrossRef]

21. Ahmad, P.; Jaleel, C.A.; Salem, M.A.; Nabi, G.; Sharma, S. Roles of enzymatic and nonenzymatic antioxidants in plants during abiotic stress. *Crit. Rev. Biotechnol.* **2010**, *30*, 161–175. [CrossRef]

22. Stolarski, M.J.; Warmiński, K.; Krzyżaniak, M.; Tyśkiewicz, K.; Olba-Zięty, E.; Graban, Ł.; Lajszner, W.; Załuski, D.; Wiejak, R.; Kamiński, P.; et al. How does extraction of biologically active substances with supercritical carbon dioxide affect lignocellulosic biomass properties? *Wood Sci. Technol.* **2020**, *54*, 519–546. [CrossRef]

23. Malm, A.; Grzegorczyk, A.; Biernasiuk, A.; Baj, T.; Rój, E.; Tyśkiewicz, K.; Dębczak, A.; Stolarski, M.J.; Krzyżaniak, M.; Olba-Zięty, E. Could supercritical extracts from the aerial parts of *Helianthus salicifolius* A. Dietr. and *Helianthus tuberosus* L. be regarded as potential raw materials for biocidal purposes? *Agriculture* **2021**, *11*, 10. [CrossRef]

24. Ostolski, M.; Adamczak, M.; Brzozowski, B.; Wiczkowski, W. Antioxidant activity and chemical characteristics of supercritical CO_2 and water extracts from willow and poplar. *Molecules* **2021**, *26*, 545. [CrossRef]

25. Singleton, V.L.; Rossi, J.A. Colorimetry of total phenolics with phosphomolybdic-phosphotungstic acid reagents. *Am. J. Enol. Vitic.* **1965**, *16*, 144–158.

26. Lamaison, J.L.; Carnat, A. The amount of main flavonoids in flowers and leaves of *Crataegus monogyna* Jacq. and *Crataegus laevigata* (Poiret) DC. (*Rosaceae*). *Pharm. Acta Helv.* **1990**, *65*, 315–320.

27. Blois, M.S. Antioxidant determinations by the use of a stable free radical. *Nature* **1958**, *181*, 1199–1200. [CrossRef]

28. Rasouli, H.; Farzaei, M.H.; Mansouri, K.; Mohammadzadeh, S.; Khodarahmi, R. Plant cell cancer: May natural phenolic compounds prevent onset and development of plant cell malignancy? A literature review. *Molecules* **2016**, *21*, 1104. [CrossRef]

29. Quideau, S.; Deffieux, D.; Douat-Casassus, C.; Pouységu, L. Plant polyphenols: Chemical properties, biological activities, and synthesis. *Angew. Chem. Int. Ed.* **2011**, *50*, 586–621. [CrossRef]

30. Aggarwal, S.; Johnson, S.; Hakovirta, M.; Sastri, B.; Banerjee, S. Removal of water and extractives from softwood with supercritical carbon dioxide. *Ind. Eng. Chem. Res.* **2019**, *58*, 3170–3174. [CrossRef]

31. Trubetskaya, A.; Budarin, V.; Arshadi, M.; Magalhães, D.; Kazanç, F.; Hunt, A.J. Supercritical extraction of biomass as an effective pretreatment step for the char yield control in pyrolysis. *Renew. Energy* **2021**, *170*, 107–117. [CrossRef]

32. Attard, T.M.; Bukhanko, N.; Eriksson, D.; Arshadi, M.; Geladi, P.; Bergsten, U.; Budarin, V.L.; Clark, J.H.; Hunt, A.J. Supercritical extraction of waxes and lipids from biomass: A valuable first step towards an integrated biorefinery. *J. Clean. Prod.* **2018**, *177*, 684–698. [CrossRef]

33. Showkat, M.M.; Falck-Ytter, A.B.; Strætkvern, K.O. Phenolic acids in Jerusalem artichoke (*Helianthus tuberosus* L.): Plant organ dependent antioxidant activity and optimized extraction from leaves. *Molecules* **2019**, *24*, 3296. [CrossRef] [PubMed]

34. Balcerek, M.; Rąk, I.; Majtkowska, G.; Majtkowski, W. Antioxidant activity and total phenolic compounds in extracts of selected grasses (*Poaceae*). *Herba Pol.* **2009**, *55*, 215–220.

35. Parveen, I.; Wilson, T.; Donnison, I.S.; Cookson, A.R.; Hauck, B.; Threadgill, M.D. Potential sources of high value chemicals from leaves, stems and flowers of *Miscanthus sinensis* 'Goliath' and *Miscanthus sacchariflorus*. *Phytochem* **2013**, *92*, 160–167. [CrossRef] [PubMed]

36. Uwineza, P.A.; Waśkiewicz, A. Recent advances in supercritical fluid extraction of natural bioactive compounds from natural plant materials. *Molecules* **2020**, *25*, 3847. [CrossRef] [PubMed]

37. Tyśkiewicz, K.; Konkol, M.; Rój, E. The application of supercritical fluid extraction in phenolic compounds isolation from natural plant materials. *Molecules* **2018**, *23*, 2625. [CrossRef]

38. Da Porto, C.; Decorti, D.; Natolino, A. Water and ethanol as co-solvent in supercritical fluid extraction of proanthocyanidins from grape marc: A comparison and a proposal. *J Supercrit. Fluids* **2014**, *87*, 1–8. [CrossRef]

39. He, J.-Z.; Shao, P.; Liu, J.-H.; Ru, Q.-M. Supercritical carbon dioxide extraction of flavonoids from pomelo (*Citrus grandis* (L.) Osbeck) peel and their antioxidant activity. *Int. J. Mol. Sci.* **2012**, *13*, 13065–13078. [CrossRef]

40. Miguez, F.E.; Villamil, M.B.; Long, S.P.; Bollero, G.A. Meta-analysis of the effects of management factors on *Miscanthus × giganteus* growth and biomass production. *Agric. For. Meteorol.* **2008**, *148*, 280–1292. [CrossRef]

41. Lemos, M.F.; Lemos, M.F.; Pacheco, H.P.; Endringer, D.C.; Scherer, R. Seasonality modifies rosemary's composition and biological activity. *Ind. Crops Prod.* **2015**, *70*, 41–47. [CrossRef]

42. Cao, Y.; Fang, S.; Fu, X.; Shang, X.; Yang, W. Seasonal variation in phenolic compounds and antioxidant activity in leaves of *Cyclocarya paliurus* (Batal.) Iljinskaja. *Forests* **2019**, *10*, 624. [CrossRef]

43. Scognamiglio, M.; D'Abrosca, B.; Fiumano, V.; Golino, M.; Esposito, A.; Fiorentino, A. Seasonal phytochemical changes in *Phillyrea angustifolia* L.: Metabolomic analysis and phytotoxicity assessment. *Phytochem. Lett.* **2014**, *8*, 163–170. [CrossRef]

44. Pacifico, S.; Galasso, S.; Piccolella, S.; Kretschmer, N.; Pan, S.-P.; Marciano, S.; Bauer, R.; Monaco, P. Seasonal variation in phenolic composition and antioxidant and anti-inflammatory activities of *Calamintha nepeta* (L.) Savi. *Food Res. Int.* **2015**, *69*, 121–132. [CrossRef]

45. Bujor, O.C.; Le Bourvellec, C.; Volf, I.; Popa, V.I.; Dufour, C. Seasonal variations of the phenolic constituents in bilberry (*Vaccinium myrtillus* L.) leaves, stems and fruits, and their antioxidant activity. *Food Chem.* **2016**, *213*, 58–68. [CrossRef]

46. Dela, G.; Or, E.; Ovadia, R.; Nissim-Levi, A.; Weiss, D.; Oren-Shamir, M. Changes in anthocyanin concentration and composition in 'Jaguar' rose flowers due to transient high-temperature conditions. *Plant Sci.* **2003**, *164*, 333–340. [CrossRef]

47. Fu, B.; Ji, X.; Zhao, M.; He, F.; Wang, X.; Wang, Y.; Liu, P.; Niu, L. The influence of light quality on the accumulation of flavonoids in tobacco (*Nicotiana tabacum* L.) leaves. *J. Photochem. Photobiol. B. Biol.* **2016**, *162*, 544–549. [CrossRef]

48. Rathee, J.S.; Hassarajani, S.A.; Chattopadhyay, S. Antioxidant activity of *Nyctanthes arbor-tristis* leaf extract. *Food Chem.* **2007**, *103*, 1350–1357. [CrossRef]

49. Zhang, L.; Gao, Y.; Zhang, Y.; Liu, J.; Yu, J. Changes in bioactive compounds and antioxidant activities in pomegranate leaves. *Sci. Hortic.* **2010**, *123*, 543–546. [CrossRef]

50. Vazquez-Olivo, G.; López-Martínez, L.X.; Contreras-Angulo, L.; Heredia, J.B. Antioxidant capacity of lignin and phenolic compounds from corn stover. *Waste Biomass Valori.* **2019**, *10*, 95–102. [CrossRef]

51. Vanderghem, C.; Jacquet, N.; Richel, A. Can lignin wastes originating from cellulosic ethanol biorefineries act as radical scavenging agents? *Aust. J. Chem.* **2014**, *67*, 1693–1699. [CrossRef]

52. Coates, J. Interpretation of infrared spectra. In *A Practical Approach*; Meyers, R.A., Ed.; Encyclopedia of Analytical Chemistry; John Wiley & Sons, Ltd.: Chichester, UK, 2006; pp. 1–23. [CrossRef]

53. Hayat, J.; Akodad, M.; Moumen, A.; Baghour, M.; Skalli, A.; Ezrari, S.; Belmalha, S. Phytochemical screening, polyphenols, flavonoids and tannin content, antioxidant activities and FTIR characterization of *Marrubium vulgare* L. from 2 different localities of Northeast of Morocco. *Heliyon* **2020**, *6*, e05609. [CrossRef] [PubMed]

54. Huck, C.W. Advances of infrared spectroscopy in natural product research. *Phytochem. Lett.* **2015**, *11*, 384–393. [CrossRef]

55. Nogales-Bueno, J.; Baca-Bocanegra, B.; Rooney, A.; Hernández-Hierro, J.M.; Byrne, H.J.; Heredia, F.J. Study of phenolic extractability in grape seeds by means of ATR-FTIR and Raman spectroscopy. *Food Chem.* **2017**, *232*, 602–609. [CrossRef] [PubMed]

56. Rodrigues, V.H.; de Melo, M.M.R.; Portugal, I.; Silva, C.M. Extraction of *Eucalyptus* leaves using solvents of distinct polarity. Cluster analysis and extracts characterization. *J. Supercrit. Fluids* **2018**, *135*, 263–274. [CrossRef]

57. Thummajitsakul, S.; Samaikam, S.; Tacha, S.; Silprasit, K. Study on FTIR spectroscopy, total phenolic content, antioxidant activity and anti-amylase activity of extracts and different tea forms of *Garcinia schomburgkiana* leaves. *LWT* **2020**, *134*, 110005. [CrossRef]

MDPI

St. Alban-Anlage 66

4052 Basel

Switzerland

Tel. +41 61 683 77 34

Fax +41 61 302 89 18

www.mdpi.com

Agriculture Editorial Office

E-mail: agriculture@mdpi.com

www.mdpi.com/journal/agriculture

www.ingramcontent.com/pod-product-compliance
Lightning Source LLC
LaVergne TN
LVHW071602170726
843515LV00008B/2321